# Surface Modification and Functionalization of Textiles

**Scrivener Publishing**
100 Cummings Center, Suite 541J
Beverly, MA 01915-6106

*Publishers at Scrivener*
Martin Scrivener (martin@scrivenerpublishing.com)
Phillip Carmical (pcarmical@scrivenerpublishing.com)

# Surface Modification and Functionalization of Textiles


Edited by

**Aminoddin Haji**

*Department of Textile Engineering, Yazd University, Yazd, Iran*

and

**Çağlar Sivri**

*Department of Engineering Management, Faculty of Engineering and Natural Sciences, Bahçeşehir University, Beşiktaş, İstanbul, Türkiye*



Scrivener Publishing

WILEY

This edition first published 2026 by John Wiley & Sons, Inc., 111 River Street, Hoboken, NJ 07030, USA and Scrivener Publishing LLC, 100 Cummings Center, Suite 541J, Beverly, MA 01915, USA

For more information about Scrivener publications please visit www.scrivenerpublishing.com.



**Wiley Global Headquarters**
111 River Street, Hoboken, NJ 07030, USA

For details of our global editorial offices, customer services, and more information about Wiley products visit us at www.wiley.com.

The manufacturer's authorized representative according to the EU General Product Safety Regulation is Wiley-VCH GmbH, Boschstr. 12, 69469 Weinheim, Germany, e-mail: Product_Safety@wiley.com.



***Library of Congress Cataloging-in-Publication Data***

ISBN 978-1-394-17460-7

Cover image: Generated with AI using Adobe Firefly
Cover design by Russell Richardson

Set in size of 11pt and Minion Pro by Manila Typesetting Company, Makati, Philippines

Printed in the USA

10 9 8 7 6 5 4 3 2 1

# Contents

# Preface

The textile industry is an important component of global trade, and textiles are an important part of our daily lives. However, the textile industry must find more sustainable and environmentally friendly processes to produce and to finish textiles in a time when the planet is now facing imminent threats from environmental factors. This book introduces state-of-the-art technologies to help the textile industry facilitate a more sustainable future by offering solutions that enhance the functional properties of fabrics while minimizing negative environmental impacts.

An extensive array of surface-modifying processes is discussed throughout this book, each of which holds promise to help transform textile materials into smarter and more sustainable products. Some processes will effectively replace existing production methods that use excessive resources with those that are potentially more efficient and environmentally sustainable, while still adding cosmetics to fabric.

The first chapter introduces the key concepts of surface modification of textiles and the accompanying environmental aspects, setting the stage for the subsequent chapters on specific techniques in applied surface science, such as plasma treatment, sputter coating and sol gel technology. Chapters 2 to 7 illustrate how appropriate textile construction can provide essential functional properties, such as antimicrobial properties, hydrophobicity, flame retardancy, etc. Chapter 8 demonstrates how laser technology can accurately modify textile surfaces, while Chapter 9 discusses conservation and modern methods of modifying wool.

Chapters 10 and 11 explore the exciting prospects of nanofinishing and microencapsulation, demonstrating how these advancements can allow textiles to do things we never thought possible, like provide self-cleaning surfaces or thermal regulation in a textile. Chapter 12 introduces chemical vapor deposition (CVD) for textiles to obtain high-performance properties; Chapter 13 focuses on electroless deposition for conductive textiles; and Chapter 14 covers enzyme treatment and immobilization as a more sustainable way to finish textiles.

In Chapter 15, ozone treatment is presented as a sustainable option for textile treatments. Ozone is a cleaner process for textile treatment than we have previously discussed. Chapter 16 looks at coating treatments that can include important attributes such as water repellency, UV resistance, and breathability; and Chapter 17 investigates skincare treatments for textiles, which have the potential to create additional value to the textiles in addition to function or aesthetics. Chapter 18 reflects upon the possibilities of the future and direction of surface modification practices in textiles.

This book shows that these technologies will not just impact the way we produce textiles but redefine what they can be. New sustainable production processes and integration of smart functionalities are on the verge of transforming the textile industry and all applications, ranging from health to electronics. This book is designed for researchers in textile technology, industry professionals, and undergraduate/postgraduate students who are determined to understand the developing landscape of textile functionalization. We thank all the contributors for their knowledge and insight, and hope this work is a useful resource for anyone interested in the future of textiles. We are sincerely grateful to our colleagues for their support. Finally, we thank Martin Scrivener and the team at Scrivener Publishing for their commitment to publishing this work.

*Dr. Aminoddin Haji*
*Dr. Çağlar Sivri*
April 2026

# 1

# An Introduction to Surface Modification Technologies for Functionalization of Textiles

**Aminoddin Haji**

*Department of Textile Engineering, Yazd University, Yazd, Iran*

***Abstract***

This chapter gives a comprehensive overview of the main surface treatment technologies that allow functional properties to be exerted on textile materials. The chapter discusses a variety of advanced physical and chemical processes such as plasma treatment, sputtering, sol-gel process, layer-by-layer assembly, grafting, ultraviolet (UV) and laser processing, and chemical vapor deposition. The chapter also discusses the use of nanomaterials, microcapsules, enzymes, and ozonated-water treatments as contemporary and sustainable means of applying functional finishing to textiles. Every technique has a critical review on how it works, its benefits and limitations, and how it acts as a technical catalog addressing the application of each technique in enabling textiles better functionality: hydrophobicity, antimicrobial activity, flame retardancy, UV protection, and conductivity. The chapter focuses on the transition from traditional wet-chemical processes toward environmentally friendly, dry, and efficient technologies, underlining the wide applications and implications of surface engineering as key to the ongoing evolution of smart and technical textiles.

***Keywords*:** Plasma technology, textile functionalization, smart textiles, sustainable finishing, surface modification

*Email*: ahaji@yazd.ac.ir

Aminoddin Haji and Çağlar Sivri (eds.) Surface Modification and Functionalization of Textiles, (1–22) 

## 1.1 Introduction

The modification of textiles by means of functionalization by surface modification is a key part of modern materials science and engineering, which has changed traditional textiles into high-performance intelligent materials. The textile industry has been using traditional wet-chemical processes to impart characteristic properties (e.g., hydrophobicity, flame retardancy, and antimicrobial properties) to textiles for many decades. Unfortunately, these processes can have a very negative impact on the environment due to fact that they consume a large amount of water and energy and generate hazardous effluents. As a result of these issues and the growing demand for textiles with additional characteristics, this sector is transitioning to more sustainable, dry and efficient surface modification technologies.

This transition is critical for the development of smart and technical textiles that will impact an increasingly broad group of sectors, including medicine, defense, sportswear, and electronics. The following sections of this chapter will explore these new developments. The various advanced physical or chemical processes can be summarized as plasma treatment, sputtering, sol-gel processing, layer-by-layer (LbL) assembly, chemical vapor deposition (CVD), etc. The chapter will delve into the operational principles, benefits, and limitations of each method while also examining the use of nanomaterials and biotechnological approaches as modern and sustainable pathways

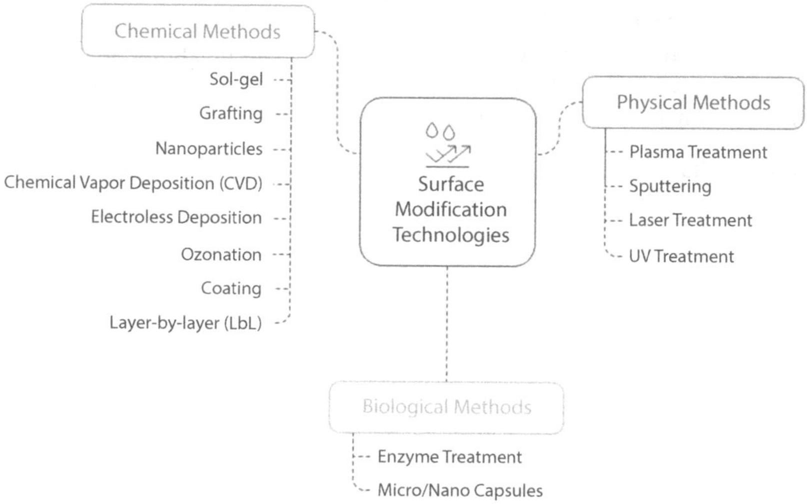


**Figure 1.1** Classification of various textile surface modification technologies.

to textile functionalization. Figure 1.1 shows a common classification of the various technologies for surface modification of textiles.

## 1.2 Plasma Treatment

Plasma, as the fourth state of matter, is a mixture of charged and neutral highly energetic particles. This mixture is composed of ions, electrons, free radicals, and photons, generated usually by applying an electric or electromagnetic field to a gas. Plasmas are generally classified into two main categories: hot (thermal) plasmas and cold (non-thermal) plasmas. Considering the sensitivity of textiles to high temperatures, only cold plasma can be practically applied in the textile industry. As an environmentally friendly and dry process, plasma technology offers a promising alternative to conventional wet processes in textile industry. It significantly reduces water and chemical consumption and eliminates hazardous effluents [1].

Unlike common textile wet processes, the effect of plasma treatment is restricted to the surface layer of textiles (to depths of less than 100 nm) without altering the bulk properties of the fibers. Upon the collision of energetic plasma species with textile surfaces, four primary effects can take place including cleaning, etching, activation, and polymerization. Plasmas of oxygen, argon, etc., are able to clean the surface of textiles by the removal of contaminants and sizing agents from the fiber surface. Energetic species can even etch the surface of the fibers in nanometer scale and increase the effective surface area by creating microscopic roughness. Surface activation occurs by breaking of covalent bonds and the formation of free radicals on the fiber surface. These radicals can subsequently react with other active plasma species or other molecules in the ambient air and convert to new functional groups [2, 3].

The type and extent of the created functional groups depend on the gas used, and various functional groups such as hydroxyl, carboxyl, or amino groups can be introduced. Employing gases such as oxygen, nitrogen, or argon, plasma surface modification enhances the hydrophilicity of textile surfaces and improves the moisture-related properties and significantly enhances dyeing and printing processes by increasing dye uptake and enabling coloration at lower temperatures. When polymerizable gases such as fluorocarbons or silanes are employed under precisely controlled condition, superhydrophobic and oleophobic surfaces suitable for self-cleaning and stain-resistant textiles can be obtained. Improved fiber surface adhesion is another result of plasma treatment that enhances the bonding of coatings and final layers in multilayer textiles, composites, and technical apparel.

Other important applications of this technology include enhancing the efficiency of antimicrobial, flame retardant, and ultraviolet (UV) resistance finishing processes [4, 5].

Plasma can be generated in two principal types: low-pressure (vacuum) and atmospheric-pressure. Low-pressure systems, despite their higher cost and batch operation, offer superior treatment uniformity. Atmospheric-pressure systems such as dielectric barrier discharge and atmospheric pressure plasma jets are more suitable for mass production due to their lower costs and continuous processing capability. However, challenges such as the stability of plasma effects (aging phenomena) and the uniform penetration of plasma into the three-dimensional porous structure of fabrics remain significant obstacles to broader industrial adoption. Despite these challenges, the substantial potential of plasma technology in creating added value and advancing smart and functional textiles underscores a promising future for this field [6, 7].

## 1.3 Sputtering

Surface modification of textiles with the aim of imparting advanced functional properties is an important domain within modern textile engineering. Among the available methods, great attention has been paid to physical vapor deposition (PVD) techniques due to their numerous advantages. Sputtering is an important and widely applied PVD technique. In this process, a target material (cathode) placed in a vacuum chamber is bombarded with high-energy ions. Due to the collision of the high energy ions, atoms or polymeric fragments are ejected from the target and subsequently deposited as a thin film on the textile substrate placed in the chamber. Sputtering, especially magnetron sputtering, is highly suitable for textile modification owing to benefits such as low-temperature processing, environmental compatibility, excellent coating adhesion, and rapid operation. The thickness and purity of the deposited nanolayers can be precisely controlled in this technique. Furthermore, pre-treatment of textile surfaces using oxygen or argon plasma significantly enhances coating adhesion. Various kinds of materials, including metals (such as copper, silver, titanium, and aluminum), metal oxides (such as $TiO_2$ and ZnO), and polymers, can be used as target and deposited on various textile substrates by sputtering method. Any kind of fibers like polyester, cotton, and polypropylene in all forms of textiles including woven, knitted, and non-woven fabrics can be coated by sputtering method [8, 9].

This technology possesses diverse applications in the textile industry. For example, a common application is metallization of textiles with the aim of imparting electrical properties. Deposition of a very thin layer of metals such as copper and silver produces conductive textiles with anti-static behavior and potential use in textronics. It also imparts electromagnetic interference (EMI) shielding property to the textile coated [9]. Another significant application of sputtering is the development of textiles with antibacterial properties. Sputter coating of textiles with metals such as silver and copper results in durable antibacterial activity against bacteria including *Staphylococcus aureus* and *Escherichia coli* [10].

Another recently explored application of sputtering is generating structural colors on textiles, which is arising from thin-film interference. It offers an environmentally friendly alternative to traditional dyeing processes. The thickness of the metal oxide layers like $TiO_2$ on a metallic underlayer determines the final color [10].

Beyond coloration, thermal management is another property which can be achieved by this technology. Photothermal heating coatings utilize materials capable of absorbing solar energy and converting it into heat. Metals (such as Au, Ge, and W), metal oxides (such as $WO_{3-x}$), metal carbides (like ZrC), and metal sulfides (such as CuS) are examples of coatings that are able to absorb sunlight and are suitable for heating applications in cold environments. Sunlight-reflecting coatings such as Ag, Ti, Al, $TiO_2$, ZnO, and aluminum-doped zinc oxide can provide cooling effect in warm climates. By simultaneous engineering of visible and infrared optical properties, it is possible to creation textiles exhibiting identical colors while having different thermal behaviors (cooling or heating) [11].

Modification of surface properties such as hydrophobicity and hydrophilicity is another crucial application of this technique [12]. Besides significant advantages and progress, sputtering still faces challenges including limited coating durability, abrasion resistance, and color stability, which necessitate further research. Sputtering technology has the potential of producing smart and multifunctional textiles for developing the next generation of textile products with technical functionalities and innovative designs.

## 1.4 Sol-Gel Technology

Sol-gel technology has been known as one of the most significant advancements in materials science in recent decades. It offers extensive possibilities for surface modification of textiles as well as various other substrates. High-purity thin-film coatings can be fabricated on the surface of textiles fibers by sol-gel technology. The sol-gel process consists of two main reactions including hydrolysis and condensation of precursors such as metal alkoxides like tetraethyl orthosilicate. The resultant colloidal suspension (sol) is subsequently converted into a three-dimensional porous network called gel [13, 14].

A major advantage of this technology for the textile industry is that it operates under mild thermal conditions, thereby preventing damage to heat-sensitive fibers. Another advantage is that sol-gel coatings can be applied to fabrics using conventional textile finishing techniques such as padding, dipping, or spraying and there is no need for new machinery. The coatings fabricated by sol-gel technology are thin and transparent with good adhesion that preserves inherent textile properties such as softness and breathability. Furthermore, by combining organic and inorganic precursors, hybrid polymeric coatings with dual characteristics of ceramics (hardness) and polymers (flexibility) can be produced [15–17].

This technology can be applied in imparting functional properties to textiles. For example, development of water- and oil-repellent surfaces is one of the applications of sol-gel technology on textiles. Low-surface-energy compounds, such as long-chain alkyl silanes or fluorinated agents, can be employed for modification of the sol-gel formulation to achieve hydrophobic and even superhydrophobic surfaces. Incorporating nanoparticles to create microscale roughness leads to the lotus effect, thus self-cleaning properties can be obtained [13, 16].

Another key area includes the production of antimicrobial textiles. Biocides like silver, copper, and zinc oxide (ZnO) nanoparticles and quaternary ammonium salts can be fixed on the textile surface by using sol-gel matrix. These coatings are of good wash durability. UV protection is yet another key application. By incorporating semiconductor nanoparticles such as titanium dioxide ($TiO_2$) or zinc oxide (ZnO) into the sol-gel matrix, textiles with high UV protection factors (UPFs) are achieved. In contrast to organic absorbers, these inorganic particles are of exceptional photostability and do not degrade even after extended exposure to light. Photocatalytic self-cleaning functionality can also be achieved by using coatings that contain the anatase phase of $TiO_2$, which degrade organic contaminants on the textile surface under UV exposure [13, 15, 17].

Imparting flame retardancy to textiles is another capability of this technique. If phosphorus- or boron-containing compounds are added to the sol, then flame resistance of both cellulosic and synthetic textiles can be improved by forming a protective ceramic char layer that hinders flame propagation. Abrasion resistance can also be enhanced by hybrid coatings reinforced with nanoparticles like alumina ($Al_2O_3$). The sol-gel matrix can also serve as a binder for pigments and dyes to enhance light fastness and wash fastness. Despite these benefits, one of the main challenges to the industrial exploitation of this technology is the utilization of organic solvents, which calls for specialized recovery and safety equipment. Research has been initiated for creating environmentally friendly water-based sol-gel systems to remove this drawback. Overall, sol-gel technology is a convenient and flexible technology that has offered access to new horizons in the design of multifunctional and intelligent textiles. This new technology has huge potential in manufacturing the next generation of high added-value textile materials with specialty and niche applications [13, 14, 16].

## 1.5 Layer-by-Layer Assembly

Surface modification of textiles to impart new functional properties has become a major area of study in textile engineering in recent decades. The LbL self-assembly method is one of the newer approaches that has garnered a lot of interest as a viable and sustainable strategy. This technique, first reported by Decher and co-workers in the 1990s, is based on alternate adsorption of oppositely charged entities on a support. LbL assembly is typically performed by sequential dipping of the support into aqueous polyelectrolyte solutions—polyanions and polycations—with washing phases between consecutive deposition cycles. Electrostatic attraction constitutes the primary driving force of assembly, but other types of interactions such as hydrogen bonding and covalent bonds can be involved as well. One of the advantages of the proposed method is that it uses water as a solvent and, hence, can be termed as green technology. Other advantages are ease of process, absence of complex equipment, minimal cost, and room temperature viability. The technique offers nanometer-level control of thickness and architecture of coating and is capable of being applied to substrates of any size or shape [18–20].

A diverse array of materials, including both synthetic and natural polyelectrolytes like chitosan and alginate, inorganic nanoparticles such as montmorillonite clay and silica, biomolecules like phytic acid and DNA, as well as carbon-based nanomaterials like carbon nanotubes and graphene, can be incorporated into LbL coatings. This tunability is a valuable asset of LbL as a handy tool for tailoring multifunctional functionality onto a variety of textile substrates from cotton, polyester, and polyamide to their blends. Improved flame retardancy is one of the most prominent applications of LbL technology in textiles. Built-up layer coatings have the capability to inhibit ignition by forming a heat and oxygen barrier or an intumescent char layer. The incorporation of agents such as phytic acid, chitosan, and ammonium polyphosphate has exhibited significant flame retardancy on polyester and cotton fabrics [18–20].

Another important application is the generation of antimicrobial properties. Textiles can have long-lasting antimicrobial properties with coatings containing bioactive polymers, such as chitosan, or silver nanoparticles. These multilayer coatings can also serve as devices for controlled release of therapeutic and bioactive agents. The LbL method can be utilized to achieve additional functionalities as well, such as UV protection by incorporating ZnO nanoparticles, superhydrophobicity through hierarchical structuring and top layer chemical modification, and EMI shielding through the use of materials like carbon nanotubes and MXenes [18, 19].

The primary disadvantage of the LbL technique is that the assembled layers have low durability under washing conditions because they rely on weak electrostatic interactions. Options to overcome these limitations included post deposition thermal treatment, additives to covalently bond layers together, and using surface pretreatments to improve adhesion. Overall LbL assembly techniques provide new routes to develop intelligent and multifunctional textiles, and research continues in order to improve the stability and commercial scale-up of this exciting technology [18, 19].

## 1.6 Grafting

As industry and consumer demand for high-performance textiles grows, fiber surface modification has emerged as an important research area. Grafting can be an effective and versatile means of providing textiles with new functional properties. This means covalent attachment of functional molecules or polymers to the textile surface. The covalent nature of the bonding gives durability and wash-fastness to the added functionality, which is an important advantage over the conventional textile finishing processes [21].

Different grafting methods have been developed, including chemical grafting by initiators, radiation grafting (e.g., electron beam), plasma grafting, and photo-grafting (under UV irradiation). All methods seek to create reactive radical sites on the fiber surface that will allow for the polymerization of functional monomers. Grafting is generally achieved *via* two methods, the "grafting-to" method, which attaches preformed polymer chain(s) to the surface, and the "grafting-from" means where the polymer chain is grown from the substrate [21, 22].

Grafting covers nearly every area of textile functionality. Just a few examples include grafting hydrophobicity for waterproof fabrics and moisture management for athletic clothing. In the area of protective textiles, grafting phosphorus- and nitrogen-containing compounds, which is an extremely effective means of achieving flame resistance. Covalent binding of antimicrobial agents such as the N-halamines or quaternary ammonium salts produces hygienic, antimicrobial textiles for medical and everyday use [21, 23].

Besides these, grafting can create more advanced textiles, such as static cling with anti-static properties, UV protection, drug delivery for wound dressings, and materials to attract, adsorb, and remove pollutants from water. Although there remain issues of scale and environmentally friendly processes for industry, research is positive and actively striving to overcome roadblocks. Grafting technology, in addition to its ability to enhance surface engineering, has helped pave the way toward the development of smart, sustainable, and customized textile materials [21].

## 1.7 UV Treatment

The advent of UV radiation technology as a dry process has provided a novel, efficient means for modifying textile surfaces. UV technology has many advantages, such as being fast, low temperature, environmental-friendly, and so on, resulting in significant interest and acceptance of UV technology as a substitute for many of the standard wet chemical treatments. The relationship between UV radiation exposure and textiles is multifaceted and is derived either through dry surface modification or through photochemical processes including grafting and photoinitiated curing [24].

In direct surface modification, UV irradiation (short wavelength) at excitation wavelengths preferably from excimer lamp 172 nm) can greatly improve the wettability (hydrophilicity) of synthetic fibers (i.e., polyester, nylon) by increasing polar functional groups, including carbonyl and carboxyl on the fiber surface. Improvement in wettability increases moisture absorption, moisture capillary transport, and increases stain removability [25].

In contrast, grafting and photochemical curing can achieve covalent attachment of functional monomers and oligomers to fiber surfaces and, as a result, generate very durable functional properties. For example, the photo-grafting of either silicone-based or fluorinated-based monomers can produce very durable water- and oil-repellent surfaces. Another interesting usage of this technology is in developing antimicrobial properties. For instance, a route toward environmentally sustainable antimicrobial activity with reasonable wash resistance is the photo-grafting of biopolymers like chitosan on cotton fabrics. In addition to this, UV irradiation can also be used to increase the dye uptake in fibers or to create differential dyeing effects through patterned exposure of a pre-dyed textile. There is a high degree of precision available in terms of patterned exposure, especially with the use of masks, which allows these processes to be well suited to producing textiles with spatially controlled wettability and complex surface functionalization. While this can produce interesting multi-functional textiles and can provide a powerful route to surface engineering, there remains a need to carefully control the process parameters to avoid polymer chain degradation and to maintain the fabrics mechanical properties. Overall, UV radiation offers a valuable method toward surface engineering and novel multifunctional textiles [24, 26, 27].

## 1.8 Laser Treatment

Laser technology, as an advanced physical tool, emerges as a relevant and advantageous strategy for textile surface modification. This technology allows controlled and precise modifications on fiber surfaces. In this context, there are two principal approaches: pulsed laser deposition (PLD) and laser-induced surface texturing. In PLD, a high-power laser beam, typically a nanosecond-sized pulse, is used to vaporize a target material; the material subsequently deposits on the fabric surface as a thin layer. In laser engraving, the surface changes are purely physical in nature and "modifies" the surface properties by creating unique micro- and nanoscale topographical characteristics. These techniques have created promising new approaches to the fabrication of textiles with engineered functionalities [27, 28].

Pulsed UV lasers such as excimer lasers can offer a controlled roughening of fiber surfaces and are an effective method for changing the wettability of the material. Laser irradiation of polymers such as polyethylene terephthalate (PET) results in unique surface topographies. This surface roughness provides the basis for physical wetting states and can be classified under either Wenzel or Cassie-Baxter. If physical changes are coupled with a chemical treatment (for example, hydrophobic coatings), then it is possible to achieve super-repellent surfaces. Perhaps one of the greatest advantages of the focused-laser approach is that it can pattern functional surfaces. This is achievable either through masking the laser path or by precise beam scanning over designated areas [27, 29].

PLD is a PVD method whereby a high-energy laser beam ablates the target material within a vacuum chamber, producing a plasma plume. The plasma then condenses as a thin film on the textile substrate. PLD is highly versatile in the Netherlands fabrication of many different materials, such as metal oxide nanoparticles and polymers. For example, depositing zinc oxide (ZnO) nanoparticles embeds UV-blocking and antifungal characteristics into fabrics. In addition, one can control the ambient gas atmosphere, allowing one to achieve either hydrophilic or hydrophobic surfaces. For instance, PTFE films deposited in this manner produced superhydrophobic surfaces on cellulosic textiles [30]. While it has been shown to be an eco-friendly alternative to traditional chlorination, laser treatment can reduce wool felting shrinkage by etching the fiber scales. Optimized parameters in this process showed that, with less loss of fabric strength, it produced similar reductions in shrinkage and eliminated the use of hazardous chlorine derivatives [31]. While the dyeability of polyamide fabric increases during laser modification, the bursting strength of the fabric decreases as a result of increased dyeability because of reduced crystallinity and increased free amino groups [32].

Laser treatments have many advantages, such as the ability to create surfaces with complex patterns, enhance dyeability, and modification of surfaces without affecting the bulk properties of the fibers. However, the two main disadvantages of laser treatments are the potential of thermal damage and decreasing mechanical strength if the procedure parameters are not accurately controlled, along with the relatively high costs associated with equipment. The prognosis for this type of technology seems good, especially as there is expected to be a move away from traditional wet chemical processes (e.g., denim bleaching) due to environmental awareness [33, 34].

## 1.9 Nanoparticles Treatment

Nanotechnology currently has a great impact on the textile finishing industry, as it is capable of modifying materials on a molecular level. Nanoparticles are particles with dimensions less than 100 nm. Because of their unique properties, especially their large surface-to-volume ratio, they are gaining popularity in textiles. This means that nanoparticle reactivity and efficacy is therefore vastly improved over bulk materials. The major ways for nanoparticles to be applied to textile products would be as a surface finish applied to a textile or to add the nanoparticles when producing the fiber. Once added to the substrates, they will produce a nanostructured surface, which are fundamentally new and add improved functions of the textile substrates [35].

The most well-known application of nanoparticles is the generation of protective and hygienic properties. Inorganic nanoparticles, such as zinc oxide (ZnO) or titanium dioxide ($TiO_2$) are very strong and stable UV absorbers and drastically improve the UPF of certain fabrics. Nanoparticles, such as silver, or ZnO will have strong antimicrobial property against a wide range of bacterial species. These coating has also shown moderate durability to repeated wash cycles [36, 37].

In addition to protective functions, nanoparticles can produce innovative surface functionalities. For example, silica nanoparticles can form the necessary roughness of the surface for the creation of both superhydrophobic and self-cleaning textiles using the lotus effect. Nanoclays and carbon nanotubes are used to enhance flame retardant and antistatic characteristics, respectively. One of the main benefits of nanoparticles is that you can interleave several functions, e.g., UV protection and antimicrobial, in the same finish, allowing manufacturers to run a more efficient and hence cost-effective process [36, 38].

Textiles can be functionalized with nanotechnology using a variety of methods. Textiles can be functionalized by adding nanoparticles to the polymer melt or solution during fiber spinning or by treating textiles with nanoparticle solutions. In the first method, nanoparticles are physically blended with a polymer melt or spinning solution which produces nanocomposite fibers. In the second method, the modification to the textile surface involves the attachment of nanoparticles onto the surface that is currently widely used in industry. The pad-dry-cure is the simplest method used in commercial procedures for the application of nanoparticles to textiles. In this case, fabrics are put into a bath of stable nanoparticle suspension (pad), liquid is removed (dry), and then the fabric is cured (cure) by heating to ensure a secure attachment of nanoparticles to the surface of the fibers [37, 39].

Nanotechnology has the potential to provide high-performance textiles with multifunctional performance. Its key advantages are high performance with low material consumption (thanks to the high specific surface area of nanoparticles), the potential for combining several functionalities in a single step, and the retention of the intrinsic fabric properties of hand feel and comfort. However, issues still need to be resolved, especially in the long-term adhesion and durability of nanoparticles on the surface of fibers in order to retain performance over time and after washing. The long-term future of this technology lies in the establishment of more sustainable and cost-effective large-scale production routes, together with thorough evaluation of the environmental and safety aspects of nanoparticles along the product life cycle.

## 1.10 Microcapsules and Nanocapsules

The encapsulation technology, including microcapsules and nanocapsules, has emerged as an advanced method of functionalizing textile surfaces and introducing smart properties. This method involves trapping an active substance (core) inside a protective polymeric shell. Microcapsules are usually micrometer sized, whereas nanocapsules, with a diameter of less than one micrometer, have a much larger surface-area-to-volume ratio and thus a faster rate of thermal transfer and are more reactive in nature. In the preparation of capsules, techniques such as *in situ* polymerization, interfacial polymerization, and coacervation are usually employed, which gives the flexibility of choosing the core and shell materials to be used, based on the intended use [40–42].

The capsules are applied to the textile surface using standard finishing methods, like pad-dry-cure and other coating techniques. Binders, usually resins made of acrylic material, are incorporated to improve durability and wash fastness by bonding the capsules to the fabric. One of the distinct features of encapsulation technology is the ability to release the core material in response to an external stimulus, typically through, but not limited to, contact by friction, pressure, temperature, or humidity. The output of the encapsulation technology system is not only a smart and user responsive system but also have potential to prolong the active materials life on the textile surface while protecting active material from degradation in the atmosphere [41, 43, 44].

Applications of encapsulation in textiles are varied. A primary application is the use of microencapsulated phase change materials (PCMs) in thermoregulating textiles that uptake, store, and release latent heat in order to produce thermal comfort. Additionally, microencapsulated extracts, such as the plant extract of lavender oil, are useful for creating durable antimicrobial textiles and enabling aromatherapeutic qualities in cotton. Other applications are mosquito repellency, controlled vitamin delivery, and the production of long-lasting scented textiles [45].

Encapsulation technology provides an advantage in producing a smart and durable functionality through a controlled release mechanism in textiles. It also protects active agents that are sensitive to the environment to allow the production of multifunctional products such as thermoregulating textiles or antimicrobial textiles that contain PCMs or extracts of plants. Perhaps the most difficult challenge lies in achieving a secure bond of capsules to the surface of the fiber while maintaining their functionality during repeated use, washing, and agitated abrasion. Future innovations, especially with the onset of nanocapsules to replace conventional microcapsules to enhance efficacy, will contribute greatly to the next generation of intelligent and high-performance textiles [42, 43, 45].

## 1.11 Chemical Vapor Deposition (CVD)

CVD is a state-of-the-art and versatile textile surface modification technology, with gaseous precursors that react/decompose inside a reactor to create a thin solid film on the fabric surface. Although conventional CVD methods require high temperatures, which render them unusable with polymeric fibers, the use of a process called plasma-enhanced CVD (PECVD) has overcome this challenge. In PECVD, plasma provides the energy to activate the precursor and deposit a film while providing a low enough temperature to avoid damage to the textile substrate from heat [46, 47].

The principal use of CVD in textiles is through the manipulation of surface wettability. The use of organosilicon precursors such as hexamethyldisiloxane, with buildup of silica-like (SiOx) coatings, can create a very hydrophobic surface on cotton, polyester, and polyamide fabrics. These coatings are thin, uniform, and transparent; they adhere strongly to the fiber surface without altering handle or appearance, they showed adequate wash fastness, with the added benefit of fluorocarbon-based precursors to produce surfaces that repel both moisture and oils [48].

In addition to controlling wettability, the coatings may be applied to deposition barrier coatings of materials to limit gas and moisture permeation, as well as to promote the dyeability of synthetic fibers, such as polypropylene, which exhibit very low dye uptake. The recent emphasis on novel precursors that include metal-organic frameworks have significantly increased the range of coatings that could be deposited, as well as supplied the opportunity to produce functional materials which contained a porous form and catalytic function, thus increasing possible value in smart and technical textiles [47].

In conclusion, CVD technology, especially in the form of PECVD, offers clear benefits for functional textile applications in that coatings can be produced uniformly, durably, and at low processing temperatures. Nevertheless, some challenges, including the need for vacuum systems, process complexity, and associated costs, present significant challenges for industrial applications. Future development will likely focus on developing atmospheric-pressure CVD processes to reduce costs and utilizing novel precursors to create multifunctional and intelligent textile materials [46].

## 1.12 Electroless Deposition

Electroless deposition or plating is a wet chemical technique that can be used as a practical method for depositing metallic layers on non-conducting substrates such as textiles. It does so through an autocatalytic redox reaction that does not require an external power supply unlike electroplating and enables uniform coating over the complex porous structure of fabrics.

Essentially, the fabric substrate is prepared in multiple ways and then placed in a plating bath that contains metal ions and a reducing agent to produce the chemical deposition of a metal layer onto the surface of the textile [49].

To successfully utilize electroless plating on textiles, particularly synthetic fibers such as polyester, surface preparation is critical. Surface preparation generally means cleaning, roughening the surface to create mechanical adhesion, and sensitizing the textile then activating it using a catalyst. Sensitization typically utilizes an aqueous solution of stannous chloride, whereas activation uses palladium chloride. Sensitization and activation would provide active sites that would drive the reduction reaction and provide adhesive attributes between the metal layer and polymeric substrate. Uniform, high-quality coatings depend on controlling the parameters of the bath (e.g., concentrations of reactants, temperature, and pH) [49, 50].

While there are multiple applications of electroless plating, one of the most important of these is the development of conductive textiles for specific technical purposes. When applying metals such as copper or nickel, a fabric can be produced that has a high electrical conductivity and can be utilized for EMI shielding in electronic and military equipment. Conductive textiles are also being used to produce flexible sensors and electrodes, smart clothes, and Joule heating textiles. Nickel-phosphorus coatings also provide good wear and corrosion protection to the fabric [50, 51].

Despite requiring surface preparation methods and strict chemical bath management, electroless plating is one of a few methods for the production of conductive and multifunctional textiles since it allows for the even coating of complex, non-conductive surfaces without electricity. There are still barriers to overcome, as the surface preparation process is cumbersome, and the expensive or hazardous chemicals must be managed, but it is reasonable to predict that future advancements will be more environmentally friendly, taking advantage of low cost, environmentally benign or preferably reusable reducing agents and activators, as well as further expanding the potential use in smart textiles and wearable electronics [50].

## 1.13 Enzyme Treatment

Over the last few decades, enzymatic technology has become a new and sustainable biotechnology and has changed the processes of surface modification in textiles. Enzymes that have a biological catalytic action have a high specificity for the substrate, meaning they modify the targeted component of the fiber surface while leaving the fiber structure mostly intact. This specificity, together with the mild environmental conditions under which they can operate (minimal temperature and pH levels), makes enzymatic processes good alternatives to harsh chemicals and energy-wasting processes in textile finishing [52, 53].

Enzymatic processes have been applied abundantly in natural fibers. For example, cellulases are used for bio-polishing of cotton fabrics, where they hydrolyze surface microfibrils, thus removing fuzz while improving softness, luster, and pilling resistance. Similarly, for protein-based fiber such as wool, they are used by proteases to modify the cuticle layer and reduce felting, thus acting as a green alternative to chlorination. Additionally, amylases are used for the desizing of starch-sized cotton fabrics because they provide targeted treatment as well as being gentle [54, 55].

The surface modification of synthetic fibers, such as polyester (PET), which are chemically inert, has presented a unique challenge; however, lipases and cutinases that hydrolyze ester bonds on the polyester surface and create hydroxyl and carboxyl groups on the surface of the fiber. This increases the hydrophilicity of the surface, enhances dye ability, and decreases potential for pilling while maintaining mechanical integrity of the fabric. Such modifications also lead to the ability to functionally modify the polyester [56, 57].

In conclusion, enzyme treatments provide unique benefits—including lower water and energy use, environmental benefits, and the highly precise and uniform application—and will yield better products. However, the described limitations of a biological approach, such as sensitivity to processing parameters, cost of use (more expensive than conventional processes), and normally, rate of reaction (compared to conventional chemical processes), are ongoing issues. Future directions are to produce more stable engineered enzymes, to use immobilization methods so that they can be reused, and to incorporate into multi-stage and sustainable finishing methods [53].

## 1.14 Ozonation

With increasing environmental awareness, the textile industry is starting to explore alternative methods to reduce pollution from traditional textile practice. One of these ways is using ozone ($O_3$) to replace chemical treatments, since ozone is a powerful oxidant. Ozone has an oxidation potential of 2.07 eV, which is greater than hydrogen peroxide (1.77 eV); the preferred oxidizing agent used in textile processing, ozone, would have very high efficiencies in a variety of applications to textile processing, including, preparation, bleaching, dyeing, finishing, and modification to textiles surface [58, 59].

One of the more promising uses of ozone within the textile pre-treatment is preparing cotton fabrics. Studies have demonstrated that ozonation can remove sizing agents and improve water absorption in raw cotton fabrics. Ozone treatment at ambient temperature not only reduces energy costs but can accomplish a level of whiteness similar to that of conventional hydrogen peroxide bleaching while causing less fiber strength damage than that of hydrogen peroxide. Ozone treatment was not as effective to remove plant-based impurities, such as motes, than alkaline scouring technique [58].

The denim industry is leading the way to industrial-scale ozone use. Ozone is often desired to create fading effects primarily on indigo-dyed fabrics. Ozone can be used instead of traditional techniques, such as pumice stone washing or sodium hypochlorite decolorization, which produce halogenated pollutants. Important metrics influencing effects compared to industry practices include duration of treatment, moisture content of the fabric, and amount of ozone. Of those metrics, length of time has been identified as the most significant driver of effect. The presence of moisture in the fabric aids because it adds ozone solubility and aids in producing hydroxyl radicals [60].

Ozone applications are not restricted to cotton and denim; ozone has been investigated for its potential in enhancing the surface of multiple textile substrates. For example, ozone is used to clear polyester from disperse dyeing. Ozone also has been applied to modify the surface of protein fibers (wool and silk) to enhance their dyeability. The benefit of ozone treatment is that modifications primarily occur on the very surface of the fabric, allowing the fabric's feel and look (total touch feel) to be retained [59, 60].

In summary, ozone processing offers a sustainable technology with many environmental advantages, such as reductions in water, energy, and chemicals and the ability to act at room temperature. However, limitations, such as reduced efficacy in the removal of particular impurities from raw cotton, need further exploration. The future of this technology is promising, and, with growing availability of specialized equipment, ozone use is expected to expand across other areas in the textile industry and will contribute to cleaner production.

## 1.15 Coating

Coating represents a core surface-engineering method that enables conventional textiles to serve as advanced technical materials. The coating process allows for the application of a layer on the surface of the fabric to either enhance existing properties or to incorporate new functional properties. The choice of the coating material—usually polyurethane, polyvinyl chloride, or silicone—and fiber types used in the substrate, varying from natural cotton to high performance aramid, will fundamentally affect the properties and performance of the coated textile [61, 62].

The method for applying coatings includes the traditional approaches of direct and transfer coating and more advanced coating methods like sol-gel processing and plasma treatment, which can provide unique advantages with respect to repetitive control of coating thickness and uniformity while providing potential environmental benefits. The prime motivation behind the literature included in this chapter is that functional properties are being imparted to fabrics such as water repellency or breathability, xenon exposure, and flame retardancy [62, 63].

The adhesion of the coating layer to the textile substrate is an important consideration when assessing the durability and lifetime performance of the coated fabric, and it can commonly be measured using testing protocols. Recent developments are in smart textiles which employ conductive materials for sensors and wearable electronics. Nanotechnology plays a significant role in these developments and thus enables the production of surfaces that are self-cleaning, antimicrobial, superhydrophobic, etc. [62, 63].

The trajectory for textile coating is influenced by the increasing shift toward environmental sustainability, including waterborne, biodegradable, or recyclable coating systems. The move toward multifunctional textiles that can provide multiple aims at once is discussing the definition of textiles used across different markets and industries including the medical, military, sportswear, and agricultural. In summary, coating is an advanced surface modification technique that improves textiles functionality for specific applications and extreme environments [61, 62].

## 1.16 Conclusion

Textile surface modification is a vibrant and evolving area of material science and is fundamental for making advanced and smart textiles. This chapter presented a range of methods from traditional industrial processes to advanced laboratory methods; each providing a unique mechanism to modify the surface characteristics of fibers and fabrics. A common theme was the demand for durable multi-functional finishes, without sacrificing comfort or mechanical performance. There has been considerable progress, while many of the technologies reviewed continue to face problems with performance under scale, cost performance, durability, and environmental considerations. The future of textile finishing will be defined by the development of continued efficiency and greener processes and innovative materials. The emerging opportunity for advanced surface modification techniques illustrates the potential for advanced products across many sectors from medicine and defense to sports and electronics and ultimately demonstrates that, if we can imagine it, then we can explore textile potential beyond traditional textiles.

**AI disclosure statement:** The author acknowledges the use of AI-based tools for minor language editing or grammatical correction. However, all intellectual contributions, analyses, and conclusions presented in this chapter are entirely the author's own.

## References

1. Mather, R.R., 13 - Surface modification of textiles by plasma treatments, in: *Surface Modification of Textiles*, Q. Wei (Ed.), pp. 296–317, Woodhead Publishing, Cambridge, UK, 2009.
2. Zille, A., Oliveira, F.R., Souto, A.P., Plasma Treatment in Textile Industry. *Plasma Processes Polym.*, 12, 2, 98–131, 2015.
3. Saleem, M., *et al.*, Functionality and applications of non-thermal plasma activated textiles: A review. *Mater. Today: Proc.*, 47, S74–S82, 2021.
4. Haji, A. and Kan, C.-W., Chapter 19 - Plasma treatment for sustainable functionalization of textiles, in: *Green Chemistry for Sustainable Textiles*, N. Ibrahim and C.M. Hussain (Eds.), pp. 265–277, Woodhead Publishing, Duxford, UK, 2021.
5. Tudoran, C., Roşu, M.C., Coroş, M., A concise overview on plasma treatment for application on textile and leather materials. *Plasma Processes Polym.*, 17, 8, e2000046, 2020.
6. McCoustra, M.R.S. and Mather, R.R., Plasma modification of textiles: understanding the mechanisms involved. *Text. Prog.*, 50, 4, 185–229, 2018.
7. Peran, J. and Ercegović Ražić, S., Application of atmospheric pressure plasma technology for textile surface modification. *Text. Res. J.*, 90, 9–10, 1174–1197, 2020.
8. Shahidi, S., Moazzenchi, B., Ghoranneviss, M., A review-application of physical vapor deposition (PVD) and related methods in the textile industry. *Eur. Phys. J. Appl. Phys.*, 71, 3, 31302, 2015.
9. Tan, X.-Q., *et al.*, Recent progress in magnetron sputtering technology used on fabrics. *Materials*, 11, 10, 1953, 2018.
10. Yuan, X., *et al.*, Properties and application of multi-functional and structurally colored textile prepared by magnetron sputtering. *J. Ind. Text.*, 51, 8, 1295–1311, 2022.
11. Hu, Q., *et al.*, Sputtering-Deposited Thin Films on Textiles for Solar and Heat Managements: A Mini-Review. *Phys. Status Solidi (a)*, 219, 3, 2100572, 2022.
12. Huang, F., *et al.*, Surface functionalization of silk fabric by PTFE sputter coating. *J. Mater. Sci.*, 42, 19, 8025–8028, 2007.

13. Textor, T., 9 - Modification of textile surfaces using the sol-gel technique, in: *Surface Modification of Textiles*, Q. Wei (Ed.), pp. 185–213, Woodhead Publishing, Cambridge, UK, 2009.
14. Ismail, W.N.W., Sol–gel technology for innovative fabric finishing—A Review. *J. Sol-Gel Sci. Technol.*, 78, 3, 698–707, 2016.
15. Rivero, P.J. and Goicoechea, J., 4 - Sol-gel technology for antimicrobial textiles, in: *Antimicrobial Textiles*, G. Sun (Ed.), pp. 47–72, Woodhead Publishing, Duxford, UK, 2016.
16. Periyasamy, A.P., *et al.*, Progress in Sol-Gel Technology for the Coatings of Fabrics. *Materials*, 13, 8, 1838, 2020.
17. Mahltig, B., Haufe, H., Bottcher, H., Functionalisation of textiles by inorganic sol-gel coatings. *J. Mater. Chem.*, 15, 41, 4385–4398, 2005.
18. Pan, Y., *et al.*, Layer-by-Layer Self-Assembly Coating for Multi-Functionalized Fabrics: A Scientometric Analysis in CiteSpace (2005–2021). *Molecules*, 27, 19, 6767, 2022.
19. Magovac, E., *et al.*, Layer-by-Layer Deposition: A Promising Environmentally Benign Flame-Retardant Treatment for Cotton, Polyester, Polyamide and Blended Textiles. *Materials*, 15, 2, 432, 2022.
20. Gomes, A.P., *et al.*, Layer-by-Layer Assembly for Biofunctionalization of Cellulosic Fibers with Emergent Antimicrobial Agents, in: *Cellulose Chemistry and Properties: Fibers, Nanocelluloses and Advanced Materials*, O.J. Rojas (Ed.), pp. 225–240, Springer International Publishing, Cham, 2016.
21. Ghosh, J., *et al.*, Grafting modification for textile functionalization: innovations and applications. *Discov. Appl. Sci.*, 7, 1, 49, 2025.
22. Abidi, N., Surface grafting of textiles, in: *Surface Modification of Textiles*, Q. Wei (Ed.), pp. 91–107, Woodhead Publishing, Cambridge, UK, 2009.
23. Nadi, A., *et al.*, Evolution in the surface modification of textiles: a review. *Text. Prog.*, 50, 2, 67–108, 2018.
24. Ferrero, F., Migliavacca, G., Periolatto, M., UV Treatments on Cotton Fibers, in: *Cotton Research*, I.Y. Abdurakhmonov (Ed.), pp. 235–256, Intech Open, London, UK, 2016.
25. Gotoh, K. and Hayashiya, M., Improvement of Serviceability Properties of Synthetic Textile Fabrics using 172 nm Ultraviolet Excimer Lamp. *Text. Res. J.*, 78, 1, 37–44, 2008.
26. Bahners, T., *et al.*, Patterned Functionalization of Textiles Using UV-Based Techniques for Surface Modification – Patterned Wetting Behavior, in: *Advances in Contact Angle, Wettability and Adhesion*, pp. 145–165, John Wiley & Sons, Inc., Hoboken, USA, 2018.
27. Bahners, T., Excimer laser irradiation of synthetic fibres as a new process for the surface modification of textiles — a review. *Opt. Quantum Electron.*, 27, 12, 1337–1348, 1995.
28. Shahidi, S., Moazzenchi, B., Ghoranneviss, M., Improving the dyeability of polypropylene fabrics using laser technology. *J. Text. Inst.*, 104, 10, 1113–1117, 2013.

29. Kan, C., A study of laser treatment on polyester substrates. *Fibers Polym.*, 9, 2, 166–170, 2008.
30. Yazdanpanah, A., *et al.*, *In situ* synthesize of ZnO nanoparticles on cotton fabric by laser ablation method; antibacterial activities. *J. Text. Inst.*, 1–11, 2021.
31. Nourbakhsh, S., Ebrahimi, I., Valipour, P., Laser treatment of the wool fabric for felting shrinkage control. *Fibers Polym.*, 12, 4, 521–527, 2011.
32. Bahtiyari, M.I., Laser modification of polyamide fabrics. *Opt. Laser Technol.*, 43, 1, 114–118, 2011.
33. Chow, Y.L.F., Chan, A., Kan, C.-W., Effect of CO2 laser irradiation on the properties of cotton fabric. *Text. Res. J.*, 82, 12, 1220–1234, 2011.
34. Štěpánková, M., Wiener, J., Dembický, J., Properties of cotton fabric after irradiation with infrared CO2 laser. *Fibers Polym.*, 15, 10, 2072–2076, 2014.
35. Mishra, R., *et al.*, The production, characterization and applications of nanoparticles in the textile industry. *Text. Prog.*, 46, 2, 133–226, 2014.
36. Vigneshwaran, N., 8 - Modification of textile surfaces using nanoparticles, in: *Surface Modification of Textiles*, Q. Wei (Ed.), pp. 164–184, Woodhead Publishing, Cambridge, UK, 2009.
37. Mahltig, B., Nanomaterials and textiles: Review on materials and applications. *J. Eng. Fibers Fabr.*, 20, 15589250251318010, 2025.
38. Jadoun, S., Verma, A., Arif, R., Modification of Textiles *via* Nanomaterials and Their Applications, in: *Frontiers of Textile Materials*, pp. 135–152, John Wiley & Sons, Inc., Hoboken, USA, 2020.
39. Jalali, E., *et al.*, Fixation of nanoparticles on fabric: Applications in general health management. *Mater. Today Commun.*, 41, 2024.
40. Peng, X., *et al.*, Biopolymers-based microencapsulation technology for sustainable textiles development: A short review. *Case Stud. Chem. Environ. Eng.*, 7, 100349, 2023.
41. Boh Podgornik, B., Šandrić, S., Kert, M., Microencapsulation for Functional Textile Coatings with Emphasis on Biodegradability—A Systematic Review. *Coatings*, 11, 11, 1371, 2021.
42. Mohamed, H.A., *et al.*, Microencapsulation and its Application in Textile Industry. *J. Text. Color. Polym. Sci.*, 22, 1, 81–87, 2025.
43. Naz, S., *et al.*, Recent advances in polymer nanoencapsulation of essential oils for multi-functional textile finishing. *Mater. Adv.*, 6, 2460–2476, 2025.
44. Abdel-Aziz, E., *et al.*, Microencapsulation and its Application in Textile Wet Processing: A Review. *J. Text. Color. Polym. Sci.*, 19, 2, 189–202, 2022.
45. El-Wekil, A.N., *et al.*, Micro/Nano Capsulation: Textile Functional Features and Application. *J. Text. Color. Polym. Sci.*, 22, 1, 313–325, 2025.
46. Wilson, J.I.B., 6 - Textile surface functionalisation by chemical vapour deposition (CVD), in: *Surface Modification of Textiles*, Q. Wei (Ed.), pp. 126–138, Woodhead Publishing, Cambridge, UK, 2009.
47. Rosace, G., Canton, R., Colleoni, C., Plasma enhanced CVD of SiOxCyHz thin film on different textile fabrics: Influence of exposure time on the abrasion resistance and mechanical properties. *Appl. Surf. Sci.*, 256, 8, 2509–2516, 2010.

48. Zareei Pour, F., Karimi, H., Madadi Avargani, V., Preparation of a superhydrophobic and superoleophilic polyester textile by chemical vapor deposition of dichlorodimethylsilane for Water–Oil separation. *Polyhedron*, 159, 54–63, 2019.
49. Jiang, S.Q. and Guo, R.H., 5 - Modification of textile surfaces using electroless deposition, in: *Surface Modification of Textiles*, Q. Wei (Ed.), pp. 108–125, Woodhead Publishing, Cambridge, UK, 2009.
50. Guo, R.H., *et al.*, An alternative process for electroless copper plating on polyester fabric. *J. Mater. Sci.: Mater. Electron.*, 20, 1, 33–38, 2009.
51. Ali, A., *et al.*, Copper electroless plating of cotton fabrics after surface activation with deposition of silver and copper nanoparticles. *J. Phys. Chem. Solids*, 137, 109181, 2020.
52. Ghaffar, A., *et al.*, Ecological Applications of Enzymes in Plants Based Textile Dyeing, in: *Dyes and Pigments - Insights and Applications*, B. Kumar (Ed.), IntechOpen, Rijeka, 2023.
53. Nierstrasz, V.A., 7 - Enzyme surface modification of textiles, in: *Surface Modification of Textiles*, Q. Wei (Ed.), pp. 139–163, Woodhead Publishing, Cambridge, UK, 2009.
54. Ismail, S.A., *et al.*, Benign Felt-proofing of Wool Fibers Using a Keratinolytic Thermophilic Alkaline Protease. *J. Nat. Fibers*, 1–13, 2020.
55. Samant, L., *et al.*, Antimicrobial and UV Protection Properties of Cotton Fabric Using Enzymatic Pretreatment and Dyeing with Acacia Catechu. *J. Nat. Fibers*, 19, 6, 2243–2253, 2020.
56. Wu, J., *et al.*, Eco-friendly surface modification on polyester fabrics by esterase treatment. *Appl. Surf. Sci.*, 295, 150–157, 2014.
57. Lee, S.H. and Song, W.S., Surface modification of polyester fabrics by enzyme treatment. *Fibers Polym.*, 11, 1, 54–59, 2010.
58. Eren, H.A. and Ozturk, D., The evaluation of ozonation as an environmentally friendly alternative for cotton preparation. *Text. Res. J.*, 81, 5, 512–519, 2011.
59. Eren, H.A., *et al.*, Ozone: An Alternative Oxidant for Textile Applications, in: *Sustainability in the Textile and Apparel Industries : Production Process Sustainability*, S.S. Muthu and M.A. Gardetti (Eds.), pp. 81–98, Springer International Publishing, Cham, 2020.
60. Hu, L., *et al.*, Investigation of Color Fading and Fabric-Touch Test for Jeans through Ozonation. *Ozone: Sci. Eng.*, 43, 3, 276–283, 2021.
61. Ghosh, J., *et al.*, Functional coatings for textiles: advancements in flame resistance, antimicrobial defense, and self-cleaning performance. *RSC Adv.*, 15, 14, 10984–11022, 2025.
62. Meirowitz, R.E., 8 - Coating processes and techniques for smart textiles, in: *Active Coatings for Smart Textiles*, J. Hu (Ed.), pp. 159–177, Woodhead Publishing, Duxford, UK, 2016.
63. Billah, S.M.R., Textile Coatings, in: *Functional Polymers*, M.A. Jafar Mazumder, H. Sheardown, A. Al-Ahmed (Eds.), pp. 1–58, Springer International Publishing, Cham, 2018.

# 2

# Surface Modification and Functionalization of Textiles by Plasma Treatment

**Aminoddin Haji[1]* and Çağlar Sivri[2]**

[1]*Department of Textile Engineering, Yazd University, Yazd, Iran*
[2]*Department of Engineering Management, Faculty of Engineering and Natural Sciences, Bahçeşehir University, Beşiktaş, İstanbul, Türkiye*

***Abstract***
Plasma treatment is an eco-friendly technology for surface modification and functionalization of textiles. It is a surface-specific process that affects only the very surface layer of the fibers, and the bulk properties remain almost intact. Generally, there are two main types of plasma treatment devices for surface treatment of textiles including low-pressure plasma and atmospheric-pressure plasma. Various modifications including activation, polymerization, cleaning, etching, and grafting can be applied on textile surfaces by plasma treatment. The type and extent of the modification depends on the process parameters such as process type, power, gas composition, and duration. In this chapter, the basics and different types of plasma technology is introduced, and the applications of different plasma treatment processes on different textile substrates are also reviewed on the basis of the literature.

***Keywords***: Dielectric barrier discharge, cold plasma, etching, plasma grafting, surface activation

## 2.1 Introduction

Plasmas are complex ionized gases composed of positive and negative ions, electrons, free radicals, and neutral species [1]. There are two main types of plasmas available for the treatment of the materials including

**Corresponding author*: ahaji@yazd.ac.ir

Aminoddin Haji and Çağlar Sivri (eds.) Surface Modification and Functionalization of Textiles, (23–38) 

low-temperature plasma (cold plasma) and high-temperature plasma (thermal plasma) [2, 3]. When sufficient energy is applied into a neutral gas, transition to the plasma state typically occurs. The applied energy can be in the form of heating to very high temperatures or by subjecting the gas to a strong electromagnetic field. This energy input leads to stripping away the electrons from the atoms or molecules causing their ionization and creating a dynamic mixture of charged particles. Thermal plasmas are characterized by possessing local thermodynamic equilibrium. Both electrons and heavy particles (ions and neutrals) are at very high temperatures, typically well exceeding 10,000 K. Consequently, the gas in a thermal plasma is often fully ionized. These high-temperature equilibrium conditions are typically achieved by supplying substantial energy to a gas, often at or near atmospheric pressure. Common generation methods include high-intensity electric arc discharges (using direct current or alternating current), inductively coupled plasmas driven by radio-frequency fields, or microwave (MW) discharges operating at high power levels. Thermal plasma technology can be used to treat various materials including metals, ceramics, nanophase materials, and composites [4].

Electrons are highly energetic in non-thermal (or cold) plasmas and the electron temperature is similar to thermal plasmas reaching values equivalent to tens of thousands of Kelvin. However, the ions and neutral gas particles remain relatively "cold," often near ambient temperature. The degree of ionization in non-thermal plasmas is typically low to moderate, often less than 1%.

Non-thermal plasmas are produced by various types of electrical discharges specifically designed to create energetic electrons without significantly heating the bulk gas. Cold plasmas can be generated at low gas pressures (such as glow discharge) or atmospheric pressure [such as dielectric barrier discharges (DBD), corona discharges, atmospheric pressure glow discharge (APGD), and atmospheric pressure plasma jets (APPJs)]. Cold plasmas are highly chemically reactive and offer various types of surface treatments for the heat-sensitive materials such as surface etching, cleaning, activation, deposition of films, and post-treatment of coated surfaces [5].

Plasma treatment has played an important role in the surface treatment of textiles from traditional ones to nano-textiles in recent years, providing a particularly protective function rather than only dressing [6]. Plasma treatment of textile substrates allows for obtaining various novel features without changing the key textile properties. Plasma treatment of textiles provides better dyeability, adhesion strength, enhanced wettability or hydrophobicity, antibacterial performance, UV protection, and so on [6].

To give more information about various textile operations and functions obtained, the plasma treatment offers practical improvements for the dyeing of textiles with natural or synthetic dyestuffs. Haji and Naebe have pointed out that plasma treatment improves the dyeing uptake of textiles with natural dyes in an eco-friendly way [7]. The literature also stressed the importance of plasma for improving adhesion between fibers and matrices in composites, adhesion in textile lamination, and adhesion of multiple resins to the textile substrates [8]. On the other hand, plasma treatment enhances the wettability of both synthetic and natural fibers by introducing polar groups on the surface, thus increasing their hydrophilic capacity. This process can be optimized by controlling the plasma variables, such as the gas composition, discharge power, pressure, and exposure time [9]. From a health or biomedical applications perspective, the plasma treatment provides antimicrobial activity with a killing efficiency of 99.9999% for both *E. coli* and *S. aureus* strains and enables the immobilization of antibacterial components for the enhancement of biomedical applications [10]. Plasma treatment also enhances the UV protection properties of textiles. By modifying fibers at the surface level, plasma technology improves the fabric's ability to block ultraviolet radiation. This treatment increases the adhesion of UV-absorbing agents such as aluminium oxide ($Al_2O_3$), nano-silver (Ag NP), and nano-titanium dioxide ($TiO_2$), thereby improving the overall UV protective performance of the material [11]. The benefits obtained from plasma treatment can even be exploited in wastewater treatment for textile processes, it provides a highly practical and efficient oxidation, carried out in a respectively short time, available from lab to industrial scale, and, finally, the entire process sounds environmentally friendly [12]. An extensive and current summary of plasma treatment in textiles has been provided in this chapter. First, plasma techniques for textiles were explored, focusing on their application to various textile structures, including fibers, woven fabrics, knitted fabrics, and nonwoven materials. The subsequent section examines the diverse end uses of plasma technology for the textiles. Finally, the section on outlook and future work recommendations offers guiding comments for the field.

## 2.2 Types of Plasma Used in Textile Treatments

Plasma types can be classified on the basis of the pressure applied or thermal characteristics. In terms of thermal classification, plasmas are typically categorized as thermal (hot) or non-thermal (cold) plasmas. As discussed earlier (Section 2.1), the temperature in thermal plasma is far higher than the melting or decomposition temperature of common textile fibers.

Thus, this plasma type is not generally suitable for the treatment of textile materials and only cold plasma will be discussed in this chapter.

### 2.2.1 Plasma Types Based on Applied Pressure

Plasma treatments in textiles can be categorized into two main types on the basis of the pressure applied: low-pressure plasma treatments, which offer precise control over surface modifications by creating a reactive plasma environment in a vacuum chamber, and atmospheric-pressure plasma treatments, which utilize ambient air to deliver functional modifications without the need for complex equipment, making them more accessible for industrial applications. Both methods significantly enhance the properties of textile materials, including their durability, functionality, and aesthetic appeal [13].

#### *2.2.1.1 Low-Pressure Plasma Treatments*

Research on plasma treatment of textiles has primarily focused on low-pressure discharges, which are non-destructive, effective at a microscopic level, and capable of removing impurities like dust, grease, and bacteria. However, their industrial use is limited by the high cost of large vacuum systems required for operation [14]. A typical low-pressure plasma treatment equipment suitable for surface modification of textiles consists of a vacuum chamber, a power supply operating at 40 kHz, 13.56 MHz, or 2.45 GHz (based on the specific process requirements, such as etching speed, uniformity, and energy consumption), a gas distribution system with controlled gas flow, and electrodes in parallel or capacitively coupled configuration. Figure 2.1 shows a schematic representation of a low-pressure plasma treatment system [15].

#### *2.2.1.2 Atmospheric Pressure Plasma Treatments*

Atmospheric pressure plasma treatments for textiles are innovative surface modification techniques that enhance material properties. Operating under ambient pressure, these treatments use ionized gases to improve characteristics such as wettability, dyeability, adhesion, and resistance to water or microbes [16]. Atmospheric-pressure plasma systems offer a practical approach to textile treatment, categorized into four main types on the basis of their electrode configuration and voltage/frequency characteristics. Corona discharge utilizes asymmetric electrodes, typically a highly curved electrode (needle or wire) and a low-curvature electrode (plate

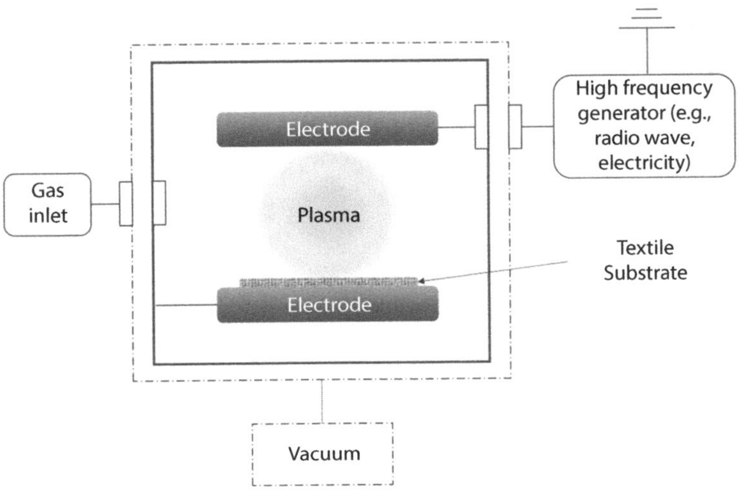


**Figure 2.1** Schematic presentation of low-pressure plasma treatment. Re-used with permission from John Wiley and Sons, License number 6063010355840 [15].

or cylinder), generating plasma in a localized region near the sharp electrode. Due to its inhomogeneity and limited treatment area, it is rarely used for textile processing. DBD employs two parallel electrodes, at least one coated with a dielectric material. This system can produce either filamentary or homogeneous plasma, with the latter preferred for uniform textile treatment. APGD operates at relatively low voltage and high frequency between bare metal electrodes, creating a uniform glow discharge. Helium is often the preferred gas due to its arc-prevention properties, although argon or nitrogen can be substituted. Finally, APPJs use concentric electrodes with the processing gas flowing through them, forming a plasma jet directed at the substrate. The ionized gas exits through a nozzle, enabling targeted surface treatment. The diversity in these systems highlights the adaptability of atmospheric-pressure plasma for various textile applications. Figure 2.2 shows a schematic representation of different atmospheric-pressure plasma treatment systems [7].

## 2.3 Applications of Cold Plasma on Textiles

In the textile industry, cold plasma technology can perform various processes while preserving the quality of textile materials. It is important for plasma generators to operate under conditions that meet specific industrial requirements. This ensures that cold plasma can be used for finishing

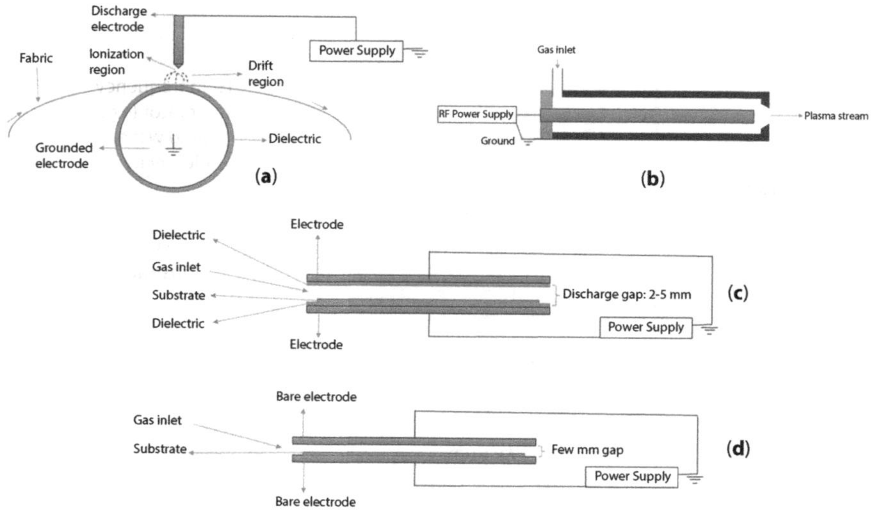


**Figure 2.2** Schematic presentation of atmospheric-pressure plasma treatment systems (a: corona discharge, b: APPJ, c: DBD, and d: APGD). Re-used with permission from Elsevier, License number 6064181242848 [7].

operations without compromising the material's integrity. Additionally, it allows for controlled modifications in quantity and structure, demonstrating that any changes to the material's integrity are carefully managed [17].

Vacuum plasma treatments in the textile industry utilize low-pressure plasma to alter the surface properties of textile materials. This eco-friendly process uses minimal amounts of chemicals. Surface activation, functional coatings, cleaning, etching, and texturing are the most common applications of the vacuum plasma treatment in the textile industry [18]. Boaretti *et al.* developed flame retardant (FR) textiles by treating polyamide 66 fabrics with MW plasma technology to reduce the amount of FR used. MW vacuum plasma, utilizing a helium/oxygen ($He/O_2$) gas mixture, activated the fabric surface, enhancing its affinity for thiourea impregnation [19]. Haji *et al.* used low-pressure plasma (oxygen and argon) and chitosan coating for surface modification of wool fabrics and improved the dyeability of wool with various natural dyes. Plasma treatment is able to partially destroy the wool hydrophobic scales, reduce the disulfide bond content, and facilitate the diffusion and fixation of the dyes on wool fibers [20–32]. Figure 2.3 shows the morphologies of untreated and plasma-treated wool fibers. The etching of surface scales is evident [28].

In another study, Haji *et al.* employed low-pressure oxygen plasma pre-treatment for surface activation to initiate grafting of acrylic acid

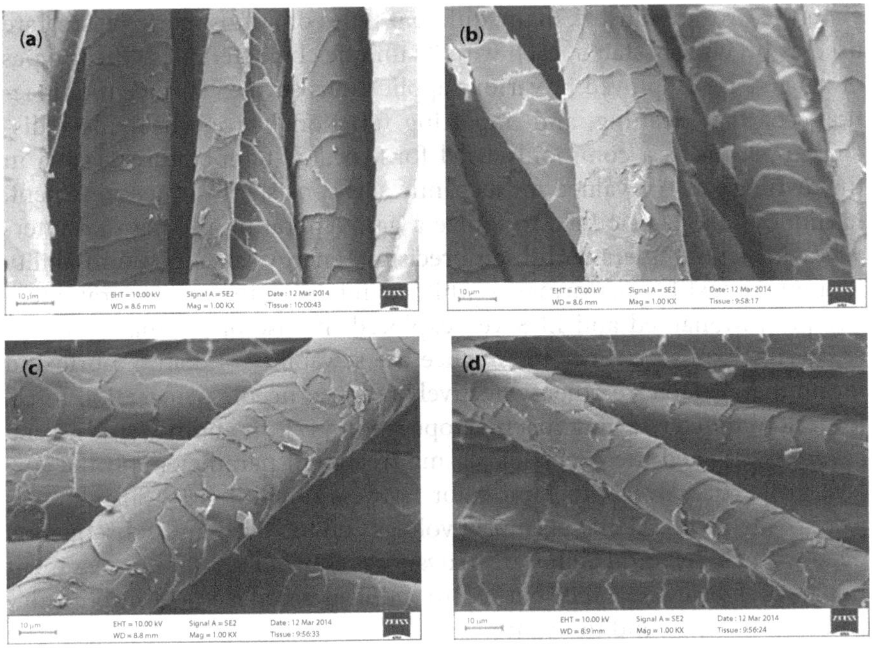


**Figure 2.3** FESEM images of wool fibers. (a) Raw, untreated fiber. (b) $O_2$ plasma-treated fiber. (c) $Ar/O_2$ plasma-treated fiber. (d) Ar plasma-treated fiber. Re-used with permission from Springer, License number 6065190247947 [28].

on polypropylene nonwoven fabric. The grafted nonwoven was used to remove cationic dye from aqueous media efficiently [33].

Haji *et al.* employed low-pressure plasma for surface functionalization of other textile fibers such as cotton, polyester, nylon, and silk. Oxygen plasma has been used for grafting of acrylic acid on cotton to facilitate dyeing with natural cationic dye as well as *in situ* synthesis of silver nanoparticles [34, 35]. Plasma treatment and subsequent chitosan coating also enhanced the attachment of titanium dioxide nanoparticles [36] and dyeability of cotton fabric with different natural and synthetic dyes [21, 37–40]. Also, water vapor plasma has been used for green *in situ* synthesis of silver nanoparticles onto cotton fabric without the use of chemical reducing agents [41].

Low-pressure oxygen plasma has been employed as a pretreatment to improve the water repellent finishing of polyester/viscose fabric using fluoroalkyl acrylic resin [42] and self-cleaning finishing of polyester/wool fabric using $TiO_2$ nanoparticles [43]. Oxygen plasma and acrylic acid also were used to improve the coating of polyester fabric with carbon nanotubes, leading to improved microwave shielding behaviour [44, 45].

In the textile industry, many different types of atmospheric plasma are used as previously stated, each with its unique characteristics. These types of plasma are employed in various applications such as surface modification, coating, cleaning, and imparting functional properties. Glow discharge plasma is a commonly used form of low-temperature plasma in the textile industry, valued for its gentle and consistent surface treatment. It employs an electric field to ionize a gas, generating plasma that interacts with textile fibers with high precision, energy efficiency, and ability to improve textiles' performance while maintaining their structural integrity [46]. Mohamed and El-Sayed explored the use of plasma treatment at varying exposure times to enhance the sewability of denim fabric and found that plasma treatment effectively reduced needle penetration force, leading to quieter sewing machine operation [47]. In another study, Feng *et al.* investigated the effects of plasma treatment parameters on the surface properties, dyeability, and color fastness of silk fabric. Silk samples were treated with APGD plasma at voltages up to 45 kV and subsequently dyed with C.I. Natural Yellow 3, analysis has shown that plasma treatment improved fabric dyeability but led to reduced color fastness in both dry and wet rubbing conditions [48].

Corona discharge applies a high-voltage, low-current electrical discharge to the textile material's surface in the presence of air or another gas, generating localized plasma. The plasma produces ozone and reactive species that induce molecular-level changes on the surface. In textile applications, corona discharge is mainly employed to enhance surface wettability, adhesion, and printability. By increasing surface energy and eliminating contaminants, it improves the bonding of dyes, inks, and coatings to both synthetic and natural fibers [49]. Nourbakhsh *et al.* explored the coating of cotton fabrics with copper and zinc oxide nanoparticles using corona discharge; corona-treated cotton displayed an uneven surface characterized by pores and holes facilitating the penetration of nanoparticles into these pores, while, on untreated cotton, the nanoparticles remained on the surface [50]. Not only for the traditional textiles or fibers, but the corona discharge has also applications on nanofibers as a powerful candidate of high-performance materials of the future. Sivri treated the surface of polyethylene oxide (PEO) nanofibers to enhance their functional capabilities with varying plasma and electrospinning parameters, including polymer solution concentrations and plasma power densities and SEM analysis revealed significant structural changes (Figure 2.4), even at low plasma frequencies, due to the nanofibers' high surface-area-to-volume ratio [51]. Abdouss *et al.* applied atmospheric pressure DBD Ar/$O_2$ plasma on nylon

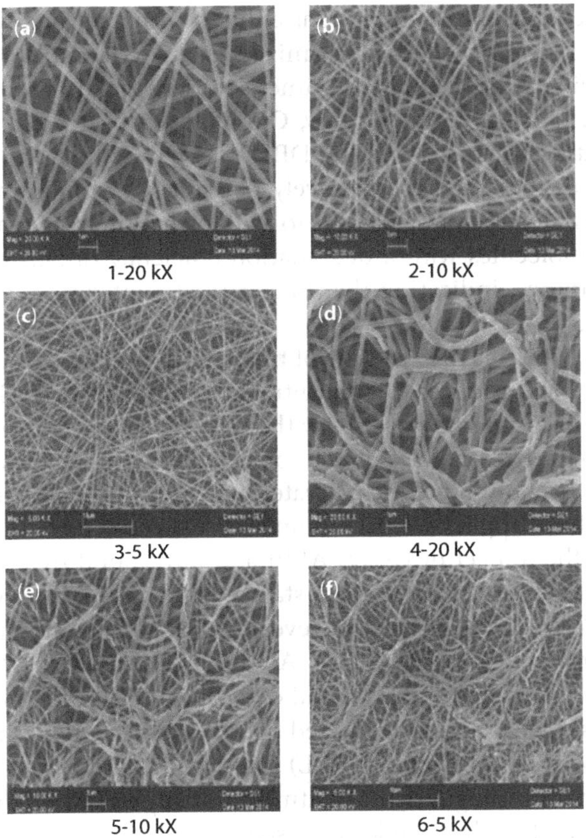


**Figure 2.4** Untreated PEO nanofibers (a, b, and c) and plasma-treated PEO nanofibers under 0.05 kW(d, e, and f) [51].

66 electrospun nanofibers and reported surface etching and increase in oxygen containing surface groups [52].

DBD plasma is a non-thermal plasma technology commonly employed for textile surface modification. Operating at atmospheric pressure, it utilizes alternating current to generate plasma between two electrodes separated by a dielectric material. This method produces energetic species, such as ions, radicals, and UV photons, which interact with textile surfaces to achieve surface functionalization and enhance properties like hydrophilicity or hydrophobicity, offering an eco-friendly processing approach [53]. Saleem *et al.* used a single-step ultrasonic biosynthesis method to synthesize and coat $TiO_2$ nanoparticles on DBD plasma-functionalized cotton fabric, and plasma treatment enhanced the cotton's self-cleaning properties. The process was optimized using a complete composite design, varying plasma exposure time, solution molarity, and ultrasonication time [54]. In another study, Oktaviana *et al.* treated wastewater from the batik industry using the DBD plasma method and found that DBD plasma active species effectively degrade the contaminants in the textile wastewater, reducing pH due to acid formation from reactions with wastewater molecules with a degradation efficiency of 54.29% achieved after 120 minutes, indicating that longer treatment times enhanced the efficiency [55].

The APPJ, which is the last type of the atmospheric plasma, is a plasma technology that produces a concentrated plasma stream. The process operates at atmospheric pressure without vacuum systems and delivers low-temperature treatment suitable for heat-sensitive textiles. It is environmentally friendly by reducing water and chemical usage and ensures precise, localized application to minimize material waste [56]. Li and Qiu investigated the impact of $He/O_2$ APPJ and ultrasound treatment on de-sizing cotton fabrics prepared from starch phosphate and polyvinyl alcohol blends. AFM and SEM analysis revealed that APPJ treatment enhances the surface roughness of the fabric. After 35 seconds of APPJ treatment followed by 20 minutes of ultrasound de-sizing at 60°C, the fiber surfaces are almost as clean as those of unsized fibers [57]. In a study carried out by Yan *et al.*, the polycaprolactone (PCL) solution was pretreated using APPJ prior to electrospinning. Plasma treatment notably enhanced the solution's conductivity and acidity, while the viscosity initially increased and then decreased. Plasma treatment also affected the molecular chain segments of PCL and modified the fiber's crystal structure. XPS analysis showed that the surface elements and the chemical bonding between carbon and oxygen remained unchanged after plasma treatment [58].

## 2.4 Conclusion

Plasma treatment, as a clean and dry alternative to many textile wet processes, has the potential to alter physical and chemical properties of textile surfaces. Several properties including but not limited to wettability, hydrophilicity, hydrophobicity, adhesion, surface tension, surface charge, and roughness can be tuned by precise selection of plasma treatment parameters. Depending the required properties and the substrate characteristics, the type of plasma and the process parameters should be selected.

Low-pressure plasma provides higher uniformity compared with the atmospheric plasma type. However, while batch processing limits the applicability of low-pressure plasma for continuous textile production, atmospheric plasma is suitable for in-line, high-speed processing with lower operational costs.

**AI disclosure statement:** The author acknowledges the use of AI-based tools for minor language editing or grammatical correction. However, all intellectual contributions, analyses, and conclusions presented in this chapter are entirely the author's own.

## References


1. Bogaerts, A., *et al.*, Gas discharge plasmas and their applications. *Spectrochim. Acta, Part B*, 57, 4, 609–658, 2002.
2. Bonizzoni, G. and Vassallo, E., Plasma physics and technology; industrial applications. *Vacuum*, 64, 3–4, 327–336, 2002.
3. Samal, S., Thermal plasma technology: The prospective future in material processing. *J. Cleaner Prod.*, 142, 3131–3150, 2017.
4. Taylor, P.R. and Pirzada, S.A., Thermal plasma processing of materials: A review. *Adv. Perform. Mater.*, 1, 35–50, 1994.
5. Bárdos, L. and Baránková, H., Cold atmospheric plasma: Sources, processes, and applications. *Thin Solid Films*, 518, 23, 6705–6713, 2010.
6. Tudoran, C., Roşu, M.C., Coroş, M., A concise overview on plasma treatment for application on textile and leather materials. *Plasma Processes Polym.*, 17, 8, e2000046, 2020.
7. Haji, A. and Naebe, M., Cleaner dyeing of textiles using plasma treatment and natural dyes: A review. *J. Cleaner Prod.*, 265, 121866, 2020.
8. McCoustra, M.R.S. and Mather, R.R., Plasma modification of textiles: understanding the mechanisms involved. *Text. Prog.*, 50, 4, 185–229, 2018.
9. Ferrero, F. and Periolatto, M., Modification of Surface Energy and Wetting of Textile Fibers, in: *Wetting and Wettability*, M. Aliofkhazraei (Ed.), InTech, London, UK, 2015.
10. Morent, R., De Geyter, N., Cools, P., Plasma modified textiles for biomedical applications, in: *Advances in bioengineering*, IntechOpen, London, UK, 2015.
11. Raslan, W.M., *et al.*, Ultraviolet Protection, Flame Retardancy and Antibacterial Properties of Treated Polyester Fabric Using Plasma-Nano Technology. *J. Mater. Appl.*, 02, 10, 1432–1442, 2011.
12. Fernandes, M., *et al.*, Plasma in textile wastewater treatment, in: *Advances in Plasma Treatment of Textile Surfaces*, S. Ul Islam and A. Haji (Eds.), pp. 267–322, Woodhead Publishing, Cambridge, USA, 2024.

13. Luciu, I., *et al.*, Low and atmospheric pressure plasma treatment of natural textile fibers, in: *23rd International Symposium on Discharges and Electrical Insulation in Vacuum*, Bucharest, 2008.
14. Sun, D. and Stylios, G.K., Fabric surface properties affected by low temperature plasma treatment. *J. Mater. Process. Technol.*, 173, 2, 172–177, 2006.
15. Naebe, M., Haque, A.N.M.A., Haji, A., The Effect of Plasma Treatment on Dyeing of Natural Fibers, in: *Innovative and Emerging Technologies for Textile Dyeing and Finishing*, L.J. Rather, A. Haji, M. Shabbir (Eds.), pp. 191–212, Wiley-Scrivener Publishing, Hoboken, USA, 2021.
16. Kan, C.-W. and Man, W.-S., Parametric Study of Effects of Atmospheric Pressure Plasma Treatment on the Wettability of Cotton Fabric. *Polymers*, 10, 3, 233, 2018.
17. Jelil, R.A., A review of low-temperature plasma treatment of textile materials. *J. Mater. Sci.*, 50, 18, 5913–5943, 2015.
18. Hassabo, A.G., *et al.*, Review on Utilizing Plasma Pretreatment on some Fabrics to Enhance their Printability. *Mater. Int.*, 4, 3, 1–16, 2022.
19. Boaretti, C., *et al.*, Investigation and Optimization of Vacuum Plasma Treatment of PA66 Fabric for Reduced Fire Retardant Consumption. *ACS Omega*, 2022.
20. Haji, A., *et al.*, Plasma-Assisted Functional Finishing of Wool and Cotton Fabrics for Improved Dyeability with Natural and Synthetic Dyes, in: *4th NED International Textile Conference*, NED University of Engineering & Technology, Karachi, Pakistan, 2020.
21. Haji, A., Plasma activation and chitosan attachment on cotton and wool for improvement of dyeability and fastness properties. *Pigm. Resin Technol.*, 49, 6, 483–489, 2020.
22. Haji, A., Natural dyeing of wool with henna and yarrow enhanced by plasma treatment and optimized with response surface methodology. *J. Text. Inst.*, 111, 4, 467–475, 2020.
23. Sajed, T., *et al.*, Modification of wool protein fiber with plasma and dendrimer: Effects on dyeing with cochineal. *Int. J. Biol. Macromol.*, 107, Part A, 642–653, 2018.
24. Haji, A., Qavamnia, S.S., Nasiriboroumand, M., The use of D-optimal design in optimization of wool dyeing with Juglans regia bark. *Ind. Textila*, 69, 2, 104–110, 2018.
25. Haji, A., Mehrizi, M.K., Hashemizad, S., Plasma and chitosan treatments for improvement of natural dyeing and antibacterial properties of cotton and wool, in: *24th IFATCC International Congress (IFATCC2016)*, Pardubice, Czech Republic, 2016.
26. Haji, A., Khajeh Mehrizi, M., Sharifzadeh, J., Dyeing of wool with aqueous extract of cotton pods improved by plasma treatment and chitosan: Optimization using response surface methodology. *Fibers Polym.*, 17, 9, 1480–1488, 2016.

27. Molakarimi, M., Khajeh Mehrizi, M., Haji, A., Effect of Plasma Treatment on Antimicrobial and Dyeing Properties of Wool Fabric Dyed with Shrimp Shell as a Natural Dye, in: *International Conference on Medical Textiles and Healthcare Products (MedTex 2015)*, Lodz, Poland, 2015.
28. Haji, A. and Qavamnia, S.S., Response surface methodology optimized dyeing of wool with cumin seeds extract improved with plasma treatment. *Fibers Polym.*, 16, 1, 46–53, 2015.
29. Haji, A., A Review on Surface Modification of Wool Fibers Using Plasma Technology and Its Effect on Dyeing Properties. *J. Text. Sci. Technol.*, 5, 3, 35–43, 2015.
30. Barani, H. and Haji, A., Analysis of structural transformation in wool fiber resulting from oxygen plasma treatment using vibrational spectroscopy. *J. Mol. Struct.*, 1079, 35–40, 2015.
31. Haji, A., Amiri, Z., Qavamnia, S.S., Natural dyeing of wool with Arnebia euchroma optimized by plasma treatment and response surface methodology. *J. Biodivers. Environ. Sci.*, 5, 2, 493–498, 2014.
32. Haji, A. and Shoushtari, A.M., Natural antibacterial finishing of wool fiber using plasma technology. *Ind. Textila*, 62, 5, 244–247, 2011.
33. Haji, A., Mousavi Shoushtari, A., Abdouss, M., Plasma activation and acrylic acid grafting on polypropylene nonwoven surface for the removal of cationic dye from aqueous media. *Desalin. Water Treat.*, 53, 13, 3632–3640, 2015.
34. Haji, A., Eco-friendly dyeing and antibacterial treatment of cotton. *Cellul. Chem. Technol.*, 47, 3-4, 303–308, 2013.
35. Haji, A., Barani, H., Qavamnia, S.S., *In situ* synthesis of silver nanoparticles onto cotton fibres modified with plasma treatment and acrylic acid grafting. *Micro Nano Lett.*, 8, 6, 315–318, 2013.
36. Haji, A., Hashemizad, S., Dehkani, S., Improved attachment of titanium dioxide nanoparticles on cotton fabric using plasma treatment and chitosan coating, in: *24th IFATCC International Congress (IFATCC2016)*, Pardubice, Czech Republic, 2016.
37. Haji, A., Khajeh Mehrizi, M., Hashemizad, S., Plasma and chitosan treatments for improvement of natural dyeing and antibacterial properties of cotton and wool. *Vlak. Textil.*, 23, 3, 86–89, 2016.
38. Haji, A., Qavamnia, S.S., Bizhaem, F.K., Salt free neutral dyeing of cotton with anionic dyes using plasma and chitosan treatments. *Ind. Textila*, 67, 2, 109–113, 2016.
39. Haji, A., Improved natural dyeing of cotton by plasma treatment and chitosan coating; optimization by response surface methodology. *Cellul. Chem. Technol.*, 51, 9-10, 975–982, 2017.
40. Haji, A. and Mahmoudi Hashemi, R., Plasma and Chitosan Pre-treatments for Improvement of Color Strength and Fastness Properties in Printing of Cotton Fabric Using Direct Dyes, in: *9th International Conference of Applied Research on Textile and Materials (CIRATM)*, Springer International Publishing, Monastir, Tunisia, pp. 129–134, 2022.

41. Khatabi, H., Bidoki, S.M., Haji, A., A green approach for *In-situ* synthesis of silver nanoparticles on cotton fabric by low pressure cold plasma. *Mater. Chem. Phys.*, 290, 126548, 2022.
42. Soltani Gerdefaramarzi, M., Fattahi, S., Haji, A., Optimization of Water-Repellent Finishing of Polyester/Viscose Fabric by Plasma Pretreatment Using Response Surface Methodology. *Fibers Polym.*, 25, 6, 2043–2050, 2024.
43. Haji, A., *et al.*, RSM optimized self-cleaning nano-finishing on polyester/wool fabric pretreated with oxygen plasma. *J. Text. Inst.*, 107, 8, 985–994, 2016.
44. Haji, A., Semnani Rahbar, R., Mousavi Shoushtari, A., Improved microwave shielding behavior of carbon nanotube-coated PET fabric using plasma technology. *Appl. Surf. Sci.*, 311, 0, 593–601, 2014.
45. Haji, A., Rahbar, R.S., Shoushtari, A.M., Plasma assisted attachment of functionalized carbon nanotubes on poly(ethylene terephthalate) fabric to improve the electrical conductivity. *Polimery*, 60, 5, 337–342, 2015.
46. Yaman, N., Improvement Surface Properties of Polypropylene and Polyester Fabrics by Glow Discharge Plasma System under Atmospheric Condition. *Tekst. ve Konfeksiyon*, 1, 1, 45–51, 2009.
47. Mohamed, A.-A.H. and El-Sayed, I.M., The impact of large volume atmospheric pressure air glow discharge plasma on enhancing denim fabric (textile) sewing process. *J. Taibah Univ. Sci.*, 18, 1, 2303814, 2024.
48. Cheng, F., *et al.*, The effect of atmospheric pressure glow discharge plasma treatment on the dyeing properties of silk fabric. *Plasma Sci. Technol.*, 22, 1, 015503, 2020.
49. Takaki, K., *et al.*, Development of a Corona Discharge Ionizer Utilizing High-Voltage AC Power Supply Driven by PWM Inverter for Highly Efficient Electrostatic Elimination, in: *Modern Applications of Electrostatics and Dielectrics*, D. Xiao and K. Sankaran (Eds.), IntechOpen, Rijeka, 2019.
50. Nourbakhsh, S., Sepehrnia, H., Akbari, E., Novel corona discharge treatment of cotton fabric with Cu and ZnO nanoparticles. *J. Text. Inst.*, 111, 9, 1269–1276, 2019.
51. Sivri, C., Development of electrospun nanofibers having novel morphologies *via* corona plasma treatment. *J. Achiev. Mater. Manuf. Eng.*, 76, 1, 36–40, 2016.
52. Abdouss, M., *et al.*, Effects of Ar/O2 atmospheric pressure dielectric barrier discharge (APDBD) on the Nylon 6,6 electrospun fibres, in: *17th Iranian Chemistry Congress*, Vali-e-Asr University of Rafsanjan, Iran, 2014.
53. Reema, *et al.*, Review of the cold atmospheric plasma technology application in food, disinfection, and textiles: A way forward for achieving circular economy. *Front. Phys.*, 10, 2022.
54. Saleem, M., *et al.*, Experimental and statistical analysis of dielectric barrier discharge plasma effect on sonochemically TiO2 coated cotton fabric using complete composite design. *Curr. Appl Phys.*, 31, 158–170, 2021.

55. Oktaviana, R.I., Kusumandari, K., Saraswati, T.E., Plasma dielectric barrier discharge for degradation of textile wastewater in a continuous system. *J. Phys. Conf. Ser.*, 1912, 1, 012023, 2021.
56. Domonkos, M. and Tichá, P., Low-Temperature Atmospheric Pressure Plasma Treatment in the Polymer and Textile Industry. *IEEE Trans. Plasma Sci.*, 1–11, 2023.
57. Li, X. and Qiu, Y., The application of He/O2 atmospheric pressure plasma jet and ultrasound in desizing of blended size on cotton fabrics. *Appl. Surf. Sci.*, 258, 19, 7787–7793, 2012.
58. Yan, J., *et al.*, Effect of atmospheric pressure plasma jet treatment on the performance of needleless electrostatic spinning with low concentration of PCL solution. *J. Text. Inst.*, 116, 2, 328–337, 2025.

# 3

# Surface Functionalization of Textiles by Sputter Coating

**Avik Kumar Dhar[1]*, Piyas Halder[2], Vijay Mohakar[1] and Ivan Moldavchuk[1]**

*[1]Department of Textiles, Merchandising, and Interiors, University of Georgia, Athens GA, Georgia (GA), USA*
*[2]Department of Textile Engineering, Shyamoli Textile Engineering College, Dhaka, Bangladesh*

***Abstract***

Sputtering is one of the most predominant surface functionalization techniques for textile materials. In this process, charged gas ions strike a target (metals, metal oxides, or ceramics) and eject atoms from the target with sufficient energy to migrate and connect with the textile substrate, resulting in a functional coating. Researchers are increasingly interested in the use of sputtering technology in textile functionalization due to its numerous advantages, including its environmental friendliness, facile approach, low operating temperature, and appropriateness for coating a wide range of substrates (metal, glass, ceramics, semiconductors, polymers, textiles, etc.). Current research on sputter-coated textiles focuses on flexible and wearable electronics, electromagnetic shielding, bacterial resistance, hydrophilic and hydrophobic characteristics, structural color, conductive and multifunctional materials, and so on. Therefore, this chapter presents a review of sputtering processes, basic chemistry, types of sputtering technology, structure-property relationships, and the crucial factors influencing the performance of coated textiles. Furthermore, this chapter discusses the most recent advances in textile sputtering technology for their multi-functional applications like electromagnetic interference shielding, protective clothing, photocatalytic activity, textile coloration, wearable electronics, and medical textiles. Finally, an overview of the existing obstacles and future directions for the effectiveness of sputter-coated materials in practice will be provided.

**Corresponding author*: avik.kumar@uga.edu

Aminoddin Haji and Çağlar Sivri (eds.) Surface Modification and Functionalization of Textiles, (39–86) © 2026 Scrivener Publishing LLC

***Keywords:*** Textiles, functional coating, sputtering, mechanism, controlling parameters, structure-property relationships, multifunctional applications

## 3.1 Introduction

Surface functionalization represents a significant advancement in fabricating advanced textiles with multifunctional properties [1, 19]. Among different surface modification techniques, chemical finishing is frequently utilized to impart functionalities in textiles [2]. However, due to increased demand to ban hazardous chemicals and minimize the amount of chemical waste, several attempts have recently been made to supplant the chemical functionalization with new ecologically acceptable options [3, 4]. Thin-film coating of textile substrates is an emerging approach to imparting functionality. Thin films are nano-structural materials with a thickness of 1 μm or less. Compared to bulk materials, they have exceptional functional properties and are mostly produced using high-power operation and vacuum settings [5, 6]. Chemical vapor deposition (CVD), sol-gel coating, layer-by-layer deposition, and physical vapor deposition (PVD) are frequently used for thin-film coatings on textile materials [7–9]. In the CVD technique, a gaseous or vaporous substance reacts at a gas-solid interface to deposit on the solid material's surface. Based on the treatment method, CVD can be categorized into ultra-high vacuum, plasma-enhanced, fluidized bed, metal-organic CVD, and so on [10–13].

PVD, on the other hand, is an atomic distribution method that involves evaporating a substance from a solid or liquid source. The vapor is delivered to the substrate in a vacuum, reduced-pressure gas, or plasma, where it is compressed to produce a thin layer with a thickness ranging from a few nanometers to micrometers [14, 15]. PVD methods include cathode arc deposition, electron beam PVD, evaporative deposition, sputtering, ion plating, and enhanced sputtering [128]. PVD methods include cathode arc deposition, electron beam PVD, evaporative deposition, sputtering, ion plating, and enhanced sputtering [16, 128]. Among numerous PVD processes, sputter coating/sputtering/co-sputtering is a promising strategy for coating textiles due to its multiple advantages, such as high deposition rate, low process temperature requirement, steady quality, multi-functional applications, strong target-substrate adhesion, roll-to-roll process for continuous operation, eco-friendly nature, and solvent-free process, which has been an exciting field of study for textile functionalization in recent years [17–19]. The advantages and disadvantages of sputter-coating of textiles are depicted in Figure 3.1. Even though

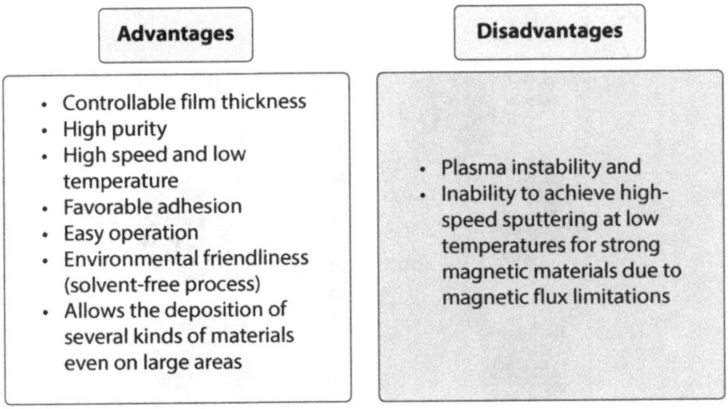


**Figure 3.1** Advantages and disadvantages of sputter coating of textiles.

there are some limitations and scopes for further applications, the benefits of this technique make this process a novel approach to fabricating multifunctional textile materials. In recent work, Zhang *et al.* developed nanoparticle (Ag)–coated aramid fibers through the magnetron sputtering technique for textile-based sensors with high detection sensitivity (detection sensitivity of $10^{-10}$ M for malachite green dye) and impressive storage stability [20]. Mrozińska *et al.* coated polylactic acid (PLA) melt-blown nonwoven with Fe by sputtering for their potential application in medical textiles. The blood coagulation mechanisms remained unaffected by PLA-Fe composites [21]. In addition to their application in medical textiles, sputter-coated textiles are extensively used in antibacterial coating, protective clothing, textile coloration, electromagnetic shielding, wastewater treatment, and so on. The versatility of sputter coating allows for tailoring textile properties for diverse functional applications (Figure 3.2). Recent statistics also demonstrate the increased trend of the applications of sputter-coated textiles. The sputter coating market size was assessed at USD 989.78 million in 2023 and USD 1,061.73 million in 2024 and is predicted to rise at a compound annual growth rate of 7.74% to reach USD 1,668.96 million by 2030 [22].

Therefore, the current chapter provides an overview of sputtering processes, basic chemistry, structure-property relationships, and factors impacting sputtering technology. Furthermore, this chapter addresses the most recent advancements in textile sputtering technology and its multifunctional applications. Finally, current advances and some future perspectives on industrial applications of sputter-coated textile materials in practice will be discussed.

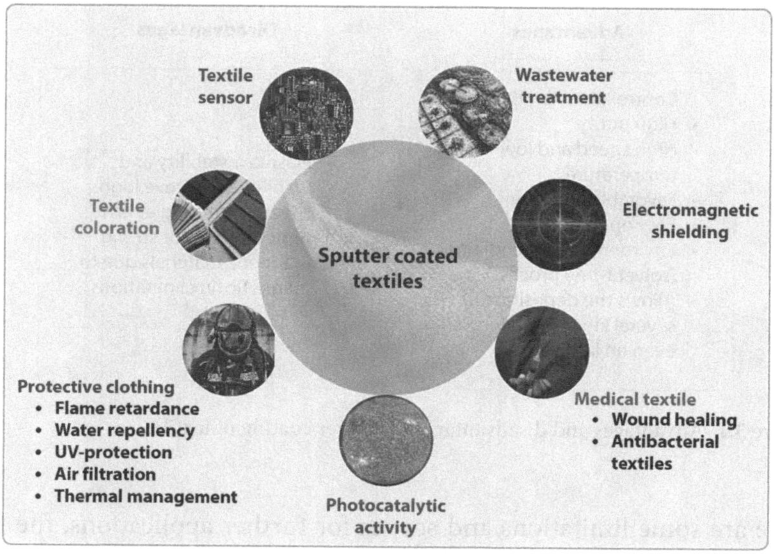


**Figure 3.2** Application areas of sputter-coated textiles.

## 3.2 Sputtering and Co-Sputtering

Sputtering is one of the most promising PVD technologies, which emerged in the late 1970s [23]. It is a typical way to deposit metal, non-metal, or their oxide layers on the substrate surface by altering the target materials and atmospheric gases [24]. This versatile technique facilitates the application of almost any kind of substrate, which makes it useful for developing smart surfaces with exceptional properties and low weight [25]. It is one of the most effective ways to achieve large-area coatings [26]. Consequently, extensive research on sputter coatings was carried out on both industry and lab scales [27]. Depending on the power sources and types of working gas, sputter coating of textiles can be classified as direct current (DC) magnetron, alternating current (AC) magnetron, radio frequency (RF) magnetron, high-power impulse magnetron sputtering (HIPIMS), reactive magnetron sputtering, and ion beam magnetron sputtering. In contrast, Co-sputtering is the extension of the conventional magnetron sputtering technique, where two or more target materials are sputtered either at once or in sequence [28]. Table 3.1 summarizes some of the recent works on sputter coating conditions of textiles for their multifunctional applications.

**Table 3.1** Recent research on the application of sputter coating of textiles for multifunctional applications.

| Materials | Metal types | Sputtering types | Apparatus used | Working gas | Sputtering conditions | Application areas | Ref. |
|---|---|---|---|---|---|---|---|
| Polyester fabrics | Ag/$TiO_2$ | DC magnetron | Sputtering unit (JPG-450 type) | Ar | - Working gas pressure - 0.6 Pa<br>- Power - 64 W<br>- Time - 10 min | Photocatalytic, conductive, anti-ultraviolet | [41] |
| | | RF reactive | Sputtering unit (JPG-450 type) | Ar+$O_2$ | - Gas flow rates-<br>▪ For Ar - 20 mL/min and<br>▪ $O_2$ - 10 mL/min<br>- Working gas pressure - 0.80 Pa<br>- Power - 300 W<br>- Time - 20 min | | |
| Cotton and lyocell-based knit textiles | Ti, Cu, Cu-doped Ti | DC magnetron | Custom-made system | Ar | - Base pressure - <$2.0 \times 10^{-4}$ Pa<br>- Current density - 75 A/$m^2$<br>- Gas flow rate - 25 sccm<br>- Work pressure - $3.0 \times 10^{-1}$ Pa<br>- Distance between target and sample - 70 mm | EMI shielding, electromechanical response | [25] |

*(Continued)*

**Table 3.1** Recent research on the application of sputter coating of textiles for multifunctional applications. (*Continued*)

| Materials | Metal types | Sputtering types | Apparatus used | Working gas | Sputtering conditions | Application areas | Ref. |
|---|---|---|---|---|---|---|---|
| PET non-woven fabrics | Ag–Cu | Roll-to-roll (R2R) magnetron | Custom-made system | Ar | - Target size - 90 × 13 $cm^2$ (Ag/Cu target at 95:5 ratio)<br>- Vacuum pressure - 5.0 × $10^{-4}$ Pa<br>- Flow rate - 400 sccm<br>- Pressure - 3.0 Pa<br>- Power supply – 200-W DC<br>- Deposition rate - 35 nm/min<br>- Time - 8.6 s<br>- Film thickness - 5 nm | Antibacterial | [42] |
| Cellulose nonwoven fabrics | Cu | DC magnetron | P.P.H. Jolex sc., Cz̨estochowa, Poland (Testbourne Ltd., Hampshire, UK) | Ar | - Target-substrate distance - 15 cm<br>- Power - 400 to 800 W<br>- Time - 8 min and 32 min | Healthcare and hygiene products | [43] |
| PET substrates | ZnO/Cu/ZnO | DC magnetron | ION'X® planar balanced magnetrons (Germany) | Ar + $O_2$ | - Target-substrate distance - 27 cm<br>- Pressure - 1.3 Pa<br>- Constant $O_2$/Ar gas flow ratio - 2/3<br>- Power - 300 W | Antiviral activity; antibacterial activity | [44] |

(*Continued*)

**Table 3.1** Recent research on the application of sputter coating of textiles for multifunctional applications. (*Continued*)

| Materials | Metal types | Sputtering types | Apparatus used | Working gas | Sputtering conditions | Application areas | Ref. |
|---|---|---|---|---|---|---|---|
| PLA–Zn composites, melt-blown nonwoven | Zn | DC magnetron | P.P.H. Jolex S.C. (Poland) | Ar | - Target-substrate distance - 15 cm<br>- Time - 5, 10, 15 min<br>- Working pressure - 0.18 Pa<br>- Power - 350–1,000 W<br>- Power density - 0.78 W/cm$^2$ | Antibacterial activity | [30] |
| Cotton woven textiles | $SiO_2$, $Al_2O_3$, Zr matrix, with Ag nanoclusters | RF co-sputtering | Keno sistec™ equipment | Ar | - Time - 60 min<br>- Power - 200, 250, and 290 W for $SiO_2$, $Al_2O_3$, and Zr, respectively, and 3, 5, and 10 W for Ag target | Antibacterial activity | [45] |
| Circular knitted fabrics (cotton and lyocell) | Ti, Cu, intermetallic Ti–Cu | DC magnetron | Custom-made system | Ar | - Base pressure - $< 2.0 \times 10^{-4}$ Pa<br>- Current density - 75 A/m$^2$<br>- Gas flow rate - 25 sccm<br>- Work pressure - $3.0 \times 10^{-1}$ Pa<br>- Distance from target - 70 mm<br>- Condition - Room temperature<br>- Time - 60 min | EMI shielding, electromechanical response | [25] |

(*Continued*)

**Table 3.1** Recent research on the application of sputter coating of textiles for multifunctional applications. (*Continued*)

| Materials | Metal types | Sputtering types | Apparatus used | Working gas | Sputtering conditions | Application areas | Ref. |
|---|---|---|---|---|---|---|---|
| Cotton textiles | $Fe_2O_3$ nanoparticles | Low-temperature plasma sputtering (Pulsed DC) | Custom-made system | $O_2$ | - Power - 200 W<br>- Time - 120 min<br>- Distance from the target - 5 cm<br>- Working pressure - 10 Pa | Antimicrobial activity | [46] |
| Non-woven PP fabrics | CN (carbon nitride) | DC-pulsed sputtering | SJ Power, Bucheon, Republic of Korea | Ar + $N_2$ | - Frequency - 60 kHz<br>- Pulse shape - square (roll-to-roll system)<br>- Target size - 0.416 m × 0.298 m<br>- Target-substrate distance - 6 cm<br>- Base pressure - $6.7 \times 10^{-4}$ Pa<br>- Working pressure - 0.23 Pa (Ar - 180 sccm and $N_2$ - 20 sccm)<br>- Power - 450 W<br>- Current density - 124 mA/$m^2$<br>- Voltage - 450 V<br>- Current - 1 A<br>- Condition - Room temperature<br>- Time - 20 s<br>- Roll speed - 100 mm/min | Antibacterial activity | [47] |

(*Continued*)

**Table 3.1** Recent research on the application of sputter coating of textiles for multifunctional applications. (*Continued*)

| Materials | Metal types | Sputtering types | Apparatus used | Working gas | Sputtering conditions | Application areas | Ref. |
|---|---|---|---|---|---|---|---|
| Cotton fabric | Zn | DC magnetron | P.P.H. Jolex S. C. (Poland) | Ar | - Target-substrate distance - 15 cm<br>- Power - 700 W<br>- Working pressure - 0.2 Pa<br>- Power density - 0.71 $W/cm^2$ | Antimicrobial activity, cytotoxicity | [48] |
| Cotton fabric, jersey fabric, and polypropylene TNT | Ag/diamond-like carbon (DLC) | Sputtering | Custom-made system | Ar | - Working pressure - $10^{-1}$ Pa<br>- Gas flow rate - 21.9 sccm<br>- Applied power - 1500 W<br>- Power - 70 W and 155 W<br>- Working pressure - 1.0 Pa and 0.8 Pa<br>- Thickness ~50 nm | Personal protective equipment | [128] |
| Collagen fabric | Ultrathin Au Films | HIPIMS | TruPlasma High pulse 4002 (G2) TRUMPF Hüttinger Sp. z o. o | Ar | - Target- rotating substrate distance - 6 cm<br>- Substrate rotational speed - 20 rpm<br>- Power - 133.1 W<br>- Peak power density - 147.5 $W/cm^2$<br>- Working pressure - 0.54 Pa<br>- Air flow rate - 150 sccm | Biomedical applications (antibacterial and antioxidant) | [40] |

(*Continued*)

**Table 3.1** Recent research on the application of sputter coating of textiles for multifunctional applications. (*Continued*)

| Materials | Metal types | Sputtering types | Apparatus used | Working gas | Sputtering conditions | Application areas | Ref. |
|---|---|---|---|---|---|---|---|
| Cotton fabric | Ag NPs | Magnetron | ZZG/JT-350; AOYI, vacuum evaporation system (ZHD 300; Beijing Technol, China) | Ar | - Power - 200 W<br>- Film thickness - 200 nm<br>- Condition - Room temperature<br>- Base pressure - $5.9 \times 10^{-4}$ Pa<br>- Working pressure - 0.25 Pa<br>- Air flow rate - 10 sccm<br>- Target-substrate distance - 10 cm<br>- Substrate rotating speed - 100 rpm | Multifunctional applications (ultraviolet (UV) blocking, IR shielding, EMI shielding, and antibacterial effect) | [49] |
| PP nonwoven | Cu/CuO | DC reactive magnetron | Model 500, Shenyang Scientific Instruments Co., Ltd., China | Ar | - Target-substrate distance - 50 mm<br>- Background vacuum pressure - $5 \times 10^{-3}$ Pa<br>- Sputtering pressure - 0.50 Pa<br>- Power - 100 W<br>- Air flow rate - 35 mL/min<br>- Time - 1 min<br>- $O_2$ was mixed with Ar to coat the CuO<br>- $O_2$ flow rate - 0 to 35 mL/min | Multifunctional applications (metallic color, UV protection) | [50] |

(*Continued*)

**Table 3.1** Recent research on the application of sputter coating of textiles for multifunctional applications. (*Continued*)

| Materials | Metal types | Sputtering types | Apparatus used | Working gas | Sputtering conditions | Application areas | Ref. |
|---|---|---|---|---|---|---|---|
| Cotton fabrics | Ag/Zn | High vacuum magnetron | JCP-350, Beijing Taikenuo Technology Co., Ltd, China | Ar + $O_2$ | - Power - 40, 50, 60 W, respectively<br>- Air flow rate - 15 sccm<br>- Ar: $O_2$ - 1:1<br>- Time - 5 min | Multifunctional Personal Protective Equipment | [51] |
| Polyester/ polyamide (90/10 %) and 100% cotton | Cu | DC magnetron | P.P.H. Jolex S. C., Poland | Ar | - Power - > 12 kW<br>- Maximum voltage - 1.2 kV<br>- Frequency - 50 Hz to 5 kHz | Antibacterial and Antiviral Protection | [52] |
| Glass fiber mat | NiO/CoO | DC magnetron | Custom-made system (Iran) | Ar | - Pressure - 26.66 kPa<br>- Voltage - 2000 V<br>- Discharge current - 220 mA<br>- Time - 60 min | Wastewater treatment, antibacterial effect | [53] |
| Cotton fiber nonwovens | Cu/Zn | DC magnetron | Custom-made system | Ar | - Base pressure - $1.5 \times 10^{-4}$ Pa<br>- Power - 100 W | Personal protective equipment | [54] |

(*Continued*)

**Table 3.1** Recent research on the application of sputter coating of textiles for multifunctional applications. (*Continued*)

| Materials | Metal types | Sputtering types | Apparatus used | Working gas | Sputtering conditions | Application areas | Ref. |
|---|---|---|---|---|---|---|---|
| PP fibers | Cu/Ag | Magnetron sputtering | JGP-450A magnetron sputtering system | Ar | - Gas flow - 15 sccm<br>- Time - 0, 15, 30, 45 min<br>- Power - 0, 10, 30, 50 W<br>- Pressure - 0, 0.5, 1, 2 Pa<br>- Substrate-target distance - 6 cm<br>- Base pressure - $5.0 \times 10^{-4}$ Pa | Multifunctional textiles (Antibacterial activity, abrasion resistant, and electrostatic materials) | [39] |
| Melt-blown polybutylene terephthalate | Zr–Cu–Ag | Co-sputtering (RF and DC power supply) | Custom-made system | Ar | - Target-substrate distance - 24.5 cm<br>- Working pressure/ atmosphere – $10^{-2}/10^{-1}$ Pa<br>- Power –<br>▪ RF for Cu - 35 W<br>▪ RF for Ag - 15 W<br>▪ DC for Zr - 300 W<br>- Film thickness - 1 μm<br>- Time - 60 min | Medical textiles | [55] |

(*Continued*)

**Table 3.1** Recent research on the application of sputter coating of textiles for multifunctional applications. (*Continued*)

| Materials | Metal types | Sputtering types | Apparatus used | Working gas | Sputtering conditions | Application areas | Ref. |
|---|---|---|---|---|---|---|---|
| PP nonwoven fabric | Nd-doped $TiO_2$ | DC magnetron | W500, China Shenyang Scientific Instrument Co., Ltd, China | Ar | - Thickness - 4 mm<br>- Base pressure - 5 × $10^{-3}$ Pa<br>- Flow rate<br>▪ Ar - 35 mL/min<br>▪ $O_2$ - 20 mL/min<br>- Working pressure - 0.5 Pa (Ti), 1 Pa (Nd)<br>- Power - 100 W (Ti), 80 W (Nd)<br>- $O_2$ was used for Ti coating | Antibacterial, UV protection | [56] |
| Cotton nonwoven | Ag/Zn | DC magnetron | Beijing Technol Science Co., Ltd. China | Ar + $O_2$ | - Power - 60 W<br>- Airflow rate -<br>▪ Ar - 15 sccm<br>▪ $O_2$ -25 sccm<br>- Time -<br>▪ 2.5 min<br>▪ 5 min<br>▪ 10 min | Hydro electroactive, antibacterial, air filter | [57] |

(*Continued*)

**Table 3.1** Recent research on the application of sputter coating of textiles for multifunctional applications. (*Continued*)

| Materials | Metal types | Sputtering types | Apparatus used | Working gas | Sputtering conditions | Application areas | Ref. |
|---|---|---|---|---|---|---|---|
| Polyethylene (PE)/Cotton/ PE dressing | Cu/Zn | Template-assisted magnetron sputtering (DC) | Custom-made system | Ar | - Vacuum pressure - $7 \times 10^{-3}$ Pa<br>- Airflow - 40 sccm<br>- Power - 100 W<br>- Time - 4 min<br>- Condition - room temperature is 25 ± 5°C | Wound dressing | [58] |
| Polyester knitwear fibers | Cu | DC magnetron | P.P.H. Jolex S. C. (Poland) | Ar | - Target-substrate distance - 15 cm<br>- Power -700 W<br>- Time - 10 min<br>- Working pressure - 0.20 Pa<br>- Power density - 0.71 W/ $cm^2$ | Multi-Functional Composite Materials | [59] |
| PLA non-woven fabrics | Cu | DC magnetron | P.P.H. Jolex s. c. (Czestochowa, Poland). | Ar | - Target-substrate - 15 cm<br>- Power - 700 W<br>- Power density - 0.72 W/ $cm^2$<br>- Working pressure - 0.2 Pa | Multi-Functional, Antimicrobial Composite Materials | [60] |
| Silk filament | Cu | Magnetron sputtering | Custom-made system | Ar | - Power - 50 W<br>- Pressure - 1.5 Pa<br>- Time - 30 min<br>- Condition - Room temperature | Fibrous humidity sensor | [61] |

### 3.2.1 Magnetron Sputtering

Magnetron sputtering is the most effective of the basic sputtering deposition processes. This technique uses powerful magnets or coils in the target holder to confine the plasma closest to the substrate. Magnetron sputtering is advantageous to other basic sputtering approaches due to its greater ion density, efficient collision of ions/atoms/electrons/gas molecules, and higher deposition rate [16]. Additionally, magnetron-sputtered films can outperform films deposited by other PVD processes and can offer the same functionality as much thicker films [26]. In addition, this method permits the integration of oxide, metallic, ceramic, and composite coatings on diverse surfaces like textiles, metals, or ceramics by regulating the process parameters and target materials [24, 29].

The key components of the magnetron sputtering of textile materials system are as follows:

- *Electric field—powered by DC, AC, or RF*
- *Sputtering target—metal, metal oxides, alloys, or ceramics*
- *Substrate (fiber, yarns, fabrics, or composites)*
- *Vacuum chamber*
- *Non-reactive gas (usually argon)*

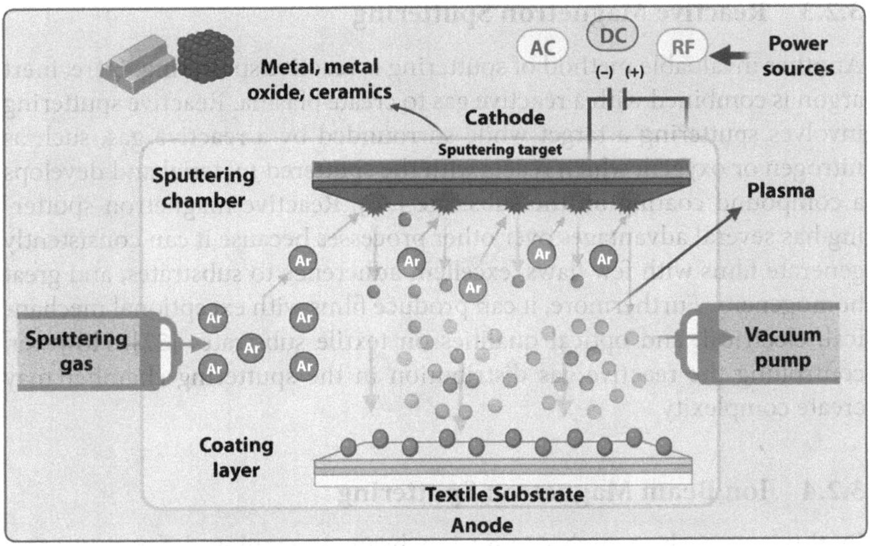


**Figure 3.3** Mechanism of magnetron sputtering coating of textiles.

The process starts when the process gas is injected into the chamber. The gas becomes ionized when an electrical voltage is applied, creating a plasma consisting of electrons and ions. The sputtering target is bombarded with process gas ions (usually argon). Atoms or ions split off from the target because of these collisions and are carried away to be deposited on the textile substrate (Figure 3.3). Process parameters such as gas pressure, applied power, process duration, and temperature can be adjusted to regulate the film's thickness and shape [30].

### 3.2.2 High-Power Impulse Magnetron Sputtering

In the HIPIMS process, a high plasma density is supplied to fabricate a denser and strongly adherent coating at a relatively low operating temperature [31]. HIPIMS exhibits more substantial peak power, faster ionization, a stronger magnetic field, large-scale production, and superior performance in comparison to conventional sputtering [32, 33]. In a recent study, a roll-to-roll HIPIMS system (flow of $O_2$/Ar, web speed 0.16 m/min) was used to develop Ti-O coatings on polyester (PET) fabric in fabricating multifunctional textiles [31]. However, a potential drawback is that the HIPIMS magnetron sputtering rate is less than what would typically be achieved with the same power input using DC magnetron sputtering because of the use of a highly peaked voltage pulse [34, 35].

### 3.2.3 Reactive Magnetron Sputtering

Another invaluable method of sputtering is reactive sputtering. Here, inert argon is combined with a reactive gas to create plasma. Reactive sputtering involves sputtering a target while surrounded by a reactive gas, such as nitrogen or oxygen, which reacts with the sputtered material and develops a compound coating on the substrate [36]. Reactive magnetron sputtering has several advantages over other processes because it can consistently generate films with few flaws, excellent adherence to substrates, and great homogeneity. Furthermore, it can produce films with exceptional mechanical, electrical, and optical qualities on textile substrates [37]. However, controlling the reactive gas distribution in the sputtering chamber may create complexity.

### 3.2.4 Ion Beam Magnetron Sputtering

In this approach, a high-energy ion beam is employed for sputtering target materials, and the resulting atoms are deposited on the substrate.

Fundamentally, this method is comparable to conventional magnetron sputtering. But it offers unique benefits related to using ion beams, namely,

- *Reduced pressures lead to a more efficient vacuum environment*
- *Targeted energy and angle of collision (bombardment angle) because of reduced ion scattering with residual gases*
- *Improved control over the sputtering process*
- *Flexible selection of target morphology and shape, and*
- *Lack of electric or thermal interaction with the cathode*

Nevertheless, the deposition rate is comparatively low, which prevents its application for large surface applications [38].

## 3.3 Factors Affecting the Performance of Sputter-Coated Textiles

The performance of the sputter-coated textiles is mostly determined by the target materials used and the surface properties of the substrate. In addition, some other factors affect the performance and film thickness during the sputtering process, like applied voltage or power, working gas type and pressure, target-substrate distance, coating time, etc. The selection of target materials and substrates is done depending on the required application areas of the coated textiles. An increase in sputtering power (up to 50 W) led to a rise in the deposition of Cu/Ag nanoparticles on the polypropylene (PP) surface, which in turn produced a composite with improved breaking strength [39]. Moreover, the target-substrate distance and deposition interval affect the film thickness. There is nearly a linear relationship between the desired content (thin-film yield) and the deposition time [30]. For example, the thickness of the Au film on the collagen substrate was reported to be 3.99 nm at 13 s and 10.37 nm after 77 s [40]. The factors affecting the performance of the sputter-coated textiles are summarized in Figure 3.4.

## 3.4 Applications for Sputter-Coated Textiles

### 3.4.1 Electromagnetic Shielding

With an increasing population and the quantity of electrical appliances used in personal, industrial, and medical settings, electromagnetic interference

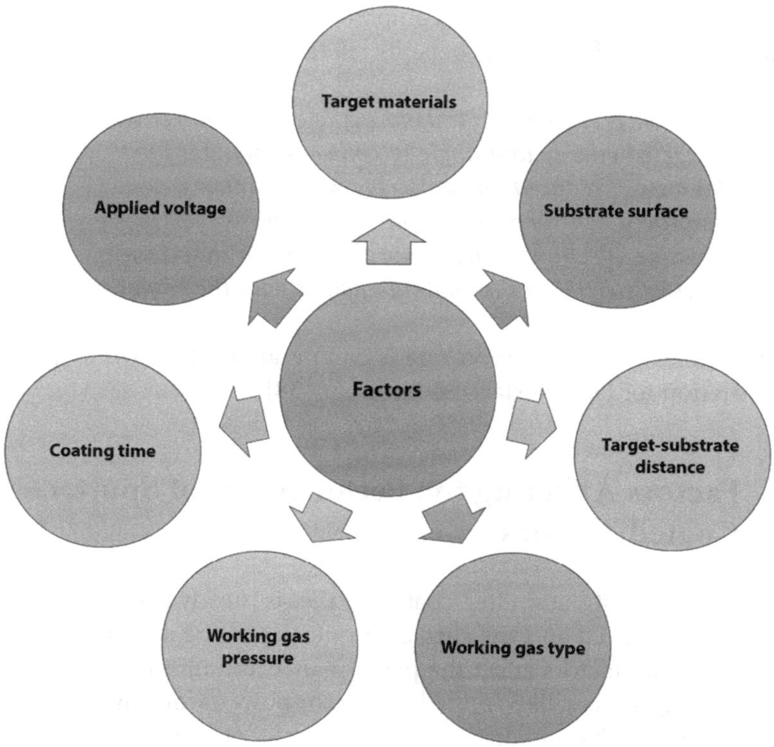


**Figure 3.4** Factors affecting the performance of sputter-coated textiles.

(EMI) issues are becoming more prevalent. Electromagnetic wave interference can cause adjacent electronic devices to malfunction, disrupting communications and potentially harming human health by producing headaches, nausea, neurological abnormalities, and other issues. To avoid the detrimental effects of these interferences, the personnel in its vicinity require a shielding substance that protects them from the adverse effects of the electromagnetic waves [62]. Metal-based materials (e.g., Ag, Cu, Au, Al, or Ni) tend to be preferred because of their excellent electrical conductivity. However, a lack of flexibility and a heavier weight make them unsuitable for portable and wearable applications. Textile-based EMI shielding materials are very versatile due to their flexibility, being soft, breathable, and easy to prepare [63, 64]. Enhancing the electrical conductivity of the materials through various techniques is a prerequisite for the EMI shielding performance of textile materials. Among the various methods, the following are noteworthy:

- *Textile coating with electrically conductive additives (e.g., carbon fibers, conductive polymer, and metal wires),*
    - *Chemical coating*
    - *PVD*
    - *CVD techniques*
- *Inclusion of organic and inorganic fillers [conductive fibers or nanoparticles, carbon-based fillers, carbon nanotube (CNT), graphene, carbon black, and MXene*
- *Multilayer structures*

However, it is frequently documented in the literature that the addition of metal or carbon additives reduces flexibility and causes poor wear and corrosion resistance of the composites [65–67].

### *3.4.1.1 Mechanism of EMI Shielding Performance*

EMI control is generally performed by reducing electromagnetic wave emissions from the source, altering their propagation path, and strengthening the susceptor's immunity. Reflection, absorption, and transmission occur when electromagnetic waves impinge on the surface. These waves are attenuated through reflection, repeated reflection, and absorption, as shown in Figure 3.5, depending on the material properties and the wave's nature. Furthermore, multiple reflections play an important role in the attenuation of electromagnetic radiation [25, 68].

The performance of EMI shielding materials is evaluated by shielding effectiveness (SE). SE evaluates how well a material restricts the EM energy of a specific frequency that passes through it. The total EMI shielding effectiveness ($SE_T$) is determined by the reflection loss ($SE_R$), absorption loss ($SE_A$), and multiple reflection loss ($SE_M$) (Equation 3.1) [68].

$$SE_T = SE_R + SE_M + SE_A \tag{3.1}$$

### *3.4.1.2 Sputter-Coated Textiles for EMI Shielding Application*

Sputter coating of textiles using metal or metal oxide (Cu, Al, Ag, Ti, Ni, and Fe) thin film offers excellent EMI shielding effectiveness. For example, polyamide fabric was sputter-coated with Cu and NiFe to improve EMI shielding performance. For sputter coating, Cu and NiFe were chosen due to their outstanding conductivity and elevated permeability [70].

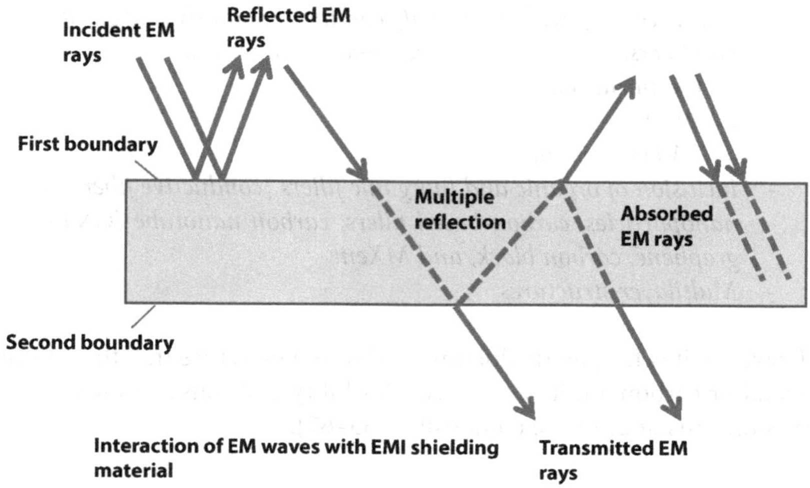


**Figure 3.5** Shielding mechanism of EM waves with a solid layer. Reproduced with permission from [69].

However, for improved performance, the developed film should demonstrate great adhesion and lower thickness, presenting an intriguing route to improving EMI shielding efficacy without compromising textile flexibility. In a recent study, thin films of Ti (309 ± 10 mm), Cu (316 ± 10 mm), and Ti-doped with Cu (359 ± 10 mm) were developed by magnetron sputtering on knit cotton (39F) and lyocell fibers (62I). For 39F, SE reached 10 dB for TiCu0.34, and nearly 15 dB for the knits functionalized with Cu (frequency range of 2-8 GHz). In contrast, for 62I knit, SE remained constant for Ti and TiCu0.34 at around 10 dB. But the Cu thin layer approached 15 dB, whereas the pristine knit textiles had an SE of 0 dB [25]. In another research, the Cu-coated bacterial cellulose (BC/Cu) composite demonstrated strong conductivity (0.026 $Sm^{-1}$), outstanding mechanical properties (41.4 MPa), and excellent EMI shielding performance (55 dB) due to the incorporation of Cu on the BC structure [71].

### 3.4.1.3 *Factors Affecting SE*

The SE of the coated fabrics depends on the number of layers of thin film. The multi-layer thin film, consisting of conductive and magnetic films, is expected to be applied to EMI shielding at high frequencies. For this reason, the SE of the 16 layers of NiFe/Cu thin film is larger than that of the eight layers of NiFe/Cu film over the frequency range from 0.7 GHz to

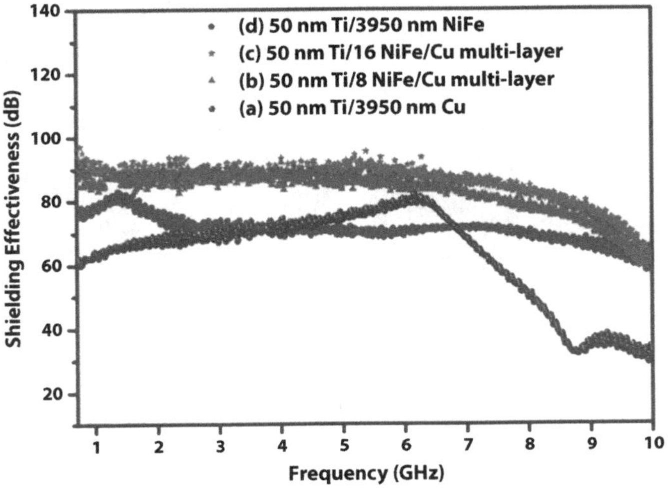


**Figure 3.6** EMI-SE of (a) Cu, (b) 8-layer NiFe/Cu, (c) 16-layer NiFe/Cu, and (d) NiFe thin film. Reproduced with permission from [70].

10 GHz (Figure 3.6). Multiple reflections at NiFe/Cu improve the SE of multilayer thin films [70].

Another approach for improving SE is to strengthen the interfacial bonding between the substrate and the thin layer during sputtering. To boost interfacial bonding strength, fibers can be etched with chemical reagents to provide a rougher surface with many pits. For example, the NaCl etching of the self-bonded natural fiber (SNFP) led to changes in the surfaces. Following the chemical etching, the fibers' surface exhibited a greater number of pits and roughness, and the amount of lignin was reduced to 52% (Figure 3.7). Increased roughness resulted in numerous mechanically locked active sites for sputtering Cu films (0.8 mm thick). After 240 min of Cu film sputtering, the SE was 25.5 dB, meaning that over 99.4% of incident signals were suppressed [72].

In another work, cotton woven fabric was padded with polyvinyl alcohol (PVA) before sputter coating with Cu, Ti, and stainless steel (SS). The PVA pretreatment improved the light transmittance of the composite structure. PVA/cotton composite coated with metal also exhibited improved electroconductivity. The Cu-coated cotton sample displayed a greater infrared reflection rate of around 25%, whereas the Cu-coated PVA/cotton exhibited a lower electrical resistivity ($5.49 \times 10^{-6}\,\Omega$ m) and a higher SE of 30 dB. Higher water contact angle (128.2°) and UPF (117.65) were revealed by the cotton fabric coated with SS [73].

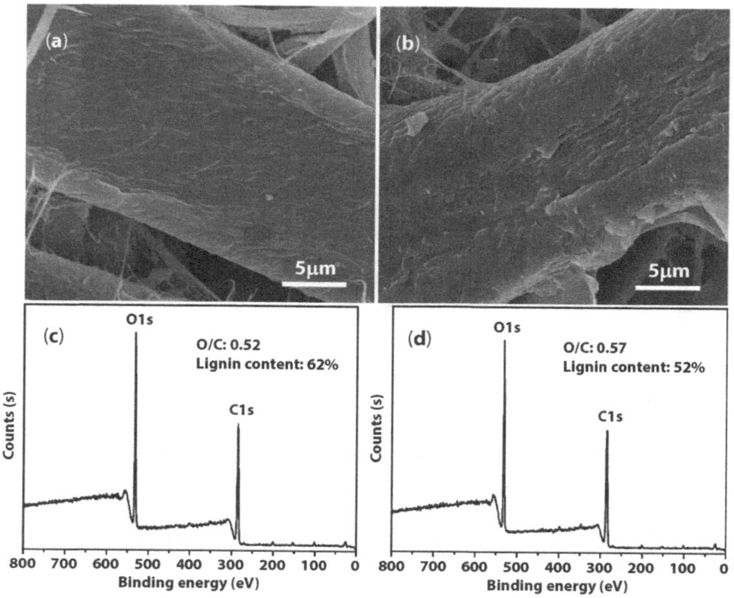


**Figure 3.7** SEM images and XPS spectra of SNFP before (a, c) and after (b, d) chemical etching. Reproduced with permission from [72].

### 3.4.2 Photocatalytic Activity

Photocatalytic technology has garnered significant research attention in a variety of sectors, particularly in semiconductor technology. Recently, polyester fabric was coated with Ag/$TiO_2$ composite films by different magnetron sputtering methods. The Ag/$TiO_2$ composite film fabricated *via* RF and DC demonstrated the closest photocatalytic activity. $TiO_2$/Ag composite films had a rougher surface than Ag/$TiO_2$ composite films, but the photocatalytic properties of Ag/$TiO_2$ composite films were superior to those of $TiO_2$/Ag composite films. The photocatalytic efficiency of the coated textiles improved regardless of whether the Ag layer was on the bottom or the surface of the $TiO_2$ films. The presence of an underlying Ag layer contributed to the increased photocatalytic performance of the Ag/$TiO_2$-coated textiles over the $TiO_2$-coated sample [41]. Yu *et al.* also explored the improved photocatalytic degradation of Rhodamine B by Ag/$Ag_2O$/$TiO_2$-coated polyester compared to $TiO_2$-coated fabrics due to the $Ag_2O$/$TiO_2$ heterojunctions [74]. Due to their photocatalytic performance, sputter-coated textiles are becoming a promising candidate for dye pollutant treatments.

### 3.4.3 Protective Clothing

Protective textiles have surged in popularity in recent decades as there has been a greater emphasis on ensuring worker safety, health, and protection from workplace risks. Therefore, it is quite vital to design protective textiles with multifunction [75].

#### *3.4.3.1 Flame Retardant Textiles*

Sputter coating is one of the novel concepts to bestow flame retardancy to diverse substrates, such as textiles, nanofibers, or plastic substrates for specialized flame retardant applications [76]. In recent work, Nomex®, basalt, and cotton fabric were coated with Al and Zr (IV) oxides to improve textiles' thermal properties using the magnetron sputtering method [77]. The modification involved applying metal (Al) and ceramic (Zr (IV) oxide) coatings with thicknesses of 1 μm and 5 μm to the fabric surface *via* magnetron sputtering. The coatings were tested for heat resistance at temperatures of 100°C and 250°C, as well as thermal radiation resistance. Thermal comfort parameters, such as thermal conductivity and heat absorption coefficient, were also evaluated. Basalt cloth coated with Zr (IV) oxide at a thickness of 5 μm achieved nearly the highest efficiency level for contact heat. All reference and modified samples showed the first level of protection against heat radiation [78].

#### *3.4.3.2 Personal Thermal Management System*

Maintaining a body temperature at the optimal metabolic rate (cooling in hot weather and heating in cold weather) is critical for survival. Because of immune system imbalances, sudden heat loss can lead to respiratory infections and heart disease. As a result, the development of thermal management fabrics is crucial in investigating personal thermal comfort [79–82].

The heat exchange mechanism of the human body is illustrated in Figure 3.8a. The coated textiles should absorb solar light and impede body radiation to fabricate an effective heating cloth that increases heat intake and decreases heat loss from the human body (Figure 3.8c). In contrast, coated textiles could function as an excellent cooling cloth by reflecting solar radiation and transmitting body radiation, which ought to improve the body's ability to release heat while suppressing heat absorption (Figure 3.8b) [80].

Therefore, depending on the application, sputter-coated textiles can operate in heating, cooling, or dual mode. Wang *et al.* fabricated a thermal management textile coated with Ag/Ti-$TiO_2$ multilayer films (thickness:

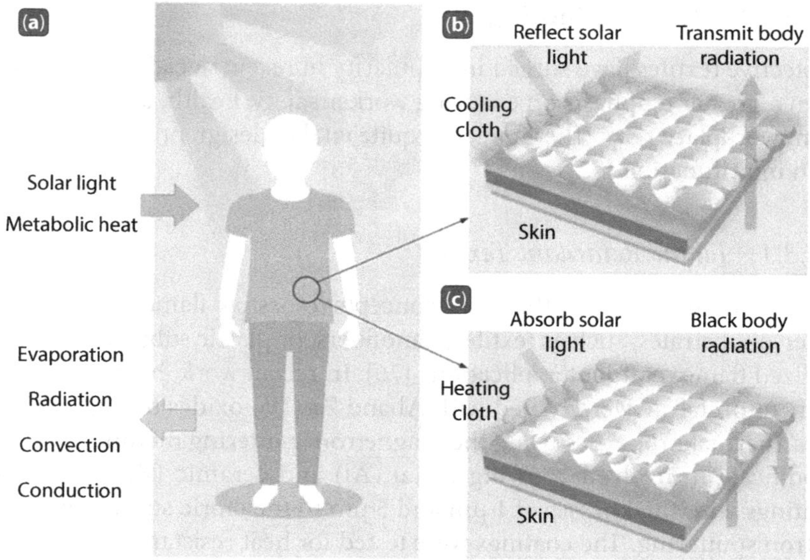


**Figure 3.8** Mechanism of personal thermal management. (a) Heat transport mechanism of the human body in the external environment. The input heat pathways consist of metabolic heat and solar light absorption (absent in the indoor environment). Contrarily, conduction convection, radiation, and evaporation are the output pathways of heat transportation. (b) The cooling fabric reflects solar light and transmits body radiation by hindering the input heat transfer channel and enhancing the output channel. (c) The heating fabric absorbs solar energy and inhibits body radiation by boosting the intake heat channel and reducing the output heat pathway for the human body. Reproduced with permission from [80].

893 nm) through magnetron sputtering. The bottom layer (Ag) demonstrated low emissivity for passive heating and high Joule heating conductivity (60.2°C at 3 V). Additionally, the Ag layer displayed good EMI shielding properties (31 dB after 5,000 stretching cycles). An impressive photothermal conversion was achieved using a nano-gradient wavelength-selecting absorber layer of Ti-$TiO_x$ (94.1°C under 1 kW/$m^2$ irradiation). Lastly, the outermost $TiO_2$ layer performed two roles:

i. It reduced surface energy for strong super hydrophobicity
ii. Acted as an anti-reflective layer to enhance solar absorption [83].

### 3.4.3.3 *Water-Repellent Fabrics*

Hydrophobic thin films coated on textiles through magnetron sputtering deposition have received substantial attention due to their potential uses in different domains, such as anti-fouling coatings, self-cleaning surfaces, protective clothing, industrial applications, and biomedical devices [84]. However, the development of water-repellent surfaces through magnetron sputtering has its challenges, especially when it comes to the production of large-area films. The main challenge, in this case, is regulating edge effects, variation in the gas flow, and substrate curvature to get good uniformity of coating on a large area [85].

The simplest test to assess the water resistance of the materials is the investigation of WCA (water contact angle). It is the main parameter that can characterize the hydrophobicity of the surface and allows us to compare materials and coating techniques in general (Figure 3.9) [84].

There are a few types of hydrophobic coatings, such as DLC films, oxides, and polytetrafluoroethylene (PTFE), which are well developed and are continuously upgraded by regulating magnetron sputtering parameters.

#### 3.4.3.3.1 Diamond-Like Hydrophobic Coatings

DLC films are thermally conductive, optically transparent materials with high hardness, low friction, corrosion resistance, and chemical stability. In terms of practical results and attempts to develop this type of coating, researchers report advanced results in terms of water contact angle (as a

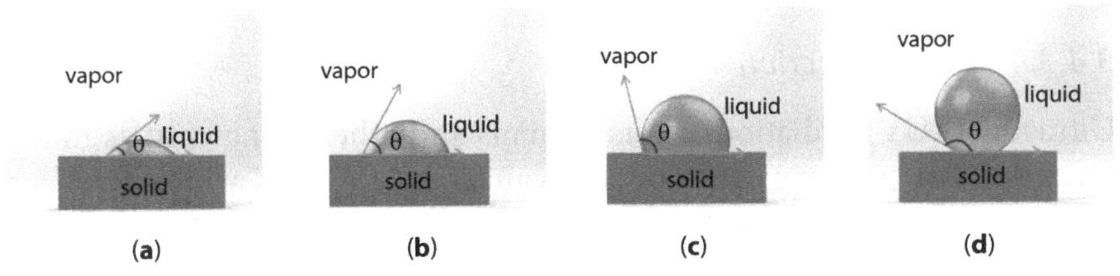


**Figure 3.9** The wetting mechanism of solid surfaces with various contact angles (solid-liquid droplet). *(a) Super hydrophilic surface: contact angle, $\theta < 5°$. (b) Hydrophilic surface; $\theta < 90°$. (c) Hydrophobic surface; $\theta = 90° \sim 150°$. (d) Super hydrophobic surface; $\theta = 150° \sim 180°$.* Reproduced with permission from [84].

main comparing factor). For example, Al-oxide surface modification of silicon material by utilizing HIPIMS sources allowed the achievement of WCA ~ 121°. The bias voltage used during the experiment was changeable, and its increase increased the WCA of the coating. A maximum WCA of 121° was reported at the voltage of –300V. The increase of voltage to an even higher number of 400 V, made the WCA lower on the level of 117.8° [86].

#### 3.4.3.3.2 Oxide Hydrophobic Coatings

Because of the versatility of the materials used for creating this type of hydrophobic coating, such as $HfO_2$, $CeO_2$, and ZnO, magnetron sputtering allows the achievement of perfect oxide thin coatings with excellent chemical and thermal stability as well as optical properties. For example, DC magnetron sputtering of ZnO material on glass at 10.67-Pa working pressure can lead to WCA up to 120°. Some other parameters besides the pressure, such as power and time, were 60 W and 20 min, respectively [87].

#### 3.4.3.3.3 Polytetrafluoroethylene Hydrophobic Coatings

PTFE is ideal for coating, particularly magnetron sputtering, due to its high thermal stability, lower dynamic friction coefficient, poor dielectric constant, and chemical inertness. For example, the $ZnO/SiO_2/PTFE$ sandwich nanostructure superhydrophobic coating made by the RF magnetron sputtering method in Ar-plasma allowed for get WCA of ~167°. The coated substrate was glass, and some other parameters, such as pressure, power, and time were: 1.5 Pa, 100W, and 30 min [88].

### *3.4.3.4 UV Protection*

Ultraviolet (UV) radiation affects humans and other living entities in the universe. The primary cause of skin cancer is UV light, which acts as a free radical carrier medium. Furthermore, it causes environmental deterioration, destroys pigments and materials, and impairs their physical properties. Therefore, the development of composites that can withstand UV radiation is required to save all creatures [89, 90]. The mechanism of UV protection of sputter-coated textiles is depicted in Figure 3.10.

The performance of UV-protective clothing is evaluated by ultraviolet protection factor (UPF) and UV transmission (%) (Equations 3.2 & 3). The UPF measures the UV radiation blocked by the fabric (UV-A and UV-B).

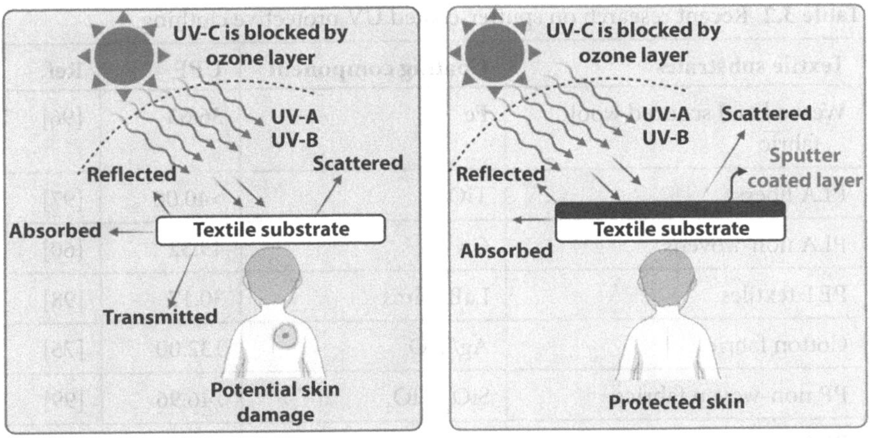


**Figure 3.10** UV protection performance of sputter-coated textiles.

A higher UPF value indicates that more UV light is filtered. In terms of UV protection, fabrics are classified as follows:

- *Good UV protection:* UPF value in the range of 15–24
- *Very good UV protection*: UPF value in the range of 25–39
- *Excellent UV protection*: UPF value of 40 or higher. There is no rating assigned if the UPF > 50 [91].

$$UPF = \frac{\int_{\lambda_1}^{\lambda_2} E(\lambda)S(V)d\lambda}{\int_{\lambda_1}^{\lambda_2} E(\lambda)S(\lambda)T(\lambda)d\lambda} \tag{3.2}$$

$$\%\, UV\ transmission = \frac{\sum_{\lambda_1}^{\lambda_2} T(\lambda)}{\lambda_2 - \lambda_1} \tag{3.3}$$

here,

*E(λ) - relative erythemal spectral effectiveness, which is proportional to its harmfulness,*
*S(λ) - solar spectral irradiance in W/m²/nm,*
*T(λ) - spectral transmission of specimen, and*
*dλ - wavelength increment in nm* [92, 93].

**Table 3.2** Recent research on sputter-coated UV protective clothing.

| Textile substrates | Coating component | UPF | Ref. |
|---|---|---|---|
| Weaved and scoured wool fabric | Fe | 36.63 | [96] |
| PLA fibers | $TiO_2$ | >40.00 | [97] |
| PLA non-wovens | Cu | 49.32 | [60] |
| PET textiles | $LaB_6$ films | 30.17 | [98] |
| Cotton fabric | $Ag/TiO_2$ | 132.00 | [75] |
| PP non-woven fabrics | $SiO_2/TiO_2$ | 746.96 | [99] |
| PET woven fabrics | $SiO_2/TiO_2$ | 1,124.00 | [99] |
| Polyester fabrics | $Ag/TiO_2$ | 30.00 | [100] |
| Bi-layered UV-EMI proof ultra-elastic e-textile | AgNPs | 335.56 | [101] |
| Polyester fabrics | Ag | 90.1 | [102] |

Sputter-coated textiles demonstrate excellent UV-blocking properties. For instance, Ag-sputtered cotton fabric had a UPF of 264 (optimum gas pressure of 6.67 Pa and sputtering time of 5 min) [94]. Another study demonstrated that sputter-coated bacterial cellulose/Cu/ZnO nanocomposite illustrated excellent UV resistance with a UPF of 1,850.33 ± 2.12 [95]. Table 3.2 summarizes some recent works on sputter-coated UV-protective clothing.

### 3.4.3.5 *Air Filtration*

Personal protective masks have evolved into an efficient and practical tool for safeguarding people from dangerous air pollutants and respiratory infections. Lately, a bilayer structured fabric was manufactured by recombining Ag/Zn two-sided sputtering cotton (SAg/ZnC) and PLA/polycaprolactone composite nanofibers (CNF). The resulting CNF@SAg/ZnC fabric had a high filtration efficiency of $PM_{0.3}$ (>99% at 85 L/min airflow rate), commendable antibacterial activity, favorable cell viability, excellent bactericidal activity against *E. coli* (99.33%) and *S. aureus* (99.19%), and high durability (>12 hours) [51]. In another research, Ag/cotton/Zn nonwoven air filters with breath-activated electroactivity were fabricated by magnetron sputtering technology. Air filters fabricated by incorporating Ag/cotton/Zn layers with PP melt-blown nonwovens or PLA nanofibers had a filtration efficiency of 96.8% with lower pressure drop [57].

### 3.4.4 Textile Coloration

Traditional textile dyeing methods are energy-intensive, generate an extensive amount of trash, and cause major environmental contamination. As a result, new environmentally friendly textile coloration procedures are urgently needed. The magnetron sputtering process has proven to be suitable for imparting structural or metallic colors by depositing metals or their oxide layers on the textile materials [103, 104]. The performance of the colored textiles is evaluated through the color parameter (K/S, CIE $L^*a^*b^*$, and color difference - ΔE) using a standard observer.

$$\frac{K}{S}=\frac{(1-R)^2}{2R} \tag{3.4}$$

Here,

*K - the Kubelka-Munk absorption coefficient of the object,*
*S - scattering coefficient, and*
*R - the reflectivity (%).*

A higher K/S number indicates more color depth and improved dyeing ability. In contrast, a smaller K/S number denotes a lower concentration of colored components [105].

In textile coloration through sputter coating, Cu is a commonly used target, which provides the textile substrates with a metallic finish to improve their aesthetics. A single-layer Cu and CuO film on PP nonwoven fabric demonstrated multifunctional properties with a metallic color. The uncoated fabric was grey. However, the Cu-coated fabric was orange-red, and CuO-coated fabrics were dark red and dark green [50].

#### *3.4.4.1 Factors Affecting Textile Coloration*

Several factors affect the coloration of textiles through sputter coating, like coating materials (metal or metal oxides), sputtering parameters (current, supplied gas), coating thickness, and finally substrate properties. For example, in a recent study, metallic coloration of polyester fabric was done with Cu and $Cu_xO_y$ film *via* a DC magnetron sputtering. The color of the Cu-coated fabrics ranged from pale yellow to yellow, whereas the color of $Cu_xO_y$-coated fabrics shifted from dark brown to dark red, then to dark green. Furthermore, $Cu_xO_y$-coated fabric had superior color fastness to rubbing (grade 3) [105].

Another important factor affecting textile coloration is sputtering current. A higher sputtering current leads to more target atoms deposited on the fabric surface and thus produces a thicker film, a larger K/S value, and a greater color depth (Figure 3.11). However, visible light absorption and reflection also have an impact on the K/S value besides film thickness [105]. Therefore, the hue of the samples is determined by the film composition, structural properties, surface morphology, and bandgap of the films that affect the absorption and reflection of visible light [106].

The change of color during sputter coating also depends on the reaction gas flow rate. Huang *et al.* explored that, when the flow of $O_2$ increased during sputtering, the film's Cu concentration reduced and transformed into $Cu_2O$, which then progressively converted to CuO. The color became

| **Current (A)** | 0 | 1 | 2 | 3 | 4 | 5 | 6 | 7 |
|---|---|---|---|---|---|---|---|---|
| **Thickness (nm)** | 0 | 45.4 | 56.5 | 66.1 | 79.1 | 87.0 | 100.2 | 122.2 |
| **K/S** | 0.21 | 0.53 | 0.68 | 0.92 | 0.99 | 0.99 | 1.02 | 1.10 |

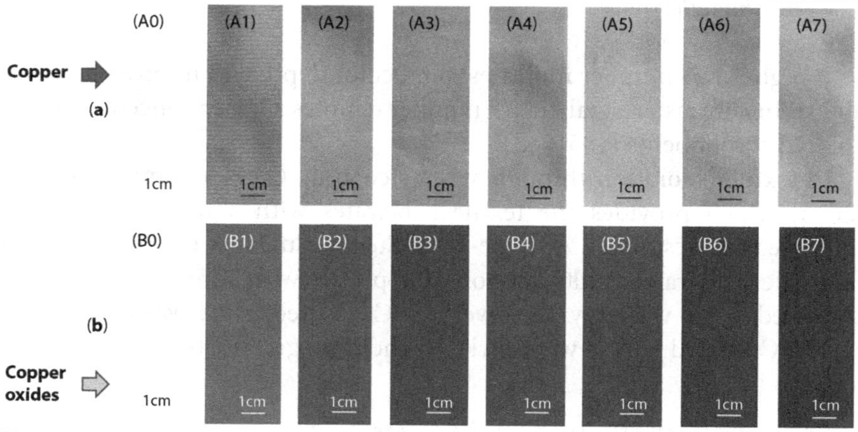


| **Current (A)** | 0 | 1 | 2 | 3 | 4 | 5 | 6 | 7 |
|---|---|---|---|---|---|---|---|---|
| **Thickness (nm)** | 0 | 34.3 | 39.2 | 45.2 | 50.5 | 56.6 | 65.2 | 78.0 |
| **K/S** | 0.21 | 3.15 | 4.29 | 4.32 | 4.52 | 5.82 | 7.58 | 8.67 |

**Figure 3.11** Colored polyester fabrics coated with (a) Cu and (b) CuO film. Film thickness and color depth (K/S) increased with the increase of applied current. Reproduced with permission from [105].

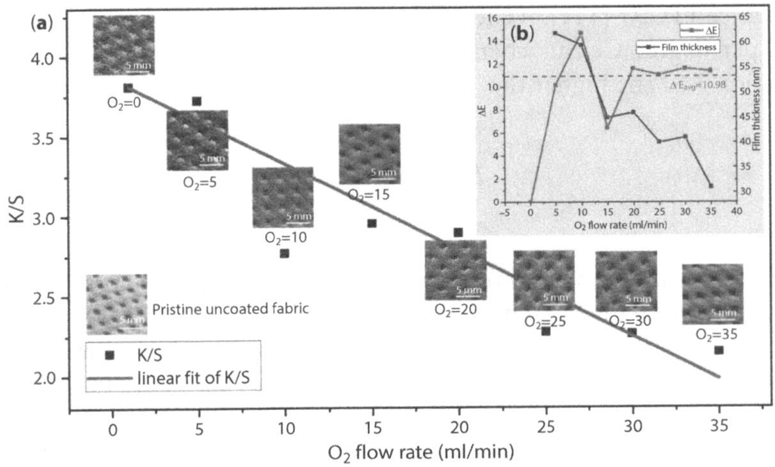


**Figure 3.12** (a) Color appearance, K/S; (b) color difference, and film thickness. *The uncoated fabric was grey, the Cu-coated fabric was orange (without* $O_2$ *flow), and the CuO-coated fabric was dark red and dark green. CuO-coated fabrics exhibited lighter and brighter colors with the increase of the* $O_2$ *flow rate (K/S value decreased). A patina green color was observed in CuO-coated fabric at an* $O_2$ *flow rate of 10 mL/min in the presence of a humid environment. Reproduced with permission from* [50].

lighter as the color brightness increased, whereas the K/S value fell as the CuO content increased. In addition, as the $O_2$ flow rate increased, the CuO film thickness was gradually reduced (Figure 3.12) [50].

## 3.4.5 Textile Sensors/Flexible and Wearable Electronics

### *3.4.5.1 Temperature Sensors*

Smart, flexible wearable sensors have emerged as the central component of modern flexible electronic systems because flexible electronic devices have seen an accelerated expansion in recent years. These sensors are essential and have numerous applications in a variety of industries, including medical treatment and health monitoring [107, 108]. Because of their inherent rigidity, flexible temperature sensors based on alloys are far from achieving the comfort requirements of wearable electronics. Consequently, there is increasing interest in textile-based temperature sensors that integrate metal nanostructures [109, 110].

A digital multimeter is used to measure the resistance change to assess the performance of temperature sensors. Moreover, the temperature characteristics are frequently explained by the temperature coefficient of resistance (TCR) [111]. TCR represents the relative change in resistance with per degree Celsius change in temperature and can be presented by Equation 3.5. The higher the TCR, the more sensitive the sensors [110].

$$TCR = \frac{(R - R_o)}{R_o \Delta T} \tag{3.5}$$

Here, R is the measured resistance, $R_0$ is the initial resistance (at 25°C), and ΔT is the change in temperature [112].

Ag nanoparticles (temperature-sensing layer) coated thermoplastic polyurethane (TPU)–polyester fabric demonstrated a TCR of 0.00234°C$^{-1}$ (sputtering power of 90 W and sputtering pressure of 0.8 Pa). After heat treatment at 160°C, the TCR improved to 0.00262°C$^{-1}$, indicating an improvement in the crystallinity of Ag particles [110]. Appiagyei *et al.* fabricated a lightweight, inexpensive, and flexible temperature sensor by depositing Ni, NiO, and Ni/NiO bilayer thin films on a PET substrate. A TCR of 3.8 × 10$^{-3}$°C$^{-1}$ was demonstrated by the developed Ni/NiO bilayer sensors, which outperformed Ni and NiO thin films, which had TCRs of 3.2 × 10$^{-3}$°C$^{-1}$ and 3.1 × 10$^{-3}$°C$^{-1}$, respectively. Even after 16,000 repeated mechanical bendings and exposure to various solvents (acid, alkali, and detergents), the Ni/NiO bilayer sensor maintained a high TCR [113].

### *3.4.5.2 pH Sensors*

Because most chemical and biological interactions in materials, biology, and environmental sciences are influenced by pH levels, pH sensors are among the most important flexible and stretchable electrochemical sensors. Traditional glass pH electrodes are unsuitable for wearable devices because of their lack of bending capability [114, 115]. Therefore, sputter-coated textile-based pH sensors are getting more attention. For instance, the sputter-coated (Au) electrospun polyurethane-based pH sensor requires only 1 μL of sweat for measurement and has good resolution. There was no statistically significant change in performance even after 35 days. Furthermore, the sensors also demonstrated rapid sweat absorption and high reproducibility [116].

### *3.4.5.3 Humidity Sensors*

During breathing, water molecules, which are vital components, disperse from the human body. Therefore, using a non-invasive humidity sensor to

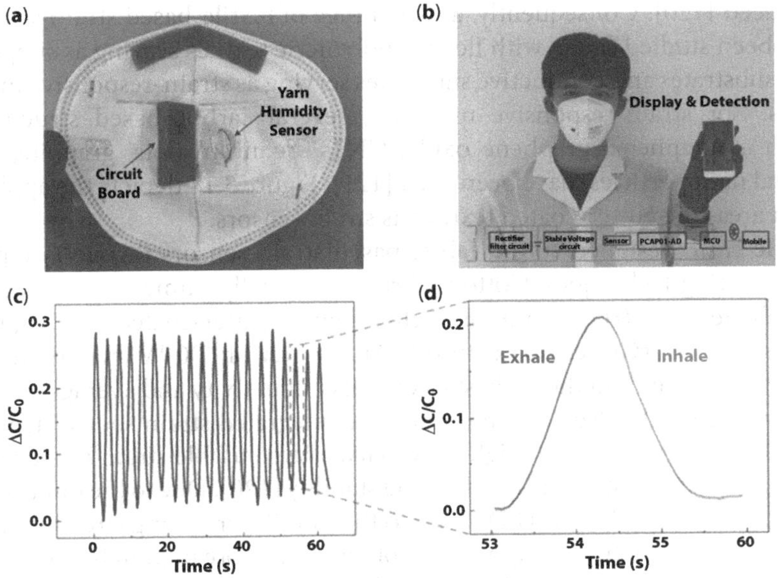


**Figure 3.13** Human respiration monitoring (real-time). (a) Developed smart mask, (b) respirator testing system with circuit diagram, and (c) respiratory signals during normal breathing conditions. While exhaling, the capacitance increases as indicated by the red curve. Still, when inhaling, it decreases, as seen by the blue curve (d) single respiratory cycle (amplified signal). Reproduced with permission from [61].

track water molecules, or the relative humidity surrounding the mouth or nose, would be a helpful method for examining respiratory health [117]. Using a domestic winding facility, coating, and sputtering technique, Ma *et al.* designed a capacitance-type humidity sensor. The study investigated the sensor's frequency characteristics and humidity response within the frequency range of 500 Hz to 300 kHz (Figure 3.13). Under the 5-kHz frequency testing condition, the sensor demonstrated good linearity, repeatability, responsiveness, and reasonably high sensitivity, making it the ideal option for the future generation of wearable sensors [61].

### *3.4.5.4 Strain Sensors*

Strain sensors exhibit considerable promise for applications in illness diagnostics, healthcare, rehabilitation, wearable textiles, and robot control [118, 119]. Because of their great flexibility, textile materials have been employed extensively for developing flexible strain sensors. However, because regular textile fabrics are electrically insulating, their susceptibility to strain is reduced [120]. Consequently, a wide range of textile-based strain sensors has been studied so far, with flexible polymeric textiles serving as supporting substrates and conductive substrates serving as strain-responsive materials. For strain-responsive materials, several carbon-based structures, such as graphene, graphene oxide, CNTs, quantum dots, graphite, and metal nanoparticles, have been used [121]. Figure 3.14 depicts the application areas of sputter-coated textiles as strain sensors.

He *et al.* designed cotton fabric-based strain sensors through dipping and coating techniques. Cotton fabric was initially immersed in reduced graphene oxide (rGO) thin film before being sputter-coated with Ag thin films. The Ag/rGO-coated cotton fabric (power of 200 W and sputtering time of 25 min) had the highest electrical conductivity and average surface resistance (2.71 Ω/sq). Furthermore, the integrated strain sensor demonstrated the advantages of high sensitivity, a large usable strain range (0%–20%), rapid response, and outstanding stability [120]. Due to the infiltration of polydimethylsiloxane (PDMS) into the multilayer during stretching, the core-sheath structure composed of polyurethane core materials, multilayer graphene nanosheets (GNSs)/Au/GNSs sheath, and PDMS wrapping layer allowed the composite sensor to simultaneously achieve excellent flexibility and stretchability, superior sensitivity, a broad range of strain sensitivity, and effective waterproof properties. After about 10,000 stretch/release cycles at a 50% applied strain, the yarn strain sensor revealed outstanding

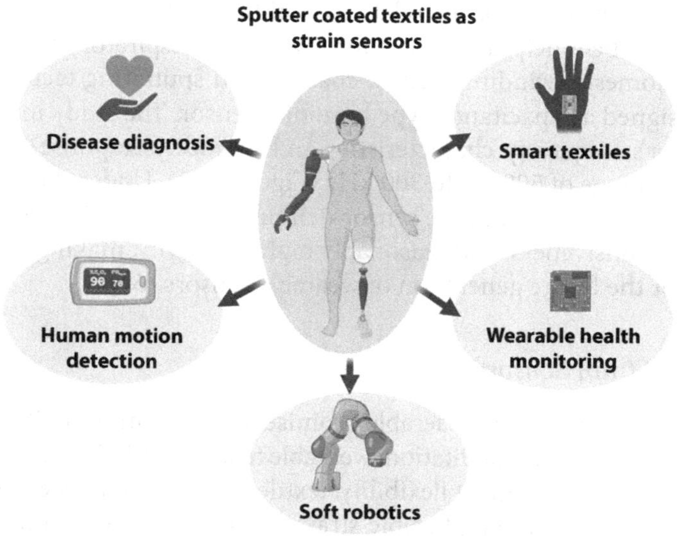


**Figure 3.14** Application areas of sputter-coated textile-based strain sensors.

stability. Furthermore, it might be easily incorporated into textiles (bandages and gloves) to detect human actions and successfully control a hand robot [118].

#### 3.4.5.5 *Flexible Capacitive Pressure Sensors*

Flexible capacitive sensors have numerous applications, including E-skin, human motion, health evaluation, and human-machine interaction. Various intricate conductive nanomaterials are commonly coated on textiles utilizing a conventional coating process to fabricate conductive electrodes of flexible capacitive pressure sensors [122]. Recently, Ag-plated conductive fabric (twill fabric) was constructed using the magnetron sputtering technique, and its conductivity reached 0.268 Ω cm (specific resistance). The flexible wearable pressure sensor was designed by combining the conductive fabric with an Ecoflex elastomer middle layer [123].

The primary components of capacitive pressure sensors are dielectric layers and electrodes. Capacitive pressure sensors often have a "sandwich" shape, with the top and lower electrodes sandwiched around the dielectric layer. The capacitance (C) is calculated by Equation 3.6 [124]:

$$C = \frac{\varepsilon_0 \varepsilon_r A}{d} \tag{3.6}$$

here,

$\varepsilon_0$ - *dielectric constant (vacuum) vacuum dielectric constant*
$\varepsilon_r$ - *dielectric constant of the dielectric layer (relative)*
$A$ - *effective area (upper and lower electrode), and*
$d$ - *distance between the two electrode plates.*

### 3.4.6 Wastewater Treatment

The application of magnetic nanoparticles (NPs) in environmental remediation is becoming more popular. They can address a wide range of issues concerning inadequate water quality. A new type of magnetic glass fiber with NiO/CoO NPs composite coatings was recently developed using plasma sputtering. Coating NiO and CoO on a glass mat resulted in significant antibacterial and photocatalytic activity. In addition, the concentration of methylene blue after exposure to UV light was dramatically reduced in the presence of the developed magnetic glass mat [53]. Although UV light photocatalysis is often faster, employing visible light-responsive materials may be less expensive due to the endless solar energy source. A unique

HIPIMS approach was used to yield nano-sized iron oxides ($FeO_x$) on polyester fabrics. After five oxidation cycles, the $FeO_x$/polyester composite was shown to be highly durable and reusable in hospital effluent [125].

### 3.4.7 Medical Textiles

#### *3.4.7.1 Antibacterial Materials*

Antimicrobial materials are greatly sought after in the production of smart packaging and personal protective equipment to safeguard food and medical personnel from pathogenic microbes [126]. Magnetron sputtering avoids the significant expenses associated with the manufacturing of nanoparticles used to modify textile surfaces. Thus, it provides a resource-efficient approach to developing antimicrobial textiles. Furthermore, the modification of textile materials by sputter coating does not require any chemicals and can be performed in a single process cycle [127].

The selection of microorganisms for antibacterial testing is related to their applicability. For instance, *S. aureus* is frequently found on the skin, *K. pneumoniae* causes various types of infections related to healthcare (respiratory infections), and *E. coli* is the most prevalent, well-studied bacteria identified in the human intestine, making it a more relevant indicator organism to use for antibacterial studies [128].

A multi-functional, antimicrobial composite material was fabricated by depositing Cu films on PET knitted textiles by the magnetron sputtering technique. The Cu-coated modified knitted textile (PET) composites demonstrated significant antibacterial activity against *S. aureus* and *E. coli* bacteria and antifungal activity against *Chaetomium globosum* [59]. In another study, Cu-coated PET substrates exhibited significant inhibitory efficacy against *E. coli* and *S. aureus* (>95%) bacteria, as well as SFV (Semliki Forest Virus) and MS2 (bacteriophage) viruses [44]. However, sputtering power and process duration have a substantial influence on the performance of the coated textiles [43]. Recent works on the antibacterial performance of sputter-coated textiles are summarized in Table 3.3.

#### *3.4.7.2 Wound Dressing*

Template-assisted magnetron sputtering offers a facile and sustainable method for developing functional wound dressings [30, 130]. Cu/Zn strips were deposited onto PE/cotton/PE substrates by template-assisted magnetron sputtering to develop an electroactive antibacterial wound dressing. The antibacterial activity of Cu/Zn@PE/Cotton/PE was based on the

**Table 3.3** Antibacterial performance of sputter-coated textiles.

| Materials | Active agent | Method | Bacterial strain used | Performance | Ref. |
|---|---|---|---|---|---|
| Cu-PLA non-woven fabrics | Cu | PN-EN ISO 20645:2006 | *E. coli* and *S. aureus* | • Better inhibition for *E. coli* | [129] |
| Cu-PET knitwear | Cu | EN ISO 20645:2006 | *E. coli* and *S. aureus* | • Better inhibition for *E. coli* | [59] |
| Ag/DLC on textile | Ag/DLC | Agar diffusion method | *S. aureus* and *K. pneumoniae* | • Inhibited the growth of both bacteria | [128] |
| CN-PP non-woven fabrics | CN | KS K 0693 test | *S. aureus* and *K. pneumoniae* | • The reduction rate was<br>▪ 89.0% for *S. aureus* and<br>▪ 91.6% for *K. pneumoniae* | [47] |
| Ag–Cu coated PET nonwoven | Ag–Cu | ASTM-E2149 and ISO 18184:2029 (E) | *E. coli* and *S. aureus* | • Antibacterial rates -<br>▪ 99.996% for *E. coli* and<br>▪ 99.978% for *S. aureus*<br>• Antiviral activity rate was 99.02%<br>• Significantly reduced the titer of the H1N1 influenza virus | [42] |
| Cu-coated cellulose nonwoven | Cu | PN-EN ISO 20645:2006 | *E. coli* and *S. aureus* | • Average inhibition zone -<br>▪ *E. coli* = 3 mm and<br>▪ *S. aureus* = 2 mm | [43] |

microscale electric fields of Cu/Zn galvanic couples. The microscale fields killed the bacteria by swiftly raising intracellular reactive oxygen species concentration and disrupting the bacterial metabolic process. However, further research is required to examine the wound healing effectiveness and *in vivo* compatibility of the composites [58]. Liu *et al.* developed a cotton-based wound dressing with a template-assisted magnetron sputtering technique. The *in vitro* and *in vivo* (mouse model) tests demonstrated that Ag/Zn/cotton can substantially expedite the wound healing process, which is associated with improved cell migration *via* electrical stimulation and significant antibacterial action (continuous release of $Ag^{+}/Zn^{2+}$), good cytocompatibility (low metal coating area percentage of 17.54%), and excellent physical properties [130]. Wang *et al.* developed a multilayer electroactive composite cotton dressing (Ag/Zn/cotton/paraffin) that includes exudate-activated electrical stimulation. In a mouse model, Ag/Zn/cotton/paraffin enhanced wound healing by triggering re-epithelialization, reducing inflammation, enhancing collagen deposition, and angiogenesis. Furthermore, the composite showed outstanding antibacterial performance and cytocompatibility [131].

## 3.5 Conclusion and Future Studies

The magnetron sputtering technique is a novel way for surface modification of textiles and has found extensive applications in industrial applications such as EMI shielding, protective clothing (UV, flame, water repellent, temperature management system, etc.), air filtration, textile coloration, wearable sensors, and medical textiles. The performance of the sputter-coated textiles depends on the applied voltage, target type, substrate surface, target-substrate distance, coating time, working gas type, and gas pressure. This chapter represents an overview of sputter coating technology and the basic chemistry of textile materials. Different factors affecting the performance of the coated textiles were summarized. Furthermore, this chapter discusses the most recent advances in their application in diverse fields. The findings of this chapter can be used to develop new materials for multifunctional applications. However, there are still numerous issues that require more investigation:

- ✓ Further research is required to systematically explore the impact of deposition factors, such as substrate temperature and pressure, gas flow rate, and gas flow pressure on film characteristics.
- ✓ Most application trials are conducted on a lab scale. As a result, a substantial study is required for the industrialization of the produced materials for mass manufacturing. From this perspective, suitable production equipment needs to be developed.

✓ Coating adhesion and mechanical properties of the coated textiles should be explored extensively.
✓ The stability of coated films against different environmental conditions is hardly explored in the literature.
✓ Because metal or metal oxides are extensively used for sputter coating, disposal of the coated materials should be a matter of concern for the environment.
✓ The adsorption behavior of the coated textiles could be explored for their potential application in water purification.
✓ The impact of metal or metal oxide-coated materials on human skin needs to be further investigated. There are a few studies in this area.
✓ Sputter coating is a water- and solvent-free textile finishing process that reduces the environmental impact. However, a systematic study is needed on the life cycle analysis, techno-economic analysis, and environmental impact assessment of the developed materials, concerning the conventional materials.

**AI disclosure statement:** The author acknowledges the use of AI-based tools for minor language editing or grammatical correction. However, all intellectual contributions, analyses, and conclusions presented in this chapter are entirely the author's own.

## References

1. Xu, J., *et al.*, AgNps-PVA–coated woven cotton fabric: Preparation, water repellency, shielding properties and antibacterial activity. *J. Ind. Text.*, 48, 10, 1545–1565, 2018.
2. Sfameni, S., *et al.*, Inorganic Finishing for Textile Fabrics: Recent Advances in Wear-Resistant, UV Protection and Antimicrobial Treatments. *Inorganics*, 11, 1, 19, 2023.
3. Berradi, M., *et al.*, Textile finishing dyes and their impact on aquatic environs. *Heliyon*, 5, 11, e02711, 2019.
4. Şimşek, B. and Karaman, M., Initiated chemical vapor deposition of poly(hexafluorobutyl acrylate) thin films for superhydrophobic surface modification of nanostructured textile surfaces. *J. Coat. Technol. Res.*, 17, 2, 381–391, 2020.
5. Oke, J.A. and Jen, T.-C., Atomic layer deposition and other thin film deposition techniques: from principles to film properties. *J. Mater. Res. Technol.*, 21, 2481–2514, 2022.

6. Mwema, F.M., Jen, T.-C., Zhu, L., *Thin film coatings: properties, deposition, and applications*, CRC Press, Boca Raton, 2022.
7. Lomov, S.V., *et al.*, Compressibility of nanofibre-grafted alumina fabric and yarns: Aligned carbon nanotube forests. *Compos. Sci. Technol.*, 90, 57–66, 2014.
8. Klaus, E., Magnetron sputtering of transparent conductive zinc oxide: relation between the sputtering parameters and the electronic properties. *J. Phys. D: Appl. Phys.*, 33, 4, R17, 2000.
9. Fang, S.K., *et al.*, Cellulose acetate superhydrophobic coatings for efficient oil–water separation using a combination of electrostatic spraying and chemical vapor deposition. *Polym. Eng. Sci.*, 64, 1, 254–263, 2024.
10. Chen, K., *et al.*, Deposition of polycrystalline Si and SiGe by ultra-high vacuum chemical molecular epitaxy. *J. Vac. Sci. Technol. A: Vac. Surf. Films*, 18, 4, 1196–1201, 2000.
11. Zheng, Y., *et al.*, Ultra-smooth and hydrophobic nitrogen-incorporated ultranano-crystalline diamond film growth in CHON gas phase system *via* microwave plasma CVD. *Surf. Coat. Technol.*, 374, 409–417, 2019.
12. Gupta, C.K. and Sathiyamoorthy, D., *Fluid bed technology in materials processing*, CRC Press, Boca Raton, 1998.
13. Manasevit, H.M., Single-crystal gallium arsenide on insulating substrates. *Appl. Phys. Lett.*, 12, 4, 156–159, 1968.
14. Noč, L. and Jerman, I., Review of the spectrally selective (CSP) absorber coatings, suitable for use in SHIP. *Sol. Energy Mater. Sol. Cells*, 238, 111625, 2022.
15. Selvakumar, N. and Barshilia, H.C., Review of physical vapor deposited (PVD) spectrally selective coatings for mid-and high-temperature solar thermal applications. *Sol. Energy Mater. Sol. Cells*, 98, 1–23, 2012.
16. Shahidi, S., Moazzenchi, B., Ghoranneviss, M., A review-application of physical vapor deposition (PVD) and related methods in the textile industry. *Eur. Phys. J. Appl. Phys.*, 71, 3, 31302, 2015.
17. Yip, J., Jiang, S., Wong, C., Characterization of metallic textiles deposited by magnetron sputtering and traditional metallic treatments. *Surf. Coat. Technol.*, 204, 3, 380–385, 2009.
18. Wei, Q., *et al.*, Evaluation of the interfacial bonding between fibrous substrate and sputter coated copper. *Surf. Coat. Technol.*, 202, 19, 4673–4680, 2008.
19. Hu, Q., *et al.*, Sputtering-deposited thin films on textiles for solar and heat managements: a mini-review. *Phys. Status Solidi (A)*, 219, 3, 2100572, 2022.
20. Zhang, S., *et al.*, Flexible silver nanoparticle@aramid nanofiber SERS substrates fabricated by radio frequency magnetron sputtering for rapid detection of toxic substances. *Mater. Today Commun.*, 39, 109263, 2024.
21. Mrozińska, Z., *et al.*, Poly(Lactide) Nonwoven Fabric with Iron Coating and Its Biological Properties. *Coatings*, 14, 8, 1050, 2024.
22. Global Sputter Coaters Market by Type (Dielectric Sputter Coatings, Magnetic Sputter Coatings, Metal Sputter Coatings), End-Users (Aerospace & Aviation, Automotive Industry, Electronics & Semiconductor) - Forecast 2024-2030. 2024 [cited 2024.

23. Wei, Q., Xu, Y., Wang, Y., 3 - Textile surface functionalization by physical vapor deposition (PVD), in: *Surface Modification of Textiles*, Q. Wei (Ed.), pp. 58–90, Woodhead Publishing, Cambridge, UK, 2009.
24. Tan, X.-Q., *et al.*, Recent progress in magnetron sputtering technology used on fabrics. *Materials*, 11, 10, 1953, 2018.
25. Lopes, C., *et al.*, EMI Shielding and Conductive Textiles Functionalized with (Ti,Cu) Nanomaterials for Biomedical Applications. *ACS Appl. Mater. Interfaces*, 15, 33, 39872–39882, 2023.
26. Kelly, P.J. and Arnell, R.D., Magnetron sputtering: a review of recent developments and applications. *Vacuum*, 56, 3, 159–172, 2000.
27. Xia, C., *et al.*, Natural fiber composites with EMI shielding function fabricated using VARTM and Cu film magnetron sputtering. *Appl. Surf. Sci.*, 362, 335–340, 2016.
28. Sittinger, V., *et al.*, Rotatable serial co-sputtering of doped titania. *Vacuum*, 114, 158–161, 2015.
29. Liu, Y., *et al.*, Investigation on the properties of nano copper matrix composite *via* vacuum arc melting method. *Mater. Res. Express*, 4, 10, 106512, 2017.
30. Mrozińska, Z., *et al.*, Activity in the Field of Blood Coagulation Processes of Poly(Lactide)-Zinc Fiber Composite Material Obtained by Magnetron Sputtering. *Coatings*, 14, 6, 2024.
31. Nguyen, T.-T.-N., *et al.*, Multifunctional Ti-O coatings on polyethylene terephthalate fabric produced by using roll-to-roll high power impulse magnetron sputtering system. *Surf. Coat. Technol.*, 324, 249–256, 2017.
32. Buehler, M.J., Nature designs tough collagen: explaining the nanostructure of collagen fibrils. *Proc. Natl. Acad. Sci.*, 103, 33, 12285–12290, 2006.
33. Sánchez-López, J.C., *et al.*, Ti6Al4V coatings on titanium samples by sputtering techniques: Microstructural and mechanical characterization. *J. Alloys Compd.*, 952, 170018, 2023.
34. Mattox, D.M., Chapter 7 - Physical Sputtering and Sputter Deposition (Sputtering), in: *Handbook of Physical Vapor Deposition (PVD) Processing*, Second Edition, D.M. Mattox (Ed.), pp. 237–286, William Andrew Publishing, Boston, 2010.
35. Bohlmark, J., *et al.*, Ionization of sputtered metals in high power pulsed magnetron sputtering. *J. Vac. Sci. Technol. A*, 23, 1, 18–22, 2005.
36. Sarkar, J., Chapter 2 - Sputtering and Thin Film Deposition, in: *Sputtering Materials for VLSI and Thin Film Devices*, J. Sarkar (Ed.), pp. 93–170, William Andrew Publishing, Boston, 2014.
37. Hammadi, O.A., Khalaf, M.K., Kadhim, F.J., Farbication of UV photodetector from nickel oxide nanoparticles deposited on silicon substrate by closed-field unbalanced dual magnetron sputtering techniques. *Opt. Quantum Electron.*, 47, 3805–3813, 2015.

38. Gago, R., Jiménez, I., Albella, J.M., *Thin film growth by ion-beam-assisted deposition techniques*, vol. 63, Elsevier Amsterdam, The Netherlands, Amsterdam, 2006.
39. Chu, C., *et al.*, Surface functionalization of nanostructured Cu/Ag-deposited polypropylene fiber by magnetron sputtering. *e-Polymers*, 21, 1, 140–150, 2021.
40. Huang, S.Y., *et al.*, Nanoarchitectonics for Ultrathin Gold Films Deposited on Collagen Fabric by High-Power Impulse Magnetron Sputtering. *Nanomaterials (Basel)*, 12, 10, 2022.
41. Yuan, X., *et al.*, Photocatalytic property of polyester fabrics coated with Ag/TiO2 composite films by magnetron sputtering. *Vacuum*, 172, 2020.
42. Huang, X., *et al.*, Sputtering-Deposited Ultra-Thin Ag-Cu Films on Non-Woven Fabrics for Face Masks with Antimicrobial Function and Breath NO(x) Response. *Materials (Basel)*, 17, 7, 2024.
43. Świerczyńska, M., *et al.*, Biochemical Evaluation and Structural Characteristics of Copper Coating Cellulose Nonwovens Prepared by Magnetron Sputtering Technology. *Coatings*, 14, 7, 843, 2024.
44. Vibornijs, V., *et al.*, Analysis of Antibacterial and Antiviral Properties of ZnO and Cu Coatings Deposited by Magnetron Sputtering: Evaluation of Cell Viability and ROS Production. *Coatings*, 14, 1, 2023.
45. Luceri, A., *et al.*, Washing resistant antibacterial composite coatings on cotton textiles. *Cellulose*, 30, 15, 9877–9897, 2023.
46. Giedraitiene, A., *et al.*, Development of Antibacterial Cotton Textiles by Deposition of Fe(2)O(3) Nanoparticles Using Low-Temperature Plasma Sputtering. *Nanomaterials (Basel)*, 13, 24, 2023.
47. Sohn, Y.S., *et al.*, Antibacterial Effects of a Carbon Nitride (CN) Layer Formed on Non-Woven Polypropylene Fabrics Using the Modified DC-Pulsed Sputtering Method. *Polymers (Basel)*, 15, 12, 2023.
48. Kudzin, M.H., *et al.*, Preparation of Cotton-Zinc Composites by Magnetron Sputtering Metallization and Evaluation of their Antimicrobial Properties and Cytotoxicity. *Materials (Basel)*, 15, 8, 2022.
49. Shichen, Z. and Jiangtao, X., Multifunctional metallic cotton fabric: effect of coating methods and PVA pretreatment. *J. Mater. Sci.: Mater. Electron.*, 33, 25, 20257–20270, 2022.
50. Huang, M.-L., *et al.*, Metallic coloration with Cu/CuO coating on polypropylene nonwoven fabric *via* a physical vapor deposition method and its multifunctional properties. *J. Text. Inst.*, 113, 7, 1345–1354, 2021.
51. Cheng, Y., *et al.*, Environmentally friendly and antimicrobial bilayer structured fabrics with integrated interception and sterilization for personal protective mask. *Sep. Purif. Technol.*, 294, 2022.
52. Cieslak, M., *et al.*, Effect of Cu Modified Textile Structures on Antibacterial and Antiviral Protection. *Materials (Basel)*, 15, 17, 2022.
53. Shahidi, S., *et al.*, *In situ* deposition of magnetic nanoparticles on glass mat using plasma sputtering method. *J. Text. Inst.*, 113, 3, 349–359, 2021.

54. Zhang, S., *et al.*, Hydro electroactive Cu/Zn coated cotton fiber nonwovens for antibacterial and antiviral applications. *Int. J. Biol. Macromol.*, 207, 100–109, 2022.
55. Sharifikolouei, E., *et al.*, Generation of cytocompatible superhydrophobic Zr-Cu-Ag metallic glass coatings with antifouling properties for medical textiles. *Mater. Today Bio*, 12, 100148, 2021.
56. Huang, M.-L., *et al.*, Antibacterial and ultraviolet protective neodymium-doped TiO2 film coated on polypropylene nonwoven fabric *via* a sputtering method. *J. Eng. Fibers Fabr.*, 16, 2021.
57. Feng, Y., *et al.*, Ag/Zn Galvanic Couple Cotton Nonwovens with Breath-Activated Electroactivity: A Possible Antibacterial Layer for Personal Protective Face Masks. *ACS Appl. Mater. Interfaces*, 13, 49, 59196–59205, 2021.
58. Dong, H., *et al.*, Cu/Zn galvanic couples composite antibacterial dressings prepared by template-assisted magnetron sputtering. *Composites, Part B*, 224, 2021.
59. Kudzin, M.H., *et al.*, Deposition of Copper on Polyester Knitwear Fibers by a Magnetron Sputtering System. Physical Properties and Evaluation of Antimicrobial Response of New Multi-Functional Composite Materials. *Appl. Sci.*, 10, 19, 2020.
60. Kudzin, M.H., *et al.*, Deposition of Copper on Poly(Lactide) Non-Woven Fabrics by Magnetron Sputtering-Fabrication of New Multi-Functional, Antimicrobial Composite Materials. *Materials (Basel)*, 13, 18, 2020.
61. Ma, L., *et al.*, All-in-one fibrous capacitive humidity sensor for human breath monitoring. *Text. Res. J.*, 91, 3-4, 398–405, 2020.
62. Verma, S., *et al.*, Recent Advances for Fabricating Smart Electromagnetic Interference Shielding Textile: A Comprehensive Review. *Electron. Mater. Lett.*, 18, 4, 331–344, 2022.
63. Jagatheesan, K., *et al.*, Fabrics and their composites for electromagnetic shielding applications. *Text. Prog.*, 47, 2, 87–161, 2015.
64. Wanasinghe, D. and Aslani, F., A review on recent advancement of electromagnetic interference shielding novel metallic materials and processes. *Composites, Part B*, 176, 107207, 2019.
65. Akram, S., *et al.*, Recent advances in electromagnetic interference (EMI) shielding textiles: A comprehensive review. *Synth. Met.*, 294, 117305, 2023.
66. Rubežienė, V. and Varnaitė-Žuravliova, S., EMI shielding textile materials, in: *Materials for Potential EMI Shielding Applications*, pp. 357–378, Elsevier, 2020.
67. Nair, A.M., Zachariah, S.M., Thomas, S., *A Mini-Review On MXene Based Textiles For Electromagnetic Interference Shielding Application*, Qeios.
68. Ding, X., *et al.*, Layered cotton/rGO/NiWP fabric prepared by electroless plating for excellent electromagnetic shielding performance. *Cellulose*, 26, 8209–8223, 2019.
69. Safdar, F., *et al.*, Polymeric textile-based electromagnetic interference shielding materials, their synthesis, mechanism and applications – A review. *J. Ind. Text.*, 51, 5_suppl, 7293S–7358S, 2021.

70. Park, J., *et al.*, Electromagnetic interference shielding effectiveness of sputtered NiFe/Cu multi-layer thin film at high frequencies. *Thin Solid Films*, 677, 130–136, 2019.
71. Lv, P., *et al.*, Copper nanoparticles-sputtered bacterial cellulose nanocomposites displaying enhanced electromagnetic shielding, thermal, conduction, and mechanical properties. *Cellulose*, 23, 5, 3117–3127, 2016.
72. Wang, Q., *et al.*, Self-bonded natural fiber product with high hydrophobic and EMI shielding performance *via* magnetron sputtering Cu film. *Appl. Surf. Sci.*, 475, 947–952, 2019.
73. Jiang, S., *et al.*, Enhanced electro-conductivity and multi-shielding performance with copper, stainless steel and titanium coating onto PVA impregnated cotton fabric. *J. Mater. Sci.: Mater. Electron.*, 29, 7, 5624–5633, 2018.
74. Yu, H.-L., *et al.*, Simple fabrication of the Ag-Ag2O-TiO2 photocatalyst thin films on polyester fabrics by magnetron sputtering and its photocatalytic activity. *Appl. Surf. Sci.*, 503, 2020.
75. Qiu, P., Xu, J., Xu, Q., Protective textiles from Ag-decorated forest-like amorphous TiO2 on cotton fabric. *J. Mater. Sci.: Mater. Electron.*, 34, 13, 1099, 2023.
76. Cai, Y., *et al.*, Structure, surface morphology, thermal and flammability characterizations of polyamide6/organic-modified Fe-montmorillonite nanocomposite fibers functionalized by sputter coating of silicon. *Surf. Coat. Technol.*, 203, 3, 264–270, 2008.
77. Miskiewicz, P., *et al.*, Assessment of Coating Quality Obtained on Flame-Retardant Fabrics by a Magnetron Sputtering Method. *Materials (Basel)*, 14, 6, 2021.
78. Miśkiewicz, P., *et al.*, Evaluation of Thermal Properties of Certain Flame-Retardant Fabrics Modified with a Magnetron Sputtering Method. *Autex Res. J.*, 21, 4, 428–434, 2021.
79. Peng, Y. and Cui, Y., Advanced textiles for personal thermal management and energy. *Joule*, 4, 4, 724–742, 2020.
80. Hu, R., *et al.*, Emerging materials and strategies for personal thermal management. *Adv. Energy Mater.*, 10, 17, 1903921, 2020.
81. Peng, L., *et al.*, Review of clothing for thermal management with advanced materials. *Cellulose*, 26, 6415–6448, 2019.
82. Qiu, S., Jia, H., Jiang, S.-X., Fabrication and characterization of thermal management fabric with heating and cooling modes through magnetron sputtering. *Mater. Lett.*, 300, 130217, 2021.
83. Wang, C., *et al.*, Multifunctional wearable thermal management textile fabricated by one-step sputtering. *Nano Today*, 45, 101526, 2022.
84. Ju, Y., *et al.*, Review on Hydrophobic Thin Films Prepared Using Magnetron Sputtering Deposition. *Materials (Basel)*, 16, 10, 2023.
85. Betz, U., *et al.*, Thin films engineering of indium tin oxide: Large area flat panel displays application. *Surf. Coat. Technol.*, 200, 20, 5751–5759, 2006.

86. Ding, J.C., *et al.*, Microstructure, mechanical, and wettability properties of Al-doped diamond-like films deposited using a hybrid deposition technique: Bias voltage effects. *Diamond Relat. Mater.*, 123, 108861, 2022.
87. Malik, G., *et al.*, Optical and other physical properties of hydrophobic ZnO thin films prepared by dc magnetron sputtering at room temperature. *J. Appl. Phys.*, 122, 14, 2017.
88. Liao, R., *et al.*, Anti-icing performance of ZnO/SiO2/PTFE sandwich-nanostructure superhydrophobic film on glass prepared *via* RF magnetron sputtering. *Mater. Lett.*, 206, 109–112, 2017.
89. Parejo, P.G., Zayat, M., Levy, D., Highly efficient UV-absorbing thin-film coatings for protection of organic materials against photodegradation. *J. Mater. Chem.*, 16, 22, 2165–2169, 2006.
90. Saravanan, D., UV protection textile materials. *Autex Res. J.*, 7, 1, 53–62, 2007.
91. Shahidi, S., Novel method for ultraviolet protection and flame retardancy of cotton fabrics by low-temperature plasma. *Cellulose*, 21, 757–768, 2014.
92. Crews, P.C., Kachman, S., Beyer, A.G., Influences on UVR transmission of undyed woven fabrics. *Text. Chem. Color.*, 31, 6, 1999.
93. Kabir, S.M.M., Dhar, A.K., Koh, J., The Influence of a Natural UV Absorber (Areca catechu) on the UV Protection and Antimicrobial Properties of Silk and Nylon Fabrics. *Fibers Polym.*, 22, 2, 382–386, 2021.
94. Shahidi, S. and Ghoranneviss, M., Plasma Sputtering for Fabrication of Antibacterial and Ultraviolet Protective Fabric. *Clothing Text. Res. J.*, 34, 1, 37–47, 2015.
95. Wasim, M., *et al.*, Surface Modification of Bacterial Cellulose by Copper and Zinc Oxide Sputter Coating for UV-Resistance/Antistatic/Antibacterial Characteristics. *Coatings*, 10, 4, 2020.
96. Shahidi, S. and Moazzenchi, B., Comparison between Mordant Treatment and Plasma Sputtering on Natural Dying and UV Protection Properties of Wool Fabric. *Fibers Polym.*, 20, 8, 1658–1665, 2019.
97. Carneiro, J., *et al.*, Photocatalytic activity and UV-protection of TiO2 nano-coatings on poly (lactic acid) fibres deposited by pulsed magnetron sputtering. *J. Nanosci. Nanotechnol.*, 11, 10, 8979–8985, 2011.
98. Wu, Y., *et al.*, Surface functionalization of nanostructured LaB6-coated Poly Trilobal fabric by magnetron sputtering. *Appl. Surf. Sci.*, 384, 413–418, 2016.
99. Huang, M.-L., Wu, Y.-Z., Lu, S.-G., Structural coloration and ultraviolet protective fabrication on fabrics coated with SiO2/TiO2 multilayer films *via* a magnetron sputtering method. *J. Ind. Text.*, 52, 15280837221114636, 2022.
100. Yuan, X., *et al.*, Properties and application of multi-functional and structurally colored textile prepared by magnetron sputtering. *J. Ind. Text.*, 51, 8, 1295–1311, 2020.
101. Dong, J., *et al.*, An All-Stretchable, Ultraviolet Protective, and Electromagnetic-Interference-Free E-Textile. *Adv. Funct. Mater.*, 33, 45, 2308426, 2023.

102. Jiang, S., *et al.*, Fabrication of Ag thin film on polyester fabric by roll to roll magnetron sputtering system. *J. Mater. Sci.: Mater. Electron.*, 26, 5, 3364–3369, 2015.
103. Huang, M., *et al.*, Structural coloration and its application to textiles: a review. *J. Text. Inst.*, 111, 5, 756–764, 2020.
104. Huang, M.-L., *et al.*, Structural colouration on textile fabrics with thin-film coating *via* magnetron sputtering: A review. *Surf. Eng.*, 38, 10-12, 830–845, 2022.
105. Huang, M.-L., *et al.*, Metallic coloration on polyester fabric with sputtered copper and copper oxides films. *Vacuum*, 178, 2020.
106. Huang, M.-L., *et al.*, Metallic coloration and multifunctional preparation on fabrics *via* nitriding reactive sputtering with copper and titanium targets. *Vacuum*, 202, 2022.
107. Evangeline, C.S. and Lenin, A., Human health monitoring using wearable sensor. *Sens. Rev.*, 39, 3, 364–376, 2019.
108. Nasiri, S. and Khosravani, M.R., Progress and challenges in fabrication of wearable sensors for health monitoring. *Sens. Actuators A*, 312, 112105, 2020.
109. Tiggelaar, R.M., *et al.*, Stability of thin platinum films implemented in high-temperature microdevices. *Sens. Actuators, A*, 152, 1, 39–47, 2009.
110. Yan, Z.Y., Liu, J.Y., Niu, J.R., Research of a Novel Ag Temperature Sensor Based on Fabric Substrate Fabricated by Magnetron Sputtering. *Materials (Basel)*, 14, 20, 2021.
111. Chi, E., Kim, W., Hur, N., Nearly zero temperature coefficient of resistivity in antiperovskite compound CuNMn3. *Solid State Commun.*, 120, 7-8, 307–310, 2001.
112. Yan, C., Wang, J., Lee, P.S., Stretchable graphene thermistor with tunable thermal index. *ACS Nano*, 9, 2, 2130–2137, 2015.
113. Appiagyei, A.B., Banua, J., Han, J.I., Flexible and patterned-free Ni/NiO-based temperature device on cylindrical PET fabricated by RF magnetron sputtering: Bending and washing endurance tests. *J. Ind. Eng. Chem.*, 100, 372–382, 2021.
114. Manjakkal, L., Dervin, S., Dahiya, R., Flexible potentiometric pH sensors for wearable systems. *RSC Adv.*, 10, 15, 8594–8617, 2020.
115. Manjakkal, L., Szwagierczak, D., Dahiya, R., Metal oxides based electrochemical pH sensors: Current progress and future perspectives. *Prog. Mater Sci.*, 109, 100635, 2020.
116. Chung, M., *et al.*, Fabrication of a Wearable Flexible Sweat pH Sensor Based on SERS-Active Au/TPU Electrospun Nanofibers. *ACS Appl. Mater. Interfaces*, 13, 43, 51504–51518, 2021.
117. Ma, L., *et al.*, Full-textile wireless flexible humidity sensor for human physiological monitoring. *Adv. Funct. Mater.*, 29, 43, 1904549, 2019.

118. Li, X., *et al.*, An ultraflexible polyurethane yarn-based wearable strain sensor with a polydimethylsiloxane infiltrated multilayer sheath for smart textiles. *Nanoscale*, 12, 6, 4110–4118, 2020.
119. Yoon, S., *et al.*, Highly stretchable metal-polymer hybrid conductors for wearable and self-cleaning sensors. *NPG Asia Mater.*, 13, 1, 2021.
120. He, S., *et al.*, Flexible and highly conductive Ag/G-coated cotton fabric based on graphene dipping and silver magnetron sputtering. *Cellulose*, 25, 6, 3691–3701, 2018.
121. Gong, S. and Cheng, W., One-dimensional nanomaterials for soft electronics. *Adv. Electron. Mater.*, 3, 3, 1600314, 2017.
122. Su, M., *et al.*, Textile-Based Flexible Capacitive Pressure Sensors: A Review. *Nanomaterials*, 12, 9, 1495, 2022.
123. Wu, R., *et al.*, A facile method to prepare a wearable pressure sensor based on fabric electrodes for human motion monitoring. *Text. Res. J.*, 89, 23-24, 5144–5152, 2019.
124. Wang, J., *et al.*, A highly sensitive and flexible pressure sensor with electrodes and elastomeric interlayer containing silver nanowires. *Nanoscale*, 7, 7, 2926–2932, 2015.
125. Coulibaly, G.N., *et al.*, Nano-sized iron oxides supported on polyester textile to remove fluoroquinolones in hospital wastewater. *Environ. Sci.: Nano*, 7, 7, 2156–2165, 2020.
126. da Silva, D.J., *et al.*, Super effective antimicrobial silver-sputtered coatings on poly(lactic acid) against bacteria and omicron SARS-CoV-2. *Mater. Today Chem.*, 30, 2023.
127. Badaraev, A., *et al.*, PLLA scaffold modification using magnetron sputtering of the copper target to provide antibacterial properties. *Resour.-Effic. Technol.*, 3, 2, 204–211, 2017.
128. Antunes, J., *et al.*, Physical Vapor Deposition Technology in Personal Protective Equipment Production: Improved Antibacterial and Hydrophobic Character of Textiles. *Coatings*, 12, 10, 1399, 2022.
129. Kudzin, M.H., *et al.*, Deposition of Copper on Poly(Lactide) Non-Woven Fabrics by Magnetron Sputtering—Fabrication of New Multi-Functional, Antimicrobial Composite Materials. *Materials*, 13, 18, 3971, 2020.
130. Liu, S., *et al.*, Template-Assisted Magnetron Sputtering of Cotton Nonwovens for Wound Healing Application. *ACS Appl. Bio Mater.*, 3, 2, 848–858, 2020.
131. Wang, S., *et al.*, Cotton fiber-based dressings with wireless electrical stimulation and antibacterial activity for wound repair. *Int. J. Biol. Macromol.*, 256, Pt 2, 128496, 2024.

4

# Sol-Gel Technology for Textile Modification

**Enfal Kayahan**

*Pamukkale University, Department of Textile Engineering, Kınıklı Campus, Denizli, Türkiye*

***Abstract***

The sol-gel method is a chemical process that relies on gelable solutions or suspensions. The sol-gel procedure, which involves the development of non-organic frameworks *via* the transformation of a colloidal suspension (sol) and its subsequent transition into a gel state to establish a network within a uniform fluid environment (gel), has become an essential approach across multiple industrial fields. This method enables the high-purity and efficient production of thin film coatings, fibers, nanopowders, and solid structures, along with their application to textile processes, offering advantages such as low cost, minimal chemical consumption, and functional and effective production. Utilizing this technology makes it feasible to impart multifunctional properties to textile surfaces, including antibacterial, anti-mosquito, oil repellent, anti-wrinkle, flame retardant, ultraviolet protection, and water repellency, self-cleaning one-step dye uptake in the same bath at reduced concentration. Research in the textile industry continues to investigate the contribution of sol-gel technology to producing materials with functional properties and reducing environmental wastewater in textile wet processes. This chapter presents an overview of sol-gel technology by examining in detail current technical reports and peer-reviewed publications on the subject and focuses on the applications of this technology in the textile industry.

***Keywords*:** Sol-gel, textile surface modification, nanosol coatings, finishing

*Email*: ekayahan@pau.edu.tr

Aminoddin Haji and Çağlar Sivri (eds.) Surface Modification and Functionalization of Textiles, (87–124) © 2026 Scrivener Publishing LLC

## 4.1 Introduction

Today, modern textile products are globally utilized in various applications, including the clothing sector, household fabrics, upholstery, medical products, and manufacturing processes. The characteristics of textiles, such as adhesion and wetting properties, largely depend on their surface structures. Therefore, modifying the surface of textiles provides an opportunity to improve functional properties, allowing them to fulfill needs and needs across various applications.

In the textile sector, sol-gel technology is crucial for creating nanocomposite polymer films on fiber surfaces. This process imparts unique mechanical, optical, electrical, and biological properties to textiles, which are unattainable with conventional finishing methods. This process is commercial importance for producing of technical fabrics and is particularly important for functional textiles used in medical, protective, and engineering contexts [1].

Surface modification involves using a variety of technologies to modify the surface characteristics of textiles, thereby providing the textile product with desired properties. Recently, modifying textile surfaces has provided numerous advantages, such as increased adhesion, wettability, and chemical affinity or resistance. These modifications can be accomplished through various methods, spanning from conventional solution treatments to cutting-edge biological methods [2, 3]. Conventional textile wet processing is composed of three key stages: preparation, color application (through dyeing or printing), and final finishing. The pretreatment phase, which includes processes such as scouring and bleaching, is carried out initially to prepare the textiles for the subsequent dyeing stage. Following the dyeing process, textiles are finished based on their intended use or specific consumer needs [4, 5]. This finishing process is the final step that adds value to textiles by imparting special functional properties, features like antimicrobial properties, water repellency, waterproofing, flame resistance, and stain or soil resistance [6]. Textile finishing can be categorized by the method (chemical wet processing or mechanical dry methods, including brushing, shearing, and raising) and by the desired outcome, whether aesthetic (altering the fabric's hand or drape) or functional. Chemical finishing generally involves adding various chemicals tailored to the textile's intended use. Among the different finishing methods, chemical finishing is extensively researched due to evolving trends and consumer demands for advanced, high-performance applications [7].

Over the past 20 years, the sol-gel–assisted method for textile finishing has become crucial in developing new applications aimed at enhancing the fundamental characteristics of textiles. Sol-gel technology is a chemical method employed for modifying surfaces. Textile finishing using sol-gel techniques provides significant benefits compared to traditional methods [8]. The primary advantages include being environmentally friendly, using fewer chemicals, enabling low-temperature treatments, posing low toxicity risks to health of humans while maintaining the natural structural and functional aspects of fabrics, allowing for adjustable coating thickness, and providing long-lasting properties to finished fabrics. It also represents a novel approach to preparing composite materials. Modification of textile fabrics can be achieved either through the application of nanosol films or by depositing silica nanoparticles [9]. The sol-gel method encompasses a broad range of processes used to produce gels from sols, or colloidal suspensions. This process involves creating inorganic matrices by forming colloidal suspensions, followed by the gelation of sols into gels, which then transform into xerogels (dry gels) after drying [10].

In most cases, the sol-gel process includes the hydrolysis and condensation of a metal-organic precursor, such as tetra n-butyl titanate, in a suitable solvent like ethanol, either with or without a catalyst [11]. Sol-gel processes achieve the formation of oxide networks from molecular precursors through consecutive hydrolysis and condensation reactions [12, 13]. During this process, alkoxide groups in the alcohol-water solution are gradually replaced by hydroxyl groups *via* hydrolysis, in the presence of acidic or basic catalysts, leading to the formation of –M–O–M– bonds. Gelation occurs as the polymer networks grow and aggregate, eventually a network that extends throughout the entire volume of the solution. At the gelation point, both the viscosity and the elastic modulus of the solution increase steadily. The gel can subsequently be dried to produce xerogels through evaporation or converted into aerogels by using supercritical fluid extraction [10, 11].

## 4.2 Synthesis of Sol-Gels and General Principles

The sol-gel approach involves the transition of a liquid "sol" phase, resulting in the formation of a solid "gel" phase [14]. Inorganic sol-gels are typically synthesized directly from chemical reactants dissolved in a liquid medium. In this process, the reactant containing a metal (M) cation is referred to as a chemical precursor within the inorganic sol or gel [15, 16]. The chemical transformation of this structure involves a complex process, and the transition from sol to gel involves intricate molecular-level reactions. The fundamental reactions involved in sol-gel formation are outlined below [17].

### 4.2.1 Solvation

The ionizable precursor-solvent complex $M(H_2O)_n^{z+}$ (where z denotes the valence of the M ions) is formed when the metal cation $Mz^+$ from the metal salt attracts water molecules. As a result of its strong tendency to maintain its coordination number, a proton ($H^+$) is released [18].

$$M(H_2O)_n^{z^+} \rightarrow M(H_2O)_{(n-1)}(OH)^{(z+1)}_{+}H^+ \tag{4.1}$$

### 4.2.2 Hydrolysis

In the hydrolysis process, the forward reaction involves the transformation of alkoxy groups into hydroxyl groups, whereas the reverse reaction is known as esterification. In this step, non-ionized molecular precursor, such as a metal alkoxide M(OR)n (where n is the valence of the metal M and R represents an alkyl group), reacts with water [19].

$$M(OR)_n + xH_2O \rightarrow M(OH)x(0R)_{nx} + xROH \tag{4.2}$$

### 4.2.3 Condensation

The condensation process can occur through two distinct pathways. In the initial stage of the first pathway (3) [19], a reaction between two hydroxyl groups leads to the formation of an oxygen bridge with the concurrent release of a water molecule. The reverse of this reaction is hydrolysis. In the second mechanism (4), condensation between a metal alkoxide and an alkoxy group forms a bridging oxygen, accompanied by the production of an alcohol molecule. The reverse process is known as alcoholysis [19–22].

$$-M - OH + OH - M \rightarrow -M - 0 - M + H_2O \tag{4.3}$$

$$-M - OH + RO - M \rightarrow M - 0 - M + ROH \tag{4.4}$$

This reaction can progressively create a large, three-dimensional polymer network through polymerization [18]. The sol-gel network's final configuration is shaped by the comparative progression of hydrolysis versus condensation, which is affected by variables including reaction temperature, water content, catalyst type, solvent characteristics, and the nature of the alkoxide group.

Sol-gel coating applications often rely on the use of aggressive organic solvents that can be harmful to the environment, particularly in large-scale industrial operations [23, 24]. These solvents, while capable of releasing volatile organic compounds, are subject to federal regulations enforced by the Environmental Protection Agency, with some states implementing even more rigorous limits. In contrast, the aqueous sol-gel technique predominantly employs water as the medium, providing a safer and environmentally responsible alternative for material processing [25, 26]. The water-based sol-gel approach primarily uses water as the solvent, but it may also include small quantities of organic co-solvents to aid in the hydrolysis of sol-gel precursors.

## 4.3 Modification of Textile Surfaces Using Sol-Gel Technology

A common approach to modifying fabrics involves dipping or spraying sol-gel formulations onto the fabric substrates [27, 28]. Typically, the sol serves as a medium or binding layer that enables the incorporation of an additional functional component, thereby imparting the characteristics of that component to the fabric. An illustrative scheme of the sol-gel process can be seen in Figure 4.1.

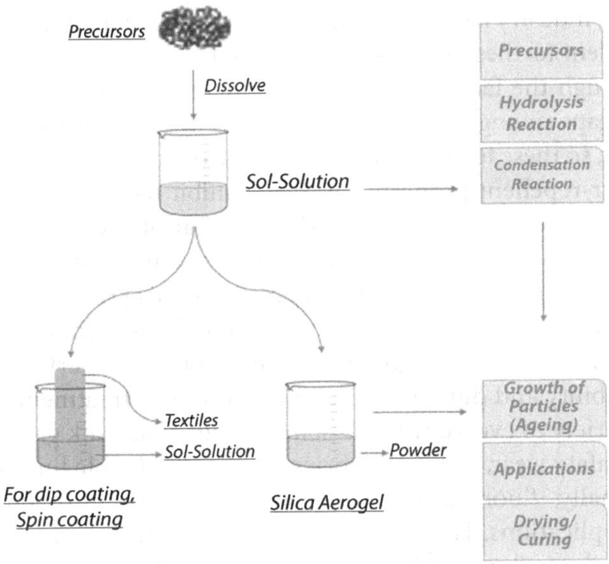


**Figure 4.1** Stages involved in the sol-gel process.

The sol-gel process is a widely used method for the chemical modification of textile fibers. In this chemistry-based surface tailoring method, the fabric substrate is exposed to liquid reactants that attach reactive groups onto its outer layer. An alternative approach for these modifications is grafting, where the initial stage consists of creating radicals on the fabric interface, followed by copolymerization with a range of monomer units. Traditional wet processes in textiles often necessitate prolonged residence times, high concentrations of aggressive chemicals, and substantial volumes of water and are typically conducted at elevated temperatures, resulting in significant energy consumption.

The sol-gel reaction is a simple technique and does not require specific conditions or high temperatures.

This chapter offers a comprehensive overview of the modification of textile materials through sol-gel technology, with in-depth analysis of current technical reports and peer-reviewed publications, and a focus on their applications in the textile industry.

## 4.4 Enhancing Textile Products with Functional Properties through the Sol-Gel Method

### 4.4.1 Water Repellent

Recently, the industry has been focusing on water-resistant protective apparel. Water-repellent textiles differ from waterproof textiles in that they allow air to pass through the fabric while blocking water. In contrast, in waterproof fabrics, the gaps among warp and weft yarns are sealed with a uniform hydrophobic layer, so these textiles block the passage of both water and air [29].

The water-repellent property can help inhibit biological settlement on the surface, thereby preventing the degradation of the textile. This characteristic directly aids in preventing biological fouling and the growth of bacteria, primarily due to the nature of biomaterials [30]. For realizing the targeted moisture-resistant property, water-repelling chemicals are applied to textile fibers, ensuring negligible influence on other performance traits such as pliability and durability. These hydrophobic treatments are widely used in a variety of everyday textiles, including luxury silk garments, work uniforms, safety wear, filtration fabrics, carpets, and table linens.

Traditionally, fluorocarbon compounds have been used for water-repellent applications. However, because of ecological and health-related issues linked to fluorocarbons, many studies have focused on non-fluorinated sol-gel coatings as alternatives [31]. According to Wenzel models, highly water-repellent fabrics can be obtained through sol-gel finishing by generating a roughened fiber surface and reducing surface energy. Typically, measurements of droplet-surface angles are employed to evaluate the wettability and efficiency of water-repellent textiles [32]. Wettability describes how a liquid spreads on a solid or liquid interface. By depositing small droplets with known surface tension onto the substrate and recording the contact angle, the surface characteristics can be determined. Surfaces with water contact angle (WCA) below 90° are regarded as hydrophilic, whereas those above 90° are hydrophobic [33].

A colloidal sol composed of $SiO_2$ particles, 3-aminopropyltriethoxysilane (APTES), and hexadecyltrimethoxysilane (HDTMS) was deposited onto cotton, blended cotton/polyester, and pure polyester textiles. The textile samples were treated with the nano-sol until complete absorption and then air-dried at room temperature overnight. The coated textiles exhibited superhydrophobic properties, with WCAs exceeding 150°. Increasing the concentration of HDTMS or silica inclusions resulted in higher contact angles and lower surface energy compared to untreated samples. Both cotton and polyester fabrics preserved their superhydrophobic characteristics after five washing and drying cycles, and hydrophobicity was retained following 600 cycles of abrasion [34].

A coating of polydimethylsiloxane (PDMS) was applied onto polyester fabric, followed by a knife-coating method using an alcoholic solution of TEOS and CTMS. The water-repellent behavior of PDMS and silica particle-coated textiles resulted in contact angles above 90°. Textiles treated with TEOS and HDTMS showed extreme water-repellency, with contact angles of 162.5° and sliding angles of 5.4°. These samples tolerated hydrostatic pressures of up to 38.6 kPa, with slight increases in sliding angles observed after additional coatings. The surface wettability and moisture repellency of the fabric remained stable even after exposure to basic and acidic solutions or ultraviolet (UV) light at different pH levels. Nonetheless, adhesion between the coating and textile substrate proved inadequate, leading to a reduction in hydrostatic resistance. Following a 10-cycle abrasion test, the fabric's tolerance to water pressure decreased to 10 kPa [35].

A technique was implemented using commercially accessible precursors that integrate both organic and inorganic components, namely fluoroalkyl-modified siloxane (FAS) and 3-(trimethoxysilyl)-propyldimethyloctadecyl ammonium chloride (SiQAC), serving as finishing agents to confer multiple functional traits to textile surfaces [36].

Both compounds were prepared in aqueous solutions at specified concentrations and deposited onto cotton textiles using dip-coating and pad-dry-cure techniques. The hybrid sols were processed through two distinct approaches: a one-step approach and a two-step approach. For the one-step procedure, a mixture of the two starting materials (FAS-SiQAC) was applied simultaneously, whereas, in the two-step approach, the SiQAC sol was first introduced, followed by the FAS sol (SiQAC + FAS). The single-step process proved to be more efficient, although both techniques successfully endowed the cotton fabrics with superhydrophobic, oleophobic, and antibacterial properties [36].

Superhydrophobic materials are achieved by engineering a surface with hierarchical roughness and implementing a chemical modification that reduces surface energy [37, 38]. A long-lasting, highly water-repellent wool textile was successfully fabricated through the employment of non-fluorinated organosilane compounds. Superhydrophobicity was achieved by applying a nanocomposite composed of HDTES oligomers and HDTES-functionalized silica particles, prepared *via* a modified Stöber method [39]. The HD-silica/oligomer nanocomposite was deposited on wool textiles using a basic dip-coating procedure, employing toluene as the solvent and triethylamine as the catalyst. Ultrasonic processing was conducted to enhance abrasion durability, environmental stability, and superhydrophobic performance. The treated samples showed excellent mechanical durability, even after being scratched with a scalpel, while maintaining their superhydrophobic properties intact after 10 washes [40].

Superhydrophobic cotton fabric was fabricated using functionalized silica hydrosols. MTMS and SDBS were dispersed in 100 mL of water and thoroughly mixed to form a uniform emulsion. Ammonium hydroxide ($NH_3 \cdot H_2O$) was added as a catalyst, followed by hexadecyltrimethoxysilane (HDTMS), with continuous stirring. The surface-modified $SiO_2$ hydrosols were applied to cotton textile samples *via* a single-step dip-pad-cure coating process. $SiO_2$ nanoparticles, with sizes ranging from 110 nm to 140 nm, were observed, and their dimensions were tuned by varying the concentrations of ammonium hydroxide and surfactants. The treated cotton textiles demonstrated outstanding superhydrophobic performance, resulting from the combined influence of $SiO_2$ nanoparticle–induced surface texture and the reduced surface energy provided by HDTMS modification [41].

Cotton fabrics were modified using a bifunctional fluorosilsesquioxane containing various functional groups, resulting in superhydrophobic characteristics. The modified material achieved a contact angle of 151° and demonstrated durability after undergoing 10 washing cycles. Superhydrophobic cotton fabrics were developed by applying a sol-gel coating containing $TiO_2$ particles. This method not only enhanced the superhydrophobicity of the fabric but also improved its UV shielding properties [42].

A water-repellent layer on polyester textiles was developed using a straightforward aqueous sol-gel approach. In this study, a novel hydrophobic coating composed of hexyltrimethoxysilane (HTMS) was successfully synthesized for polyester fabric *via* a single-step aqueous sol-gel process conducted under acidic conditions. A total of five distinct molar ratios of hydroxyl-terminated methyl siloxane (HTMS) to water were prepared (1:40, 1:30, 1:20, 1:12, and 1:3). The resulting solution, prepared at a concentration of 5% (w/v), was deposited on polyester textiles *via* a simple dip-pad-cure procedure. This mixture was continuously stirred at 30°C on a magnetic stirrer until a transparent solution was obtained, after which it was applied to the fabric using the same dip-pad-cure method. The influence of water proportion on water-repellency was investigated through practical testing of the treated polyester samples. The hydrophobic properties were further assessed by measuring the WCA. The treated polyester textile demonstrated a WCA of 136.2° with particle sizes around 115.3 μm. The sol-gel solution with a molar ratio of 1:3 was selected as the optimal composition, as it exhibited the highest water repellency in the hydrophobicity assessment.

The HTMS sol-gel coating produced a uniform surface structure without detectable cracks, consequently increasing the textural features of the treated polyester textile. HTMS was selected as the precursor for this research because of its robust adhesion to the substrate, potentially leading to an increased WCA. Using HTMS without combining it with other precursors led to a rough surface structure, which is a result of the low surface energy associated with the long-chain alkyl silane in the HTMS coating. Additionally, HTMS acts as a crosslinker, creating a denser and less hydrophilic surface on the substrate, which enhances its hydrophobic properties. The treated fabric maintained its softness and smooth texture, ensuring comfort for the wearer. The successful execution of the sol-gel process was confirmed by Fourier transform infrared spectroscopy (FTIR), which identified the Si-O-Si linkage. This observation was further supported by energy-dispersive X-ray spectroscopy (EDX). Scanning electron microscopy (SEM) images revealed a well-defined microstructure of the consistent, thin xerogel film, with no detectable cracks. The use of HTMS alone produced a textured surface. This texturing, resulting from the reduced surface energy of the long-chain alkyl silane in the HTMS layer, provided efficient water-repellent properties to the treated polyester textile [43].

Water-based $SiO_2$-$TiO_2$ aerogels were successfully developed by incorporating $TiO_2$ into a methyl trimethoxy silane (MTMS)–based $SiO_2$ sol, utilizing a surfactant to facilitate the process. MTMS served as the precursor, whereas cetyltrimethylammonium bromide (CTAB) acted as a catalyst to facilitate even dispersion of the aqueous phase with the organic components. The solution was carefully blended with $TiO_2$ to form $SiO_2$-$TiO_2$ alcogels, which were subsequently immersed in distilled water at 60°C for a duration of 12 hours. Subsequently, the mixture was dried at room pressure across varying temperatures for two hours, resulting in the formation of the final $SiO_2$-$TiO_2$ aerogels. To prepare the coating suspension, $SiO_2$-$TiO_2$ aerogel powder was combined with polydimethylsiloxane and its crosslinker in 20 mL of ethanol. The resulting sol was applied onto cotton textiles *via* a dip-immersion method. The treated fabric demonstrated strong water-repellent behavior, reaching a WCA of 152°. Additionally, it offered antifouling, self-cleaning, and liquid-repelling features, together with photocatalytic functionality [44].

An environmentally friendly, water-based coating method was developed to create a polyethylene terephthalate (PET) fabric that exhibits exceptional water-repellent superhydrophobic properties. In the related study, the researchers stated that their main goal was to produce and utilize nanoparticles to enhance surface roughness. In this study, octadecyltrichlorosilane (OTS) solutions, free from fluorine and containing different concentrations of hollow silica (HS) nanoparticles, were used to produce a superhydrophobic surface *via* a straightforward dip-coating technique. The results indicated that the inclusion of HS nanoparticles significantly elevated, with higher concentrations resulting in improved water repellency and increased surface roughness. The treated fabrics exhibited an impressive WCA of 152.4°, showcasing their superhydrophobic fiber surface. Additionally, the durability of these superhydrophobic properties was tested through a laundering procedure, which revealed that the fabrics retained their water repellency even after 20 washing cycles. EDX and X-ray diffraction (XRD) analyses validated the assessments of morphology did not show any significant structural changes. The prepared fabric was then rigorously tested, including laundering in a washing machine, to assess its durability. Importantly, the fibers preserved their strength and resilience during the testing process, with only minimal loss of hollow $SiO_2$ nanoparticles. It was emphasized that, this eco-friendly and cost-effective method presents significant potential for application in the apparel sector and additional fields, offering an efficient solution for resisting water stains and enhancing functionality across various uses [45].

Previous studies reported the fabrication of multifunctional cotton textiles through the sol-gel approach employing organosilicon ionic liquids (Si-ILs). A key advantage of this method is the ability to modify the anionic part of the Si-ILs by immersing the material in a medium with the targeted anion. The sol-gel process was used to anchor methylimidazolium or pyridinium ionic liquids onto the cotton surface. Through modification of the ILs' anionic component *via* metathesis, eight variants of functionalized cotton textiles with $PF_6$–, $(CF_3SO_2)2N$–, $BF_4$–, and $CH_3CO_2$– anions were successfully produced [46].

In another investigation, multifunctional cotton fabrics exhibiting enhanced hydrophobicity, water repellency, oil-water separation, and self-cleaning ability were developed *via* the sol-gel process incorporating ionic liquids. This work focused on sols synthesized with 1-methylimidazolium chloride propyltriethoxysilane and 1-pyridinium chloride propyltriethoxysilane, which were applied to a cellulose substrate using a pad-dry-cure technique. The process of functionalizing textiles with organosilanes derived from ionic liquids involves the following: (i) preparing organosilicon ionic liquids from 1-methylimidazole and pyridine, (ii) generating sols from these ionic salts, (iii) depositing them onto textiles *via* an impregnation-dry-cure approach, and (iv) executing anionic metathesis to confer desired functionalities. The treated fabrics were subsequently immersed in diluted solutions of hexafluorophosphoric acid ($HPF_6$), lithium bis(trifluoromethanesulfonyl)imide [$Li(CF_3SO_2)2N$], sodium tetrafluoroborate ($NaBF_4$), and sodium acetate ($NaCH_3CO_2$), resulting in incorporation of $PF_6$–, $(CF_3SO_2)2N$–, BF4–, and $CH_3CO_2$– anions on the textile surfaces. SEM analysis revealed a very thin coating layer on the cotton fibers. The analysis of the treated fabrics revealed a microstructured surface containing silica and fluorine molecules. Fabrics incorporating fluorine exhibited significant hydrophobicity, excellent water repellency, and self-cleaning properties. Additionally, these highly hydrophobic fabrics demonstrated the capability to separate mixtures of petroleum ether/water, hexane/water, cyclohexane/water, toluene/water, chloroform/water, dichloromethane/water, and diesel/water using a simple filtration method. This exceptional performance highlights their potential for practical and industrial applications [47].

A study focused on finding substitutes for conventional techniques and hazardous fluorinated substances was carried out, emphasizing the creation of materials featuring nanoscale roughness and reduced surface energy to produce hydrophobic surfaces. To address challenges such as synthesizing water-compatible, low-interfacial-energy materials and selecting suitable adhesives, the authors explored the sol-gel approach. The results indicate that a TEOS-PDMS hybrid layer constitutes an effective route to enhance the hydrophobicity and surface behavior of cotton and polyester textiles. In this work, aqueous hydrophobic coatings for cotton and polyester were produced *via* a single-step sol-gel route; the coatings were evaluated in terms of wettability, interfacial energy, microstructure, elemental composition, durability, breathability and mechanical performance [48]. A hybrid precursor was synthesized by co-condensing TEOS with PDMS, and the formulation was deposited onto cotton and polyester substrates using a pad-dry-cure finishing. Treated samples exhibited markedly improved water repellency, with droplets beading on the surface without penetrating the fibers. Maximum WCA values were recorded at a TEOS:PDMS molar ratio of 1:0.25, reflecting excellent hydrophobicity. FTIR analysis showed Si–O–Si linkages in the coating, whereas SEM-EDX characterization demonstrated robust adhesion of the layer to the textile surfaces, more pronounced on cotton. The sol-gel coating caused substantial reductions in air permeability (≈96.63% for cotton and ≈55.43% for polyester) and led to decreases in tensile strength and load-bearing capacity for both substrates, with cotton experiencing a smaller decline. Although hydrophobic performance remained acceptable under prolonged moderate heating, water repellency diminished after several laundering cycles, indicating a need to improve long-term stability. On the positive side, the TEOS-PDMS films exhibited excellent abrasion resistance, making them attractive for applications that require durable, water-repellent finishes [48].

A novel water-repellent finish for textiles was developed using a sol-gel process. They combined 3-glycidyloxypropyltrimethoxysilane (KH560) and gamma-methacryloxypropyltrimethoxysilane (KH570) as precursors, which were then modified with branched silicone through sol-condensation. This approach led to the creation of a fluorine-free, next-generation polysiloxane-based water repellent with enhanced water-repellent properties. Various parameters and conditions were tested during the chemical preparation to optimize performance. The prepared chemicals were characterized by utilizing various analytical methods, such as FTIR, thermogravimetric analysis, particle size analysis, and water-repellent testing. Particle size analysis revealed that the samples had a homogeneous distribution, facilitating efficient contact and penetration with textile fibers. The study confirmed the presence of functional groups that contribute to effective water repellency, with a particularly high performance observed for the coupling agent ratio KH570 = 0.02 mol:0.04 mol. Overall, the research successfully synthesized a new water repellent by combining KH560 and KH570, which, when modified with branched silicone, formed a film on the fabric's surface, demonstrating excellent water-repellent properties [49].

In another study, Allaca *et al.* (2023) investigated various practical applications for frontline worker clothing, focusing on essential features such as water repellency (hydrophobic properties) and absorption, with the needs of the Covid-19 outbreak in mind. A sol-gel route was used to synthesize a titania gel precursor for dip-coating. The objective was to fabricate cotton and silk textiles that both repel water and reflect harmful UV and infrared radiation. Considering the high cost of scaling such materials industrially, the team developed a cost-effective, titania-based, wet-chemical sol-gel dip-coating approach to produce superhydrophobic cotton and silk substrates. XRD of the coated samples confirmed the presence of anatase-phase $TiO_2$ (no rutile detected) and showed no significant alteration of the fabrics' cellulosic component. Wettability tests reported contact angles of 146° for cotton and 136° for silk (sessile-drop measurements yielded 147° and 132°, respectively) [104]. Diffuse reflectance spectroscopy measured reflectivities of ~84% (UV) and ~92% (IR). Coatings were applied to locally sourced textiles from Narsapuram (Andhra Pradesh, India) using a low-cost dip-coater assembled from repurposed computer parts. Comparative diffuse reflectance measurements versus uncoated controls indicated roughly a 30% increase in reflectance. Overall, the results suggest that a single titania layer can simultaneously act as an effective thermal shield and a hydrophobic coating for cotton and silk. As a result, it was concluded that the coated fabrics exhibited excellent hydrophobic and reflective properties, making them suitable for various maintenance-free apparel applications [50].

Polyester fibers have lower wettability compared to natural fibers. In this context, micro-nano structured hybrid coatings were investigated to advance research on developing safer, fluorine-free alternatives to harmful hydrophobic treatments for polyester fabrics. Their work aims to eliminate the need for pretreatments, for example, laser exposure techniques designed to modify surface features. This work presents a simple sol-gel approach for producing breathable polyester fabrics with superhydrophobic properties. The method involves the hydrolysis and condensation of (3-glycidyloxypropyl)trimethoxysilane (GPTMS). This compound is interconnected with long-chain alkylalkoxysilanes to form a functional network. Using GPTMS, these alkoxysilanes, and a catalyst, a moisture-compatible superhydrophobic layer was applied to polyester fabrics. The process is straightforward, fluorine-free, and employs the dip-pad-dry-cure technique. During the procedure, GPTMS underwent co-condensation with non-fluorinated silanes, including triethoxy(ethyl)silane (C2), triethoxy(octyl)silane (C8), and hexadecyltrimethoxysilane (C16). HCl and 1-methylimidazole served as catalysts to facilitate the opening of the epoxy ring. The resulting material formed a hybrid, cross-linked polyalkylsiloxane network. Finally, this network was deposited on polyester surfaces using the dip-pad-dry-cure method, yielding durable and functional superhydrophobic coatings. The surface properties of the treated static and dynamic water repellency tests were conducted on polyester fabrics. Furthermore, the fabrics were evaluated for their endurance against common aqueous liquids and mechanical wear, as well as for their uptake of moisture and breathability. The micro- and nano-architecture of the modified fabric surface was analyzed using scanning electron microscopy. SEM images at high magnification revealed numerous grooves and protrusions distributed along the polyester filaments after treatment. This minute surface structuring contributes to sustaining their superhydrophobic character. The resulting superhydrophobic coatings demonstrated high resistance to water-based stains and maintained excellent hydrophobicity even after multiple abrasion cycles. The treated fabrics demonstrated outstanding superhydrophobicity, highlighting their potential for use in sustainable, fluorine-free water-repellent coatings across various sectors, including technical textiles, protective workwear, automotive applications, construction, biomedical fields, furniture, and high fashion industries [51].

### 4.4.2 Flame Retardancy

The burning process in flames is primarily driven by oxidative reactions that proceed in the gas phase. For a polymer to engage in flaming combustion, it must first undergo degradation, leading to the production of combustible volatile species. These volatile species subsequently mix with atmospheric oxygen, facilitating the maintenance of the flame. Owing to the strong energy liberation during flaming, heat reaching the material surface accelerates its decomposition and sustains the ongoing burning cycle.

Research aimed at imparting flame-retardant properties to products such as nylon, polyester, polypropylene, cotton, and wool, which exhibit high flammability in the textile industry, have been reported in the literature. In recent years, the sol-gel process has also been extensively explored for its potential to reduce the flammability of textile products [52–55]. This method effectively applies both organic and inorganic components with diverse functionalities to various textile products. For example, silica-based sol-gel coatings have been shown to suppress polymer dripping during combustion, which represents a critical safety risk in synthetic materials. In addition, they contribute to prolonging the stability and protective efficiency of complementary flame-retardant surface treatments [56, 57]. Such advancements present promising and innovative opportunities for the functional finishing of textile products, contributing to reduced flammability and increased durability under standard processing conditions.

The progress in nanotechnology, driven by the incorporation of mineral-based additives, surface-engineered fabrics with fire-resistant features, and tailored polymer systems, has enabled the development of textiles with lowered ignition susceptibility [52]. Over the last ten years, the creation of purely mineral-based or combined organic–mineral ceramic networks *via in situ* sol-gel processing has been recognized as a promising approach for fire protection in diverse macromolecular materials, such as bulk thermoplastics, thermosets, fibers, textiles, and foamed structures [58]. Studies have shown that the incorporation of sol-gel–derived phases into a polymer matrix not only enhances mechanical and thermal properties but also improves resistance to flame spread and exposure to radiative heat flux. The resulting ceramic or hybrid phases function as protective barriers, reducing heat, oxygen, and mass transfer between the degrading material and its environment. Specifically, they operate primarily in the condensed phase, providing thermal protection to the polymer matrix while promoting the formation of a stable carbonaceous residue, commonly referred to as char [59]. The sol-gel method can substantially lower the high chemical concentrations typically required in conventional processes (usually around 300–500 g/L) or replace halogen-based chemicals with more eco-friendly alternatives for flame retardant treatments in traditional textile finishing. This results in a more environmentally friendly and cost-effective approach to producing flame-retardant finished fabrics. In contrast to their organic counterparts, inorganic flame retardants are considered more environmentally friendly and non-polluting, as they do not release harmful chemicals into ecosystems or pose risks to human health during combustion [60]. From a technical perspective, certain flame retardants containing halogen atoms—including polychlorinated biphenyls as well as deca- or pentabromodiphenyl ethers—have demonstrated high effectiveness, although their toxicity remains a concern. As a consequence, the use of these substances has been prohibited, which has accelerated research in both scientific and industrial domains to develop environmentally friendly alternatives [61, 62]. In this context, additives composed solely of phosphorus or in combination with nitrogen have been formulated as replacements for halogen-containing compounds [63].

For cotton flame retardancy, effective commercial solutions that are halogen-free include flame retardants specifically designed for coatings and back-coated fabrics, along with products such as Proban® and Pyrovatex®, which utilize hydroxymethylphosphonium salts and N-methylol phosphonopropionamide derivatives, respectively. While these derivatives exhibit high flame-retardant efficiency, they also present significant drawbacks. For instance, Pyrovatex® can bind approximately 50% of the flame retardant to the fabric surface, resulting in substantial loss during the first washing cycle. In the case of Proban®, its application necessitates a specialized industrial process, and there is a risk of formaldehyde release associated with the use of the flame-retardant material [64].

An acidic precursor combined with nitrogen-containing agents was utilized to deposit flame-resistant coatings onto cotton fabric. The study focused on employing straightforward, low-toxicity, and safe chemicals for developing fire-protective layers. Key process parameters, such as immersion duration, thermal treatment, and curing conditions, were systematically optimized to form N-Si network sol-gel coatings on the cotton substrate. The immersion period ranged from 30 minutes to 14 hours, whereas thermal treatment at around 90°C varied from ambient drying to 15 hours. Curing was conducted at 165°C. The coating's microstructure was examined using low-vacuum scanning electron microscopy (LV-SEM), while the chemical composition of the treated cotton was determined by FTIR and EDS analyses. Overall, the fire-resistance performance of untreated cotton was markedly improved. Samples exposed to prolonged immersion and extended heating (≈15 hours) showed notable alterations in color and mechanical integrity. In contrast, fabrics treated with shorter immersion times of 0.5 hours and air-dried at room temperature exhibited minimal negative impacts from the acidic constituents of the sol-gel formulation. The DI-0.5 h-15 (burning rate = 4.98 $cm^2/s$ and burning time = 36 s) treatment resulted in the highest char residue, approximately 32.93%, by effectively resisting degradation in the early stages. These findings were further corroborated by flame spread and burning rate analyses, which indicated the lowest flame spread rate for the DI-5 h-RT sample (burning rate = 6.49 $cm^2/s$ and burning time = 30 s) [65].

Phosphorus-based flame retardants applied to textiles *via* the sol-gel approach have attracted considerable attention because of their remarkable fire-resistance performance and eco-friendly characteristics. In 2021, a silicon-phosphorus-nitrogen flame-retardant, phenylphosphinyl- and amine-functionalized silanol (WPAS), was synthesized through the sol-gel route to produce cotton fabrics with both flame-retardant and self-cleaning properties. Initially, 20 g of 3-Hydroxyphenyl phosphinyl-propanoic acid (CEPPA) was dispersed in 75 mL of water and stirred for 30 minutes. Subsequently, 30 g of APTES was incorporated into the mixture along with 75 mL of ethanol, followed by 60 minutes of stirring to ensure complete reaction and uniform dispersion. The WPAS product was obtained by drying the solution at 60°C for 24 hours. Prior to treatment, the cotton substrate was rinsed with distilled water and conditioned at 80°C for 2 hours. For the preparation of the flame-retardant solution, 20 g of CEPPA and 30 g of APTES were mixed in a 1:1 water/ethanol solution (totaling 150 g) and stirred for 90 minutes to yield a 25% WPAS dispersion. The solution pH was adjusted to 5–6 with ammonia and stirred for an additional 30 minutes. Cotton fabrics were immersed in the prepared solution for 1 hour, then passed through a squeezing step to remove excess liquid, and finally heat-treated at 160°C for 5 minutes. The results showed that adding 28.4 wt% WPAS increased the cotton's limiting oxygen index to 30.4%, induced self-extinguishing behavior under heat exposure, and decreased the overall heat release by 37% [66]. The improved flame resistance was linked to the phosphorus radicals scavenging in the gas phase and the protective effect of the condensed char layer [65, 66].

A flame-retardant coating solution was developed using a water/ethanol solvent system. A hybrid coating referred to as tri-elemental synergistic network polymer (TSNP) was formulated by combining diphenyl phosphate with APTES through the sol-gel technique, and this formulation was subsequently applied to cotton fabric to improve its fire resistance. At the beginning of the experiment, 20 g of diphenylphosphoric acid was added to 80 mL of water and stirred continuously until a uniform solution was achieved. Meanwhile, 20 g of APTES was dissolved in 80 mL of pure ethanol and gradually introduced into the aqueous solution, with stirring maintained for two hours to achieve thorough mixing. The resulting solution was dried at 60°C for 24 hours. Subsequently, the cotton fabric was thoroughly rinsed with deionized water and dried under the same conditions. To prepare a solution with 20% TSNP by weight, 20 g each of diphenylphosphoric acid and APTES were combined in 160 g of a water-ethanol mixture at a 1:1 ratio and stirred for two hours. The fabric was immersed in this solution for a full 24 hours, and any surplus liquid was removed by gentle pressing. Finally, the treated cotton was heat-treated at 160°C for 5 minutes and then dried at 60°C for 24 hours, resulting in flame-retardant cotton textiles. The authors highlighted that the interaction between silicon, phosphorus, and nitrogen in the coating creates a robust protective layer during combustion, greatly enhancing the fabric's resistance to flames. In addition, the study discovered that the siloxane component participates in a hydrolysis-condensation reaction, resulting in the creation of a hydrophobic silicone layer on the fabric's surface, which significantly improves its water resistance. Furthermore, the flame-retardant coating solution developed in this study has a long shelf life, and a single immersion is sufficient to impart flame resistance to the cotton fabric, eliminating the need for multiple applications such as layer-by-layer self-assembly. This method is described as straightforward, rapid, and effective [67].

### 4.4.3 Self -Cleaning and UV Protection

Controlled exposure to sunlight can be beneficial for health; however, prolonged or intense exposure to UV radiation may have detrimental effects.

Both UVA (320–400 nm) and UVB (280–320 nm) wavelengths are capable of initiating cellular mechanisms that contribute to premature skin aging, erythema, hyperpigmentation, DNA lesions, and an elevated likelihood of skin cancer. Fabrics can serve as an effective barrier against these harmful UV rays. Studies indicate that even textiles designed primarily for aesthetic purposes can offer measurable skin protection. Given that UV radiation possesses significantly higher energy than visible light, it can induce chemical changes that jeopardize human health and degrade textile fibers. Moreover, UV-protective treatments, commonly known as UV-blocking finishes, are among the key chemical coatings applied to textiles, providing protection for both the wearer and the material itself against UV-induced damage.

The level of UV protection provided by fabrics depends on various factors, including the type of material, porosity, thickness, and color, with white summer clothing generally offering low UV protection. Additionally, the type of textile material affects UV protection; for example, cotton fabrics are more susceptible to UV light compared to polyester, which has an aromatic structure that enhances UV absorption. Factors such as fabric thickness and thread density also play a significant role in determining UV-protection properties [68]. However, the use of laundry detergents with optical brighteners can enhance a fabric's ability to block UV radiation. As a result, research into developing UV-protective treatments for textiles is growing.

Tetraethyl orthosilicate (TEOS) is an inorganic precursor frequently used to deposit uniform silica layers on textile surfaces, favored for its low-temperature processing and cost-effectiveness. The gelation of TEOS, catalyzed by acids or bases, has been widely characterized through its viscoelastic response during the sol-gel transformation. In this work, the influence of dibutyltin diacetate (DBTA) as a polycondensation promoter for TEOS was investigated using a multi-stage coating method on cotton fabrics, resulting in a robust surface layer. Two different TEOS concentrations, 0.03 M and 0.3 M, were tested to assess the effect of DBTA. Coatings formed with the lower concentration (0.03 M) displayed better mechanical behavior than untreated cotton and samples coated with the higher concentration (0.3 M). While the higher TEOS content improved thermal stability and laundering performance, it created stiff surface layers that restricted fiber mobility, leading to reduced tensile strength, elongation, and flexibility. Inclusion of DBTA enhanced abrasion resistance and further improved both thermal stability and washing durability [53].

The effectiveness of inorganic sol liquids in protecting cellulose-based polymers in cultural heritage textiles has been demonstrated, offering a promising alternative to existing materials [69]. However, further investigation is required into the mechanism of the DBTA catalyst, and it should be replaced with more environmentally friendly options. Another study reported that sustainable $SiO_2$ nanoparticle–chitosan coatings, exhibiting a mean particle diameter of approximately 150 nm, successfully conferred multiple functional improvements to polyester and viscose textiles with different fabric constructions, including warp and weft arrangements [70]. In terms of mechanical performance, the nanocomposite enhanced the overall elongation and tensile properties as the proportion of $SiO_2$ nanoparticle–chitosan increased. Nevertheless, applying excessive amounts of the nanocomposite led to reductions in strength and stretchability, as nanoparticles penetrated not only the fabric surface but also the interior of the fibers.

Some textile polymers, such as cellulose, polyester, poly(p-phenylene benzobisoxazole), and p-aramid fibers, exhibit limited resistance to UV radiation and can degrade over time. However, depending on the textile's intended use, additional UV resistance may be necessary beyond the fabric's inherent durability. For instance, while polyester's light stability is adequate for clothing, fibers used in textile roofing exposed to prolonged sunlight can degrade, resulting in reduced mechanical strength over time [71].

To enhance UV resistance, certain textile polymer fibers can be modified prior to spinning using UV-absorbing pigments such as titanium dioxide ($TiO_2$) or by incorporating organic molecules that function as radical scavengers. As a result, coatings that block UV radiation are considered among the key chemical finishing treatments for textiles, safeguarding both users and fabrics from UV-induced damage. The use of nanosols combined with the sol-gel approach offers a promising route for fabricating efficient UV-blocking layers. These nanosols typically incorporate inorganic UV-absorbing compounds such as zinc oxide (ZnO) [72] or titanium dioxide [73, 74] which effectively capture UV energy. Their UV absorption capability is particularly strong in the 200–400 nm range, due to the inherently high extinction coefficients of ZnO and $TiO_2$ for UV light.

$TiO_2$ and ZnO nanoparticles, incorporated into textiles using a sol-gel method, are known to offer high UV protection. Because most summer clothing is light and colorless, choosing textiles with high UV protection is crucial. The utilization of $TiO_2$ and ZnO nanoparticles not only offers UV protection but also harnesses the photocatalytic properties of $TiO_2$, providing additional benefits such as self-cleaning and antibacterial functions through photocatalytic reactions.

ZnO nanoparticles were immobilized on cotton fabrics using a sol-gel spin-coating procedure. The coating solution was composed of zinc acetate dihydrate as the precursor, 2-methoxyethanol as the solvent, and monoethanolamine as a stabilizing agent. SEM observations revealed that the ZnO nanoparticles were approximately spherical, with an average diameter around 15 nm, and were homogeneously distributed over the textile surface. Due to their wide bandgap, these nanoparticles exhibited photocatalytic activity under UV illumination. EDX analysis confirmed the elemental composition, showing zinc and oxygen contents of 63.97% and 30.30%, respectively, indicating high purity. Both the standalone ZnO nanoparticles and the ZnO-coated cotton fabrics were evaluated as photocatalysts for degrading methyl orange dye in aqueous solutions under sunlight and a 200 W Philips lamp. After 7 hours, the nanoparticles achieved 93% dye removal under sunlight and 26.4% under lamp illumination. Moreover, the ZnO-coated fabrics showed notable self-cleaning behavior, with 73% dye decomposition under sunlight and 30.7% under artificial light [75].

A simple, environmentally friendly, and efficient approach was employed, using folic acid to facilitate simultaneous synthesis and surface modification of nanostructures on cotton textiles. This safe protocol allowed *in situ* formation of $Cu_2O$/ZnO nanostructures on cotton under mild conditions. Acting as a biotemplate, folic acid promoted the stabilization and reduction of nanoparticles while minimizing particle aggregation. The influence of folic acid on the treated textiles was examined in terms of UV shielding, water absorption behavior, crease recovery, hand feel, thickness, and mechanical performance. Results showed that the biotemplate significantly increased UV-blocking capacity, shortened water droplet absorption times, improved bending flexibility, and enhanced both wrinkle recovery and tensile properties. SEM images revealed that folic acid contributed to the formation of finer nanoparticles on the fabric surface, with particle sizes averaging approximately 48 nm. XRD analysis confirmed the crystal structures of ZnO (hexagonal wurtzite) and $Cu_2O$ (cubic), showing no notable differences due to the presence of folic acid. Furthermore, the increased strength of the cotton textiles indicated effective incorporation of the nanostructures into the fiber matrix. Overall, the presence of folic acid improved UV protection, water affinity, fabric handle, wrinkle resistance, and mechanical integrity of the treated cotton [76].

The UV protection factor (UPF) of fabrics treated with $TiO_2$ nanoparticles synthesized through an *in situ* process was determined to precisely evaluate the textile's inherent properties. The cotton-polyester twill fabric blend (70–30%) with a weight of 246.67 g/m$^2$ was coated with $TiO_2$ nanoparticles directly during the synthesis. The *in situ* coating procedure consisted of multiple phases: cleaning the fabrics, generating nanoparticles, infusing them into the fabric, and drying the treated material $TiO_2$ nanoparticles were formed directly on the cotton fabric during this process: fabric samples (4 × 4 cm) were initially cleaned with ethanol in a shaker, rinsed with distilled water, and then dried in an oven at 80°C. A 4.2 mol/L NaOH solution was utilized for the preparation of the nanoparticles. The solution underwent ultrasonic cleaning for 15 minutes in an ultrasonic bath.

A mixture of 1 mL of ethanol, 4.5 mL of titanium isopropoxide (TTIP), and an additional 1 mL of ethanol was gradually introduced into a fabric solution in an ultrasonic bath for 5 hours. Afterward, the fabrics were rinsed with distilled water and oven-dried at 80°C for 2 hours. Gravimetric analysis indicated a weight increase from 0.4297 g to 0.4745 g, reflecting a deposition of roughly 10.42% titanium dioxide nanoparticles. Imaging and elemental mapping confirmed the direct adherence of nanoparticles to the fibers, whereas X-ray diffraction verified their formation on the textile surface. The UPF of untreated fabrics was 3.67, which increased to 55.82 after coating, demonstrating enhanced UV protection. Crucially, the fabrics' breathability and moisture-handling properties remained largely unchanged. These results indicate that titanium dioxide nanoparticles can substantially improve UV shielding in workwear textiles without compromising wearer comfort. Fabrics with UPF below 20 are generally considered insufficient for UV protection according to global standards. However, following modification, the UPF values of these fabrics exceeded 50, demonstrating "excellent" UV protective properties according to these standards [77].

Variations in fiber composition and their interactions with functional coatings have been shown to influence the UV protection performance of textiles. In particular, untreated cotton fabrics or those treated solely with $TiO_2$ nanoparticles often fail to reach the criteria for "excellent UV protection," whereas cotton-polyester blends generally exhibit superior UV-blocking efficiency [78]. Consistently, the present investigation demonstrates that incorporating $TiO_2$ nanoparticles onto textile surfaces can markedly enhance the UPF, improving the fabrics' ability to shield against harmful UV radiation. This improvement is dependent on the nanoparticle content, as higher $TiO_2$ loadings correlate with greater UV attenuation [79]. The enhancement is also attributable to the intrinsic UV-absorbing properties of $TiO_2$. As a semiconductor with a bandgap around 3.2 eV, $TiO_2$ can absorb photons with energies above its bandgap, effectively mitigating UV exposure and providing substantial protection to the underlying textile [80].

Most clothing worn during peak UV exposure in summer consists of light-colored, lightweight materials. Therefore, when the goal is to achieve maximum shielding, the nature of the textile substrate must be taken into account. An effective strategy involves embedding titanium dioxide ($TiO_2$) and zinc oxide (ZnO) nanoparticles into fabrics through sol-gel processing. Among these, $TiO_2$ is particularly noted for its strong capacity to mitigate both UVA and UVB exposure by absorbing UV radiation [81, 82]. Beyond their UV-shielding role, titanium dioxide in a specific crystalline phase exhibits photo-activated catalytic behavior, delivering added functions such as autonomous stain-degradation and antimicrobial activity. When the coated surface is illuminated by UV photons whose energy meets or exceeds $TiO_2$'s bandgap, electron-hole pairs are generated within the oxide lattice and drive surface redox chemistry. These reactions produce oxygen-derived reactive intermediates (for example, singlet oxygen, superoxide species and hydroxyl radicals), which oxidize organic pollutants and compromise microbial envelopes, thereby underpinning the observed self-cleaning and biocidal effects [83]. Liu *et al.* [84] utilized titanium butoxide and TEOS as coating precursors, with 1,2,3,4-butanetetracarboxylic acid (BTCA) serving as a crosslinker and additives such as citric acid, a triethanolamine salt and tributyl citrate included to modify the formulation; these systems were applied to two wool substrates (untreated/raw wool and Kroy-processed wool) to evaluate the approach. During the chlorination treatment, the scales of the Kroy-processed wool fabrics were removed. Wool samples treated with carboxylic acid at an optimal concentration of 60 g/L were immersed in a $TiO_2/SiO_2$ colloid for 5 minutes and then dried at 50°C for 30 minutes. To enhance the stability of the $TiO_2/SiO_2$ films, the samples were boiled for an additional 20 minutes. After boiling, the samples were baked at 120°C for 3 minutes, followed by thorough washing with cold water and air drying. Three different molar ratios of $TiO_2/SiO_2$ sol (30:70, 50:50, and 70:30) were tested to determine the optimal ratio for achieving the best self-cleaning properties on treated wool fabrics. Both types of wool fabrics were successfully coated with a 60 g/L carboxylic acid solution and a 50:50 $TiO_2/SiO_2$ gel. It was observed that the treated wool exhibited higher UV absorbance after coating with $TiO_2/SiO_2$ nanoparticles. While silica ($SiO_2$) itself has a low refractive index and no UV absorption, its presence enhances the interaction between light and the $TiO_2$ particles, leading to higher absorption. The UPF value of raw wool fabric treated with the 50:50 $TiO_2/SiO_2$ sol reached 356, marking a 102% increase compared to untreated raw wool. Similarly, Kroy-processed wool treated with the same ratio saw a 566% increase in UPF value compared to its untreated counterpart. From the analysis of the SEM images, this can be attributed to the increased deposition of $TiO_2/SiO_2$, likely due to the removal of scales and enhanced adhesion between surfaces, resulting from the larger specific surface area. These findings suggest that $TiO_2/$

$SiO_2$ nanoparticles effectively scatter UV radiation, as their particle size is smaller than the UV wavelength. Consequently, it can be concluded that the combination of high photocatalytic efficiency and small particle size provides $TiO_2/SiO_2$ with excellent UV-blocking properties [84].

Cotton fabrics were treated with $TiO_2$ nanoparticles *via* the sol-gel method to enhance UV protection, antibacterial properties, and flame retardancy. The nanoparticles were synthesized from 5 g of natural rutile sand through acid extraction using the sol-gel process. In the study, colloidal silica and $TiO_2$ nanoparticle-embedded colloidal silica solutions were prepared using the sol-gel method, followed by sonication. PEG was incorporated into the prepared composites to enhance the biocompatibility and stability of the solution. The particle size of the colloidal $TiO_2/SiO_2$ sol was measured. The solution was applied to the surface of cotton fabric using the pad-dry-cure method. Cotton fabric samples, each measuring 15 × 15 cm, were rinsed twice with a 1 wt% NaOH solution and deionized water under ambient conditions. After cleaning, the fabrics were dried in an oven at 303 K for 5 minutes. The fabric substrate was then dipped four times into a uniform solution of silica sol and $TiO_2/SiO_2$ hybrid solution, using the pad-dry-cure technique with appropriate pressure (4 $kg/cm^2$) and speed (5 rpm). Finally, the coated fabric was heated in an oven at 333 K for 3 minutes under dry heat to eliminate any remaining impurities. The nanoparticles deposited on the fabric surface demonstrated higher stability both before and after washing, as confirmed by SEM–EDS analysis. The washing durability tests after 5 and 10 washes showed that the nanoparticles adhered strongly to the fabric surface. The $TiO_2/SiO_2$-coated fabric exhibited improved UV resistance, flame retardancy, and antibacterial activity against Gram-positive *Staphylococcus aureus* and Gram-negative *Escherichia coli* compared to uncoated fabrics, both before and after washing [85].

Building on previous research, Rashid *et al.* aimed to develop an innovative method for functionalizing textiles through the *in situ* synthesis of a $TiO_2$/dimethyloctadecyl [3-(trimethoxysilyl) propyl] ammonium chloride (SiQAC)/Ag nanocomposite. This approach focused on creating durable, multifunctional textiles with enhanced antibacterial, photocatalytic self-cleaning, and UV protection properties. In this study, a straightforward sol-gel/hydrothermal reaction method was employed for the *in situ* synthesis of an Ag-doped $TiO_2$/SiQAC nanocomposite coating on cellulose fibers. The process was carried out under mild and industry-compatible hydrothermal conditions, representing an innovative strategy for generating a multifunctional hybrid layer containing $TiO_2$ on cellulosic materials through a green and affordable sol-gel–hydrothermal route. The working assumption was that the polysiloxane framework formed by SiQAC could effectively immobilize Ag-doped $TiO_2$, thereby yielding a robust multifunctional material with improved antimicrobial efficacy, achieved through the combined action of passive and active bio-barriers together with a controlled release of silver Additionally, the study focused on examining the interactions between the alkoxy groups of SiQAC, the hydroxyl groups on the $TiO_2$ surface, and those on the cellulose fibers. The photocatalytic mechanism of the $TiO_2$-based hybrid nanocomposite film was also proposed and discussed; particular, it was examined the interactions between $TiO_2$/SiQAC/Ag. In this study, two different sols were prepared at room temperature: TTIP and TTIP/SiQAC. The $TiO_2$ sol was created by hydrolyzing a 4% dilution of TTIP precursor in ethanol, whereas the TTIP/SiQAC sol was formulated as an equivalent molar mixture of 4% TTIP and SiQAC (1:1 molar ratio) in absolute ethanol. In both sols, hydrolysis was initiated by adding acidified water. For the application process, cotton samples were fully immersed in TTIP or TTIP/SiQAC solutions at 20°C, followed by padding with a two-roller machine to achieve a wet pickup of 80 ± 5%. The samples then underwent hydrothermal treatment to synthesize $TiO_2$, by immersion in boiling distilled water containing AgNO3 solutions at various concentrations (0.5, 1.0, and 2.0 mM) for silver doping. After the treatment, the textile samples were dried at 100°C.

This work demonstrates the successful creation of a multifunctional hybrid layer incorporating $TiO_2$ on cellulose-based textiles, prepared *via* an environmentally sustainable and cost-efficient sol-gel/hydrothermal route. The presence of silver (Ag) as a dopant was found to enhance the photocatalytic activity of $TiO_2$, with higher $AgNO_3$ concentrations resulting in even greater activity. The treated cotton demonstrated strong antimicrobial performance arising from both release-mediated biocidal action and surface-contact antibacterial mechanisms contributed by the coating components. Although repeated laundering diminished the surface-contact activity associated with SiQAC, the fabric's overall self-sterilizing capability remained effective, producing bacterial reductions greater than 6.7 log for both organisms tested. The hybrid nanocomposite film also conferred substantial, long-term UV shielding, albeit with a slight reduction in UV-blocking efficiency when SiQAC was present. The synergistic action of the film's constituents was further reflected in its photocatalytic self-cleaning: $TiO_2$ + SiQAC + Ag2.0 treated samples achieved roughly 80% degradation of methylene blue [86].

$TiO_2$ nanoparticles were synthesized using the sol-gel method, following a previously reported procedure [87]. To impart self-cleaning and antibacterial properties, Cu(II) was introduced into sol-gel–derived $TiO_2$ nanoparticles, which were then applied onto cotton fabrics. The sol-gel mixture was prepared by combining TTIP (25 mL, 97%) and acetic acid (5 mL) in 500 mL of distilled water, followed by 3.5 mL of $HNO_3$. After maintaining the mixture at 80°C for 30 minutes, the $TiO_2$ suspension was magnetically stirred for 2 hours. Copper sulfate pentahydrate was subsequently added as a Cu precursor, and various Cu(II) concentrations (0.2%–5.5% by weight) were obtained by mixing the solutions in appropriate ratios [88].

Commercial 100% cotton fabrics were obtained from a manufacturer and initially treated with a non-ionic detergent to remove wax, grease, stains, and other surface impurities. The fabrics were then cut into 5 cm × 5 cm pieces. Titanium dioxide nanoparticles were applied onto the fabric using a dip-coating approach. Following application, the samples were rinsed, left to air-dry, and subsequently heated at 100°C to ensure the nanoparticle layer remained stable. To investigate self-cleaning behavior, the treated fabrics were stained with coffee, and changes in stain intensity were monitored through K/S measurements under sunlight exposure. Whereas previous studies often employ synthetic dye solutions to assess the self-cleaning efficiency of $TiO_2$ nanoparticles [75, 89], this study utilized coffee to better represent practical conditions. The fabric incorporating 1% Cu(II) initially showed a stronger stain coloration; however, 78% of the coffee stain was removed after 1 hour of sunlight, and the stain completely disappeared after 2 hours. $TiO_2$-functionalized fabrics demonstrated photocatalytic self-cleaning and antibacterial effects when exposed to light, whereas these properties were not observed in dark conditions [88].

Self-cleaning fabrics can repel dirt and oil while also functioning as disinfectants [90]. This makes them ideal for use in the production of medical and sportswear, uniforms, and various household items such as curtains. These fabrics are particularly beneficial for military personnel operating in extreme environments. Beyond the functional enhancements provided to textile products, self-cleaning textiles offer a potential solution to the challenges posed by water scarcity due to climate change [91]. These textiles, developed with advanced technology, require less frequent washing and wet processing.

### 4.4.4 Antibacterial Finishes

The sol-gel method has proven effective for creating antibacterial finishes, offering several advantages such as eco-friendliness, reduced chemical use, lower temperature requirements, minimal toxicity to humans, and long-lasting antibacterial properties. Different sol-gel formulations have shown the ability to inhibit bacterial growth or provide antimicrobial activity. Strong antibacterial finishes can be obtained by incorporating anatase-phase $TiO_2$ along with metallic elements such as Ag, Cu, Zn, or Ni.

A study was conducted to evaluate the antibacterial effectiveness of textiles treated with coatings of titanium dioxide ($TiO_2$), zinc oxide (ZnO), and a combination of both. Initially, the research focused on synthesizing $TiO_2$ and ZnO nanoparticles—both separately and as a composite—at the laboratory scale using sol-gel techniques. The goal was to identify the most effective coating formulation for antimicrobial performance. In this study, titanyl sulfate, zinc nitrate hexahydrate, and distilled water were used to prepare the solution. For the lab-scale experiments, specific amounts of salt were dissolved in 125 mL of water, whereas 37.5 L was used for roll-to-roll experiments. The mixture was stirred at room temperature until fully dissolved. Once the solution became transparent, the remaining water was added and stirring continued for another two minutes. To avoid early nanoparticle formation, the solution was kept at 20°C until applied. White cotton samples (10 cm × 10 cm) were used as the textile substrate, and *Escherichia coli* bacteria were tested for antibacterial activity. At the conclusion of the study, it was emphasized that textiles coated with a combination of $TiO_2$ and ZnO nanoparticles exhibited stable antibacterial properties, suppressing over 99.99% of *Escherichia coli* without requiring UV light activation. Conversely, it was stated that textiles coated solely with $TiO_2$ nanoparticles showed no significant antibacterial activity in the absence of UV radiation [92].

A study was conducted to impart antibacterial properties to cotton fabrics by coating them with a zinc-containing silica sol. For this purpose, a zinc-based sol was developed in combination with the precursor GPTMS. The resulting coatings exhibited significant antibacterial activity against *Staphylococcus aureus* and *Klebsiella pneumoniae*. These effects were sustained for up to 10 washing cycles, although visible bacterial colonies began to appear after 20 washes [93]. Textiles with fluorescent, repellent, or antimicrobial properties are increasingly being used in various everyday applications. The development of multifunctional coatings is gaining significant attention, especially for applications in areas such as signaling and the medical field. Numerous studies have focused on surface modifications using nanosols to enhance the performance of these specialized textiles, improving properties such as color, fluorescence longevity, self-cleaning, or antimicrobial capabilities.

Multifunctional coatings were successfully produced on cotton fabrics by applying nanosols through sol-gel reactions. These coatings consisted of hybrid materials constructed from a host matrix formulated using TEOS combined with organosilanes such as dimethoxydimethylsilane or dimethoxydiphenylsilane in equal mass proportions. The multifunctional modification of cotton fabric surfaces using nanosols involved integrating two curcuminoid dyes into siloxane matrices, resulting in textiles that possess fluorescent, antimicrobial, and water-repellent properties. The sol-gel process created textiles with fluorescent hybrids under low-temperature conditions, which demonstrated excellent dye absorption and color stability. This dyeing approach was environmentally sustainable, as the fluorescent hybrids were water-insoluble, allowing for straightforward removal through filtration. By incorporating yellow or red α, β-diketone derivatives during the nanogel coating process, textiles with yellow-green or orange-red fluorescence were created, making them suitable for applications in signaling or social media engagement. Cotton fabrics treated with organic-inorganic composite coatings showed contact angles ranging from 129° to 145°, demonstrating their hydrophobic nature and enabling self-cleaning surfaces. The sol-gel coatings adhered effectively to the textile base, maintaining remarkable durability even after 15 washing cycles and abrasion tests. With controlled adjustment of the siloxane scaffold, curcumin-based compounds gained higher antimicrobial efficiency, resulting in fabrics with improved resistance to microbial growth. The treated fabrics exhibited a hydrophobic surface along with fluorescent and antimicrobial features, and their color was responsive to changes in pH levels [94].

As noted in the literature, most publications on this topic concentrate on viruses and bacteria, with only a few studies addressing fungi and the application of silver (Ag) and zinc oxide (ZnO) [95, 96]. However, considering the growing impact of fungal infections on human health, such as diseases aspergillosis caused by species of the *Aspergillus* genus, research should increasingly focus on antifungal fabrics. In the research conducted by Igal *et al.*, the sol-gel technique was utilized to create modified silica by integrating silver (Ag) and zinc oxide (ZnO), along with carbon. These modified silica materials served as antimicrobial agents in the development of fabrics designed to manage diseases and inhibit damage from various infectious organisms, including fungi and bacteria. During the synthesis and impregnation process, fabric samples measuring 4 cm × 4 cm were submerged in the prepared solution for one minute, after which they were taken out and left to dry for a week. This method involved using TEOS as a precursor, in combination with ammonium hydroxide as a catalyst to facilitate basic hydrolysis. The mixture was prepared by first adding 30 mL of the solvent, ethanol ($CH_3CH_2OH$), into a beaker while maintaining a nitrogen ($N_2$) environment. Subsequently, 23.5 mL of ammonium hydroxide ($NH_4OH$) was introduced as a catalyst, and finally, 34 mL of the TEOS precursor [$Si(OC_2H_5)_4$] was incorporated into the solution.

To conclude, an additional 13.5 mL of ethanol and 10 mL of distilled water were incorporated into the mixture. This process was conducted within a fume hood under controlled conditions. Carbon sourced from discarded batteries, along with high-purity ZnO (99.99%) and silver nitrate ($AgNO_3$, 99.99%), was mixed in while stirring. To evaluate the durability of the coating on the fabric, multiple washing cycles were performed. The antimicrobial effectiveness against fungal and bacterial strains was subsequently assessed. The results confirmed that the green sol-gel method is effective in creating coatings that prevent the growth of fungi and bacteria. Furthermore, it was highlighted that the additives exhibited significant persistence, maintaining their inhibitory effects even after up to 20 washing cycles [97].

### 4.4.5 Anti-Felting Finishing on Wool

A significant drawback of wool is its inherent tendency to felt during washing, which is caused by the arrangement of the cuticle scales on the fibers. This issue can lead to the swelling of wool fibers during wet processing and contributes to shrinkage. Although only a limited number of studies have explored this area, there is potential for utilizing sol-gel technology to achieve shrink resistance in wool. A sol prepared using a PPT $[Si(OEt)_3]_3$ precursor was applied to wool fibers through a pad-dry-cure method to impart functional properties. The researchers discovered that an anti-felting effect could be achieved with curing temperatures as low as 120°C [98]. A thermal storage composite was developed by incorporating phase change materials into the matrix through a sol-gel coating method. This approach resulted in enhanced thermal storage capabilities, making the material suitable for use in both commercial and residential buildings [99].

Initially, the wool textile underwent a pre-treatment using the serine protease Esperase 8.0L in combination with sodium sulfite, followed by immersion in a sol-gel–derived polymer matrix. To improve the longevity of the sol-gel coating on the wool, two distinct alkoxysilanes, one bearing an epoxy functional group and the other a mercapto group, were introduced into the formulation. The synergy between the enzymatic pre-treatment and the sol-gel application produced a lightweight wool material exhibiting both dimensional stability and water-repellent characteristics while preserving the original fiber coloration. Among the two silanes tested, the mercapto-functionalized alkoxysilane delivered superior outcomes in terms of polymer deposition on the fibers, shrinkage control, fabric brightness, and resistance to repeated laundering, compared to the epoxy-functionalized counterpart [100].

### 4.4.6 Wrinkle Resistance

Silane coupling agents have been utilized to enhance the wrinkle resistance of textile fabrics without the need for formaldehyde, thereby preserving the base properties of the textiles, such as whiteness and strength [101].

A wrinkle-resistant finish was applied to pre-treated linen fabrics using the sol-gel method. The study examined three distinct pretreatment methods: bleaching, bleaching combined with caustic treatment, and bleaching followed by mercerization. Four different sol-gel formulations for wrinkle-resistant finishing were evaluated, with the concentration of GPTMS being the sole variable influencing the results. The formulations included several consistent chemicals, specifically 98% TEOS, hydrochloric acid (HCl), and ethanol. The solutions were blended at ambient temperature for a duration of 24 hours before being diluted with distilled water at a ratio of 1:10. Subsequently, these diluted solutions were applied to the linen samples with a laboratory padding machine, achieving a 90% wet pick-up rate while operating at a speed of 30 m/min and a pressure of 3 bar. The samples underwent a drying process at 100°C for 5 minutes, followed by curing at 150°C for 30 minutes using a laboratory stenter machine. The findings indicated that the sol-gel technique significantly enhanced the wrinkle resistance of the linen fabrics, with mercerization yielding superior results compared to the other pretreatment methods. Furthermore, the application of the sol-gel wrinkle-resistant finish did not considerably change the fabric's physical characteristics, such as strength and texture, nor did it impact visual aspects like color. Characterization analyses confirmed changes in the chemical structure of the linen. The study also highlighted the ecological benefits of the sol-gel method, as it did not release formaldehyde, unlike conventional wrinkle-resistant finishing agents [102].

The application of various metal alkoxides, including aluminum isopropoxide (AIP), titanium tetra-isopropoxide, and zirconium tetrabutoxide, along with hydrophobic trialkoxysilanes such as methyltriethoxysilane (MTEOS) and octyltriethoxysilane (OTES), to cotton fabrics was explored using the sol-gel method. In this study, these applications were combined with nanosols containing GPTMS. While treatments with metal alkoxides alone did not enhance wrinkle resistance, the addition of GPTMS was found to improve the wrinkle resistance of the fabric. Incorporating AIP, OTES, or MTEOS into the GPTMS solution further increased the dry wrinkle recovery angle to 289°, 290°, and 294°, respectively.

The treatment of the samples with GPTMS alone resulted in a 10% reduction in tensile strength compared to the untreated specimens.

Furthermore, the addition of AIP to the GPTMS solution led to an additional 10% decrease in tensile strength. In contrast, incorporating OTES or MTEOS into the GPTMS and AIP solution improved the fabric's tensile strength by 10%. Additionally, when the sample was treated with GPTMS containing AIP or with combinations of AIP, Dynasylan F8815 (a fluorinated alkyl-functional water-based oligosiloxane), MTEOS, and OTES, the contact angle shifted to 112°, 147°, 145°, and 136°, respectively, indicating a reduction in hydrophilicity. Moreover, pretreatment with BTCA enhanced the wrinkle resistance of the fabrics treated with GPTMS solutions; however, this also resulted in a decrease in tensile strength [103].

## 4.5 Conclusion

The global textile market is largely driven by advancements in high-performance textile products, with sol-gel–assisted textile finishing being one of the key contributors to maintaining a competitive edge.

The sol-gel procedure is a novel and effective approach for modifying fabrics, offering simplicity and numerous advantages. It presents an exciting opportunity for exploring the potential of using sol-gel technology to modify the surfaces of various textile products. The primary reasons for textile materials functionalization are the preservation and enhancement of existing properties, as well as the development of new material characteristics. The sol-gel technique, as an advanced method for creating multifunctional materials, allows for the coating of textile surfaces with chemically or physically modified silica sols. This process results in textiles that exhibit features such as water, oil, and soil repellency, wrinkle resistance, photochromic and thermochromic color-changing properties, and antimicrobial functionality. This chapter highlights recent findings on the application of modified silica for textile surface modification.

While sol-gel–based textile process offers promising benefits, there are still many considerations to address. The water-based sol-gel method is continually evolving, with efforts focused on developing innovative and sustainable products. Improving this process will decrease the dependence on chemicals, such as ethanol and isopropanol, by substituting them with water during synthesis. This shift not only lowers operational and environmental costs but also, in the long term, benefits investors by reducing overall production expenses.

**AI disclosure statement:** The author acknowledges the use of AI-based tools for minor language editing or grammatical correction. However, all intellectual contributions, analyses, and conclusions presented in this chapter are entirely the author's own.

## References

1. Schindler, W.D. and Hauser, P.J., *Chemical finishing of textiles*, CRC, Woodhead, 2004.
2. Mahltig, B., Haufe, H., Böttcher, H., Functionalisation of textiles by inorganic sol–gel coatings. *J. Mater. Chem.*, 15, 4385, 2005.
3. Ismail, W.N.W., Sol–gel technology for innovative fabric finishing—A review. *J. Sol-Gel Sci. Technol.*, 78, 698–707, 2016.
4. Periyasamy, A.P., Ramamoorthy, S.K., Lavate, S.S., Eco-friendly denim processing, in: *Handbook of Ecomaterials*, pp. 1–21, Springer, Cham, 2019.
5. Periyasamy, A.P. and Venkatesan, H., Eco-materials in textile finishing, in: *Handbook of Ecomaterials*, pp. 1461-1482. Springer, Cham, 2019.
6. Schindler, W.D. and Hauser, P.J., *Chemical finishing of textiles*, Woodhead Publishing Ltd, Cambridge, 2004.
7. Shabbir, M. and Sheikh, J.N., Introduction to textiles and finishing materials, in: *Frontiers of Textile Materials*, pp. 1–11, Wiley, 2020.
8. Mahltig, B., Knittel, D., Schollmeyer, E., Öttcher, H.B., Incorporation of triarylmethane dyes into sol-gel matrices deposited on textiles. *J. Sol-Gel Sci. Technol.*, 32, 219–222, 2004.
9. Elkashouty, M., Elsyaed, H., Twaffiek, S., Salem, T., Elhadad, S.S.M., An overview: Textile surface modification by using sol-gel technology. *Egypt. J. Chem.*, 63, 3301–3311, 2020.
10. Aurobind, S.V., Amirthalingam, K.P., Gomathi, H., Sol-gel based surface modification of electrodes for electro analysis. *Adv. Colloid Interface Sci.*, 121, 1–7, 2006.
11. Mackenzie, J.D. and Bescher, E.P., Chemical routes in the synthesis of nanomaterials using the sol-gel process. *Acc. Chem. Res.*, 40, 795–803, 2007.
12. Pierre, A.C., *Introduction to sol-gel processing*, Springer International Publishing, Cham, 2020.
13. Livage, J., Beteille, F., Roux, C., Chatry, M., Davidson, P., Sol–gel synthesis of oxide materials. *Acta Mater.*, 46, 743–750, 1998.
14. Gupta, R. and Chaudhury, N.K., Entrapment of biomolecules in sol–gel matrix for applications in biosensors: problems and future prospects. *Biosens. Bioelectron.*, 22, 2387–2399, 2007.
15. Locher, M., Romano, V., Weber, H.P., Rare-earth doped sol–gel materials for optical waveguides. *Opt. Lasers Eng.*, 43, 341–347, 2005.

16. Pierre, A.C., *Introduction to sol-gel processing*, Springer International Publishing, Cham, 2020.
17. Livage, J., Basic principles of sol-gel chemistry, in: *Sol-Gel Technologies for Glass Producers and Users*, pp. 3–14, Springer US, Boston, MA, 2004.
18. Danks, A.E., Hall, S.R., Schnepp, Z., The evolution of 'sol–gel' chemistry as a technique for materials synthesis. *Mater. Horiz.*, 3, 91–112, 2016.
19. Periyasamy, A.P., Venkataraman, M., Kremenakova, D., Militky, J., Zhou, Y., Progress in sol-gel technology for the coatings of fabrics. *Materials*, 13, 1838, 2020.
20. Bokov, D., *et al.*, Nanomaterial by sol-gel method: synthesis and application. *Adv. Mater. Sci. Eng.*, 2021, Article ID 5102014, 2021.
21. Walcarius, A. and Collinson, M.M., Analytical chemistry with silica sol-gels: traditional routes to new materials for chemical analysis. *Annu. Rev. Anal. Chem.*, 2, 121–143, 2009.
22. Yuan, X., Xu, L., Pan, H., Shen, Y., Wang, L., Xie, M., Eco-friendly approach for preparation of water-based superhydrophobic silica aerogels *via* ambient pressure drying. *Mater. Res. Express*, 8, 015021, 2021.
23. Loy, D.A., Sol–gel processing, in: *Encyclopedia of Physical Science and Technology*, pp. 257–276, Elsevier, 2003.
24. Ciriminna, R. and Pagliaro, M., Open challenges in sol–gel science and technology. *J. Sol-Gel Sci. Technol.*, 101, 29–36, 2022.
25. Anand, S.S., Philip, B.K., Mehendale, H.M., Volatile organic compounds, in: *Encyclopedia of Toxicology*, pp. 967–970, Elsevier, Oxford, 2014.
26. Xu, L., Zhuang, W., Xu, B., Cai, Z., Superhydrophobic cotton fabrics prepared by one-step water-based sol–gel coating. *J. Text. Inst.*, 1–9, 2011.
27. Camlibel, N.O. and Arik, B., Sol-gel applications in textile finishing processes, in: *Recent Applications in Sol-Gel Synthesis*, InTech, Rijeka, 2017.
28. Mäkinen, H. and Jussila, K., Coldprotective clothing: types, design and standards, in: *Protective Clothing*, pp. 3–38, Elsevier, 2014.
29. Ielo, I., Giacobello, F., Castellano, A., Sfameni, S., Rando, G., Plutino, M.R., Development of antibacterial and antifouling innovative and eco-sustainable sol–gel-based materials: from marine areas protection to healthcare applications. *Gels*, 8, 26, 2021.
30. Hossain, M.R., Newby, S., Ahmed, M.R., Optimised fluorine free polysiloxane-based water repellent for cellulosic fabric. *J. Text. Inst.*, 116, 670–679, 2025.
31. Venkataraman, M., *et al.*, Preparation of electrosprayed microporous membranes. *IOP Conf. Ser.: Mater. Sci. Eng.*, 460, 012017, 2018.
32. Wenzel, R.N., Resistance of solid surfaces to wetting by water. *Ind. Eng. Chem.*, 28, 988–994, 1936.
33. Zhao, Q., Wu, L.Y.L., Huang, H., Liu, Y., Ambient-curable superhydrophobic fabric coating prepared by water-based non-fluorinated formulation. *Mater. Des.*, 92, 541–545, 2016.
34. Xue, C.H., Li, M., Guo, X.J., Li, X., An, Q.F., Jia, S.T., Fabrication of superhydrophobic textiles with high water pressure resistance. *Surf. Coat. Technol.*, 310, 134–142, 2017.

35. Simončič, B., *et al.*, Multifunctional water and oil repellent and antimicrobial properties of finished cotton: influence of sol–gel finishing procedure. *J. Sol-Gel Sci. Technol.*, 61, 340–354, 2012.
36. Crick, C.R. and Parkin, I.P., Preparation and characterisation of super-hydrophobic surfaces. *Chem. Eur. J.*, 16, 3568–3588, 2010.
37. Wang, Y., Zhu, Y., Zhang, C., Li, J., Guan, Z., Transparent, superhydrophobic surface with varied surface tension responsiveness in wettability based on tunable porous silica structure for gauging liquid surface tension. *ACS Appl. Mater. Interfaces*, 9, 4142–4150, 2017.
38. Han, J.T., Kim, S.Y., Woo, J.S., Lee, G., Transparent, conductive, and super-hydrophobic films from stabilized carbon nanotube/silane sol mixture solution. *Adv. Mater.*, 20, 3724–3727, 2008.
39. Zhang, J., Li, B., Wu, L., Wang, A., Facile preparation of durable and robust superhydrophobic textiles by dip coating in nanocomposite solution of organosilanes. *Chem. Commun.*, 49, 11509, 2013.
40. Xu, L., Zhuang, W., Xu, B., Cai, Z., Superhydrophobic cotton fabrics prepared by one-step water-based sol–gel coating. *J. Text. Inst.*, 1–9, 2011.
41. Przybylak, M., Maciejewski, H., Dutkiewicz, A., Preparation of highly hydrophobic cotton fabrics by modification with bifunctional silsesquioxanes in the sol-gel process. *Appl. Surf. Sci.*, 387, 163–174, 2016.
42. Za'im, N.N.M., Yusop, H.M., Wan Ismail, W.N.S., Synthesis of water-repellent coating for polyester fabric. *Emerg. Sci. J.*, 5, 747–754, 2021.
43. Wan, J., Xu, L.H., Pan, H., Wang, L.M., Shen, Y., Green water-based fabrication of $SiO_2$–$TiO_2$ aerogels with superhydrophobic and photocatalytic properties and their application on cotton fabric. *J. Porous Mater.*, 28, 1501–1510, 2021.
44. Sheraz, M., Choi, B., Kim, J., Enhancing textile water repellency with octadecyltrichlorosilane (OTS) and hollow silica nanoparticles. *Polymers (Basel)*, 15, 4065, 2023.
45. Boukhriss, A., Gmouh, S., Hannach, H., Roblin, JP., Cherkaoui, O., Boyer, D., Treatment of cotton fabrics by ionic liquid with $PF_6^-$ anion for enhancing their flame retardancy and water repellency. *Cellulose*, 23, 3355–3364, 2016.
46. Bentis, A., Boukhriss, A., Zahouily, M., Manoun, B., Gmouh, S., Functionalization of cotton fabrics by solgel method using ionic liquids with highhydrophobic, excellent water repellent, oil/water separation, and self-cleaning properties. *Cellulose*, 30, 6719–6740, 2023.
47. Abu Bakar, N.H., Wan Ismail, W.N., Mohd Yusop, H., Mohd Zulkifli, N.F., Synthesis of a water-based TEOS–PDMS sol–gel coating for hydrophobic cotton and polyester fabrics. *New J. Chem.*, 48, 933–950, 2024.
48. Hossain, M.R., Newby, S., Ahmed, M.R., Optimised fluorine free polysiloxane based water repellent for cellulosic fabric. *J. Text. Inst.*, 1–10, 2024.
49. Allaka, G. and Yepuri, V., Inexpensive hydrophobic and infrared reflective coatings on the cotton and silk fabrics using sol-gel dip coating technique. *Mater. Today Proc.*, 1–4, 2023.

50. Sfameni, S., Lawnick, T., Rando, G., Visco, A., Textor, T., Plutino, M.R., Super-hydrophobicity of polyester fabrics driven by functional sustainable fluorine-free silane-based coatings. *Gels*, 9, 109, 2023.
51. Alongi, J., Colleoni, C., Rosace, G., Malucelli, G., Phosphorus- and nitrogen-doped silica coatings for enhancing the flame retardancy of cotton: Synergisms or additive effects? *Polym. Degrad. Stab.*, 98, 579–589, 2013.
52. Colleoni, C., Donelli, I., Freddi, G., Guido, E., Migani, V., Rosace, G., A novel sol-gel multi-layer approach for cotton fabric finishing by tetraethoxysilane precursor. *Surf. Coat. Technol.*, 235, 192–203, 2013.
53. Guido, E., *et al.*, Thermal stability and flame retardancy of polyester fabrics sol–gel treated in the presence of boehmite nanoparticles. *Polym. Degrad. Stab.*, 98, 1609–1616, 2013.
54. Liang, S., Neisius, N.M., Gaan, S., Recent developments in flame retardant polymeric coatings. *Prog. Org. Coat.*, 76, 1642–1665, 2013.
55. Davesne, A.-L., Jimenez, M., Samyn, F., Bourbigot, S., Thin coatings for fire protection: An overview of the existing strategies, with an emphasis on layer-by-layer surface treatments and promising new solutions. *Prog. Org. Coat.*, 154, 106217, 2021.
56. Levchik, S.V., Introduction to flame retardancy and polymer flammability, in: *Flame Retardant Polymer Nanocomposites*, pp. 1–29, Wiley, 2007.
57. Chiang, C., Ma, C.M., Wu, D., Kuan, H., Preparation, characterization, and properties of novolac-type phenolic/$SiO_2$ hybrid organic–inorganic nano-composite materials by sol–gel method. *J. Polym. Sci. A Polym. Chem.*, 41, 905–913, 2003.
58. Shen, J., Liang, J., Lin, X., Lin, H., Yu, J., Wang, S., The flame-retardant mechanisms and preparation of polymer composites and their potential application in construction engineering. *Polymers (Basel)*, 14, 82, 2021.
59. Wang, Y., Cao, X., Zhang, L., Effects of cellulose whiskers on properties of soy protein thermoplastics. *Macromol. Biosci.*, 6, 524–531, 2006.
60. Hull, T.R., Law, R.J., Bergman, A., Environmental drivers for replacement of halogenated flame retardants, in: *Polymer Green Flame Retardants*, pp. 119–179, Elsevier, 2014.
61. Weijs, L., Dirtu, A.C., Malarvannan, G., Covaci, A., Bioaccumulation and biotransformation of brominated flame retardants, in: *Comprehensive Analytical Chemistry*, pp. 433–491, Elsevier, Amsterdam, 2015.
62. Hörold, S., Phosphorus-based and intumescent flame retardants, in: *Polymer Green Flame Retardants*, pp. 221–254, Elsevier, 2014.
63. Samanta, K., Basak, S., Chattopadhyay, S.K., Gayatri, T.N., Water-free plasma processing and finishing of apparel textiles, in: *Handbook of Sustainable Apparel Production*, pp. 18–53, CRC Press, Boca Raton, FL, 2015.
64. Rehman, Z.U., Khan, L., Hwain, L., Chiho, Y., Koo, B.H., Si-N matrix as an effective fire retardant source for cotton fabric, prepared through sol–gel process. *Fire*, 7, 69, 2024.

65. Fu, C., *et al.*, Self-cleaning cotton fabrics with good flame retardancy *via* one-pot approach. *Polym. Degrad. Stab.*, 192, 109700, 2021.
66. Yan, X.-Y., *et al.*, Designing durable self-cleaning nanofiltration membranes *via* sol-gel assisted interfacial polymerization for textile wastewater treatment. *Sep. Purif. Technol.*, 289, 120752, 2022.
67. Ma, Y., Wang, Y., Ma, L., Zhu, Z., Fabrication of hydrophobic and flame retardant cotton fabric *via* sol-gel method. *Cellulose*, 30, 11829–11843, 2023.
68. Osterwalder, U., Schlenker, W., Rohwer, H., Martin, E., Schuh, S., Facts and Fiction on Ultraviolet Protection by Clothing. *Radiat. Prot. Dosim.*, 91, 255–259, 2000.
69. Trovato, V., Rosace, G., Colleoni, C., Sfameni, S., Migani, V., Plutino, M.R., Sol-gel based coatings for the protection of cultural heritage textiles. *IOP Conf. Ser.: Mater. Sci. Eng.*, 777, 012007, 2020.
70. Attia, N.F., El Ebissy, A.A., Morsy, M.S., Sadak, R.A., Gamal, H., Influence of textile fabrics structures on thermal, UV shielding, and mechanical properties of textile fabrics coated with sustainable coating. *J. Nat. Fibers*, 18, 2189–2196, 2021.
71. Mahltig, B. and Textor, T., *Nanosols and textiles*, World Scientific, 2008.
72. Textor, T., Bahners, T., Schollmeyer, E., Modern approaches for intelligent surface modification. *J. Ind. Text.*, 32, 279–289, 2003.
73. Daoud, W.A. and Xin, J.H., Low temperature sol-gel processed photocatalytic titania coating. *J. Sol-Gel Sci. Technol.*, 29, 25–29, 2004.
74. Xing, Y. and Ding, X., UV photo-stabilization of tetrabutyl titanate for aramid fibers *via* sol–gel surface modification. *J. Appl. Polym. Sci.*, 103, 3113–3119, 2007.
75. Shaban, M., Abdallah, S., Khalek, A.A., Characterization and photocatalytic properties of cotton fibers modified with ZnO nanoparticles using sol–gel spin coating technique. *Beni-Suef Univ. J. Basic Appl. Sci.*, 5, 277–283, 2016.
76. Noorian, S.A., Hemmatinejad, N., Bashari, A., One-pot synthesis of $Cu_2O$/ZnO nanoparticles at present of folic acid to improve UV-protective effect of cotton fabrics. *Photochem. Photobiol.*, 91, 510–517, 2015.
77. Rabiei, H., Farhang Dehghan, S., Montazer, M., Khaloo, S.S., Koozekonan, A.G., UV protection properties of workwear fabrics coated with $TiO_2$ nanoparticles. *Front. Public Health*, 10, 2022.
78. Kathirvelu, S., Multifunctional finishing of cotton and blended fabrics treated with titanium dioxide nano-particles. *Indian J. Sci. Technol.*, 1, 1–9, 2008.
79. Chaudhari, S.B., Mandot, A.A., Patel, B.H., Effect of nano $TiO_2$ pretreatment on functional properties of cotton fabric. [Online]. Available: www.ijerd.com, 2012.
80. Palaskar, S.S., Desai, A.N., Shukla, S.R., Development of multifunctional cotton fabric using atmospheric pressure plasma and nano-finishing. *J. Text. Inst.*, 107, 405–412, 2016.

81. Shabbir, M., Ahmed, S., Sheikh, J.N., *Frontiers of Textile Materials*, Wiley, Hoboken, NJ, 2020.
82. Shahid, M. and Adivarekar, R., *Advances in Functional Finishing of Textiles*, Springer Singapore, Singapore, 2020.
83. Inamuddin, A.M., Asiri, A.M., Lichtfouse, E., *Nanophotocatalysis and Environmental Applications: Detoxication and Disinfection*, Springer International Publishing, 2020.
84. Liu, S., Zhang, Q., Xu, Z., Yang, S., Liu, H., Surface modification of $TiO_2/SiO_2$ composite hydrosol stabilized with polycarboxylic acid on Kroy-process wool fabric. *J. Adhes. Sci. Technol.*, 31, 1209–1228, 2017.
85. Dhineshbabu, N., Arunmetha, S., Manivasakan, P., Karunakaran, G., Rajendran, V., Enhanced functional properties of cotton fabrics using $TiO_2/SiO_2$ nanocomposites. *J. Ind. Text.*, 45, 674–692, 2016.
86. Chehade, G., Demir, M.E., Dincer, I., Yuzer, B., Selcuk, H., Experimental investigation and analysis of a new photoelectrochemical reactor for hydrogen production. *Int. J. Hydrogen Energy*, 43, 12049–12058, 2018.
87. Rashid, M.M., Tomšič, B., Simončič, B., Jerman, I., Štular, D., Zorc, M., Sustainable and cost-effective functionalization of textile surfaces with Ag-doped $TiO_2$/polysiloxane hybrid nanocomposite for UV protection, antibacterial and self-cleaning properties. *Appl. Surf. Sci.*, 595, 153521, 2022.
88. Yuzer, B., Akyildiz, H., Kahraman, M.V., Aktaş, A., Baysal, M., Photocatalytic, self-cleaning and antibacterial properties of Cu(II) doped $TiO_2$. *J. Environ. Manage.*, 302, 114023, 2022.
89. Zahid, M., Multifunctional and responsive textile nanocomposites for high value applications, University of Genova–Istituto Italiano di Tecnologia Bio Nanotechnology, Genova, 2018.
90. Saad, S.R., Mahmed, N., Abdullah, M.M.A.B., Sandu, A.V., Self-cleaning technology in fabric: A review. *IOP Conf. Ser.: Mater. Sci. Eng.*, 133, 012028, 2016.
91. Xu, B., Ding, Y., Qu, S., Cai, Z., Superamphiphobic cotton fabrics with enhanced stability. *Appl. Surf. Sci.*, 356, 951–957, 2015.
92. Abramova, A.V., Melnikov, G.Y., Ershov, S.E., Sviridova, T.A., Abramov, A.V., A sol-gel method for applying nanosized antibacterial particles to the surface of textile materials in an ultrasonic field. *Ultrason. Sonochem.*, 60, 104788, 2020.
93. Poli, R., Colleoni, C., Calvimontes, A., Polášková, H., Dutschk, V., Rosace, G., Innovative sol–gel route in neutral hydroalcoholic condition to obtain antibacterial cotton finishing by zinc precursor. *J. Sol-Gel Sci. Technol.*, 74, 151–160, 2015.
94. Raduly, F.M., Dobre, R., Vlase, L., Bostanaru, A.C., Todea, M., Balcu, I., Multifunctional finishing of cotton fabric with curcumin derivatives coatings obtained by sol–gel method. *Gels*, 9, 369, 2023.
95. El-Naggar, M.E., Khattab, T.A., Abdelrahman, M.S., Aldalbahi, A., Hatshan, M.R., Development of antimicrobial, UV blocked and photocatalytic self-cleanable cotton fibers decorated with silver nanoparticles using silver carbamate and plasma activation. *Cellulose*, 28, 1105–1121, 2021.

96. Jin, L., Yuen, C.W.M., Jiang, Y., Kan, C.W., Wang, X., Development of novel segmented-pie microfibers from copper-carbon nanoparticles and polyamide composite for antimicrobial textiles application. *Text. Res. J.*, 92, 3–14, 2022.
97. Igal, K., Zanotti, K., Zuin, V.G., Vazquez, P., Sol-gel technology for greener and more sustainable antimicrobial textiles that use silica matrices with C, and Ag and ZnO as biocides. *Curr. Res. Green Sustain. Chem.*, 4, 100177, 2021.
98. Yi, H. and Yan, K., Polyurethane modified with 3-aminopropyltriethoxysilane as wool antifelting agent. *J. Appl. Polym. Sci.*, 109, 2169–2175, 2008.
99. Chen, C., Wang, A., Xu, G., Ni, C., High-performance phase change composite of acetamide/silica-network for thermal storage. *Nanosci. Nanotechnol. Lett.*, 5, 84–88, 2013.
100. Shen, J., Smith, E., Chizyuka, M., Prajapati, C., Development of durable shrink-resist coating of wool with sol-gel polymer processing. *Fibers Polym.*, 18, 1769, 2017.
101. Montazer, M. and Harifi, T., *Nanofinishing of Textile Materials*, Woodhead Publishing, Duxford, 2018.
102. Arik, B., Characterization and Wrinkle Resistance Enhancement by Sol-Gel Method of Variously Pretreated Linen Fabrics. *Fibers Polym.*, 21, 82, 2020.
103. Schramm, C., Binder, W.H., Tessadri, R., Durable Press Finishing of Cotton Fabric with 1,2,3,4-Butanetetracarboxylic Acid and TEOS/GPTMS. *J. Sol-Gel Sci. Technol.*, *29*, 155, 2004.
104. Allaka, G., Yepuri, V., Inexpensive hydrophobic and infrared reflective coatings on the cotton and silk fabrics using sol-gel dip coating technique. *Materials Today: Proceedings,* 2023.

5

# Layer-by-Layer (LbL) Deposition Method for Surface Modification of Textiles

**Rasool Shabanloo[1]* and Aminoddin Haji[2]**

[1]*Textile Engineering Department, Amirkabir University of Technology (Tehran Polytechnic), Tehran, Iran*
[2]*Department of Textile Engineering, Yazd University, Yazd, Iran*

***Abstract***
The layer-by-layer (LbL) assembly method has emerged as a versatile technique for the surface modification of textiles, offering precise control over the deposition of nanoscale materials. This chapter delves into the LbL method, highlighting its ability to enhance textile properties through the sequential adsorption of oppositely charged polyelectrolytes. We explore the methodology's applications in imparting functionalities such as antimicrobial, UV protection, and hydrophobicity to fabrics. The chapter also addresses the environmental and economic aspects of LbL, proposing sustainable practices and forecasting future advancements in smart textiles. Through this comprehensive overview, the chapter aims to provide a foundational understanding of LbL in textile innovation.

***Keywords***: Layer-by-layer assembly, polyelectrolyte multilayers, functional textiles, antimicrobial finishing, UV protective fabrics, hydrophobic textile coatings, smart and sustainable textiles

## 5.1 Introduction

### 5.1.1 Overview of Textile Surface Modification

Recent years have witnessed a significant change in surface modification techniques in textile engineering. This change has been driven by the search for textiles that not only meet functional requirements for different applications but also adhere to sustainable manufacturing practices.

**Corresponding author*: rasulshabanloo1864@gmail.com

Aminoddin Haji and Çağlar Sivri (eds.) Surface Modification and Functionalization of Textiles, (125–158) 

Nanotechnology has played a crucial role, bringing in nanoscale materials that give textiles remarkable properties like self-cleaning abilities and improved mechanical strength [1]. There are other methods for modification increasingly recognized for their capacity to apply inorganic coatings, providing properties such as durability and UV shielding, all while prioritizing environmental responsibility through the use of eco-friendly chemicals and efficient application methods [2].

In addition to these advancements, plasma treatment has emerged as a sustainable, dry alternative to traditional wet chemical treatments. It allows for precise modifications at the molecular level, enhancing dyeability and printability [3]. Electrospinning and microencapsulation techniques have become popular. Electrospinning is used to create non-woven fabrics with high surface area-to-volume ratios for barrier applications, whereas microencapsulation is utilized to incorporate active agents into textiles for controlled release functions [4–6].

Smart materials and finishes integrated into textile substrates have become a frontier of innovation in technological advancements. These materials and methods are used to react to environmental stimuli like changes in temperature, variations in pH, or exposure to light. This marks the arrival of a new era of adaptive and interactive textiles [7, 8]. The progress made in this area does not just improve the performance and features of textiles but also helps make the industry more sustainable by reducing the need for resource-heavy finishing methods. As the sector develops further, it is expected to introduce even more advanced innovations that change the potential of textile materials.

### 5.1.2 Importance of LbL Technique in the Textile Industry

The textile industry faces significant challenges related to sustainability, including excessive water usage, chemical consumption, and energy expenditure. Traditional dyeing processes contribute to environmental pollution by generating large amounts of untreated hazardous waste. In this context, the layer-by-layer (LbL) deposition method emerged as a promising solution to address these issues, firstly in 1966 for the deposition of oppositely charged colloidal particles [9].

#### *5.1.2.1 Reducing Chemical, Energy, and Water Consumption*

The production of textiles involves multiple stages, including the creation of fibers, yarn, and fabric, as well as the processes of dyeing and finishing. These processes all require considerable amounts of water and energy. For example, the production of cotton fibers uses around 1,559 kg of fresh water per kg, whereas polyester fiber production consumes approximately 108 kWh of electricity per kg. Additionally, during subsequent processes like bleaching, dyeing, printing, and finishing, the consumption of clean water can reach as high as 200 L/kg. Utilizing LbL surface modification techniques presents opportunities to decrease chemical usage, energy consumption, and water usage, thereby contributing to sustainability [10].

The LbL method for surface modification of textiles represents a significant advancement in sustainable textile engineering. This technique involves the sequential deposition of oppositely charged materials to create thin films on textile surfaces, which can impart various functional properties [10, 11]. The use of natural resources can be decreased by the LbL method. For example, it reduces the need for harmful chemicals typically used in conventional textile finishing processes. LbL supports a more eco-friendly production cycle by utilizing water as a solvent and allowing active substances to be applied at very low concentrations. Additionally, it cuts down on energy usage by enabling textile treatment at room temperature and atmospheric pressure, as opposed to traditional methods that often require high temperatures and pressure. Finally, the LbL technique significantly reduces water usage by eliminating the need for large volumes of water in rinsing and washing steps, which are common in standard dyeing and finishing operations [10].

Flame-retardant treatment for cotton, polyester, and polyamide textiles, enhancing safety while being environmentally friendly [11], sustainable dyeing and finishing of cotton fabrics, achieving coloration and functional properties with efficient resource usage [12], antibacterial and UV protection functions for cotton fabrics through nano-ZnO composite film deposition [13], and incorporation of graphene oxide nanosheets in textiles, demonstrate the versatility of LbL technique in creating advanced high performance, environmentally friendly materials [14].

The examples highlight how the LbL method contributes to the promotion of environmentally friendly practices in the textile industry, establishing it as a fundamental element for future advancements in sustainable textile production.

## 5.2 Fundamentals of LbL Deposition

### 5.2.1 Basic Principles of LbL Deposition

LbL deposition presents an effective and precise method for producing multilayer coatings by sequentially adsorbing oppositely charged materials. This technique utilizes electrostatic interactions, hydrogen bonding, and other forces to construct thin films with regulated thickness and composition. The essential concepts of LbL deposition include material selection, the deposition procedure, and the mechanisms underlying the development of multilayer structures [15, 16].

### 5.2.2 Choice of Materials

The selection of materials for LbL deposition is crucial for achieving the desired film properties. Typically, LbL deposition involves the use of polyelectrolytes, which are polymers with ionizable groups that dissociate in water to form charged molecules. These polyelectrolytes can be classified as cationic (positively charged) or anionic (negatively charged) [17, 18]. Cationic polyelectrolytes, such as chitosan, which contain positive $-NH_3$ functional groups [19], are one of the most well-known polymers that can associate with polyanionic polymers. Polycationic and polyanionic can make durable assemblies by electrostatic interactions [20]. In addition to polyelectrolytes, various functional additives can be incorporated into the LbL process *via* different methods such as encapsulation in the case of organic and inorganic additives [21], spin coating [22], spray-assisted alignments [23], spin-spray [24], and cross-linking [25] for incorporation of nanoparticles (NPs). These functional additives can assist in improving electrochemical and chemical properties [26], UV protection properties [27], as well as mechanical properties [28]. Green chemicals were also used in LbL surface modification. For instance, the surface of linen fabric was functionalized using green tea extract (GTE) and chitosan. The linen fabric exhibited over 75% antibacterial activity, up to 96% antioxidant properties, and more than 22.5 UV protection factor, demonstrating the imparting of functional properties [29]. The surface of polyamide 66 (PA66) textiles was modified using chitosan and sodium alginate (SA) in combination with phytic acid (PA) through LbL deposition. This method of surface modification, assisted by microcrystalline cellulose, opens up possibilities for a more environmentally friendly treatment of PA66 textiles, leading to improved hydrophilic and flame-retardant properties [30].

### 5.2.3 Deposition Process

The LbL deposition process can involve the alternate immersion of the substrate into solutions containing oppositely charged materials. As shown in Figure 5.1, each immersion step allows for the adsorption of a monolayer of material onto the substrate, followed by rinsing to remove excess material and prevent cross-contamination between layers.

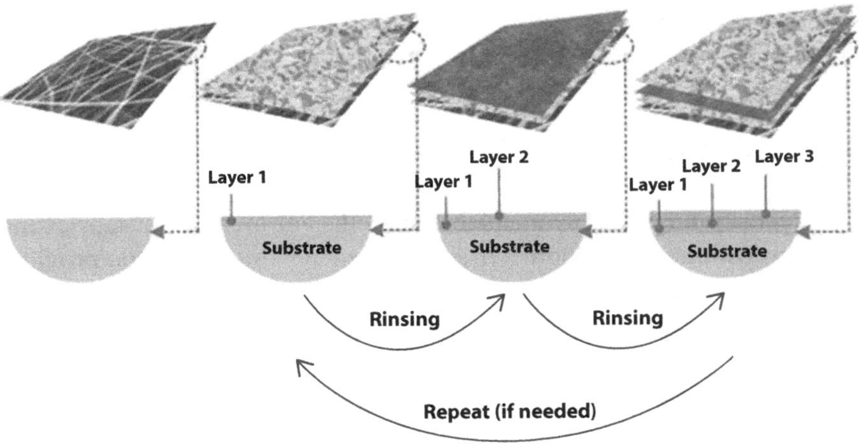


**Figure 5.1** Assembly of layers in three steps of deposition [31].

### 5.2.4 Mechanisms of Layer Formation

The principal mechanism responsible for LbL assembly is the electrostatic attraction of species with opposing charges. Depending on the materials employed, covalent and hydrogen bonding may also be formed. These processes are all spontaneous. Therefore, the process of layer generation at the nanoscale through these binding forces—which produces laminar composites with desirable properties—is referred to as "self-assembly" [32].

The LbL deposition process can involve the alternate immersion of the substrate into solutions containing oppositely charged materials. The substrate is first immersed in a solution containing a cationic polyelectrolyte, resulting in the adsorption of a positively charged layer. After rinsing, the substrate is then immersed in an anionic polyelectrolyte solution, leading to the formation of a negatively charged layer. This alternating adsorption process is repeated to build up the desired number of layers [33]. Different parameters can affect the building up of the layers, such as pH, type of salt, reaction temperature, ionic strength, and chemical properties of polyelectrolytes [15].

In addition to electrostatic adsorption, layers can be assembled *via* hydrogen bonding [34] and covalent bonds [35]. It is proposed that hydrogen bonds form between the lone electron pairs on nitrogen and oxygen atoms in the pyrrole rings of polypyrrole molecules and the amino acid residues in wool proteins [36]. Furthermore, carbon fiber surfaces can accept graphene oxide (GO) and silica ($SiO_2$) *via* covalent bonding, leading to the improvement of mechanical properties and wettability [31].

## 5.3 Advantages of LbL Deposition

LbL deposition is a highly adaptable and efficient method for fabricating multilayer films with nanometer-scale accuracy. This technique provides several benefits, making it an appealing option for various applications, especially in the enhancement of textile surfaces.

### 5.3.1 Precise Control over Layer Thickness and Composition

One of the most significant advantages of LbL deposition is the precise control that it provides over the thickness and composition of the deposited films [37]. Adjusting the number of deposition cycles, the concentration of solutions, temperature, pH of the solution, salt concentration, the polyelectrolytes' molecular weight, rinsing time, and the immersion time, researchers can finely tune the thickness of each layer, often with nanometer precision [38–40]. Considering the number of layers deposited on substrates, each cycle typically adds a nanometer-scale layer to the film, allowing for the creation of coatings with thicknesses ranging from a few nanometers to several micrometers [41, 42]. An increase in the number of layers in the composite fabric can enhance some properties, such as conductivity [43]. This level of control is crucial for applications requiring specific layer properties, such as UV protection and sensing [44], antibacterial efficiency [45], and conductivity of E-textiles [46].

### 5.3.2 Versatility in Material Selection

The selection of appropriate materials can greatly enhance the functional properties of textiles, such as their antimicrobial efficacy [47], food packaging [48], and flame retardancy [49]. For instance, incorporating bioactive agents into LbL assemblies can impart antimicrobial qualities to textiles, rendering them particularly advantageous for use in medical and healthcare environments [50]. In addition, the choice of suitable materials can significantly improve the durability and performance of textiles [51, 52]. It is crucial to reduce the dependence on detrimental chemicals, along with minimizing energy and water consumption. The use of natural or biodegradable materials in LbL processes can greatly diminish the environmental impact in comparison to traditional textile finishing methods [53].

### 5.3.3 Simplicity and Cost-Effectiveness

The LbL technique is notably straightforward and economical when compared to alternative surface modification techniques. It does not necessitate advanced equipment or harsh processing environments. The fundamental procedure entails the sequential immersion of a substrate in aqueous solutions containing the desired materials, accompanied by rinsing steps [54]. This method simply enables the creation of multilayer structures on various substrates, including colloidal particles [55], quartz [56], and micelles [57], through the alternating electrostatic deposition of oppositely charged polyelectrolytes. In addition, this procedure is recognized for its convenience, versatility, and eco-friendliness, allowing precise control over the thickness and surface properties of the deposited layers without requiring expensive equipment. Furthermore, it permits the direct use of biodegradable or biocompatible polymers, such as chitosan (CS) and polyacrylic acid (PAA), as polyelectrolytes for chemical deposition [58]. Thin films composed of poly(vinyl alcohol) and graphene nanoplatelets, which were stabilized using poly(4-styrenesulfonic acid) (PSS), can also be fabricated through a straightforward and economical LbL method to enhance the flame resistance of cotton fabric [59].

## 5.4 Methods of LbL Deposition

### 5.4.1 Dip-Coating Method of LbL Deposition

#### *5.4.1.1 Process Overview*

The dip-coating method is commonly employed in LbL deposition due to its ease of use, adaptability, and efficiency. In LbL deposition, dip coating typically consists of several stages: immersion, adsorption, rinsing, and drying. Initially, the substrate is submerged in a solution containing the first polyion, allowing the charged components to adhere to the substrate surface through electrostatic interactions. After a specific duration, the substrate is taken out of the solution and thoroughly rinsed with a solvent, often deionized water, to eliminate loosely bound or excess material. Subsequently, the substrate is dried, either through air-drying or using a gentle stream of nitrogen. This process is then repeated with a solution containing the oppositely charged poly-ion, and the cycle continues until the desired number of layers or thickness is attained [60, 61]. Figure 5.2 depicts the process of LbL dip coating.

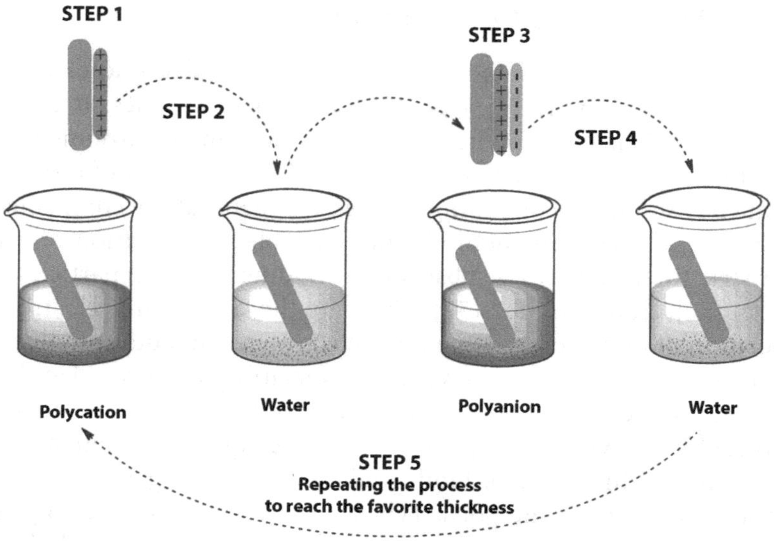


**Figure 5.2** The process of LbL dip coating [62].

### 5.4.1.2 *Advantages of Dip Coating*

One of the primary advantages of dip coating in LbL deposition is its simplicity and ease of implementation. The process does not require complex or expensive equipment, making it accessible for both research and industrial applications. Additionally, dip coating allows for excellent control over film thickness by adjusting parameters such as immersion time, solution concentration, and the number of deposition cycles [63]. Dip coating is also highly versatile. It can be applied to substrates of various shapes, sizes, and materials, including flat surfaces, fibers, and three-dimensional objects [64].

### 5.4.1.3 *Challenges and Considerations*

Despite its advantages, the dip-coating method in LbL deposition also presents some challenges. One of the main issues is the potential for uneven coating, especially on highly porous substrates, due to the rapid absorption of coating solution, which can have negative effects on the performance of LbL-coated materials [65]. Moreover, air bubbles created during submerging or rinsing can result in adverse impacts [66, 67]. In addition, the rinsing step can be challenging in achieving uniform coatings and preventing defects in the multilayer structure [68]. Therefore, achieving uniform coverage requires careful optimization of the deposition parameters during immersion to ensure even adsorption.

#### 5.4.1.4 Applications of Dip Coating in LbL Deposition

Dip coating is widely used in various applications due to its ability to produce multilayers with a rather good quality film. In the textile industry, it is employed to impart functional properties such as UV resistance and hydrophobicity [69], antimicrobial activity [70], and antifouling finishing of cotton textiles [71]. In the biomedicine field, dip-coated LbL films are utilized for the creation of drug delivery systems. Oppositely charged polyelectrolytes, for example, are employed to assist textiles in regulating drug delivery. The precise management of layer thickness and composition permits the development of coatings that can either release therapeutic agents in a controlled manner or offer a surface suitable for cell growth. Additionally, the incorporation of multiple layers led to an increase in the antibacterial effectiveness of cotton against *Escherichia coli* (*E. coli*) and *Staphylococcus aureus* (*S. aureus*) [72, 73]. In addition, dip coating can be used in electromagnetic shielding because it plays a significant role in public contamination. Therefore, LbL dip coating is a method of defense against electrical and electromagnetic signals. Chitosan is employed as a polycation together with graphene, whereas poly (sodium 4-styrene sulfonate) (PSS) acts as a polyionic material and can be applied to cotton fabric. Consequently, this process allows for the attainment of electrical and electromagnetic shielding [74].

## 5.5 Spin Coating Method of LbL Deposition

Another widely used technique in LbL deposition is the spin coating method, which has distinct benefits in the production of uniform and thin films. This approach entails swiftly spinning a substrate while dispensing a solution onto its surface, utilizing centrifugal force to evenly distribute the solution and create a thin, uniform layer.

### 5.5.1 Process Overview

The spin coating in the LbL deposition process consists of dispensing, spinning, and drying. Initially, a small amount of solution with the initial polyion is dispensed onto the center of the substrate. Then, the substrate is rapidly spun at a high speed, usually ranging from a few hundred to several

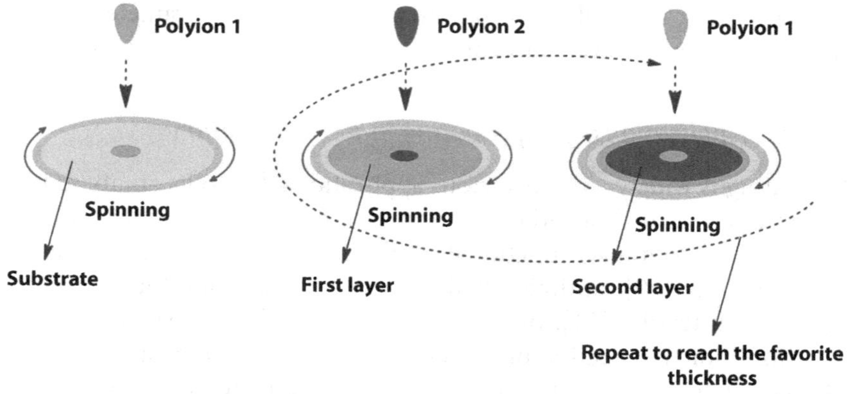


**Figure 5.3** The procedure of LbL spin coating [76].

thousand revolutions per minute. The centrifugal force generated by the spinning spreads the solution across the substrate, creating a thin, uniform layer. Any excess solution is pushed out from the substrate's edges. After the spinning is finished, the substrate is dried, and the process is repeated with a solution containing the oppositely charged polyion. This sequence is then repeated until the desired number of layers is attained [75]. Figure 5.3 illustrates the procedure of LbL spin coating.

### 5.5.2 Advantages of Spin Coating

The deposition of thin films often involves the use of spin coating, a technique known for its numerous advantages. The simplicity and reproducibility of the method make it appropriate for high-throughput manufacturing processes. Additionally, spin coating is cost-effective and capable of depositing films over large areas with minimal material wastage [77]. The method can be used with a variety of materials, such as polymers, inorganic compounds, and organic-inorganic hybrids, showcasing its versatility [78]. Moreover, it can create precise, smooth, and uniform films. This level of precision is attainable by adjusting parameters like spin speed, spin time, and solution concentration, enabling the production of films ranging from a few nanometers to several micrometers in thickness. The qualities mentioned above are what render spin coating an essential instrument in scenarios where flawless, high-quality films are necessary, such as in microelectronics, optics, and protective coatings [79, 80]. In addition, the spin coating can be widely used to disperse the high-viscosity materials on the planar substrate and create highly organized multilayers [81–83].

### 5.5.3 Challenges and Considerations

The spin-coating technique used for LbL deposition has the advantage of generating consistent thin films, but it comes with various challenges and limitations. A major obstacle is the struggle to achieve even coatings on non-flat or irregularly shaped textile surfaces due to the preference for centrifugal force for flat substrates during spin-coating [79]. Another significant challenge is the technical difficulties of uniformly coating 2D substrates with irregular shapes and the inability to apply films to 3D substrates [84]. For instance, drug-releasing thin films created through spin coating have been developed successfully; however, their application is restricted to coatings on 3D substrates [85]. It is important to note that in some cases larger volume of solution is usually needed compared to other methods due to the expulsion of excess material during the spinning process. This could pose a limitation, particularly when dealing with costly or limited materials, especially at industrial levels [86].

### 5.5.4 Applications of Spin Coating in LbL Deposition

Spin coating is an essential part of the LbL deposition method for textile materials, offering various applications to enhance the functionality and performance of fabrics. The first use of spin coating is to create coatings on textiles, providing long-lasting protection against bacterial and fungal growth. Furthermore, it is used to improve the fire resistance of fabrics through flame-retardant treatment [11, 87]. In addition, spin coating is employed to aid in the administration of superhydrophobic layers, bestowing water-repellent qualities to textiles. It is also used to produce UV-protective coatings that guard fabrics against harmful ultraviolet radiation. The spin coating method is beneficial for producing conductive textiles, which are crucial for wearable electronics and smart textiles. Furthermore, it is utilized to formulate self-cleaning coatings that enable textiles to resist dirt and stains. Additionally, the technique is valuable for creating antistatic coatings, minimizing static electricity accumulation on fabrics. It is also used for applying biocompatible coatings on medical textiles to improve their compatibility with human tissues. Moreover, spin coating contributes to the advancement of thermo-regulating textiles that can adjust to temperature fluctuations, offering comfort in different environmental conditions. Finally, it is vital to produce multifunctional coatings that integrate several of the aforementioned properties, resulting in comprehensive enhancements to textile materials [88, 89].

## 5.6 Spray Coating Method of LbL Deposition

The spray coating method in LbL deposition is an innovative and versatile technique that allows for the rapid and uniform application of multilayer films over large and complex surfaces. This method involves the use of a spray nozzle to atomize the solution into fine droplets, which are then deposited onto the substrate [90].

### 5.6.1 Process Overview

In LbL deposition, the spray coating process involves several crucial stages: atomization, deposition, rinsing, and drying. Initially, a spray nozzle is used to atomize a solution containing the first polyion into fine droplets. The substrate is positioned to receive the sprayed solution, resulting in the formation of a thin and uniform layer upon contact with the droplets. Subsequently, the substrate is rinsed to eliminate excess material and then dried. This sequence is then repeated using a solution containing the oppositely charged polyion, and the process is reiterated until the desired number of layers is achieved. The atomization process plays a key role in ensuring the uniform distribution of droplets, enabling even coverage of the substrate [91]. Figure 5.4 illustrates the schematic of this method of layer deposition.

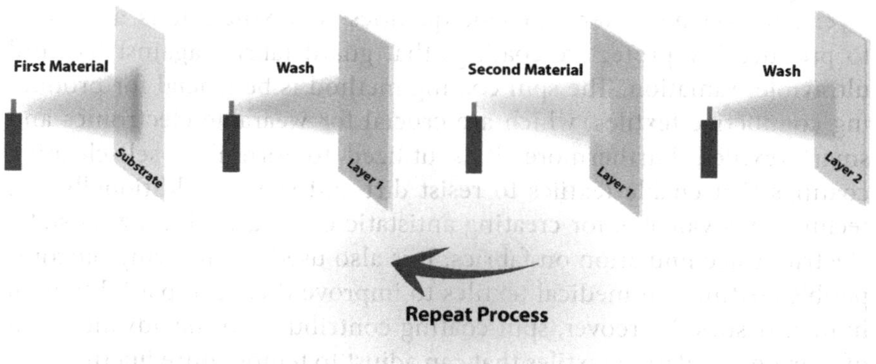


**Figure 5.4** The schematic of spray coating LbL deposition [92].

### 5.6.2 Advantages of Spray Coating

The spray coating method has numerous advantages, making it an appealing choice for LbL deposition. A key advantage is its capability to uniformly coat large and irregularly shaped surfaces. Unlike dip or spin coating, spray coating does not necessitate the substrate to be immersed or spun, making it well-suited for applications involving complex geometries and large areas, such as textile fabrics [93]. Additionally, it is simple to use and can be adapted to create flame-retardant fabric, antibacterial fabric, and other functional fabrics. The LbL spraying technique can efficiently construct functional coatings based on interactions (often electrostatic or covalent interactions) between the chosen reagents [94, 95]. Control over the morphology of the deposited layer is achievable through adjustable spraying parameters. Furthermore, composite materials can be fabricated by LbL deposition of multilayered films, offering a convenient method for assembling multicomponent films with customizable morphology [96, 97]. The method is also highly efficient and suitable for high-throughput production, significantly reducing the overall deposition time compared to traditional methods [98].

### 5.6.3 Challenges and Considerations

Spray coating for LbL deposition on textiles presents several challenges and limitations. One significant issue is the non-uniformity of the coating, which can result from the difficulty in controlling the spray parameters, such as droplet size and distribution. This non-uniformity can lead to inconsistent surface properties and performance of the coated textiles. Additionally, the process often requires multiple passes to achieve the desired thickness, which can be time-consuming and inefficient. Another limitation is the potential for clogging of the spray nozzles, especially when using solutions with high viscosity or particulate matter. This can interrupt the coating process and necessitate frequent maintenance. Furthermore, the adhesion of the layers can be compromised due to the rapid evaporation of the solvent, leading to weak interlayer bonding. Environmental concerns also arise from the use of volatile organic compounds in some spray-coating formulations, which can pose health risks and require proper ventilation and handling [11, 12, 65, 99].

### 5.6.4 Applications of Spray Coating in LbL Deposition

Spray coating is widely used in various applications due to its ability to quickly and uniformly apply multilayer films. In the textile industry, spray-coated LbL films are used to impart functional properties such as water repellency [100], UV protection [101], and anti-reflection [102]. Moreover, flame retardancy of polyester fabrics can be achieved by spray-coating silica-based nanomaterials [103]. These achievements are similar when poly (lactic acid) surface modified by chitosan-based spray coating. Spray coating is also widely used for biomedical applications. Controlled drug release is the result of spray coating surface modification of textiles, which makes it a proper choice for biomedical uses [104]. Graphene and other conductive polymers have been functionalized with groups like carboxylic acid and incorporated with materials like *N,N,N*-cetyltrimethylammonium bromide for application on cotton fabric. This process has led to enhancements in both electromagnetic protection and electrical conductivity of the fabric [105].

## 5.7 Comparison of Dip Coating, Spin Coating, and Spray Coating Methods in LbL Deposition

LbL deposition utilizes various techniques, such as dip coating, spin coating, and spray coating, each with its unique characteristics, advantages, and limitations, making them suitable for different applications. This section provides a comprehensive evaluation of these three methods, focusing on their underlying principles, benefits, and challenges.

When comparing LbL deposition methods such as dip coating, spin coating, and spray coating, it is important to recognize their unique advantages and limitations. Dip coating is effective for coating complex shapes and porous substrates, but it can be time-consuming due to the multiple steps involved. Spin coating is well-known for producing uniform and smooth coatings with precise thickness control, making it suitable for flat substrates but less so for irregular shapes or large areas. Spray coating offers an efficient way to apply coatings to large surfaces and substrates that are hard to coat through immersion or spinning. However, achieving uniformity in the coating can be challenging, especially with intricate shapes, requiring optimization of spray parameters. The choice of coating method depends heavily on specific application requirements, such as substrate type, desired coating properties, and production efficiency. Understanding the differences between these coating techniques is crucial for selecting the most suitable approach for LbL deposition on textiles or other materials to achieve optimal performance and quality in the final product. Dip coating is also easily scalable for industrial applications, allowing for efficient coating of large surface areas, which is advantageous for mass production.

On the other hand, spin coating excels in controlling film thickness by adjusting parameters such as spin speed, solution concentration, and spin time. This method can produce films with exact thicknesses. This level of control is crucial for applications requiring highly smooth and uniform films, such as in optical and electronic devices. Additionally, spin coating offers high reproducibility and consistency in film properties, ensuring that each layer meets the desired specifications. This reproducibility is particularly important in research and development settings where consistency is key. Spray coating is recognized for its efficiency and rapid application process, making it suitable for large-scale and high-throughput production. This method is ideal for coating large, irregularly shaped, or textured surfaces, providing a level of flexibility that is unmatched by the other two methods. Its ability to quickly cover large areas also makes it an attractive choice for applications where time and efficiency are critical.

In summary, while dip coating offers versatility and scalability, making it suitable for a wide range of substrates and industrial applications, spin coating provides excellent control over film thickness and smoothness, essential for high-precision applications. Spray coating, with its rapid application process and ability to coat complex surfaces, is ideal for large-scale production and substrates with diverse shapes.

Each method has its unique strengths, and the choice of technique will depend on the specific requirements of the application and the nature of the substrate being coated.

## 5.8 Characterization of LbL Modified Textiles

Textile materials have been modified with the LbL method to enhance their different properties by depositing alternating layers of materials such as NPs, biomolecules, and a wide range of polymers. Analyzing the surface of LbL-modified textiles with different methods, such as scanning electron microscopy (SEM), water contact angle (WCA), atomic force microscopy (AFM), and X-ray photoelectron spectroscopy (XPS) are imperative part of understanding the structure and morphology of the coatings. These methods offer detailed images that expose the surface topography and composition of the modified textiles. In addition, mechanical properties can play a significant role in the end uses of LbL-modified textiles.

Here are the most important and applicable characterization methods used to consider the crucial properties of LbL surface-modified textiles.

### 5.8.1 Surface Morphology

#### *5.8.1.1 Scanning Electron Microscopy (SEM)*

The surface morphology of textiles that have been modified by the LbL method can be easily studied using SEM. This technique provides valuable insights into the structural and topographical characteristics of the modified surface. The SEM can be used to investigate the smoothness and uniformity of the surface, which are altered by the LbL deposition method. For instance, a study was developed to consider the surface roughness of LbL-coated and uncoated mercerized cotton textiles using SEM images. The SEM images showed that the surface roughness increased as the number of bilayers deposited increased. The morphology of mercerized cotton textiles before and after LbL modification demonstrated agglomeration that occurred after layer deposition, which may be influenced by multiple factors such as the polyelectrolyte and nanoformulations, the type and strength of the polyelectrolyte solution (whether it is cationic or anionic), the duration of immersion in deionized water, as well as the repeated drying process at each stage [106, 107].

Moreover, cotton fabric, coated with 17 layers of graphene, showed electron transfer efficiency and more roughness on the surface with no change in fabric structure in the first deposited layers [108]. In another structural characterization study, researchers investigated the modification of cotton fabrics with chitosan and polyelectrolytes using the LbL method. The study involved a thorough surface characterization using SEM, which uncovered the interaction between various layers and the overall surface appearance [109].

Anatase titanium dioxide ($TiO_2$) NPs with opposite charges to PAA were used to coat the negatively charged cellulose acetate (CA) nanofibers through electrostatic LbL assembly. The application of five bilayers on CA nanofibers increased the Brunauer–Emmett–Teller surface area from 2.5 $m^2/g$ for uncoated nanofibers to 6.0 $m^2/g$ for coated fibers. Additionally, the average diameter of the fibers increased from 344 to 584 nm. The reason for the increase in fiber diameter may be due to the rough surface of the deposited multilayers [110].

#### *5.8.1.2 Atomic Force Microscopy (AFM)*

AFM is a highly effective technique for analyzing the surface of textiles modified through the LbL deposition process. It provides detailed three-dimensional, high-resolution images of the surface topography, allowing for in-depth analysis of the morphological properties of the LbL coatings. This approach is particularly proficient at examining the thickness, roughness, and uniformity of the deposited layers. Recent studies have demonstrated the utility of AFM in evaluating the surface characteristics of textiles subjected to LbL treatment. Demir and Türemen [109] utilized AFM to characterize the surface of cotton textiles coated with polyelectrolytes and chitosan, highlighting the complex surface morphology and interactions between multiple layers. Similarly, Li *et al.* [111] examined the roughness of cellulose fibers treated with polyelectrolyte and lignosulfonate multilayers using AFM. They discovered that the surface roughness significantly increased with the increase in the number of bilayers.

AFM phase images can also provide insights into the uniformity and distribution of coatings for applications requiring precise surface modifications. The ability of AFM to describe LbL-modified textiles is further improved by its capability to measure the mechanical properties of the surface, such as bending and adhesion [112]. Due to these attributes, AFM plays a critical role in optimizing the LbL process and ensuring the desired functional properties of the modified fabrics. The significance of AFM in surface characterization will grow as the demand for advanced textile materials increases, providing essential information for the development of high-performance textile products.

## 5.9 X-Ray Photoelectron Spectroscopy (XPS)

Analyzing LbL deposition on textiles with XPS is an effective method. The elemental composition and chemical bonding of the surface layers can be thoroughly analyzed using this technique, which is essential for comprehending the effectiveness and qualities of the LbL assembly.

In research to enhance the thermal stability and flame retardancy of cotton fabrics, XPS analysis was utilized to verify the effective application of gelatin (GEL)/adenosine 5′-monophosphate (AMP) and $SiO_2$ sol-gel coatings. The findings from XPS offered in-depth insights into the elemental compositions and functional groups present on the surface, which play a vital role in improving flame-retardant characteristics [113]. In the other study, cotton fabrics with a positive charge were treated with ZnO NPs using an LbL technique, resulting in the creation of 10 and 16 multilayer films on the fabric surface. It is clear from Figure 5.5 that analysis using XPS revealed peaks at 283.95, 530.11, and 1,033.7 eV for carbon, oxygen, and zinc, respectively. The intensity of the zinc peak increased with the addition of more layers. According to XPS, the cationized samples contained 0.8% nitrogen, whereas the 10 and 16 multilayer films had zinc content of 17.7% and 18.8%, respectively. It is anticipated that the oxygen content, originating mainly from OH groups on cotton, will decrease in the

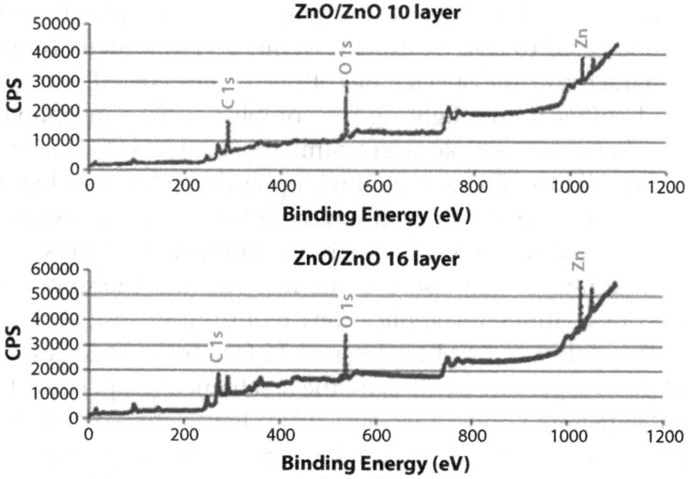


**Figure 5.5** XPS spectra of 10 and 16 multilayer films of nano-ZnO deposited on woven cotton fabric that is cationically charged [114].

10-layer films and may increase in the 16-layer films due to the presence of additional ZnO [114].

The application of poly(sodium 4-styrene sulfonate) (PSS) and poly(allylamine hydrochloride) (PAH) onto cotton fibers using an LbL approach was documented by the researchers. Upon examination using XPS analysis, specific peaks at 281.91, 398.91, and 528.91 eV were observed on the positively charged woven cotton fabric, indicating the presence of carbon, oxygen, and nitrogen in the unmodified sample. After the LbL modification, new peaks at 398.91 and 164.91 eV emerged, demonstrating the incorporation of nitrogen and sulfur from PAH and PSS layers [115].

## 5.10 Mechanical Properties

LbL deposition is a useful technique to enhance the mechanical properties of textiles. Recent research has indicated substantial advancements in the tensile strength, elasticity, and abrasion resistance of LbL-coated textiles.

One example is the use of nanocellulose and chitosan layers, which have proven to improve the tensile strength and flexibility of cotton fabrics [116]. In another investigation, a total of 15 bilayers were applied to cotton fabric using the LbL method. These bilayers consisted of poly(styrene sulfonate) (PSS) as an anionic agent and synthesized silver-loaded chitosan NPs (CS-Ag) as a cationic agent. The mechanical properties, such as tensile strength, air permeability, and bending (flexural) rigidity, were examined on coated fabrics in both the warp and weft directions, with an average of three samples analyzed. The purpose was to evaluate the impact of the coating process. The results of the tensile strength tests indicated

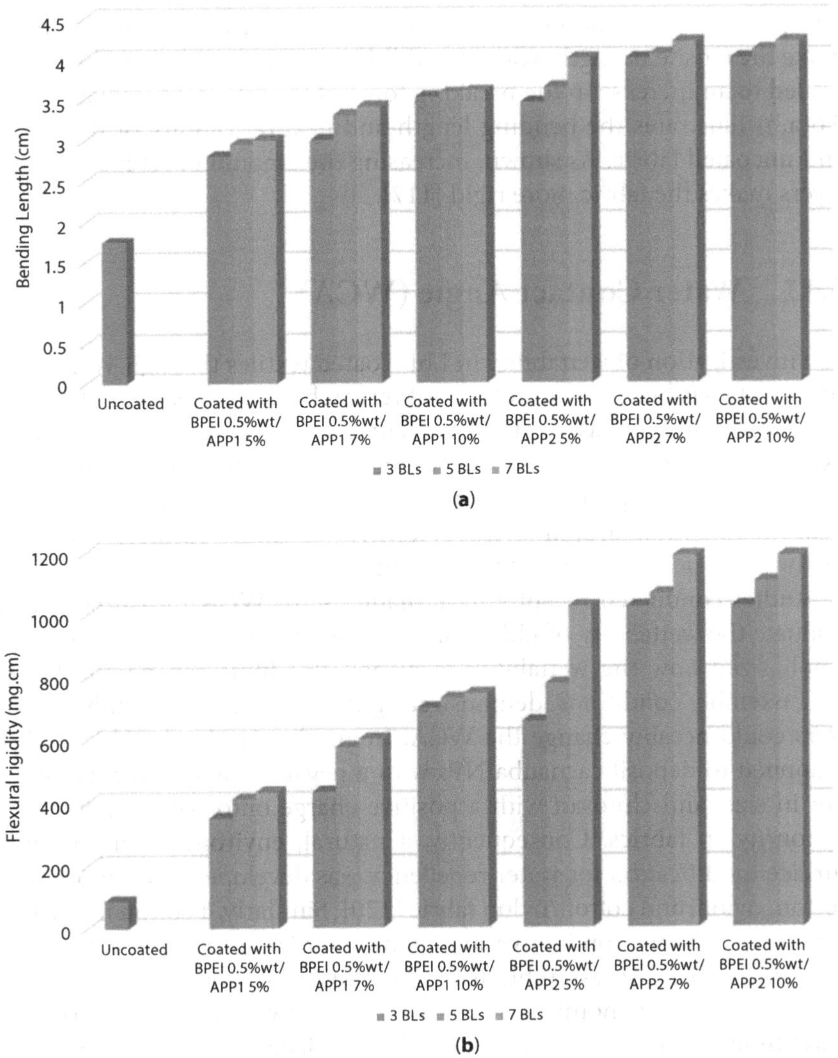


**Figure 5.6** (a) Bending length and (b) flexural rigidity of uncoated and LbL-coated polyester fabric with branched-polyethyleneimine (BPEI)/ammonium polyphosphate (APP). BLs, bilayers. Reused with permission from Sage Publications, License number: 6043710879321 [117].

that the dipping process had a negative effect on tensile strength, although these changes were not statistically significant ($p < 0.05$). Additionally, the bending tests demonstrated that the flexibility of the samples remained unchanged and was unaffected by the coating process [45].

The mechanical properties of polyester fabric were evaluated after LbL-coating with branched-polyethyleneimine (BPEI) and ammonium polyphosphate (APP) solution. Results showed that the coating process did not make the tensile strength worse. In addition, increasing the number of layers led to an increase in the breaking force of the fabric. Moreover, Figure 5.6(a,b) illustrates the bending length and flexural rigidity of the coated and uncoated fabric. As shown, increasing the amount of APP and coated layers makes the fabric more rigid [117].

## 5.11 Water Contact Angle (WCA)

The investigation of wettability in LbL-coated textiles through WCA measurements has been a central focus of research. Therefore, the LbL assembling technique, an easy and cost-effective bottom-up approach, has been extensively used to create super-wetting surfaces with micro/nanostructures, specific components, and adjustable thickness [118]. Assessing the hydrophobic or hydrophilic characteristics of textile surfaces using this method is crucial as it directly affects their practical uses.

Studies conducted recently have made use of WCA measurements to evaluate the wettability of LbL-coated textiles. For example, Li *et al.* [119] emphasized how the wettability of Nafion/PEI films can be affected by LbL assembly conditions, demonstrating that particular assembly parameters could notably change the WCA. In another study, the LbL method is applied to deposit carnauba NPs with a negative charge, averaging 290 nm in size, and chitosan with a positive charge onto cotton, nylon, and cotton/nylon fabrics. Consequently, a natural, environmentally friendly surface modification for water repellency was developed and reported for cotton, nylon, and cotton/nylon fabric [120]. Similarly, a cotton fabric with superhydrophobic, conductive, and flame-retardant properties was created by utilizing the LbL assembly technique involving poly (ethyleneimine), APP, and carbon nanotubes. This was followed by treating the fabric with poly(dimethylsiloxane) [121]. M.R. Abeywardena *et al.* [122] proposed a new, uncomplicated, and cost-efficient approach for producing superhydrophobic polyester fabric. This approach involves applying an extremely thin layer of surface-modified superhydrophobic precipitated calcium carbonate (SHPCC) using the dip-coating technique. They utilized dolomite as the source of calcium for creating SHPCC, along with stearic acid (SA) to modify the surface of calcium carbonate and bind the synthesized SHPCC to the textile. The results of the WCA tests demonstrated the superhydrophobic properties of the coated polyester fabric, possibly attributed to the incorporation of SHPCC particles that create a hierarchical structure, thereby enhancing hydrophobicity.

## 5.12 Antimicrobial Properties of LbL-Modified Textiles

Antimicrobial textiles, known as functionally active textiles, can either kill microorganisms or prevent their growth. These textiles are classified based on their ability to combat microbes, including antibacterial, antifungal, and/or antiviral properties. Some antimicrobial textiles can simultaneously combat bacteria, fungi, and viruses [123].

Chitosan, known for its natural antibacterial properties, has been applied in various ways to take advantage of its antibacterial effects. For example, when chitosan is combined with citric acid and sodium lignosulfonate solution and deposited LbL on cotton fabric using boric acid, the fabric gains UV protection, becomes flame retardant, and exhibits antimicrobial activity [124]. The fabric was treated with a multilayer nanocoating comprising magnesium lignosulfonate, chitosan, and mono ammonium phosphate (MAP) using an LbL assembly to develop a sustainable and efficient multifunctional treatment for cotton. By applying a five-bilayer coating on cotton, the combined properties of magnesium lignosulfonate, chitosan, and MAP were harnessed to offer UV protection and antimicrobial activity against *E. coli*, *S. aureus*, and *C. albican* [125]. Besides natural chitosan NPs, other NPs were also used. For example, silver NPs (Ag NPs) have been used for the introduction of antibacterial properties to textile materials against more than 600 microorganisms [126]. Moreover, extensive research has been conducted on the synergistic enhancement of antibacterial activity through the application of different methods using AgNPs and chitosan in combination [127, 128]. In another study, polyethyleneimine (PEI) and polyanionic PSS polyelectrolytes were utilized as polycation and polyanion layers, alternating with dispersions of negatively- and positively-charged $TiO_2$ particles. The textile obtained strong antibacterial properties due to the combination of PEI and $TiO_2$, which both originally possessed antibacterial properties [129]. Table 5.1 shows the other investigations concerning the antimicrobial activity of different textiles, coated with different LbL coating methods.

**Table 5.1** Different compounds on different textiles using different coating methods to impart antibacterial activity.

| Antimicrobial activity | Type of textiles | LbL coating method | Antimicrobial activity compound | References |
|---|---|---|---|---|
| Antibacterial | Cotton | Dip coating | ZnO NPs | [130] |
| Antibacterial and antiviral | Polyester | Spray coating | Copper NPs | |
| Antibacterial | Polyester | Spray coating | Ag NPs | [131] |
| Antibacterial | Wool | Dip coating | GO | |
| Antibacterial, antifungal | Cotton/ polyester | Dip coating | Chitosan | [132] |
| Antibacterial | Silk | Spin coating | $TiO_2$ NPs | |
| Antibacterial | Nylon | Dip coating | Lysozyme | [133] |
| Antibacterial | Cotton/ polyester | Dip coating | Nafion | |
| Antibacterial | Cotton | Spin coating | Silver NPs | [134] |
| Antibacterial and antifungal | | Spray coating | Chitosan/silver NPs | |

## 5.13 Future Directions of the LbL Surface Modification

The advancement of LbL surface modification in textiles faces several challenges, but significant progress is expected in this field. Overcoming these challenges will be essential for the successful integration and widespread use of LbL technologies in textile applications.

### 5.13.1 Scalability of LbL Deposition

The scalability of the LbL deposition process poses a major challenge. Although LbL coatings have been proven effective in small-scale experiments, it is a significant challenge to apply these findings to large-scale industrial production. Future research efforts should concentrate on creating automated and high-throughput LbL deposition techniques that can uphold the accuracy and consistency of coatings on a massive scale. Advancements in robotic systems and cutting-edge manufacturing technologies could be crucial in addressing these scalability challenges.

### 5.13.2 Durability and Stability of Coatings

The practical application of LbL coatings relies heavily on their ability to withstand real-world conditions, making durability and stability crucial. Enhancing the mechanical strength and environmental resistance of these coatings should be the focus of upcoming research efforts. This may include integrating cross-linking agents, creating coatings with multiple functions, and investigating new materials with better adhesion and resilience. To confirm the effectiveness of LbL-modified textiles, it will be necessary to conduct long-term field tests and accelerated aging studies.

### 5.13.3 Environmental and Economic Considerations

The textile industry is increasingly focusing on sustainability, so it is important to thoroughly assess the environmental and financial impacts of LbL surface modification. It is crucial for future research to prioritize the creation of environmentally friendly and cost-efficient LbL processes. This involves using biodegradable and non-toxic materials, as well as refining deposition techniques to reduce waste and energy usage. Conducting life cycle assessments and cost-benefit analyses will be essential for proving the practicality of LbL technologies from environmental and economic standpoints.

### 5.13.4 Innovations and Emerging Trends

The future development of LbL surface modification for textiles may be influenced by various emerging trends and advancements. By incorporating smart and responsive materials into LbL coatings, it is possible to create textiles with enhanced functions like self-cleaning, antimicrobial properties, and responsiveness to stimuli. Furthermore, the combination of LbL technology with other advanced manufacturing methods such as 3D printing and nanofabrication could pave the way for the creation of complicated and highly personalized textile surfaces. The collaboration between academia, industry, and government agencies will play a critical role in driving these innovations and ensuring their successful integration into the textile industry.

**AI disclosure statement:** The author acknowledges the use of AI-based tools for minor language editing or grammatical correction. However, all intellectual contributions, analyses, and conclusions presented in this chapter are entirely the author's own.

## References


1. Periyasamy, A.P., Militky, J., Sachinandham, A., Duraisamy, G., Nanotechnology in Textile Finishing: Recent Developments, in: *Handbook of Nanomaterials and Nanocomposites for Energy and Environmental Applications*, p. 1, Springer Nature, Switzerland, 2020.
2. Periyasamy, A.P., Venkataraman, M., Kremenakova, D., Militky, J., Zhou, Y., Progress in sol-gel technology for the coatings of fabrics. *Materials*, 13, 1838, 2020.
3. Kothari, A., Plasma Technology in Textile: A Step towards the Green Environment. *Contemp. Soc. Sci.*, 29, 2019.
4. Aghabeigi, E., Shabanloo, R., Akbari, S., Aghaji, A.G., Electrospun PLLA/PEI nanofibers for controlled drug release behaviors and antibacterial efficiency. *J. Macromol. Sci. Part A*, 61, 186, 2024.
5. Mirjalili, M. and Zohoori, S., Review for application of electrospinning and electrospun nanofibers technology in textile industry. *J. Nanostruct. Chem.*, 6, 207, 2016.
6. Shabanloo, R., Akbari, S., Mirsalehi, M., Hybrid electrospun scaffolds based on polylactic acid/PAMAM dendrimer/gemini surfactant for enhancement of synergistic antibacterial ability for biomedical application. *Biomed. Mater.*, 17, 045009, 2022.
7. Rana, N., 3. Microencapsulation Finishing in Textiles, in: *Innovation and Recent Advancement in Fashion and Textile*, p. 33, HOW Academics, Delhi, India, 2023.
8. Shabanloo, R., Karimi, N., Montazer, M., Nanomaterials and textile industry, in: *Handbook of Nanomaterials*, vol. 1, p. 309, Elsevier, Amsterdam, Netherlands, 2024.
9. Iler, R., Multilayers of colloidal particles. *J. Colloid Interface Sci.*, 21, 569, 1966.
10. Korkmaz, G., Kılınç, M., Kılınç, N., Kut, Y.D., The Role of Surface Modification Methods for Sustainable Textiles, 2023.
11. Magovac, E., Vončina, B., Jordanov, I., Grunlan, J.C., Bischof, S., Layer-by-layer deposition: a promising environmentally benign flame-retardant treatment for cotton, polyester, polyamide and blended textiles. *Materials*, 15, 432, 2022.
12. Uğur, SS., Sustainable dyeing and finishing of cotton fabrics with layer-by-layer technique. *Coatings*, 13, 1129, 2023.

13. Gomes, A.P., Mano, J.F., Queiroz, J.A., Gouveia, I.C., Layer-by-layer deposition of antibacterial polyelectrolytes on cotton fibres. *J. Polym. Environ.*, 20, 1084, 2012.
14. Ariga, K., Ahn, E., Park, M., Kim, B.S., Layer-by-layer assembly: recent progress from layered assemblies to layered nanoarchitectonics. *Chem. Asian J.*, 14, 2553, 2019.
15. Iqbal, M.H., Kerdjoudj, H., Boulmedais, F., Protein-based layer-by-layer films for biomedical applications. *Chem. Sci.*, 15, 9408, 2024.
16. Mishra, A., Bhuyan, N.N., Xu, H., Sharma, G.D., Advances in layer-by-layer processing for efficient and reliable organic solar cells. *Mater. Adv.*, 4, 6031, 2023.
17. Peng, M., Zhang, X., Xiao, X., Dong, M., Zhao, G., Liu, P., Chen, Y., Wang, C., Polyelectrolytes fabrication on magnesium alloy surface by layer-by-layer assembly technique with antiplatelet adhesion and antibacterial activities. *J. Coat. Technol. Res.*, 16, 857, 2019.
18. Percival, S.J., Small, L.J., Spoerke, E.D., Rempe, S.B., Polyelectrolyte layer-by-layer deposition on nanoporous supports for ion selective membranes. *RSC Adv.*, 8, 32992, 2018.
19. Souza, P.R., de Oliveira, A.C., Vilsinski, B.H., Kipper, M.J., Martins, A.F., Polysaccharide-based materials created by physical processes: from preparation to biomedical applications. *Pharmaceutics*, 13, 621, 2021.
20. Vilsinski, B.H., de Oliveira, A.C., Souza, P.R., Martins, A.F., Polysaccharide-based polyelectrolyte multilayers fabricated *via* layer-by-layer approach: From preparation to applications. *Prog. Org. Coat.*, 196, 108720, 2024.
21. Szabó, T., Péter, Z., Illés, E., Janovák, L., Talyzin, A., Stability and dye inclusion of graphene oxide/polyelectrolyte layer-by-layer self-assembled films in saline, acidic and basic aqueous solutions. *Carbon*, 111, 350, 2017.
22. Danglad-Flores, J., Eftekhari, K., Skirtach, A.G., Riegler, H., Controlled deposition of nanosize and microsize particles by spin-casting. *Langmuir*, 35, 3404, 2019.
23. Hu, H., Pauly, M., Felix, O., Decher, G., Spray-assisted alignment of Layer-by-Layer assembled silver nanowires: a general approach for the preparation of highly anisotropic nano-composite films. *Nanoscale*, 9, 1307, 2017.
24. Gittleson, F.S., Hwang, D., Ryu, W.-H., Hashmi, S.M., Hwang, J., Goh, T., Taylor, A.D., Ultrathin nanotube/nanowire electrodes by spin–spray layer-by-layer assembly: A concept for transparent energy storage. *ACS Nano*, 9, 10005, 2015.
25. Zhou, Y., Cheng, M., Zhu, X., Zhang, Y., An, Q., Shi, F., A facile method for the construction of stable polymer–inorganic nanoparticle composite multilayers. *J. Mater. Chem. A*, 1, 11329, 2013.
26. Ma, W., Zhang, Y., Li, F., Kou, D., Lutkenhaus, J.L., Layer-by-layer assembly and electrochemical study of alizarin red s-based thin films. *Polymers*, 11, 165, 2019.

27. Abd El-Hady, M., Sharaf, S., Farouk, A., Highly hydrophobic and UV protective properties of cotton fabric using layer by layer self-assembly technique. *Cellulose*, 27, 1099, 2020.
28. Srivastava, S. and Kotov, N.A., Composite layer-by-layer (LBL) assembly with inorganic nanoparticles and nanowires. *Acc. Chem. Res.*, 41, 1831, 2008.
29. Saini, S., Gupta, A., Singh, N., Sheikh, J., Functionalization of linen fabric using layer by layer treatment with chitosan and green tea extract. *J. Ind. Eng. Chem.*, 82, 138, 2020.
30. Kundu, C.K., Song, L., Hu, Y., Micro crystalline cellulose aided surface modification and deposition of green polyelectrolytes for the improved hydrophilicity and flame retardancy of polyamide 66 fabric. *Int. J. Biol. Macromol.*, 254, 127610, 2024.
31. Zhang, X., Xu, Y., Zhang, X., Wu, H., Shen, J., Chen, R., Xiong, Y., Li, J., Guo, S., Progress on the layer-by-layer assembly of multilayered polymer composites: Strategy, structural control and applications. *Prog. Polym. Sci.*, 89, 76, 2019.
32. Cai, M., Yang, J., Lu, X., Lu, X., Layer-by-Layer Self-Assembly Strategies of Atomically Thin Two-Dimensional Nanomaterials: Principles, Methods, and Functional Applications. *ACS Appl. Nano Mater.*, 2024.
33. Huang, S.-H., Lin, G.-L., Gallardo, M.R., Chu, Y.-T., Wang, C.-H., Millare, J.C., Lee, K.-R., Incorporation of graphene oxide in the layer-by-layer self-assembly of polyacrylic acid and poly (diallyldimethylammonium chloride) to fabricate nanocomposite membrane for forward osmosis. *Mater. Chem. Phys.*, 315, 128989, 2024.
34. Stockton, W.B. and Rubner, M.F., Molecular layer processing of polyaniline *via* the use of hydrogen bonding interactions. *MRS Online Proc. Lib. (OPL)*, 369, 587, 1994.
35. Kang, M., Chen, S., Yang, R., Li, D., Zhang, W., Fabrication of an eco-friendly clay-based coating for enhancing flame retardant and mechanical properties of cotton fabrics *via* LbL assembly. *Polymers*, 14, 4994, 2022.
36. Maity, S., Singha, K., Pandit, P., Conductive polymer-coated wool composites for novel applications, in: *Wool Fiber Reinforced Polymer Composites*, p. 423, Woodhead Publishing, Elsevier Ltd, Sawston, Cambridge, UK, 2022.
37. Sahebalzamani, M., Ziminska, M., McCarthy, H.O., Levingstone, T.J., Dunne, N.J., Hamilton, A.R., Advancing bone tissue engineering one layer at a time: a layer-by-layer assembly approach to 3D bone scaffold materials. *Biomater. Sci.*, 10, 2734, 2022.
38. Lavalle, P., Picart, C., Mutterer, J., Gergely, C., Reiss, H., Voegel, J.-C., Senger, B., Schaaf, P., Modeling the buildup of polyelectrolyte multilayer films having exponential growth. *J. Phys. Chem. B*, 108, 635, 2004.
39. Porcel, C., Lavalle, P., Decher, G., Senger, B., Voegel, J.-C., Schaaf, P., Influence of the polyelectrolyte molecular weight on exponentially growing multilayer films in the linear regime. *Langmuir*, 23, 1898, 2007.

40. Wang, W., Xu, Y., Backes, S., Li, A., Micciulla, S., Kayitmazer, A.B., Li, L., Guo, X., von Klitzing, R., Construction of compact polyelectrolyte multilayers inspired by marine mussel: effects of salt concentration and pH as observed by QCM-D and AFM. *Langmuir*, 32, 3365, 2016.
41. Leake, K., Eberbach, T., Stensland, A., Watts, L., Yochum, H., Flexible automated system for laser modified layer by layer assembly. *Rev. Sci. Instrum.*, 95, 2024.
42. Wang, M., Song, Y.-J., Jiang, W., Fornasiero, F., Urban, J.J., Mi, B., Layer-by-layer Assembly of Nanosheets with Matching Size and Shape for More Stable Membrane Structure than Nanosheet-Polymer Assembly. *ACS Appl. Mater. Interfaces*, 16, 26568, 2024.
43. He, J., Li, R., Gu, F., Preparation of polyaniline/nylon conducting fabric by layer-by-layer assembly method. *J. Appl. Polym. Sci.*, 128, 1673, 2013.
44. Cui, Y., Zheng, G., Jiang, Z., Zhou, Y., Wang, Q., Zhou, M., Wang, P., Yu, Y., Fabrication of stretchable PEDOT: PSS coated cotton fabric *via* LBL electrostatic self-assembly and its UV protection and sensing properties. *Cellulose*, 29, 2699, 2022.
45. Gadkari, R., Ali, S.W., Joshi, M., Rajendran, S., Das, A., Alagirusamy, R., Leveraging antibacterial efficacy of silver loaded chitosan nanoparticles on layer-by-layer self-assembled coated cotton fabric. *Int. J. Biol. Macromol.*, 162, 548, 2020.
46. Kim, S.H. and Ko, H.C., Robust and Highly Conductive PEDOT: PSS: Ag Nanowires/Polyethyleneimine Multilayers Based on Ionic Layer-by-Layer Assembly for E-Textiles and 3D Electronics. *ACS Appl. Electron. Mater.*, 4, 2413, 2022.
47. Zheng, Z., Li, X., Dai, X., Du, C., Layer-by-layer Construction of Antibacterial and Anticoagulant Blood Contacting Materials. *Macromol. Biosci.*, 2400015, 2024.
48. Zhang, J., Guan, B., Zhang, Y., Hu, J., Sun, T., Dong, T., Yun, X., Development of high barrier poly (l-actic acid)/chitosan/graphene oxide flexible films for meat packaging by layer-by-layer. *Food Biosci.*, 60, 104304, 2024.
49. Malucelli, G., Nanostructured Flame-Retardant Layer-by-Layer Architectures for Cotton Fabrics: The Current State of the Art and Perspectives. *Nanomaterials*, 14, 858, 2024.
50. Gomes, A.P., Mano, J.F., Queiroz, J.A., Gouveia, I.C., Layer-by-layer assembly for biofunctionalization of cellulosic fibers with emergent antimicrobial agents, in: *Cellulose Chemistry and Properties: Fibers, Nanocelluloses and Advanced Materials*, p. 225, Springer International Publishing (Springer), Cham, Switzerland, 2016.
51. Chung, I.J., Kim, W., Jang, W., Park, H.-W., Sohn, A., Chung, K.-B., Kim, D.-W., Choi, D., Park, Y.T., Layer-by-layer assembled graphene multilayers on multidimensional surfaces for highly durable, scalable, and wearable triboelectric nanogenerators. *J. Mater. Chem. A*, 6, 3108, 2018.

52. Levitt, A., Zhang, J., Dion, G., Gogotsi, Y., Razal, J.M., MXene-based fibers, yarns, and fabrics for wearable energy storage devices. *Adv. Funct. Mater.*, 30, 2000739, 2020.
53. Eid, B.M. and Ibrahim, N.A., Recent developments in sustainable finishing of cellulosic textiles employing biotechnology. *J. Cleaner Prod.*, 284, 124701, 2021.
54. Shakir, I., Ali, Z., Bae, J., Park, J., Kang, D.J., Layer by layer assembly of ultrathin V 2 O 5 anchored MWCNTs and graphene on textile fabrics for fabrication of high energy density flexible supercapacitor electrodes. *Nanoscale*, 6, 4125, 2014.
55. Park, M.-K., Onishi, K., Locklin, J., Caruso, F., Advincula, R.C., Self-assembly and characterization of polyaniline and sulfonated polystyrene multilayer-coated colloidal particles and hollow shells. *Langmuir*, 19, 8550, 2003.
56. Zhang, L. and Sun, J., Layer-by-layer deposition of polyelectrolyte complexes for the fabrication of foam coatings with high loading capacity. *Chem. Commun.*, 3901, 2009.
57. Kim, B.-S., Park, S.W., Hammond, P.T., Hydrogen-bonding layer-by-layer-assembled biodegradable polymeric micelles as drug delivery vehicles from surfaces. *ACS Nano*, 2, 386, 2008.
58. De Villiers, M.M., Otto, D.P., Strydom, S.J., Lvov, Y.M., Introduction to nanocoatings produced by layer-by-layer (LbL) self-assembly. *Adv. Drug Delivery Rev.*, 63, 701, 2011.
59. Jang, W., Chung, I.J., Kim, J., Seo, S., Park, Y.T., Choi, K., Improving fire resistance of cotton fabric through Layer-by-Layer assembled graphene multilayer nanocoating. *J. Korean Phys. Soc.*, 72, 1052, 2018.
60. Babaeipour, S., Nousiainen, P., Kimiaei, E., Tienaho, J., Kohlhuber, N., Korpinen, R., Kaipanen, K., Österberg, M., Thin multifunctional coatings for textiles based on the layer-by-layer application of polyaromatic hybrid nanoparticles. *Mater. Adv.*, 2024.
61. Li, J., Li, X., Liu, J., Hu, B., Cao, X., Wu, Z., Li, B., Robust, breathable and antibacterial superamphiphobic cotton fabrics prepared by layer-by-layer dip coating. *Appl. Surf. Sci.*, 669, 160582, 2024.
62. Kotoulas, K.T., Campbell, J., Skirtach, A.G., Volodkin, D., Vikulina, A., Surface modification with particles coated or made of polymer multilayers. *Pharmaceutics*, 14, 2483, 2022.
63. Gang, D., Park, S., Yoo, J.M., Gu, M., Electrochemical Reforming of Methanol using Multilayer Nanoarchitectonics with Ligand Exchange-Induced Layer-by-Layer Assembly toward Electrocatalytic Hydrogen Production. *ACS Appl. Energy Mater.*, 2024.
64. Ojstršek, A., Jug, L., Plohl, O., A review of electro conductive textiles utilizing the dip-coating technique: their functionality, durability and sustainability. *Polymers*, 14, 4713, 2022.
65. Sahebalzamani, M., Hashemi, T.S., Nejad, Z.M., Agarwal, S., McCarthy, H.O., Levingstone, T.J., Dunne, N.J., Deposition of multilayer coatings onto

highly porous materials by Layer-by-Layer assembly for bone tissue engineering applications using cyclic mechanical deformation and perfusion. *Mater. Adv.*, 5, 2316, 2024.
66. Carosio, F., Cuttica, F., Di Blasio, A., Alongi, J., Malucelli, G., Layer by layer assembly of flame retardant thin films on closed cell PET foams: Efficiency of ammonium polyphosphate versus DNA. *Polym. Degrad. Stab.*, 113, 189, 2015.
67. Pan, Y., Zhan, J., Pan, H., Wang, W., Tang, G., Song, L., Hu, Y., Effect of fully biobased coatings constructed *via* layer-by-layer assembly of chitosan and lignosulfonate on the thermal, flame retardant, and mechanical properties of flexible polyurethane foam. *ACS Sustainable Chem. Eng.*, 4, 1431, 2016.
68. Narkhede, M., Thota, S., Mosurkal, R., Muller, W.S., Kumar, J., Layer-by-layer assembly of halogen-free polymeric materials on nylon/cotton blend for flame retardant applications. *Fire Mater.*, 40, 206, 2016.
69. Xiong, M., Ren, Z., Liu, W., Fabrication of UV-resistant and superhydrophobic surface on cotton fabric by functionalized polyethyleneimine/SiO 2 *via* layer-by-layer assembly and dip-coating. *Cellulose*, 26, 8951, 2019.
70. Tambunlertchai, S., Srisang, S., Nasongkla, N., Development of antimicrobial coating by later-by-layer dip coating of chlorhexidine-loaded micelles. *J. Mater. Sci. - Mater. Med.*, 28, 1, 2017.
71. Javaid, S., Mahmood, A., Nasir, H., Iqbal, M., Ahmed, N., Ahmad, N.M., Layer-by-layer self-assembled dip coating for antifouling functionalized finishing of cotton textile. *Polymers*, 14, 2540, 2022.
72. Junthip, J., Tabary, N., Chai, F., Leclercq, L., Maton, M., Cazaux, F., Neut, C., Paccou, L., Guinet, Y., Staelens, J.N., Layer-by-layer coating of textile with two oppositely charged cyclodextrin polyelectrolytes for extended drug delivery. *J. Biomed. Mater. Res. Part A*, 104, 1408, 2016.
73. Shirvan, A.R., Nejad, N.H., Bashari, A., Antibacterial finishing of cotton fabric *via* the chitosan/TPP self-assembled nano layers. *Fibers Polym.*, 15, 1908, 2014.
74. Tian, M., Du, M., Qu, L., Chen, S., Zhu, S., Han, G., Electromagnetic interference shielding cotton fabrics with high electrical conductivity and electrical heating behavior *via* layer-by-layer self-assembly route. *RSC Adv.*, 7, 42641, 2017.
75. Naveas, N., Pulido, R., Torres-Costa, V., Agulló-Rueda, F., Santibáñez, M., Malano, F., Recio-Sánchez, G., Garrido-Miranda, K.A., Manso-Silván, M., Hernández-Montelongo, J., Antibacterial films of silver nanoparticles embedded into carboxymethylcellulose/chitosan multilayers on nanoporous silicon: a layer-by-layer assembly approach comparing dip and spin coating. *Int. J. Mol. Sci.*, 24, 10595, 2023.
76. Liu, C.Y., Bittinger, S.C., Bose, A., Meyer, A., Schlicke, H., Vossmeyer, T., Tuning the Interfacial Chemistry of Nanoparticle Assemblies *via* Spin-Coating: From Single Sensors to Monolithic Sensor Arrays. *Adv. Mater. Interfaces*, 11, 2301058, 2024.

77. Grosso, D., Boissière, C., Faustini, M., Thin film deposition techniques, in: *The Sol-Gel Handbook*, p. 277, Wiley-VCH Verlag GmbH & Co. KGaA (an imprint of Wiley), Weinheim, Germany, 2015.
78. Chaudhary, K.T., Thin film deposition: Solution based approach, in: *Thin Films*, IntechOpen, London, United Kingdom, 2021.
79. Das, R. and Chanda, A., Fabrication and properties of spin-coated polymer films, in: *Nano-size polymers: Preparation, properties, applications*, p. 283, Springer International Publishing (Springer), Cham, Switzerland, 2016.
80. Mendhe, A.C., Spin Coating: Easy Technique for Thin Films, in: *Simple Chemical Methods for Thin Film Deposition: Synthesis and Applications*, p. 387, Springer, Cham, Switzerland, 2023.
81. Richardson, J.J., Björnmalm, M., Caruso, F., Technology-driven layer-by-layer assembly of nanofilms. *Science*, 348, aaa2491, 2015.
82. Shim, B.S., Podsiadlo, P., Lilly, D.G., Agarwal, A., Lee, J., Tang, Z., Ho, S., Ingle, P., Paterson, D., Lu, W., Nanostructured thin films made by dewetting method of layer-by-layer assembly. *Nano Lett.*, 7, 3266, 2007.
83. Tan, W.S., Cohen, R.E., Rubner, M.F., Sukhishvili, S.A., Temperature-induced, reversible swelling transitions in multilayers of a cationic triblock copolymer and a polyacid. *Macromolecules*, 43, 1950, 2010.
84. Keeney, M., Jiang, X., Yamane, M., Lee, M., Goodman, S., Yang, F., Nanocoating for biomolecule delivery using layer-by-layer self-assembly. *J. Mater. Chem. B*, 3, 8757, 2015.
85. Yu, L., Lee, Y.Y., Tay, F.E., Iliescu, C., Spray coating of photoresist for 3D microstructures with different geometries. *J. Phys. Conf. Ser., IOP Publishing*, 34–937, 2006.
86. Guzmán, E., Ortega, F., Rubio, R.G., Layer-by-layer materials for the fabrication of devices with electrochemical applications. *Energies*, 15, 3399, 2022.
87. Zhang, X., Ou-Yang, S., Wang, J., Liao, L., Wu, R., Wei, J., Construction of antibacterial surface *via* layer-by-layer method. *Curr. Pharm. Des.*, 24, 926, 2018.
88. Liu, X., A study of the layer-by-layer films with properties of self-healing and UV-protection, 2020.
89. Silva, A.C., Silvestre, A.J., Freire, C.S., Vilela, C., Modification of textiles for functional applications, in: *Fundamentals of natural fibres and textiles*, p. 303, Elsevier, United Kingdom, 2021.
90. Zhao, S., Caruso, F., Dähne, L., Decher, G., De Geest, B.G., Fan, J., Feliu, N., Gogotsi, Y., Hammond, P.T., Hersam, M.C., The future of layer-by-layer assembly: a tribute to ACS nano associate editor Helmuth Mohwald. *ACS Nano*, 13, 6151, 2019.
91. Fan, Q., Ma, J., Xu, Q., An, W., Qiu, R., Multifunctional coatings crafted *via* layer-by-layer spraying method. *Prog. Org. Coat.*, 125, 215, 2018.
92. Krywko-Cendrowska, A., di Leone, S., Bina, M., Yorulmaz-Avsar, S., Palivan, C.G., Meier, W., Recent advances in hybrid biomimetic polymer-based films: from assembly to applications. *Polymers*, 12, 1003, 2020.

93. Zope, I.S., Foo, S., Seah, D.G., Akunuri, A.T., Dasari, A., Development and evaluation of a water-based flame retardant spray coating for cotton fabrics. *ACS Appl. Mater. Interfaces*, 9, 40782, 2017.
94. Carosio, F., Fontaine, G., Alongi, J., Bourbigot, S., Starch-based layer by layer assembly: efficient and sustainable approach to cotton fire protection. *ACS Appl. Mater. Interfaces*, 7, 12158, 2015.
95. Mateos, A.J., Cain, A.A., Grunlan, J.C., Large-scale continuous immersion system for layer-by-layer deposition of flame retardant and conductive nanocoatings on fabric. *Ind. Eng. Chem. Res.*, 53, 6409, 2014.
96. Al-Hussein, M., Herzig, E.M., Schindler, M., Löhrer, F., Palumbiny, C.M., Wang, W., Roth, S.V., Müller-Buschbaum, P., Comparative study of the nanomorphology of spray and spin coated PTB7 polymer: Fullerene films. *Polym. Eng. Sci.*, 56, 889, 2016.
97. Roth, S., Herzog, G., Körstgens, V., Buffet, A., Schwartzkopf, M., Perlich, J., Kashem, M.A., Döhrmann, R., Gehrke, R., Rothkirch, A., *In situ* observation of cluster formation during nanoparticle solution casting on a colloidal film. *J. Phys.: Condens. Matter*, 23, 254208, 2011.
98. Tang, H., Zhang, G., Ji, S., Rapid assembly of polyelectrolyte multilayer membranes using an automatic spray system. *AlChE J.*, 59, 250, 2013.
99. Butt, M.A., Thin-film coating methods: a successful marriage of high-quality and cost-effectiveness—a brief exploration. *Coatings*, 12, 1115, 2022.
100. Mahadik, S., Pedraza, F., Mahadik, S., Comparative studies on water repellent coatings prepared by spin coating and spray coating methods. *Prog. Org. Coat.*, 104, 217, 2017.
101. Kashem, M.N.H., Liu, X., Ding, Z., Li, W., Spin-spray-assisted layer-by-layer assembly of thick polymer films with self-healing, UV-protection, and anti-fog properties. *J. Polym Sci.*, 61, 1040, 2023.
102. Kwon, S.-I., Kyung, K.-H., Park, J.-Y., Lee, I.-S., Kim, J.-H., Kim, S.-H., Shiratori, S., Uniform anti-reflective films fabricated by layer-by-layer ultrasonic spray method. *Colloids Surf. Physicochem. Eng. Asp.*, 580, 123785, 2019.
103. Carosio, F., Di Blasio, A., Cuttica, F., Alongi, J., Frache, A., Malucelli, G., Flame retardancy of polyester fabrics treated by spray-assisted layer-by-layer silica architectures. *Ind. Eng. Chem. Res.*, 52, 9544, 2013.
104. Wang, X., Zhou, Y., Johnson, M., Milne, C., Li, Y., Wang, W., Zhang, N., Xu, Q., A novel hydrophilic, antibacterial chitosan-based coating prepared by ultrasonic atomization assisted LbL assembly technique. *J. Funct. Biomater.*, 14, 43, 2023.
105. Nooralian, Z., Gashti, M.P., Ebrahimi, I., Fabrication of a multifunctional graphene/polyvinylphosphonic acid/cotton nanocomposite *via* facile spray layer-by-layer assembly. *RSC Adv.*, 6, 23288, 2016.
106. Rivero, P.J., Urrutia, A., Goicoechea, J., Arregui, F.J., Nanomaterials for functional textiles and fibers. *Nanoscale Res. Lett.*, 10, 1, 2015.
107. Ostrander, J.W., Mamedov, A.A., Kotov, N.A., Two modes of linear Layer-by-Layer growth of nanoparticle– polylectrolyte multilayers and different

interactions in the Layer-by-layer deposition. *J. Am. Chem. Soc.*, 123, 1101, 2001.

108. Badawi, N.M., Batoo, K.M., Hussain, S., Agrawal, N., Bhuyan, M., Bashir, S., Subramaniam, R., Kasi, R., Design and Characterization of Electroconductive Graphene-Coated Cotton Fabric for Wearable Electronics. *Coatings*, 13, 1601, 2023.
109. Demir, A. and Türemen, M., Structural characterization of LbL assembled multilayers by using different polyelectrolytes on cotton fabrics. *Fibers Polym.*, 18, 2298, 2017.
110. Ding, B., Kim, J., Kimura, E., Shiratori, S., Layer-by-layer structured films of TiO2 nanoparticles and poly (acrylic acid) on electrospun nanofibres. *Nanotechnology*, 15, 913, 2004.
111. Li, H., Fu, S., Peng, L., Zhan, H., Surface modification of cellulose fibers with layer-by-layer self-assembly of lignosulfonate and polyelectrolyte: effects on fibers wetting properties and paper strength. *Cellulose*, 19, 533, 2012.
112. Lee, C., Wei, X., Kysar, J.W., Hone, J., Measurement of the elastic properties and intrinsic strength of monolayer graphene. *Science*, 321, 385, 2008.
113. Zhang, X., Li, D., You, F., Li, G., Zhou, C., Cheng, D., Pan, Y., Wang, J., Ma, J., Improved thermal stability and flame retardancy levels of cotton fabrics treated by GEL/AMP layer-by-layer assembly and silica gel composite coatings. *Cellulose*, 30, 6695, 2023.
114. Uğur, SS., Sarıışık, M., Aktaş, A.H., Uçar, MC., Erden, E., Modifying of cotton fabric surface with nano-ZnO multilayer films by layer-by-layer deposition method. *Nanoscale Res. Lett.*, 5, 1204, 2010.
115. Hyde, K., Rusa, M., Hinestroza, J., Layer-by-layer deposition of polyelectrolyte nanolayers on natural fibres: cotton. *Nanotechnology*, 16, S422, 2005.
116. Asta, N., Loist, M., Reid, M.S., Wågberg, L., Model systems for clarifying the effects of surface modification on fibre–fibre joint strength and paper mechanical properties. *Cellulose*, 1, 2024.
117. Wattanatanom, W., Charuchinda, S., Potiyaraj, P., Flame behavior and mechanical properties of polyester fabrics coated with intumescent coatings *via* layer-by-layer assembly. *Text. Res. J.*, 89, 4691, 2019.
118. Li, X., Du, X., He, J., Self-cleaning antireflective coatings assembled from peculiar mesoporous silica nanoparticles. *Langmuir*, 26, 13528, 2010.
119. Sung, C. and Kang, M., Wettability studies of layer-by-layer films of Nafion/Polyethylenemine/SiO2 nanoparticles. *J. Polym. Res.*, 30, 323, 2023.
120. Bashari, A., Salehi K, A.H., Salamatipour, N., Bioinspired and green water repellent finishing of textiles using carnauba wax and layer-by-layer technique. *J. Text. Inst.*, 111, 1148, 2020.
121. Xue, C.-H., Wu, Y., Guo, X.-J., Liu, B.-Y., Wang, H.-D., Jia, S.-T., Superhydrophobic, flame-retardant and conductive cotton fabrics *via* layer-by-layer assembly of carbon nanotubes for flexible sensing electronics. *Cellulose*, 27, 3455, 2020.

122. Abeywardena, M., Yashomala, M., Elkaduwe, R., Karunaratne, D., Pitawala, H., Rajapakse, R., Manipura, A., Mantilaka, M., Fabrication of water-repellent polyester textile *via* dip-coating of *in-situ* surface-modified superhydrophobic calcium carbonate from dolomite. *Colloids Surf. A Physicochem. Eng. Asp.*, 629, 127397, 2021.
123. Iyigundogdu, Z.U., Demir, O., Asutay, A.B., Sahin, F., Developing novel antimicrobial and antiviral textile products. *Appl. Biochem. Biotechnol.*, 181, 1155, 2017.
124. Safi, K., Kant, K., Bramhecha, I., Mathur, P., Sheikh, J., Multifunctional modification of cotton using layer-by-layer finishing with chitosan, sodium lignin sulphonate and boric acid. *Int. J. Biol. Macromol.*, 158, 903, 2020.
125. Petkovska, J., Mladenovic, N., Marković, D., Radoičić, M., Vest, N.A., Palen, B., Radetić, M., Grunlan, J.C., Jordanov, I., Flame-retardant, antimicrobial, and UV-protective lignin-based multilayer nanocoating. *ACS Appl. Polym. Mater.*, 4, 4528, 2022.
126. Ali, S.W., Rajendran, S., Joshi, M., Synthesis and characterization of chitosan and silver loaded chitosan nanoparticles for bioactive polyester. *Carbohydr. Polym.*, 83, 438, 2011.
127. Li, Q., Jiang, S., Jia, W., Wang, F., Wang, Z., Cao, X., Shen, X., Yao, Z., Novel silver-modified carboxymethyl chitosan antibacterial membranes using environment-friendly polymers. *Chemosphere*, 307, 136059, 2022.
128. Nawaz, R., Naqvi, S.T.R., Fatima, B., Zulfiqar, N., Farooq, M.U., Haq, M.N.U., Hussain, D., Javeed, A., Rasul, A., Jafri, L., Cost-effective fabrication, antibacterial application and cell viability studies of modified nonwoven cotton fabric. *Sci. Rep.*, 12, 2493, 2022.
129. Carré, G., Garnier, L., Moeller-Siegert, J., Gies, J.-P., Keller, V., André, P., Keller, N., Antibacterial textiles functionalized by layer-by-layer assembly of polyelectrolytes and TiO 2 photocatalyst. *RSC Adv.*, 5, 38859, 2015.
130. Gulati, R., Sharma, S., Sharma, R.K., Antimicrobial textile: recent developments and functional perspective. *Polym. Bull.*, 79, 5747, 2022.
131. Novi, V.T., Gonzalez, A., Brockgreitens, J., Abbas, A., Highly efficient and durable antimicrobial nanocomposite textiles. *Sci. Rep.*, 12, 17332, 2022.
132. Chen, X., Ding, F., Zhang, S., Liu, Y., Hou, X., Ren, X., Flame-retardant, antibacterial and hydrophobic multifunctional coatings on cotton fabrics *via* layer-by-layer self-assembly. *Cellulose*, 30, 6679, 2023.
133. Gibbons, E., Krysmann, M., Gavalas, S., Heslop, K., Kelarakis, A., Nafion-Based Layer-by-Layer Coatings with Antimicrobial Activity. *Mater. Proc.*, 14, 10, 2023.
134. Emam, H.E., Antimicrobial cellulosic textiles based on organic compounds. *3 Biotech*, 9, 1, 2019.

# 6

# Textile Fiber Grafting: Strategies and Functionality

**Md. Reazuddin Repon[1]*, Shubhajit Dutta[1] and Shumaila Kiran[2]**

*[1]Department of Bioproducts and Biosystems, School of Chemical Engineering, Aalto University, Espoo, Finland*
*[2]Department of Applied Chemistry, Government College University, Faisalabad, Pakistan*

***Abstract***
The interest in multifunctional textiles has increased phenomenally, and diverse functional properties can be endowed through grafting polymerization. This chapter covers different grafting techniques on natural and synthetic textile fibers with functional properties like antibacterial, flameproof, hydrophobic, UV shielding, and mechanically durable. Chemical, radiation, enzymatic, UV, and plasma-assisted grafting techniques are elaborated on. Monomer type, initiation method, and processing conditions are analyzed with grafting yield and functional performance. Natural fibers such as cotton, silk, wool, and jute, as well as synthetic fibers like polypropylene, polyester, and nylon, are responsive to grafting treatments and are therefore studied. This chapter also discusses concerns in textile grafting and highlights areas for further development.

***Keywords*:** Antibacterial activity, chemical grafting, flame resistance, natural fibers, oil–water separation

## 6.1 Introduction

In recent years, the global textile industry has transformed because of the need for functional and high-performance materials. While textiles were traditionally woven and knit for strength and durability to be used for their

**Corresponding author*: reazmbstu.te@gmail.com

Aminoddin Haji and Çağlar Sivri (eds.) Surface Modification and Functionalization of Textiles, (159–174) © 2026 Scrivener Publishing LLC

look, feel, or strength, now, they require antimicrobial, flame-retardant, ultraviolet (UV), conductive, and water-resistant properties. Looking good and feeling comfortable is not enough. Meeting these demands has led to new techniques in focus surface modification processes that add special features while maintaining the inherent value of the original fibers [1–3].

Grafting polymerization is one of the best processes for altering textile fibers to bestow other functional properties. At the fiber surface, polymer chains are chemically linked to the fiber to create graft copolymers [4–6]. These copolymers graft new functions while retaining the original mechanical or structural properties of the fiber. This process is very flexible and can be used with a wide variety of materials, such as natural fibers of cotton, silk, and wool, to synthetic ones like polypropylene (PP), rayon, and nylon [7, 8]. Consequently, grafting is utilized in textiles: in clothing, home textiles, technical textiles, and medical textiles, among others [9–11]. To perform grafting, several procedures have been developed, including chemical grafting, radiation-induced techniques, grafting by enzymes, UV grafting, plasma treatments, and light-induced processes. All of these techniques have varying degrees of grafting effectiveness in terms of ecological impact and grafting with different kinds of fibers. Modifications done through grafting, in comparison to conventional surface coating techniques, are tougher and are more resistant to washing [4, 12–14].

This chapter focuses on grafting modifications of textile fibers and covers the different groups of fibers that can be altered through grafting, such as natural fibers and synthetic fibers. This chapter looks at the different and numerous functional properties that can be integrated through grafting, such as antibacterial properties, hydrophobic and oleophobic properties, flame retardancy, and UV protection. This chapter is important because it looks at how different conditions of grafting, such as the type of monomer, the method of initiation, or reaction time, influence the properties of the textile. The chapter also covers the challenges that still need to be tackled and the future perspectives in textile grafting.

## 6.2 Grafting Strategies and Functionality Analysis of Textile Fibers

### 6.2.1 Grafting Strategies and Functionality of Natural Fibers

Natural fibers are garnering more focus because they are renewable, biodegradable, low-cost, and eco-friendly [15, 16]. However, they are not as versatile as other materials due to heating instability, low strength, and high moisture absorption. However, these challenges can be addressed using grafting modification, emerging as a viable solution [17]. The grafting methods of natural fibers and their functional properties are included in Table 6.1.

The study focuses on developing multifunctional properties, such as superhydrophobicity, antibacterial activity, and enhanced flame resistance, to cotton fabric with 3-aminopropyltriethoxysilane (APTES) [1]. With a focus on concentration and reaction time, it was found that optimal conditions to 14% APTES for 10 hours. With these parameters, APTES modified the cotton fabric such that the treatment provided a water contact angle of 155.52°, indicative of superhydrophobicity. The treated cotton fabric not only self-cleaned exceptionally well but also demonstrated high efficiency in oil–water separation, achieving over 93.33% separation for varying oils. Thermally, the modified fabric demonstrated far better burn and charcoal formation than untreated cotton, which enhanced its flame-retardant properties. The grafted cotton also has strong antibacterial effects against *E. coli* and *S. aureus*.

In a grafting-from approach described by Zhao, Okada, and Sugimoto, methyl methacrylate (MMA) is graft-polymerized onto cotton fibers using a Phen DEZ initiator, which is a diethylzinc–1,10 phenanthroline complex in an aqueous emulsion system at room temperature. This results in high graft yields and ultra-high molecular weight PMMA chains (>$10^6$ Da) being attached to cotton cellulose [18]. Using this organozinc initiator made grafting easy, solvent-free, and possible at room temperature, which got around a lot of the problems that come with most radical-initiated methods. The modified cotton was less hydrophilic and worked better with organic solvents like MMA.

Arioli *et al.* come up with a way to covalently attach glycine-based polyamidoamine oligomers (MGLY) to cotton fibers using a water-based, radical graft polymerization method [19]. First, MGLY was made by adding N,N′-methylenebisacrylamide and glycine together, which made oligomers with two acrylamide ends. Then, aqueous MGLY solutions (10 or 20 wt%) with potassium persulfate as an initiator were soaked into cotton strips. When persulfate broke down in heat, it made radicals that oxidized cellulose. This made active sites that started the polymerization of MGLY terminal acrylamide groups and caused covalent grafting at levels of 2%–15%. The grafted cotton did not hold up to heat and oxidation as well as cotton that was only adsorbed with MGLY, but it did burn more slowly and was more stable when exposed to UV light. Researchers look into grafting hydroxyethyl methacrylate (HEMA) onto cotton fabrics using a radical graft copolymerization method in another study [20]. Cotton cellulose

**Table 6.1** Summary of textile fibers used, grafting agents, grafting methods, and their functional properties developed during the modifications.

| Textile materials | Grafting agent | Grafting method | Functional properties | Ref. |
|---|---|---|---|---|
| Cotton | 3-Aminopropyl trimethoxysilane | Chemical grafting | Antibacterial activity, wettability, flame resistance, oil–water separation | [1] |
| Cotton | Methyl methacrylate | Chemical grafting | Decreased hydrophilicity, increased compatibility | [18] |
| Cotton | Acrylamide | UV-induced grafting | Increased hydrophilicity, swelling behavior, and electrical conductivity | [33] |
| Cotton | Polyamidoamine oligomer | Chemical grafting | UV resistance, flame resistance | [19] |
| Cotton | Hydroxyethyl methacrylate | Chemical grafting | Durable finishing, wrinkle recovery | [20] |
| Cotton | Methyl methacrylate | Radiation-induced grafting | Improved strength | [24] |
| Cotton | Fatty acid | Microwave plasma-assisted grafting | Excellent water repellency | [21] |
| Cotton | Fatty acid | Microwave plasma-assisted grafting | Hydrophobicity | [22] |
| Cotton | Urethane acrylate | Light-induced grafting | Water repellency | [34] |
| Silk | Methyl methacrylate | Radiation-induced grafting | Improved strength | [24] |

*(Continued)*

**Table 6.1** Summary of textile fibers used, grafting agents, grafting methods, and their functional properties developed during the modifications. (*Continued*)

| Textile materials | Grafting agent | Grafting method | Functional properties | Ref. |
|---|---|---|---|---|
| Jute | Hydroxyethyl methacrylate | UV-induced grafting | Durability | [23] |
| Silk | Glycidyl methacrylate | Chemical grafting | UV resistance, water repellency | [25] |
| *Bombyx mori* silk | Methyl methacrylate | Chemical grafting | Improved dye uptake, thermal stability | [26] |
| Silk | Polyethyleneimine | Enzyme-mediated grafting | Antibacterial activity | [27] |
| Kosa silk | Vinyl benzyl trimethylammonium chloride | Chemical grafting | Antibacterial activity | [28] |
| Silk | Hydroxyethyl methacrylate | Radiation-induced grafting | Improved whiteness, mechanical resilience, and strength | [29] |
| Silk | Hydroxyethyl methacrylate | Radiation-induced grafting | Enhanced mechanical stability | [30] |
| Wool | Phenolic monomer | Enzyme-mediated grafting | Improved dye uptake, color variation | [31] |
| Wool | 2-hydroxyethyl methacrylate | Chemical grafting | Improved shrink resistance, washing, and rubbing fastness | [32] |
| Polypropylene | Methacrylated tannic acid | Photo-induced grafting | Antioxidant and antibacterial | [35] |
| Polypropylene | Phenolic monomer | Enzyme-induced grafting | Increased dyeability, hydrophilicity | [36] |

(*Continued*)

**Table 6.1** Summary of textile fibers used, grafting agents, grafting methods, and their functional properties developed during the modifications. (*Continued*)

| Textile materials | Grafting agent | Grafting method | Functional properties | Ref. |
|---|---|---|---|---|
| Polypropylene | Acrylamide | UV-induced grafting | Increased hydrophilicity, swelling behavior, and electrical conductivity | [37] |
| Poly (ethylene terephthalate) | Gaseous Cl2 | UV-induced grafting | Improved dye uptake | [38] |
| Nylon | - | UV-induced grafting | Antibacterial activity | [8] |
| Polyester | - | UV-induced grafting | Antibacterial activity | [8] |
| Polyester | Chitosan | Radiation-induced grafting | Antibacterial activity | [39] |
| Polyester | Chitosan and collagen | Radiation-induced grafting | Antibacterial activity | [40] |
| Poly (ethylene terephthalate | PEGDA and METAC | UV-induced grafting | Increased moisture absorption | [41] |
| Rayon | Methyl methacrylate | Radiation-induced grafting | Improved strength | [24] |
| Rayon | Ceric ammonium nitrate | Chemical grafting | Improved dye uptake | [42] |

fibers are first put in contact with water-based HEMA solutions. Then, under controlled conditions, chemical radicals start the graft polymerization process. After the grafting, the fabrics are treated with polycarboxylic acids to give them a long-lasting press finish. Cotton-grafted fabrics are better at keeping their wrinkle recovery and mechanical strength than fabrics that are not grafted.

In a method that does not use solvents, cotton fabric is put in low-pressure microwave plasma with gases like $N_2$, $O_2$, or Ar. This generates reactive sites on the fiber surfaces [21]. These active centers make it possible to covalently graft other fatty acids, such as lauric or stearic acid, onto cellulose, which gives it new functions. The results showed that hydrophobic modification and alterations to the surface morphology worked, making materials that are hydrophobic and strong at the same time. Another work on cotton fabric employing microwave plasma-assisted graft copolymerization revealed the same patterns of behavior [22].

A UV-induced graft copolymerization method used to change natural jute yarn by starting polymer grafting right under UV light [23]. The grafting agent is HEMA, which is introduced to a solution to soak into the yarn. Then, UV light is used to start radical grafting on the fiber surface. The method gets a high grafting yield, frequently more than 80%, which makes jute yarns stronger. The approach works well for making long-lasting changes to the surface of jute fibers through UV-activated radical polymerization.

Using γ-radiation–induced graft copolymerization, MMA was added to cotton, rayon, and silk yarns to improve their characteristics [24]. A cobalt-60 gamma source and methanol (MeOH) as a swelling solvent were used for the grafting. The best grafting yield came from a total dose of 500 krad and a dosing rate of 200 krad/h. MMA was the grafting agent, and γ-rays were used to graft the two things at the same time. Silk had the highest graft yield (35%), followed by rayon (22%) and cotton (17%). The study also looked at how other additives, like N-vinylpyrrolidone, tripropylene glycol diacrylate, trimethylolpropane triacrylate (TMPTA), sulfuric acid, lithium nitrate, and urea, affected the amount of grafting. Some additions, like urea and lithium nitrate, made graft yields much higher, notably in silk. Others, like TMPTA, had steric hindrance effects.

A study shows how to use *in-situ* graft copolymerization to improve the functional performance of silk fabric [25]. Ascorbic acid and hydrogen peroxide were utilized to start the process, and glycidyl methacrylate (GMA) was used as a grafting monomer to change silk. The conditions were changed to make it more UV-resistant and resistant to aging. The modified silk's whiteness index went from 26.71% to 88.23% at the best grafting rate of 8.8%. Its breaking strength went from 174.6 N to 211.5 N, and it became more water-repellent, with a water contact angle over 90°, compared to untreated silk. This study shows that this eco-friendly chemical change can make silk materials far more resistant to UV light, more durable, and more water-resistant. Using a diethylzinc–1,10 phenanthroline complex (Phen DEZ) as an initiator, another study reports a new one-step emulsion grafting of MMA onto *Bombyx mori* silk fibers. The authors were able to get graft yields of more than 200% under the best conditions by changing the temperature of the reaction and the ratio of initiator to monomer [26]. The modified silk had far better thermal stability, with decomposition temperatures starting around 40°C higher than those of untreated fibers, according to thermogravimetric analysis. Tests showed that the MMA grafted silk not only took up bright colors, but also held onto them better after washing. This occurred because PMMA filled in the empty spaces and made the fibers freer.

Researchers have come up with a way to use enzymes to graft silk materials that have fast and long-lasting antibacterial properties [27]. The process uses the enzyme laccase to catalyze the enzymatic graft copolymerization of p-hydroxyphenylacetic acid-modified polyethyleneimine (mPEI) onto silk fibroin. This efficiently binds mPEI to tyrosine residues on the surface of the fabric. After grafting, silver nanoparticles are deposited *in situ* within the mPEI network to make an antibacterial layer. The fabric that comes out has a surface charge that changes a lot (zeta potential goes from −32 mV to +21.7 mV) and a silver content of roughly 1.81%. The treated silk kills 99.9% of *Escherichia coli* and *Staphylococcus aureus* within 30 minutes of contact. The activity stays over 99.9% even after 18 hours, including after 10 wash cycles. Another study says that Kosa silk fibers can be changed by graft copolymerization with vinyl benzyl trimethylammonium chloride (VBT) as the grafting agent. This makes the silk surfaces antibacterial [28]. The silk fiber is treated first, and then it is put into a solution of VBT monomer and other chemicals, including thiourea, in a controlled environment to speed up the grafting reaction. The amount of grafts that grow increases with the amount of thiourea, but only up to a certain point (around $10.0 \times 10^{-3}$ M). Characterization shows that cationic polymer chains are covalently attached to silk, which makes it more effective against microbes.

Researchers looked into how electron beam irradiation could cause HEMA to polymerize onto silk fabric [29]. They looked at two methods: pre-irradiation (where silk is first irradiated and then HEMA is added) and co-irradiation (where both silk and monomer are irradiated at the same time). When grafted silk fabrics were tested, they had different attributes than untreated samples, such as better whiteness, mechanical strength, and breaking strength. Another study looked into using electron-beam irradiation to graft methacrylamide (MAA) onto silk fibroin. This is known as radiation-induced grafting/crosslinking [30]. Silk is exposed to a high-energy electron beam, which creates reactive radicals on the silk surface and in the environment where the monomers are. This allows crosslinking and graft copolymer synthesis to happen at the same time. Methacrylamide sticks to activated silk surfaces to make them more stable and change the structure of the polymer.

Using laccase to polymerize and graft a range of plant-based phenolic monomers (including gallic acid, syringic acid, catechin, catechol, and ferulic acid) into polymeric dye materials is an enzyme-catalyzed grafting method that has been used on wool fibers [31]. The oxidative coupling that laccase helps with lets these phenols make both homo- and hetero-polymers, which then color yak wool. The phenolic chemicals and their binary combinations reacted to make a wide range of stable hair-coloring polymers. For example, gallic/syringic acids, catechin/catechol, and ferulic/syringic acids made brown, black, and red colors, respectively. Using a chemically induced free radical technique with benzoyl peroxide (BPO), 2-HEMA was grafted onto wool yarns in another study [32]. We looked at how temperature, duration, monomer concentration, and initiator concentration affected grafting yield to find the best circumstances for grafting. The best conditions were 50°C, 3 hours, 0.25 M HEMA, and 0.06 M BPO, which gave a yield of 55%. The grafting made the wool less likely to swell, but it made it a little more resistant to shrinking and washing, and rubbing without changing the way it took up dye.

### 6.2.2 Grafting Strategies and Functionality of Synthetic Fibers

Synthetic fibers are very strong, last a long time and do not break down in nature. But their surfaces are generally hydrophobic, which makes them less useful, especially in situations where better dyeability, adhesion, and wettability are needed. Grafting modification is a good way to change the surface properties of synthetic fibers by adding functional monomers or polymers. Table 6.1 shows the different ways that synthetic fibers can be grafted and what they can do.

Researchers have discovered that chemically adding tannic acid (TA) to PP fabrics in two steps gives them high antibacterial and antioxidant characteristics [35]. The first step is to change the chemical structure of TA by adding GMA to make methacrylated tannic acid (TA_acr). In the second phase, photoinduced grafting under UV light (using a photoinitiator DMPA) covalently attaches TA_acr to both raw PP and PP that has already been coated with chitosan. This makes the bond very strong and adds weight, even after washing with ethanol. The textiles made from this process have a high antioxidant activity (RSA ~ 95%) and are very effective against *E. coli* and *S. aureus*, notably the chitosan-coated PP grafted with TA_acr. Figure 6.1 shows a diagram of the two-step procedure used to functionalize PP fabrics with methacrylated TA. A different study talks about using laccase to covalently functionalize PP surfaces through enzymatic grafting [36]. First, the PP substrates are treated with plasma to create

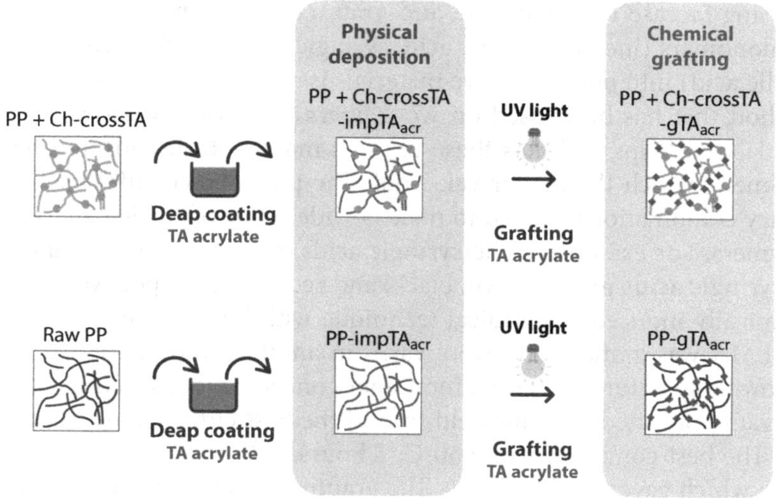


**Figure 6.1** Schematic illustration of the two-step process used to functionalize polypropylene fabrics with methacrylated tannic acid [35].

reactive surface functional groups and make the polymer more wettable. Then, the enzyme laccase speeds up oxidative coupling processes with phenolic monomers (also known as functional phenolic grafting agents). This makes it possible for covalent connections to form between the phenols and the activated PP surface. The technique changes the surface in a way that lasts without using harsh chemicals or high temperatures. It also makes functionalized PP with better qualities, like being more dyeable and water-loving. A different study used a process called radiation-induced post-irradiation grafting to attach acrylamide (AAm) to PP films. First, PP films are treated to gamma radiation (pre-irradiation stage) to make reactive radical sites. Then, they are put in an acrylamide solution in water under nitrogen at controlled temperatures to help graft copolymerization. The study looks at how the pre-irradiation dose (kGy), monomer concentration, reaction time, solvent, and inhibitor (Mohr's salt) concentration affect the yield of grafting. Under the best conditions, like a moderate dosage of around 40 kGy, 25 wt% acrylamide, and about 35°C, the degree of grafting reaches several percent, making PP g PAAm films. Compared to untreated PP, the changed films are more hydrophilic, swell more, and conduct electricity better [37]. Another investigation also demonstrated that cotton acted the same way [33].

When gaseous chlorine is present, poly (ethylene terephthalate) (PET) textiles are exposed to UVC irradiation (≈254 nm). This causes UV-assisted photo grafting (chlorination) by adding chlorine atoms to the PET surface (>4 wt%) without changing the characteristics of the bulk material [38]. The results verified that the surface had been changed and that chlorine had been added. The dyeing experiments also showed that the material had a much stronger affinity for cationic dyes. Another study looked at the photo-initiated radical grafting of two vinyl monomers—poly(ethylene glycol) diacrylate (PEGDA) and [2-methacryloyloxy ethyl]trimethylammonium chloride (METAC)—onto PET woven fabric to make it more water-loving and better at handling moisture [41]. When exposed to UV light, the activated surface allows the monomers to bond covalently, changing the wettability and lowering the water contact angle without weakening the fabric. The results show that the grafting layers are even, that they can absorb more moisture, and that they can spread moisture more easily. Also, $^{60}Co$ $\gamma$ rays were used to pre-irradiate the polyester fibers, which made reactive radical sites on their surfaces. The fibers were then graft copolymerized using acrylic acid (AAc) and N-vinylformamide (NVF) after being exposed to radiation [39]. The grafts containing NVF were broken down even further by water to make primary amino groups, and then they were chemically linked to chitosan by ester to make surfaces that kill bacteria. The materials that had chitosan added to them were very good at killing *E. coli* and *Pseudomonas aeruginosa*. In a different study, collagen (COL) was stuck to the chitosan-grafted surfaces to make them more biocompatible and help cells stick to them [40]. The changes made to the PET fibers made them very effective at killing *Staphylococcus aureus*, *Escherichia coli*, and *Pseudomonas aeruginosa*.

It was found in a study that changed rayon fibers by grafting them with acrylonitrile (AN) using two methods: a chemical redox method that used ceric ammonium nitrate (CAN) in nitric acid ($HNO_3$) and a gamma radiation-induced mutual grafting method [42]. The chemical approach got a maximum grafting yield of 160.01% at 45°C in 120 minutes by employing the best amounts of CAN, $HNO_3$, and AN in 20 mL of water. The radiation approach, on the other hand, got a maximum Pg of 90.24% at room temperature with a dose of 3.456 kGy/h and 10 mL of water. When put in solvents such as water, methanol, ethanol, DMF, and acetone, the grafted fibers swelled differently from ungrafted rayon.

## 6.3 Current Challenges of the Fiber Grafting Strategy

i. Some grafting procedures need a lot of steps, very specific reaction conditions, or sophisticated tools like radiation sources or plasma reactors. These things make processing prices go up.

ii. High-energy grafting methods that use radiation or powerful chemical initiators can break the fiber or make it weaker. Adding new features without making the fiber weaker is hard.
iii. It is hard to get a high grafting efficiency without making homopolymers or changing the surface unevenly. It is important to be able to control the grafting density, chain length, and surface dispersion, but it is not easy to do so.
iv. Some grafting techniques involve hazardous monomers, initiators, and solvents, which are bad for the environment and safety. For industrial uses, waste treatment and emissions control are quite significant.
v. Grafting can make things last longer than surface coatings, but some grafted functions still wear out over time because of washing and abrasion. It is still hard to improve long-term stability when things are used in the actual world.

## 6.4 Future Perspectives of Fiber Grafting Strategy

i. New grafting technologies that are better for the environment, such as enzyme-based reactions, water-based systems, and bio-based monomers, could be used. These strategies try to keep effectiveness while lowering toxicity and harm to the environment.
ii. For improved control over grafting design, advanced methods such as controlled radical polymerization could be used. This can assist in making surfaces that have certain, predictable qualities and are more consistent.
iii. Grafting is likely to be a key part of making smart fabrics that react to things like temperature, pH, wetness, or electric impulses. Using grafting with nanoparticles or responsive polymers could make new things like medical textiles, sensors, and electronics that you can wear.

## 6.5 Conclusion

Grafting modification is a useful and flexible way to improve the functional qualities of textile fibers without changing their natural features. Many different grafting methods have been used on both natural and synthetic fibers to provide them with important qualities like resistance to microbes, water, and oil, as well as better mechanical durability. Even though there have been a lot of improvements, it is still hard to move grafting procedures from the lab to the factory since they need to be cost-effective and environmentally friendly. Additionally, getting consistent grafting with exact architectural control and long-term stability in real-world situations is still a technical challenge. The development of eco-friendly grafting methods, including enzyme-mediated procedures and the use of bio-based monomers, together with advances in controlled polymerization and nanotechnology, could lead to the creation of smart textiles for the next generation.

**AI disclosure statement:** The author acknowledges the use of AI-based tools for minor language editing or grammatical correction. However, all intellectual contributions, analyses, and conclusions presented in this chapter are entirely the author's own.

## References


1. Alassod, A., *et al.*, Functionalization of aminoalkylsilane-grafted cotton for antibacterial, thermal, and wettability properties. *RSC Adv.*, 12, 32, 20906–20918, 2022.
2. Zhang, Y.Y., *et al.*, Recent advances in the surface modification strategies to improve functional finishing of cotton with natural colourants-A review. *J. Cleaner Prod.*, 335, 2022.
3. Ghasemian, R., Barani, H., Khazaei, F., Surface modification and performance of wool fibers after combined plasma and enzyme treatments. *Sci. Rep.*, 15, 1, 2025.
4. Bhattacharya, A. and Misra, B.N., Grafting: a versatile means to modify polymers - Techniques, factors and applications. *Prog. Polym. Sci.*, 29, 8, 767–814, 2004.
5. Slagman, S., Zuilhof, H., Franssen, M.C.R., Laccase-Mediated Grafting on Biopolymers and Synthetic Polymers: A Critical Review. *Chembiochem*, 19, 4, 288–311, 2018.
6. Wei, L.Q. and McDonald, A.G., A Review on Grafting of Biofibers for Biocomposites. *Materials*, 9, 4, 2016.
7. Shavandi, A. and Ali, M.A., Graft polymerization onto wool fibre for improved functionality. *Prog. Org. Coat.*, 130, 182–199, 2019.
8. Michael, J.K. and Zhengmao, Z., Practical Anti-Microbial Surfaces on Nylon and Polyester by UV Photochemistry, in: *Conference on Lasers and Electro-Optics (CLEO)*, IEEE, Baltimore, MD, USA, 2007.
9. Silva, A.C.Q., *et al.*, Modification of textiles for functional applications, in: *Fundamentals of Natural Fibres and Textiles*, M.I.H. Mondal (Ed.), pp. 303–365, Woodhead Publishing, Duxford, United Kingdom, 2021.
10. Meshram, M.W., *et al.*, Graft copolymers of starch and its application in textiles. *Carbohydr. Polym.*, 75, 1, 71–78, 2009.
11. Princi, E., *et al.*, Grafting polymerization on cellulose based textiles: A13 C solid state NMR characterization. *Eur. Polym. J.*, 41, 6, 1196–1203, 2005.

12. Abidi, N., Surface grafting of textiles, in: *Surface Modification of Textiles*, Q. Wei (Ed.), pp. 91–107, Woodhead Publishing, Cambridge, United Kingdom, 2009.
13. Solpan, D., Torun, M., Güven, O., Comparison of pre-irradiation and mutual grafting of 2-chloroacrylonitrile on cellulose by gamma-irradiation. *Radiat. Phys. Chem.*, 79, 3, 250–254, 2010.
14. Sangermano, M. and N.R., Light induced grafting-from strategies as powerfull tool for surface modification. *eXPRESS Polym. Lett.*, 13, 2, 135–145, 2019.
15. Liang, Z.H., *et al.*, Multi-functional silk fibers/fabrics with a negligible impact on comfortable and wearability properties for fiber bulk. *Chem. Eng. J.*, 415, 2021.
16. Dev, B., *et al.*, Mapping the progress in natural fiber reinforced composites: Preparation, mechanical properties, and applications. *Polym. Compos.*, 44, 7, 3748–3788, 2023.
17. Debashish, R., *et al.*, Cellulose modification by polymer grafting: a review. *Chem. Soc. Rev.*, 38, 7, 2046–2064, 2009.
18. Zhao, C., Okada, H., Sugimoto, R., Chemical modification of cotton by methyl methacrylate *via* emulsion polymerization. *Polym. J.*, 51, 10, 1023–1028, 2019.
19. Arioli, M., *et al.*, Enhancing Cotton Fabrics Through Grafting of Glycine-Based Polyamidoamine. *Polymers*, 17, 12, 2025.
20. Xu, W.L. and Shyr, T.W., Durable press finishing of cotton fabrics by polycarboxylic acids after graft copolymerization with hydroxyethyl methacrylate. *Text. Res. J.*, 70, 1, 8–10, 2000.
21. Abidi, N. and Hequet, E., *Cotton fabric graft copolymerization using microwave plasma. I. Universal attenuated total reflectance-FTIR study. J. Appl. Polym. Sci.*, 93, 1, 145–154, 2004.
22. Abidi, N. and Hequet, E., *Cotton fabric graft copolymerization using microwave plasma. II. Physical properties. J. Appl. Polym. Sci.*, 98, 2, 896–902, 2005.
23. Uddin, M.K., Khan, M.A., Ali, K.M.I., Modification of jute yarn by graft-copolymerization with ultraviolet radiation. *Radiat. Phys. Chem.*, 48, 4, 511–517, 1996.
24. Bashar, A.S., Khan, M.A., Ali, K.M.I., Modification of Cotton, Rayon and Silk Fibers by Radiation-Induced Graft Co-Polymerization. *Radiat. Phys. Chem.*, 45, 5, 753–759, 1995.
25. Ahmed, S., *et al.*, Performance analysis of functional properties of silk fabric enhanced *via in situ* grafting modification. *Polym. Eng. Sci.*, 65, 3, 1474–1482, 2025.
26. Zhao, C., Okada, H., Asakawa, M., Sugimoto, R., Yoshino, K., Grafting Methyl Methacrylate onto Silk *via* Emulsion Graft Copolymerization Using a Diethylzinc Complex Initiator. *Green Sustainable Chem.*, 9, 4, 135–142, 2019.

27. Wang, F.Y., *et al.*, Rapid Antibacterial Effects of Silk Fabric Constructed through Enzymatic Grafting of Modified PEI and AgNP Deposition. *ACS Appl. Mater. Interfaces*, 13, 28, 33505–33515, 2021.
28. Verma, A., *et al.*, Surface Modification of Kosa Silk Fiber by Graft Copolymerization with Vinyl Benzyl Trimethylammonium Chloride (VBT) and Its Antibacterial Study. *Russ. J. Appl. Chem.*, 97, 5, 515–525, 2024.
29. Shao, J.Z., Liu, J.Q., Chen, Z.L., Grafting of silk with electron beam irradiation. *Color. Technol.*, 117, 4, 229–233, 2001.
30. Liu, J.Q., Shao, J.Z., Zheng, J.H., Radiation grafting/crosslinking of silk using electron-beam irradiation. *J. Appl. Polym. Sci.*, 91, 3, 2028–2034, 2004.
31. Jeon, J.R., *et al.*, Laccase-catalysed polymeric dye synthesis from plant-derived phenols for potential application in hair dyeing: Enzymatic colourations driven by homo- or hetero-polymer synthesis. *Microb. Biotechnol.*, 3, 3, 324–335, 2010.
32. Nuralin, F. and Pulat, M., Grafting of 2-Hydroxy Ethyl Methacrylate onto Wool Yarns to Improve Their Shrink-Resistance and Fastness Properties. *Tekst. ve Konfeksiyon*, 28, 1, 27–33, 2018.
33. Shukla, S.R. and Athalye, A.R., Ultraviolet-Radiation Induced Graft-Copolymerization of Hydroxyethyl Methacrylate onto Cotton Cellulose. *J. Appl. Polym. Sci.*, 44, 3, 435–442, 1992.
34. Ferrero, F., *et al.*, Water-repellent finishing of cotton fabrics by ultraviolet curing. *J. Appl. Polym. Sci.*, 107, 2, 810–818, 2008.
35. Fouilloux, J., *et al.*, Chemical Modification of Commercial Fabrics by Photoinduced Grafting Tannic Acid to Produce Antioxidant and Antibacterial Textiles. *Sustainability*, 16, 11, 2024.
36. Schroeder, M., *et al.*, Laccase-Induced Grafting on Plasma-Pretreated Polypropylene. *Biomacromolecules*, 9, 10, 2735–2741, 2008.
37. Shukla, S.R. and Athalye, A.R., Ultraviolet Radiation-Induced Graft-Copolymerization of 2-Hydroxyethyl Methacrylate onto Polypropylene. *J. Appl. Polym. Sci.*, 51, 9, 1567–1574, 1994.
38. Kordoghli, B., *et al.*, UV irradiation-assisted grafting of poly(ethylene terephthalate) fabrics. *Colloids Surf., A*, 441, 606–613, 2014.
39. Hu, S.G., Jou, C.H., Yang, M.C., Surface grafting of polyester fiber with chitosan and the antibacterial activity of pathogenic bacteria. *J. Appl. Polym. Sci.*, 86, 12, 2977–2983, 2002.
40. Jou, C.H., *et al.*, Biofunctional properties of polyester fibers grafted with chitosan and collagen. *Polym. Adv. Technol.*, 18, 3, 235–239, 2007.
41. Bongiovanni, R., *et al.*, Moisture and Surface Properties of Radically Photo-Grafted Poly-(Ethylene Terephthalate) Woven Fabric. *Fibers*, 12, 10, 2024.
42. Inderjeet, K., Neelam, S., Vandna, K., Modification of fiber properties through grafting of acrylonitrile to rayon by chemical and radiation methods. *J. Adv. Res.*, 4, 6, 547–557, 2013.

7

# Importance of Surface Treatment with UV Radiation in Textiles

**Ayşegül Körlü[1] and Mohammad Khajeh Mehrizi[2]***

[1]*Department of Textile Engineering, Faculty of Engineering, Ege University, Izmir, Turkey*
[2]*Department of Textile Engineering, Yazd University, Yazd, Iran*

***Abstract***
Surface treatment using ultraviolet (UV) radiation significantly enhances textile properties. UV rays are part of the electromagnetic spectrum and have wavelengths shorter than visible light, categorized into three types: UV-A (320–400 nm), UV-B (280–320 nm), and UV-C (200–280 nm), with UV-C having the highest energy. This treatment improves water repellency, dye absorption, and durability while reducing reliance on harsh chemicals, making it a sustainable choice. Additionally, UV radiation boosts color fastness, ensuring that textiles stay vibrant after washing. This method also lowers energy and water usage in textile treatments. Moreover, UV treatment enhances production efficiency, offering manufacturers cost savings and consistent quality. As technology advances, UV radiation's role in textile production will likely expand, highlighting its importance. Furthermore, it allows for creative aesthetic effects, enabling unique designs that are difficult to achieve with traditional methods. In the dynamic world of textile manufacturing, UV treatment continues to influence sustainability, efficiency, and creativity.

***Keywords:*** Eco-friendly, surface modification, UV radiation, functional properties, textiles, finishing processes

## 7.1 Introduction

In today's fast-paced world, the importance of innovative methods in textile finishing is paramount. With growing concerns about sustainability and

**Corresponding author*: mkhajeh@yazd.ac.ir

Aminoddin Haji and Çağlar Sivri (eds.) Surface Modification and Functionalization of Textiles, (175–190) © 2026 Scrivener Publishing LLC

environmental impact, the textile industry must adopt practices that reduce its ecological footprint. One such method is the use of ultraviolet (UV) light in textile finishing, which not only improves fabric surface properties but also provides several environmental advantages. By utilizing UV light, manufacturers can significantly reduce the need for water and harmful chemicals traditionally used in finishing processes. This shift not only saves vital resources but also minimizes the release of toxic substances into the environment, aligning with the industry's growing commitment to sustainability [1].

Moreover, UV finishing can be applied to a wide array of fabrics, including natural fibers like cotton and synthetic materials such as polyester. This versatility allows for the development of innovative products that cater to the evolving demands of consumers who prioritize both performance and ecological responsibility. As brands increasingly seek to differentiate themselves in a competitive market, the adoption of UV technology can serve as a key selling point, appealing to environmentally conscious consumers.

In addition to its ecological benefits, UV light finishing can also enhance textiles' aesthetic qualities. Fabrics treated with UV light often exhibit improved brightness and clarity of colors, resulting in visually striking products that stand out in the marketplace. This combination of performance and visual appeal positions UV-treated textiles as a desirable choice for both manufacturers and consumers alike [2].

As the textile industry continues to change, embracing innovative methods such as UV light finishing will be essential for meeting the challenges of sustainability and environmental surveillance. By investing in these advanced techniques, manufacturers not only can reduce their ecological footmark but also can contribute to a more sustainable future for the industry as a whole. UV light in the textile industry effectively modifies the surface properties of fabrics [3–5].

This chapter discusses UV technology, its applications, contributions to sustainable textile production, and its advantages in production efficiency. It also covers the impact of UV technology on fabric properties, surface modification before coloration, the UV curing process, and pilling prevention, etc.

## 7.2 Understanding UV Radiation and Applicational Zones of UV Treatment

Electromagnetic waves are vibrations that propagate rapidly through space. The electromagnetic spectrum classifies these waves by their frequency and wavelength. The lowest frequency and longest wavelength are radio waves, whereas gamma rays represent the highest frequency and shortest wavelength. UV rays have a higher frequency than violet light, positioned between visible light and X-rays, with wavelengths longer than X-rays but shorter than visible rays. UV radiation accounts for 5% of the total sunlight reaching the Earth's surface, compared to 50% for visible light and 45% for infrared radiation. Despite its small proportion, UV radiation possesses the highest quantum energy.

UV radiation is classified into three types based on wavelength:

1. UV-A (315–400 nm): The most common type reaching the Earth's surface, UV-A rays penetrate deeply into the skin, contributing to aging and certain skin cancers.
2. UV-B (280–315 nm): Responsible for sunburn and significant in skin cancer development, UV-B rays are more intense than UV-A but less prevalent.
3. UV-C (200–280 nm): The most harmful type, UV-C radiation is primarily absorbed by the Earth's atmosphere and does not reach the surface. In addition to their varying wavelengths and effects, the different types of UV radiation play distinct roles in human health and the environment [6].

While UV-A rays are abundant and often associated with long-term damage, they are also known to generate free radicals, which can lead to oxidative stress and skin cell damage. This sustained exposure can manifest not only as skin aging but also as an increased risk of melanoma, the deadliest form of skin cancer.

UV-B rays, on the other hand, are crucial for the synthesis of vitamin D in the skin, a vital nutrient for bone health and immune function. However, their ability to cause direct DNA damage elevates their role as a significant risk factor in the development of skin cancers, particularly in fair-skinned individuals who are more susceptible to the effects of sunlight.

UV-C radiation, while the most potent in terms of damage potential, is largely filtered out by the ozone layer, which protects life on Earth from its adverse effects. However, human activities, such as the use of specific artificial sources like germicidal lamps, can expose individuals to UV-C radiation, reinforcing the need for proper safety precautions.

Understanding the characteristics of each type of UV radiation underscores the importance of protective measures such as wearing sunscreen, protective clothing and seeking shade, particularly during peak sunlight hours. By taking these precautions, individuals can better safeguard their skin and overall health against the damaging effects of UV exposure.

Due to its unique properties, UV technology is used in various fields. Some prominent usage areas are disinfection and sterilization, medical applications, industrial applications, horticulture and agriculture, cosmetics and skin care, and photography and imaging [7].

UV light serves a critical role in the disinfection of both drinking water and wastewater by inactivating a wide variety of harmful microorganisms, including bacteria, viruses, and other pathogens that can pose significant health risks. In healthcare settings such as hospitals and laboratories, UV-C lamps are utilized extensively to sterilize surfaces, medical equipment, and various tools to ensure a hygienic and safe environment. Furthermore, manufacturers of food products also employ UV technology to maintain sanitation standards that protect public health. In addition to its applications in healthcare, this technology is integrated into HVAC systems for residential and commercial buildings to combat airborne pathogens, which helps improve overall indoor air quality and creates a healthier living environment for occupants.

Beyond disinfection, UV light plays an important role in dermatology, where it is used to treat various skin conditions, including psoriasis, eczema, and vitiligo, promoting healing and relief for patients dealing with these chronic issues. For surgical procedures, the effectiveness of UV radiation in sterilizing surgical instruments and other medical devices cannot be overstated, as it ensures that these tools are free from harmful microorganisms before any procedure is performed. Moreover, in manufacturing processes, UV light is harnessed to quickly cure and harden inks, paints, and coatings, enhancing productivity and efficiency in these industries.

In the realm of personal care, UV radiation is used in tanning beds, where it induces skin tanning by stimulating melanin production, allowing individuals to achieve a desired skin tone. Additionally, this technology is notable in specialized photography, where UV light captures images that reveal details and features that are not visible under regular lighting conditions. These various applications collectively highlight the remarkable versatility and significance of UV technology across a multitude of sectors, showcasing its effectiveness not only in disinfection but also in industrial processes, agriculture, healthcare, and even consumer products, making it an invaluable tool in modern society.

## 7.3 Effects of UV Radiation on Fibers and Fabric Properties

Textiles, such as apparel and curtains, are typically thought of as protective barriers against harmful UV rays from the sun. However, UV radiation also plays a crucial role in altering the surface properties of fabrics, serving as a commercial alternative to traditional methods without altering the main characteristics of the material. During this modification, high-density surface fibers act as a shield, preventing UV rays from penetrating and protecting the underlying fibers that provide strength. For optimal results, these surface fibers must be directly exposed to UV rays, or effective photoinitiators should be used to generate reactive free radicals. The photoinitiators applied in textiles are generally odorless, non-toxic, cheap, and easily washable. Consequently, aromatic photoinitiators are commonly preferred for most UV applications. UV exposure can lead to the noticeable fading of dyes and natural colors in fibers, making them look washed out over time [8].

Prolonged UV exposure negatively impacts textiles in several ways: it weakens fiber integrity, reduces tensile strength, and increases breakage susceptibility. UV radiation can make fibers, particularly synthetic ones, brittle and less flexible while also causing surface defects like roughness and pitting. Additionally, it degrades both natural and synthetic fibers, leading to a loss of mechanical properties and diminishing elasticity. Damaged surfaces from UV exposure can attract dirt and grime, affecting aesthetic qualities. Moreover, prolonged exposure to UV light can cause dyes in fabrics to fade and change their colors [9–11].

Nevertheless, UV surface modification offers a sustainable and environmentally friendly alternative to conventional chemical processes, largely because it eliminates the need for harmful chemicals and does not generate toxic by-products, which can be detrimental to both health and the environment. When UV light is utilized within the textile industry, it demonstrates the ability to effectively change and enhance the surface properties of various fabrics, leading to potential improvements in their performance and functionality. This innovative approach not only reduces the ecological footprint of fabric treatment but also opens new avenues for developing advanced textile materials that meet modern market demands.

Ratna *et al.* pointed out that the application of UV treatment can effectively generate active sites on the surfaces of fabrics. This process enhances the fabrics' ability to react with a variety of dye agents, leading to a significant improvement in the overall dyeing process. As a result, this augmentation facilitates the production of more vibrant and intense colors, while also providing greater depth in color achieved with specific types of dye agents [12].

Zhang *et al.* conducted a comprehensive study that demonstrated how UV radiation has a beneficial impact on the colorfastness of cotton fabrics that have been dyed using reactive dyes. Their findings suggest that the application of UV processing may function as a highly effective pre-treatment step prior to the actual dyeing process, potentially enhancing the overall quality and longevity of the dye on the fabric [13].

Baran *et al.* studied the impact of UV absorption on the photocatalytic degradation of cationic and anionic dyes in TiO2 suspensions. Using spectrophotometry, they found that cationic dyes exhibited increased adsorption and faster photodegradation rates compared to anionic dyes under UV illumination. Furthermore, a linear relationship was observed between dye solution absorbance and the photodegradation rate constant [14].

In another research, Hong studied effect of UV irradiation on color and functionality of wool and cotton fabrics finished with spent coffee extract. This study focuses on using UV irradiation to enhance the coloration and functionality of wool and cotton fabrics finished with spent coffee extract. The results display that UV irradiation can meaningfully increase the color strength, as well as enrich their antimicrobial and antioxidant properties [15].

Khajeh Mehrizi *et al.* focused their research on polyester fabric dyed with the natural dye derived from Prangos ferulacea, documenting that the maximum color strength was achieved after subjecting the fabric to two hours of UV exposure at a temperature of 130°C. This treatment resulted in colorfastness that ranged from good to excellent for wash and rub tests. These results underline the intricate balance between achieving desirable dye characteristics and the durability of the treated fabrics under various conditions [16].

Altering the fiber surface significantly boosts dye absorption, enhancing uptake and fixation rates even at lower application temperatures. This improvement leads to better wettability, as shown in studies by Bhatti *et al.* Additionally, these modifications facilitate the implementation of various pattern designs. Test results indicate that UV-irradiated fabric has greater color strength than non-irradiated fabric, with maximum strength observed after 120 minutes of UV exposure, likely due to cellulose oxidation from UV light [17].

Periolatto and colleagues conducted a study in which they improved the acid dyeability of wool fibers by employing UV irradiation as a pretreatment method. They experimented with different radiation doses during the process, which allowed them to identify optimal conditions for enhancing color uptake. As a result of their work, the treated wool achieved satisfactory levels of dye fastness, maintaining its color when subjected to washing and rubbing tests. These results were evaluated according to established ISO standards, demonstrating the effectiveness of their approach in modifying wool's dyeing properties [11].

Jang and Jeong conducted a detailed study on the surface modification processes applied to PET and PTT fibers, employing a combination of UV light and ozone treatment. This innovative approach resulted in a significant alteration of the fibers' surface properties. This change in surface chemistry led to an enhanced ability to absorb dyes, particularly cationic dyes, due to the increased electrostatic interactions between the newly charged surface and the positively charged dye molecules. Such modifications finally progress the dyeability of PET and PTT fibers, showcasing the effectiveness of the UV/$O_3$ treatment method in textile applications [18].

Zuber *et al.* investigated the modification of cellulosic fibers *via* UV irradiation. The fibers were pretreated and then subjected to either conventional mercerization or UV irradiation. Mercerized fibers demonstrated better tear and tensile strength than both untreated and UV-irradiated fibers, highlighting the detrimental effect of UV radiation on mechanical properties. Furthermore, there was no weight loss in the cellulose fabric after UV exposure [19].

Eren *et al.* studied the influence of UV Light on the physical properties of polyester-blended fabrics. The result showed that the potential of UV technology can enhance both bending rigidity and hydrophilicity of the fabric [20].

In the research conducted by Karim *et al.* in 2014, a comprehensive investigation was carried out focusing on the application of UV-curable inkjet printing techniques on poly(lactic acid) fabrics. During this innovative printing process, the specific wavelength of UV radiation that was employed was 200 nm. The study included a detailed comparison between two different types of printed PLA fabrics: those that were thermally hardened through conventional inkjet printing methods and those that were treated with the UV-curable inkjet printing approach. Upon analyzing the results, it was concluded that the PLA fabrics which underwent UV-curable inkjet printing exhibited superior color fastness, demonstrating greater resistance to fading and degradation. The findings indicated that using UV-curable inkjet printing for PLA fabrics resulted in a significant enhancement in both aesthetic quality and functional behavior [21].

Mansuri *et al.* studied the simultaneous bleaching and anti-pilling of cotton fabric by applying UV radiation to hydrogen peroxide-treated knitted samples. They evaluated the impact of a photoinitiator, sodium hydroxide as an activator, sodium silicate for $H_2O_2$ stabilization, and irradiation time on whiteness, yellowness, pilling rate, and strength loss. The results show that adequate whiteness and acceptable strength loss can be achieved using UV-irradiated pad-steam processes [22].

The study focused on investigating the application of radical and cationic UV curing techniques to improve the water and oil repellency properties in various types of fabrics, specifically cotton, polyester, and polyamide. To track the progress of polymerization during the curing process, the researchers measured the weight increase of the treated fabrics over time. In addition, the effectiveness of the water and oil repellency was evaluated through multiple methods. Their evaluations provided a comprehensive understanding of how the UV curing methods impacted the performance characteristics of the fabrics in terms of their resistance to both water and oil [23].

The integration of ozone treatment and UV technology is a widely adopted method for the effective treatment of wastewater generated by the textile industry. This combination not only serves the primary purpose of purifying textile wastewater but also has the potential to promote the creation of various compounds, including UV-hydrogen peroxide and chlorinated compounds. These additional reactions lead to a range of byproducts that may have significant implications for the treatment process itself [8]. Khajeh Mehrizi and Azadi investigated the decoloration of direct dyes using multiwalled carbon nanotubes and hydrogen peroxide under UV radiation. Their findings revealed that carbon nanotubes, due to their porous structure and large surface area, are effective adsorbents for removing various organic pollutants [24]. Malakootian *et al.* studied the decolorization of Acid Red 18 textile dye using a hybrid UV/catalytic ozonation process (COP) with ZnO as a catalyst. This UV/COP hybrid advanced oxidation process demonstrated high efficiency in removing resistant dye contaminants from textile wastewater [25].

In 2015, Rahmatinejad and colleagues showed the application of UV light combined with ozone as an environmentally friendly method for modifying wool surfaces. Their study revealed that the dual-sided UV/ozone irradiation process exhibited a slightly superior performance compared to the single-sided irradiation method. This finding was used in both the pad-dry-cure process and the print-dry-cure fluorocarbon finishing techniques. The results indicate a clear advantage in utilizing the dual-sided approach for improving the characteristics of wool through these eco-conscious treatments [26].

## 7.4 Future Directions and Challenges

The changing world and life perspectives force the academy and industries to study innovative subjects, use new technologies, and produce functional products. Today, the world is progressing to a quite different future because of developing technologies, the problems caused by new technologies and solutions to the problems. The textile industry also takes its share of the transformation. The concept of textiles is different from one's concept before 20-30 years ago. Future textiles are being developed to perform specialized functions for both clothing and technical uses. Here are some key features they aim to have [27]:

- Eco-Friendly Production and Disposal: They should be made and disposed of in ways that are not harmful to the environment.
- Durability: They should resist aging and wear.
- Maintenance and Self-Cleaning: They should be easy to clean, take care of, and able to clean themselves.
- Aesthetics and Comfort: They should look good, feel lovely to touch, and be easy to shape.
- Protection: They should protect against harmful environmental factors such as UV rays, electromagnetic radiation, microorganisms, high temperatures, and chemicals.

These special functions come from choosing suitable fibers and modifying traditional textile production processes. The goal is to use materials that adapt and respond positively to changing conditions [27]. Currently, much effort is being put into fabrics that can generate, store, and provide energy to the wearer for use in daily activities [28]. Interest has already concentrated on products with high added value that knowledge-intensive technologies have aided. UV technology, which is a different technique from conventional textile production technologies, can meet the needs mentioned above. Therefore, the use of UV technology in the textile industry is a popular research subject. By conducting a comprehensive search on the Scopus database using keywords such as "textile" and "UV" or "curing" or "surface modification," excluding "wastewater treatment," we aimed to gauge the popularity of UV technology in textile research. The search yielded 1746 references between 2014 and 2024, with the highest number of publications recorded in 2022 (Figure 7.1). Despite a slight decrease in 2023, the number of publications on the subject had already reached 157 before the end of 2024, indicating a potential surpassing of 2023's figures. These references are invaluable for machine designers and UV technology users.

Research on UV technology in the textile industry has gained significant traction due to its potential to address the challenges of our time and

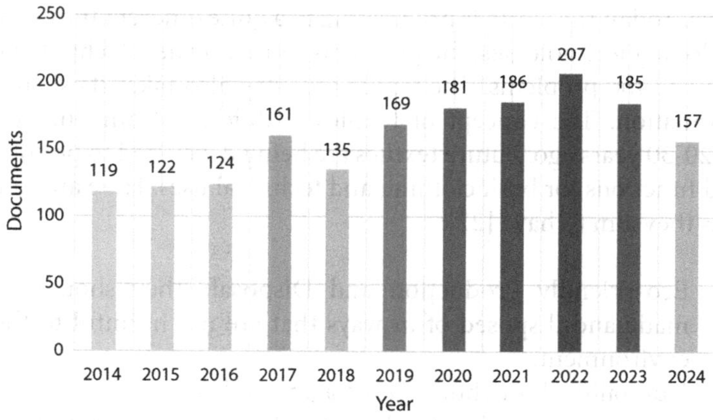


**Figure 7.1** Number of publications about textile and UV technology from 2014 to 2024 (Scopus with terms' textile and UV or curing or surface modification, out of wastewater treatment) [29].

ensure the industry's relevance in the modern era. Traditional methods consume a lot of energy and chemicals. Instead, newer methods using radiation have shown promise. These methods use less energy and fewer chemicals, reduce waste, are easy to operate, and work quickly [30, 31]. Panda *et al.* combine desizing, scouring and bleaching in one step with the aid of UV-C radiation. The process took 25 minutes, and this method used approximately 71% water, 73%, and 70% energy compared to traditional techniques. According to the life cycle analysis, the innovative one-step pretreatment reduces over 60% of environmental impacts, promoting sustainability. However, UV-C light has a harmful nature to humans [32].

The wavelength of UV has an important effect on human health. If the wavelength is short, then it will be more dangerous for humans than longer ones. Therefore, it is crucial to have a safe distance between people and light sources, which varies depending on the type of lamp and the lamp's wattage rating [33]. Moreover, UV lights can generate ozone during the process [30]. It is an oxidizing agent that can harm, particularly ventilation necessary when using strong UV-B and UV-C light [34]. On the other side, ozone can cause yellowing and a reduction in material strength. If professionals consciously design the system, then the risks will be reduced [35]. In many workplaces, UV light sources are within an enclosure, making them safe workers' conditions. However, accidents can happen. Although most artificial UVR sources have light output, they are used on and off, and workers might sometimes be in different positions

**Table 7.1** Precautions to minimize health risks of indoor UVR exposure can be met in the textile industry and research* [36].

| Source | Potential for overexposure | Hazard description | Safety precautions |
|---|---|---|---|
| Germicidal lamps for sterilization and disinfection | High | UV-C lamps sterilize work areas in hospitals, the food industry, and laboratories. | Engineering controls for reducing exposure |
| Photocuring, photohardening and etching | Medium | UV lamps in cabinets can emit radiation through openings that may exceed safe exposure limits. | Engineering controls to close openings and training |
| "Black lights" used in non-destructive testing (NDT) | Medium to low | UV-A lamps used in NDT in banking, commerce, materials inspection, and entertainment.<br>"Black lights" in insect control and entertainment are usually below exposure limits. | Engineering controls to reduce eye exposure, using PPE<br>No precautions are needed for insect control and entertainment lamps, or lower power NDT lamps. |
| UV lasers | High | Intense UV radiation at a single wavelength can exceed exposure limits in seconds for both direct and stray light. | Administrative and engineering controls, PPE, and training |
| General lighting | Low | Most lamps emit minimal UVR, and high-intensity lamps have protective coverings to absorb any UVR. However, if this covering is damaged, then overexposure to UVR can occur. | No precautions are needed under normal conditions.<br>Caution should be taken if the protective envelope is broken or cracked. |

*Overexposure varies in distance and duration of exposure. This table provides general guidance and is not exhaustive.

relative to the source. Health risks from UVR exposure depend on the main factors:

- the spectral characteristics and biologically effective UVR emissions of the source,
- the distance and orientation of personnel relative to the source, and
- the duration of exposure, taking into account skin type.

Implementing and reinforcing stringent safety measures is crucial in mitigating the risks associated with accidental UVR exposure. It is essential to ensure personnel adherence to safety protocols. Users must be trained on UV protection, and warning signs must be available at the workplace entrances [36]. Users must wear suitable personal protective equipment (PPE) like protective glasses, gloves, and garments. Table 7.1 shows health risks and precautions to minimize health risks from indoor UVR exposure.

On the other side, robots and dark factories can be a solution to the human health problem. For example, the interest in using service robots has increased, especially with the 2019 pandemic. It has led to the research and development of autonomous UV germicidal irradiation robots. Many research institutes and companies are working on and testing these UV disinfection robots. Moreover, there are commercial [37]. Therefore, in manufacturing, robots and intelligent autonomous systems are not so far today. It will be possible to use UV technology in the textile with the aid of intelligent autonomous systems in the not-so-distant future. The developments can solve some environmental problems. However, UV technology with intelligent autonomous systems can cause social challenges like reduction of employment.

## 7.5 Conclusion

The industry's evolution from ancient practices to modern techniques highlights its remarkable adaptability and innovation. As one of the largest consumer goods industries globally, textiles have an immense economic impact and provide vital employment opportunities. However, the industry's water usage and environmental footprint necessitate more sustainable practices.

UV treatment is a practical, cost-effective, and efficient method to modify the surface of polymers without making direct contact. It involves fewer steps and can serve multiple purposes, like curing photo-sensitive

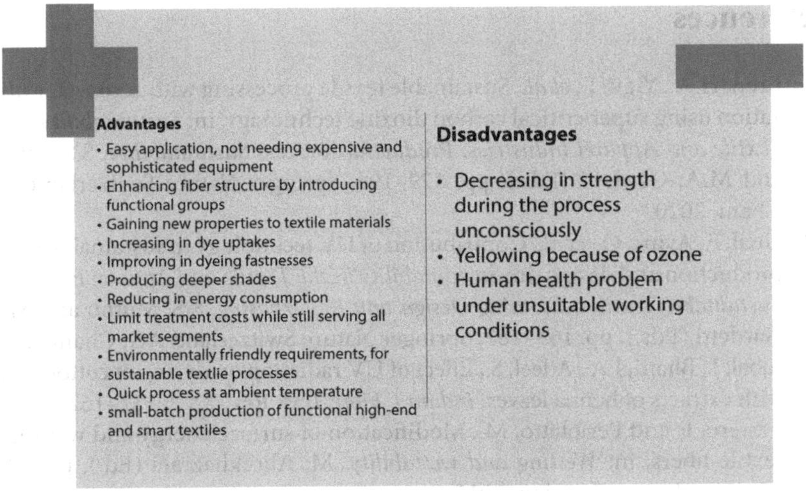


**Figure 7.2** Advantages and disadvantages of UV technology in textile industry [39–42].

polymers, increasing surface adhesiveness, and adding functional groups to the surface [38]. By reducing pollution and water consumption, UV technology supports the creation of eco-friendly and energy-efficient textile products (Figure 7.2). UV technology facilitates a variety of functional applications, such as improving functionalities and enabling the creation of materials with advanced protective features. Moving toward a future where intelligent textile autonomous systems become more, the integration of technology will undoubtedly play a crucial role.

Nevertheless, alongside its benefits, UV technology has health risks to workers and potential social challenges, such as unemployment due to automation. Therefore, it is imperative to implement robust safety measures and rigorously promote the development of that efficiency with human and environmental well-being; technology advances the textile industry, offering innovative ways for future developments. As the industry continues to evolve, the focus must remain on fostering sustainable practices and addressing the socio-economic impacts of advancements. As a result, it is estimated that UV technology in the textile industry will become common because of changing world conditions.

**AI disclosure statement:** The author acknowledges the use of AI-based tools for minor language editing or grammatical correction. However, all intellectual contributions, analyses, and conclusions presented in this chapter are entirely the author's own.

## References


1. Eren, H.A., Yiğit, I., *et al.*, Sustainable textile processing with zero water utilization using super critical carbon dioxide technology, in: *Sustainability in the Textile and Apparel Industries: Production Process Sustainability*, S.S. Muthu and M.A. Gardetti (Eds.), pp. 179–196, Springer Nature Switzerland AG, Cham, 2020.
2. Unal, F., Avinc, O., *et al.*, Contribution of UV technology to sustainable textile production and design, in: *Sustainability in the Textile and Apparel Industries: Sustainable Textiles, Clothing Design and Repurposing*, S.S. Muthu and M.A. Gardetti (Eds.), pp. 163–187, Springer Nature Switzerland AG, Cham, 2020.
3. Iqbal, J., Bhatti, I.A., Adeel, S., Effect of UV radiation on dyeing of cotton fabric with extracts of henna leaves. *Indian J. Fibre Text. Res.*, 32, 2, 157–162, 2008.
4. Ferrero, F. and Periolatto, M., Modification of surface energy and wetting of textile fibers, in: *Wetting and wettability*, M. Aliofkhazraei (Ed.), pp. 139–168, Intechopen, London, 2015.
5. Mohammed, M., Rahman, R., *et al.*, Surface treatment to improve water repellence and compatibility of natural fiber with polymer matrix: Recent advancement. *Polym. Test.*, 115, 107707, 2022.
6. Ansari, J.A., Ismail, M., Farida, M., Investigate the efficacy of UV pretreatment on thermal inactivation of Bacillus subtilis spores in different types of milk. *Innov. Food Sci. Emerg. Technol.*, 52, 387–393, 2019.
7. Ramos, C.C.R., Roque, J.L.A., *et al.*, Use of ultraviolet-C in environmental sterilization in hospitals: A systematic review on efficacy and safety. *Int. J. Health Sci. (Qassim)*, 14, 6, 52–65, 2020.
8. Perincek, S.D., Duran, K., E. Körlü, A., Bahtiyari, MI., Ultraviolet Technology. *Tekst. ve Konfeksiyon*, 17, 4, 219, 2007.
9. Zhu, Z. and Kelley, M.J., Poly (ethylene terephthalate) surface modification by deep UV (172 nm) irradiation. *Appl. Surf. Sci.*, 236, 1-4, 416, 2004.
10. Periolatto, M., Ferrero, F., Vineis, C., Rombaldoni, F., Multifunctional finishing of wool fabrics by chitosan UV-grafting: An approach. *Carbohydr. Polym.*, 98, 1, 624–629, 2013.
11. Periolatto, M., Ferrero, F., Migliavacca, G., Low temperature dyeing of wool fabric by acid dye after UV irradiation. *J. Text. Inst.*, 105, 10, 1058–1064, 2014.
12. Ratna, S.M., Bansal, A., Jain, S.K., Influence of UV radiation treatments on the dyeing performance of textiles. *Indian J. Fibre Text. Res.*, 45, 2, 223, 2020.
13. Zhang, F., Xu, W., Liu, G., Effects of UV irradiation on color fastness of reactive dyed cotton fabrics. *Int. J. Clothing Sci. Technol.*, 30, 3, 451, 2018.
14. Baran, W., Makowski, A., Wardas., W., The effect of UV radiation absorption of cationic and anionic dye solutions on their photocatalytic degradation in the presence $TiO_2$. *Dyes Pigm.*, 76, 1, 226–230, 2008.
15. Hong, K.H., Effect of UV irradiation on color and functionality of wool and cotton fabrics finished with spent coffee extract. *Cellulose*, 31, 5931–5946, 2024.

16. Khajeh Mehrizi, M., Jokar, M., Shahi, Z., The coloration of polyester fabric with Prangos ferulacea natural dye: new approaches toward a cleaner production. *Pigm. Resin Technol.*, 52, 3, 377–382, 2023.
17. Bhatti, I.A., Adeel, S., Abbas, M., Effect of radiation on textile dyeing, in: *Textile dyeing*, P.J. Hauser (Ed.), pp. 1–17, In Tech, New York, 2011.
18. Jang, J. and Jeong, Y., Nano roughening of PET and PTT fabrics *via* continuous UV/$O_3$ irradiation. *Dyes Pigm.*, 69, 3, 137–143, 2006.
19. Zuber, M., Ziaa, K.M., *et al.*, Modification of cellulosic fibers by UV-irradiation. Part II: After treatments effects. *Int. J. Biol. Macromol.*, 51, 5, 743–748, 2012.
20. Eren, S., Aşıkoğlu, M., Özenç, A.A., Influence of UV Light on the Physical Properties of Polyester-Blended Fabrics. *3 rd International Conference on Scientific and Academic Research*, Konya, Turkey, pp. 218–221, 2023.
21. Karim, N.M., Rigout, M., Yeates, S.G., Carr, C., Analysis of UV-cured and thermally-cured inkjet printed poly (lactic acid) fabrics. *NIP 30/Digital Fabrication Conference*, Philadelphia, USA, 2014.
22. Mansuri, A.H., Khoddami, A., Rezaei, D.A., Using Ultraviolet Radiation for the Bleaching and Pilling Reduction of Knitted Cotton Fabric. *Fibres Text. East. Eur*, 19, 3, 74–77, 2011.
23. El-Molla, M.M., Synthesis of polyurethane acrylate oligomers as aqueous UV-curable binder for inks of ink jet in textile printing and pigment dyeing. *Dyes Pigm.*, 74, 2, 371–379, 2007.
24. Khajeh Mehrizi, M. and Azadi, S., Direct Dye Removal by Using UV/Hydrogen Peroxide/Multiwalled Carbon Nanotubes. *Fibers Polym.*, 18, 1962–1968, 2017.
25. Malakootian, M., Smith Jr., A., *et al.*, Decoloration of textile Acid Red 18 dye by hybrid UV/COP advanced oxidation process using ZnO as a catalyst immobilized on a stone surface. *Desalin. Water Treat.*, 182, 385–394, 2020.
26. Rahmatinejad, J., Khoddami, A., Avinc, O., Innovative hybrid fluorocarbon coating on UV/ ozone surface modified wool substrate. *Fibers Polym.*, 16, 2416–2425, 2015.
27. Militký, J., Venkataraman, M., Periyasamy, A.P., Recent Trends in Textile Structures, in: *Fibrous Structures and Their Impact on Textile Design*, J. Militký, M. Venkataraman, A.P. Periyasamy (Eds.), pp. 15–77, Springer Nature Singapore Pte Ltd., Gateway East, 2023.
28. Shah, M.A., Pirzada, B.M., *et al.*, Applications of nanotechnology in smart textile industry: A critical review. *J. Adv. Res.*, 38, 55–75, 2022.
29. Scopus, https://www.scopus.com/home.uri, 2024.
30. Abou Elmaaty, T., Okubayashi, S., Elsisi, H., Electron beam irradiation treatment of textiles materials: a review. *J. Polym. Res.*, 29, 117, 2022.
31. Ferrero, F., Periolatto, M., Tempesting, L., Water and Oil Repellent Finishing of Textiles by UV Curing: Evaluation of the Influence of Scaled-Up Process Parameters. *Coatings*, 7, 5, 60, 2017.
32. Panda, S.K.B.C., Sen, K., Mukhopadhyay, S., A sustainable low temperature combined pretreatment process for cotton assisted by ultraviolet-C radiation. *Sustainable Mater. Technol.*, 37, e00664, 2023.

33. Anubrata, M., Biswajit, G., Suvro, M., Himadri, B., Some Studies on Content of Ultraviolet Rays in Conventional Light Sources vs LED. *J. Microw. Eng. Technol.*, 10, 1, 2023.
34. Ultraviolet Radiation, Environment, Health & Safety, https://ehs.lbl.gov/resource/documents/radiation-protection/non-ionizing-radiation/ultraviolet-radiation/, 2024.
35. Claus, H., Ozone Generation by Ultraviolet Lamps. *Photochem. Photobiol.*, 97, 3, 471–476, 2021.
36. ICNIRP Statement on Protection of Workers Against Ultraviolet Radiation, https://www.icnirp.org/cms/upload/publications/ICNIRPUVWorkersHP.pdf, 2024.
37. Mehta, I., Hsueh, H.Y., *et al.*, UV Disinfection Robots: A Review. *Rob. Auton. Syst.*, 161, 104332, 2023.
38. Nemani, S.K., *et al.*, Surface Modification of Polymers: Methods and Applications. *Adv. Mater. Interfaces*, 5, 1801247, 2018.
39. Islam, S.U. and Mohammad, F., High-Energy Radiation Induced Sustainable Coloration and Functional Finishing of Textile Materials. *Ind. Eng. Chem. Res.*, 54, 15, 3727–3745, 2015.
40. Gashti, M.P., *et al.*, Electromagnetic shielding response of UV-induced polypyrrole/silver coated wool. *Fibers Polym.*, 16, 585–592, 2015.
41. Ferrero, F., Migliavacca, G., Periolatto, M., UV Treatments on Cotton Fibers, in: *Cotton Research*, I.Y. Abdurakhmonov (Ed.), pp. 233–255, IntechOpen, London, 2016.
42. Seipel, S., Junchun, Y., *et al.*, Inkjet printing and UV-LED curing of photochromic dyes for functional and smart textile applications. *RSC Adv.*, 50, 28395, 2018.

8

# Laser Treatment of Textile Materials

**Mohammad Khajeh Mehrizi[1]*, Umut Kıvanç Şahin[2], Zahra Shahi[1] and İsmet Ege Kalkan[2]**

[1]*Department of Textile Engineering, Yazd University, Yazd, Iran*
[2]*Department of Textile Engineering, Istanbul Technical University, Istanbul, Turkey*

***Abstract***
In the last decade, progress has been made in surface treatments for textiles with laser systems that this technology has advantages, such as speed, precision, and versatility. Laser is a form of electromagnetic radiation that is increasingly favored in textiles for their efficiency and simplicity. Laser treatment modify textiles surfaces without the use of toxic chemicals or water. It is used for removing indigo dye, heating threads, and creating unique textures. This chapter explores the various applications of lasers and shows how this technology revolutionizes textile modification. Laser operations paving the way for a more sustainable future in textile production.

***Keywords*:** Laser, textile, surface modification, sustainability, treatment, radiation, electromagnetic

## 8.1 Introduction to Laser Technology in Textiles

### 8.1.1 Overview of Laser Technology

LASER is light amplification caused by the stimulated emission of radiation. Atoms elevated to higher energy levels emit laser light through a process called "stimulated emission" [1–3]. Laser beams can focus on small spots, producing highly intense light [4, 5]. Laser irradiation can induce physical and chemical changes on the surfaces of polymers [2]. Lasers are an effective method for the surface treatment of materials without

**Corresponding author*: mkhajeh@yazd.ac.ir

Aminoddin Haji and Çağlar Sivri (eds.) Surface Modification and Functionalization of Textiles, (191–202) © 2026 Scrivener Publishing LLC

water [6]. Synthetic textiles are compatible with laser treatment, enabling marking, engraving, and welding owing to their thermoplastic nature. By contrast, organic fabrics do not melt under laser heating [7].

### 8.1.2 Historical Background

Quantum electronics began with the development of stimulated radiation of ultrahigh frequency electromagnetic oscillations, referred to as the "Molecular generator" by N. Basov and A. Prochorov, and "MASER" (Microwave Amplification by Stimulated Emission of Radiation) by American physicists C. Towns, J. Gordon, and H. Zinger. The laser era began 6 years later (1960) when Theodore H. Maiman demonstrated the red fluorescence of Ruby crystals under green-light illumination. He created a device that called LASER (Light Amplification by Stimulated Emission of Radiation) [8].

### 8.1.3 Importance of Laser Treatment in the Textile Industry

Laser technology in the textile industry applied for removing indigo dye from denim, pattern creation for altered dyeability, surface roughening, and textile cutting [6]. In a study by Kan, the effects of laser radiation on polyester materials were investigated. The results showed significant improvements in properties, such as wettability and air permeability, and negative effects including reduced tensile strength, increased yarn abrasion, and altered bending properties. Laser irradiation led to the formation of ripple structures on the fiber surfaces that resulted in increased air space between the fibers, and the enhancement of air permeability. The ripple structure enhances the surface roughness, leading to greater unwettability and longer wetting times. The improved bending properties of laser-irradiated yarns can decrease fabric flexibility and elastic recovery, potentially impacting tailorability, draping quality, and wear. In addition, the reduction in tensile strength and elongation under high-fluence laser irradiation may be due to the deposition of debris on the polyester surface [2]. In a study by Ondogan, successful results were achieved in designing fabrics, such as drapery, 100% cotton denim, and leather, by controlling the laser beam power. Unlike sanding and stoning methods, laser textile designs do not wear down or deform fabric textures [5]. A study by Nourbakhsh found that plasma discharge and laser treatment enhanced the wettability of polyamide 66 and poly (ethylene terephthalate) fibers. Laser-treated fibers develop a uniform ripple-like structure that traps the soil, complicating washing. In contrast, plasma treatment produces an uneven surface that facilitates soil removal [6]. In a comprehensive study conducted by Hassan, copper metal particles were effectively deposited onto a knitted cotton fabric that had undergone treatment with a $CO_2$ laser *via* a process known as electroless plating. This innovative $CO_2$ laser treatment enhanced the surface roughness of the fabric, leading to significant improvements in the durability and overall performance of the metal-coated substrate. As a result of this treatment, the laser-treated fabric samples demonstrated not only enhanced electrical conductivity but also exhibited greater resistance to wear and tear when compared to fabric samples that did not undergo laser treatment. This indicates that the incorporation of $CO_2$ laser technology into the treatment process plays a crucial role in optimizing the functional properties of cotton fabrics [9]. In a study by Shahidi, the *in situ* synthesis of copper nanoparticles (40–100 nm) *via* pulsed laser ablation of a cotton fabric was investigated. This is a simple and cost-effective technique, with no by-products. Antibacterial activity tests have demonstrated a 100% reduction in bacterial survival in all treated samples [10]. In a study by Esteves, 100% polyamide 6.6 and 100% polyester fabrics were treated with a $CO_2$ pulsed laser and dyed with different commercial dyes. Laser treatment roughens the fiber surface, increases the overall surface area, and enhances dye adsorption [1].

## 8.2 Principles of Laser Treatment

### 8.2.1 Basic Principles of Laser Operation

The laser surface design machine features a wavelength-specific laser source, galvanometer with a mirror, and computer program that controls the mirrors and laser power [5]. The geometry and dimensions of laser-produced features can be precisely adjusted through process parameters, such as laser intensity, pulse duration and shape, scanning speed, and laser shot overlap. The intensity and duration of the laser beam affect the textured volume on the surface, which typically ranges from a few nanometers to micrometers [11, 12].

### 8.2.2 Types of Lasers Used in Textile Treatment

While extensive research exists on the UV laser treatment of synthetic fibers such as polyamide and polyester, less is known about the $CO_2$ laser treatment of these polymers [1]. UV lasers are more suitable for polymers because most organic materials adsorb UV radiation and cause photochemical reactions in the surface layers without damaging the bulk material [12]. Infrared lasers, such as $CO_2$, are the most powerful and appear to be advantageous because of their large beam size, high efficiency, easy operation, use of non-toxic gases, and low cost [1]. Ultrashort pulsed lasers are favored because they modify surfaces without altering the bulk properties of the material [12].

### 8.2.3 Mechanisms of Interaction between Laser and Textile Materials

When a laser beam impacts a surface, energetic photons excite the electrons in the target material. Beer Lambert's law predicts that this excitation generates heat from absorbed photon energy. This heating can melt or vaporize the material. The ablation mechanism can be chemical, thermal, or a combination of the two. UV radiation can cause covalent bonds in polymers to break, leading to photodegradation ablation, whereas in photothermal ablation, electronic excitation from UV photons thermalizes and fractures polymer bonds [13].

## 8.3 Applications of Laser Treatment in Textiles

### 8.3.1 Cutting and Engraving

Laser cutting has advantages such as high speed, precision, and ease of operation; it is used in various industries, particularly in clothing [3]. Gas lasers, particularly $CO_2$ lasers, are typically used in fabric cutting. The introduction of gases such as $N_2$ and He reduces charring and clears debris from the cutting area [7]. Laser engraving creates clean and permanent marks without physical contact, and, as a result, it eliminates wear and tear [3].

### 8.3.2 Surface Modification

Laser irradiation transforms morphology of polymer from smooth to a regular, roll-like structure, affecting particle adhesion, wetting properties, and optical appearance [6]. Furthermore, high-fluence laser irradiation of aromatic polymers leads to the deposition of debris ranging from yellow to black. This debris, composed of ionized carbon-rich substances, ultimately condenses into large aggregates, resulting in a more hydrophobic surface [2].

### 8.3.3 Marking and Branding

Laser marking is a technique for surface marking that its advantage is its remarkable speed. It has high accuracy. The clarity of marking on products and legibility that are essential for identification is another benefit. The durability of laser markings means that they can withstand wear and tear over time. This technique can be applied to various substrates, demonstrating its adaptability across different industries [7].

### 8.3.4 Cleaning and Finishing Processes

The cleaning soiled fabrics is a sensitive process. Traditional methods of cleaning textiles may cause fibers to lift, break, or swell, potentially compromising the appearance of the fabric. Laser-cleaning methods are as promising alternatives. This method has shown significant potential for effectively removing fine dirt from natural fibers without damaging the fibers. Consequently, many conservators are now considering laser cleaning as a viable solution to preserve and restore the beauty and durability of textile artifacts [14].

## 8.4 Advantages and Disadvantages of Laser Treatment

### 8.4.1 Advantages

In the laser method, there is a minimal chance of product degradation, no need for more or fewer consumables, and no issue with the disposal of harmful by-products [3]. For instance, the use of lasers in denim engraving adds value by replacing conventional denim distressing methods [15]. Nowadays, many textile businesses, including fabric production facilities and the apparel industry, use laser engraving and cutting technologies extensively. It is possible to combine laser cutting, engraving, and branding in one step. Because laser-based processes do not involve mechanical tool wear, the results are more precise. Because lasers treat materials without force, no material attachment is necessary. Because bonded edges formed, synthetic fibers do not tear the fabric. The laser-based method is lint-free and pristine [16].

### 8.4.2 Disadvantages

Significant initial expenditures pose a challenge for small- and medium-sized businesses. This requires a skilled operator, which could be a drawback in some situations [16]. It is challenging to create a natural effect, and, to give the product a natural appearance after fading, physical labor is frequently needed. Both the laser beam and released gases pose health risks. In certain situations, the maintenance and service of the laser system can be problematic [17].

## 8.5 Case Studies and Industry Examples

### 8.5.1 Fashion Industry Applications

The textile and fashion industries could benefit significantly from the use of lasers to create fresh and creative patterns. Exact and clean cuttings can be accomplished using laser therapy. Smaller businesses and couture houses still prefer to cut fabric in an old-fashioned way with scissors or a vertical vibrating blade [18, 19].

Fabric or textiles can be precisely and carefully cut, one or more layers at a time, without ragged edges or hems on the materials, using CAD technology and a laser [20]. When laser-cut synthetic fabric materials are utilized, the hem of the cloth fuses and looks well-finished, negating the need to sew the cut edge. This process has numerous applications in the production of bags, shoes, textiles, and clothing. Stitching is not necessary because lasers can cut fabrics with extreme precision without leaving ragged edges [21].

### 8.5.2 Technical Textiles

Numerous scientists have attempted to imitate organic surface patterns, such as the extensively studied lotus leaf, to achieve roughness patterns at the nano- and micrometer scales and the essential requirements of the Cassie-Baxter state. Applying hybrid coatings of nanoparticles embedded in silane or fluorinated binders to textured technical surfaces is a popular idea for simultaneously providing low surface free energy and micro-/nano-roughness at the same time. A hybrid coating for cotton fabrics made of silica nanoparticles and perfluorooctylated quaternary-ammonium silane coupling agent is an illustration of this idea, as described by Yu *et al.* [22]. Together with an increase in the oil repellence, the water contact angle reached 145°. These hybrid coatings have several problems, including poor wash stability and abrasion resistance of the particles and/or binders [23].

### 8.5.3 Home Textiles

The number of fabric layers that the laser beam can cut is a limitation of laser cutting. The best results were achieved when cutting one or a few lays; several plies did not yield the same accuracy and precision. Furthermore, there is a possibility that the cut edges fuse together, particularly in the case of synthetic materials. The use of a laser is necessary in certain situations to seal the edges of the cut patterns and sewed garment sections to prevent fraying. Similar to how multiple-layer cutting is prioritized in clothing production facilities, it appears doubtful that laser cutting will catch on. Nonetheless, it works well for cutting sails when fusing the edges of synthetic and woven textiles is preferred. Single-ply cutting is a standard method. Additionally, several areas of home furnishings use laser cutting [3].

### 8.5.4 Automotive Textiles

Carbon fiber-reinforced polymer (CFRP) parts exhibit remarkable mechanical stiffness and strength in relation to weight, making them a popular choice for structural components in the aviation and automotive industries. As with other automated production processes for CFRP components, preform cutting has not yet been fully optimized. Within the HAZ, the fibers exhibited geometric changes as the laser-cut kerfs fused together. Although laser cutting appears to be a viable substitute for the mechanical cutting procedures that are currently in use, there are still several unanswered concerns about its suitability for large-scale production [24].

## 8.6 Future Trends in Laser Treatment of Textiles

### 8.6.1 Innovations in Laser Technology

Laser-based marking and engraving are intricate physical processes of significant scientific and practical value. Owing to their technical properties, laser radiation offers high-quality processing. Laser marking is a popular technique in the fashion industry for successfully creating unique designs and artistic embellishing on any surface of textile products. They are also utilized for crucial irreversible technical data that are beneficial to both customers and manufacturers. Marks allow for tracking raw material sources, state monitoring and control, and environmental effect assessment, in addition to containing certain data about textile products [25].

### 8.6.2 Integration with Smart Textiles

Conductive tracks made of metal foil on textiles can be joined using a laser. No additional preparation processes are required when a wavelength that is well absorbed by one of the connecting partners, in this case, metal and cloth, is selected. Depending on the structure and material, a punching procedure may be required to embroider conductive tracks onto the fabrics. High-conductivity metals can be utilized immediately with a laser, negating the need for pretreatment. The laser process can also be integrated with other laser machining techniques such as cutting or engraving [26].

### 8.6.3 Sustainability and Environmental Impact

One of these new technologies, laser application, provides sustainability by reducing the amount of chemicals, energy, and wastewater used in textile production [27–29]. According to Štěpánková *et al.* [29], it also offers process flexibility and ease of design. The repeated creation of the same designs is made possible by the faster and easier transfer of the produced designs to the textile material using laser technology. As a result, producing textile items with high added value and luxurious appearance is quite simple [26, 30].

## 8.7 Conclusion

### 8.7.1 Summary of Key Points

Through "stimulated emission," lasers produce focused light that can be used to modify the surfaces of materials such as metals, polymers, and wood. Water-free laser treatments change the characteristics of textiles; they work especially well for welding, engraving, and branding of synthetic fabrics. Textile cutting, pattern generation, and removal of indigo dye were all performed using lasers. They have an impact on the quality of the fabric; for example, they may increase air permeability while decreasing the strength and flexibility. Lasers can alter the surfaces of textiles at nano- and micro-scales with great precision. The important factors that affect laser design are scanning speed, pulse duration, and intensity, rapid and accurate cutting without coming into contact with anything. common in technological fabrics and fashion.

Changes in the adherence and texture of cloth when applied to coatings and treatments. Strong and transparent perfect clothing labels. Without harming the fibers, laser cleaning aids in the removal of dirt from the textiles. With less wear and tear, precise cutting and engraving, and fewer by-products, it is environmentally beneficial. High starting costs, need for operator expertise, possible health risks, and difficulties in producing natural fabric effects. Combining smart textiles with sustainability advances and reducing energy and chemical use in textile production to lessen environmental effects.

### 8.7.2 Future Outlook for Laser Treatment in Textiles

Compared with conventional textile processes, the laser approach is completely different because it is pollution- and waste-free in its design and operation. Using a laser instead of traditional methods for cutting, engraving, embossing, fading denim, and other processes has a number of additional benefits. Additionally, compared to other methods, lasers have a lower risk of product degradation, use fewer consumables, and eliminate the need to dispose of harmful byproducts [7]. The outdated laser equipment is heavy, cumbersome, and difficult to maintain. However, modern equipment is easy to maintain, operate, and learn how to use. Additionally, older equipment required frequent gas refilling and had a number of safety concerns [3].

**AI disclosure statement:** The author acknowledges the use of AI-based tools for minor language editing or grammatical correction. However, all intellectual contributions, analyses, and conclusions presented in this chapter are entirely the author's own.

## References

1. Esteves, F. and Alonso, H., Effect of $Co_2$ Laser Radiation on Surface and Dyeing Properties of Synthetic Fibres. *Res. J. Text. Apparel*, 11, 3, 42–47, 2007.
2. Kan, C.W., Impact on textile properties of polyester with laser. *Opt. Laser Technol.*, 40, 113–119, 2008.
3. Nayak, R. and Padhye, R., The use of laser in garment manufacturing: an overview. *Fash. Text.*, 3, 5, 1–16, 2016.
4. Kan, C.W., Lam, Y.L., Siu, Y.T., Effects of laser treatment on fabric characteristics and performance. *Surf. Innovations*, 3, 4, 228–236, 2015.
5. Ondogana, Z., Pamuka, O., Ondoganb, E.N., Ozguney, A., Improving the appearance ofall textile products from clothing to home textile using laser technology. *Opt. Laser Technol.*, 37, 631–637, 2005.
6. Nourbakhsh, S. and Ebrahimi, I., Different Surface Modification of Poly (Ethylene Terephthalate) and Polyamide 66 Fibers by Atmospheric Air Plasma Discharge and Laser Treatment: Surface Morphology and Soil Release Behavior. *J. Text. Sci. Eng.*, 2, 2, 2012.
7. Jamal, Z., Yadav, D.N., Rani, S., Application of laser Technology in fashion industry. *Int. J. Home Sci.*, 4, 43–56, 2018.

8. Angelova, Y.P., Factors influencing the laser treatment of textile materials: An overview. *J. Eng. Fibers Fabr.*, 15, 1–16, 2020.
9. Hassan, Z., Atalay, O., Kalaoglu, F., Conductive cotton fabric using laser pre-treatment and electroless plating. *J. Text. Inst.*, 113, 5, 737–747, 2022.
10. Shahidi, S., Rashidian, M., Dorranian, D., Preparation of antibacterial textile using laser ablation method. *Opt. Laser Technol.*, 99, 145–153, 2018.
11. Koussi, E.K., Jung, H.J., Faure, N., Donnet, C., Mauclair, C., Sedao, X., Comparative Study of Ultraviolet and Infrared Femtosecond Laser Irradiation on Textile Polymers PET and PA66. *JLMN-J. Laser Micro/Nanoeng.*, 15, 3, 2020.
12. Sezer, H.K., Short Review on Laser Texturing and Cleaning Carbon Fibre Composites for Aerospace Applications. *J. Polytech.*, 19, 4, 623–631, 2016.
13. Atwee, T., Elshafei, N., Mostafa, H.Y., Bassynouni, F., Surface Treatments using $CO_2$ Lasers on a Variety of Synthetic Textiles. *J. Mater. Sci.*, 6, 101, 2022.
14. Chillè, C., Rossignoli, G., Nagasawa, T., Osticioli, I., Agresti, J., Siano, S., Ciofini, D., Laser cleaning of artificially soiled cotton fabric: Assessment of surface chemical effects. *J. Cult. Heritage*, 6, 1–11, 2023.
15. Kan, C.W., $CO_2$ laser treatment as a clean process for treating denim fabric. *J. Cleaner Prod.*, 66, 624–631, 2014.
16. Dudeja, J.P., Laser application in apparel industry. *Int. J. Adv. Sci. Technol. Eng. Manag. Sci.*, 4, 9, 1–5, 2018.
17. Sarkar, J. and Rashaduzzaman, Md., Laser fading technology: facts and opportunities, Textile Today, 2014.
18. Harder, F., *Fashion for Profit: A professional's complete guide to designing, manufacturing, & marketing a successful line*, 9th ed., p. 130, Harder Publications, Rolling Hills Estates, CA, 2010.
19. Wind, J. and Mahajan, V., *Digital marketing: Global strategies from the world's leading experts*, 1st ed., pp. 102–1 30, John Wiley and Sons, New York, NY, 2001.
20. Vidyasagar, P.V., *Encyclopedia of textiles - Textile automation*, 1st ed, vol. 3, Mittal Publications, 169, New Delhi, India, 2000.
21. Yuan, G., Novel surface treatment for fashion garments, 2012.
22. Yu, M., Guotuan, G., Meng, W.D., Qing, F.L., Superhydrophobic cottonfabric coating based on a complex layer of silica nanoparticles and perfluorooctylated quaternary ammonium silane coupling agent. *Appl. Surf. Sci.*, 253, 3669–3673, 2007.
23. Bahners, T. and Gutmann, J.S., Highly liquid repellent technical textiles obtained by means of combined photo-chemical and laser surface modifications. *Adv. Contact Angle Wettab. Adhes.*, *2*, 333–363, 2015.
24. Fuchs, A.N., Schoeberl, M., Tremmer, J., Zaeh, M.F., Laser cutting of carbon fiber fabrics. *Physics Procedia*, *41*, 372–380, 2013.
25. Angelova, Y., Mežinska, S., Lazov, L., Innovative laser technology in textile industry: marking and engraving, in: *Environment. Technologies. Resources. Proceedings of the International Scientific and Practical Conference*, vol. 3, pp. 15–21, 2017.

26. Qin, H., Hajiaghajani, A., Escobar, A.R., Zargari, A.H.A., Jimenez, A., Kurdahi, F., Tseng, P., Laser-Induced Graphene-Based Smart Textiles for Wireless Cross Body Metrics. *ACS Appl. Nano Mater.*, *6*, 20, 19158–19167, 2023.
27. Morgan, L., Laser enhanced dyeing of wool for textile design, in: *Re-thinking textiles and surfaces*, University of Huddersfield, UK, 2014.
28. Nourbakhsh, S., Ebrahimi, I., Valipour, P., Laser treatment of the wool fabric for felting shrinkage control. *Fibers Polym.*, 12, 4, 521–527, 2011.
29. Štěpánková, M., Wiener, J., Dembický, J., Impact of laser thermal stress on cotton fabric. *Fibres Text. East. Eur*, 18, 3, 70–73, 2010.
30. Ortiz-Morales, M., Poterasu, M., Acosta-Ortiz, S.E., Compean, I., Hernandez-Alvarado, M.R., *et al.*, A comparison between characteristics of various laser-based denim fading processes. *Opt. Lasers Eng.*, 39, 1, 15–24, 2003.

9

# Conventional and Advanced Techniques for the Physical, Chemical, and Biochemical Modifications of Wool

**Seiko Jose[1]*, Anuradha Sankaran[2], Kalaprasad Gopalan Nair[3], Delna Maria[3], Uttara Krishnan[3] and Athira. K. Ramachandran[3]**

[1]*Textile Manufacture and Textile Chemistry Division, ICAR – Central Sheep and Wool Research Institute, Avikanagar, Rajasthan, India*
[2]*Department of Physical Chemistry, University of Madras, Chennai, Tamil Nadu, India*
[3]*Department of Chemistry, Sree Narayana College, Thrissur, Kerala, India*

***Abstract***
Wool is one of the oldest and preferred textile fibers to mankind. It is a protein fiber, which has a complex and unique structure covered with unidirectional tiny overlapping scales. Chemically, wool is a complex structure of the protein in which keratin is the major component. Nowadays, apart from conventional garments, wool is extensively used in technical textiles and has diversified applications. The unique properties of wool such as crimp, high moisture content, lightweight, thermal insulation, fire retardancy, and felting make distinct wool from other natural fibers. To overcome some of the inherent drawbacks and to enhance the performance, the wool fiber surface is often modified through chemical or physical methods. Among these, radiation treatment, enzymatic treatment, steam explosion, and chlorite treatment are well established. The major objective of these treatments is surface functionalization, leading to better interfacial adhesion and consequent improvement in physico-mechanical properties. This chapter reviews the physical, chemical, and biochemical methods of modification of wool fiber and its subsequent applications.

***Keywords*:** Biopolymer, chlorite treatment, enzymes, grafting, keratin, plasma treatment, surface modification

**Corresponding author*: seiko_jose2005@yahoo.co.in

Aminoddin Haji and Çağlar Sivri (eds.) Surface Modification and Functionalization of Textiles, (203–228) 

## 9.1 Introduction

Wool is a high-quality luxury textile material derived from sheep. Wool fibers are made up of polypeptides and peptides. The unique properties of wool fiber are well exploited for textiles as well as other industries. Recently, a significant effort has been initiated to find new applications for wool through the development of unconventional wool products [1, 2]. Wool has a typical scale structure on its surface, which helps in the easy identification of wool from other fibers under the microscope. The length and thickness of the individual scale on the surface of the wool fiber may vary according to the variety of wool. Generally, fine wool has more scales per unit length than coarse wool. In fine wool, the edges of the scales are more often found to be sharp, whereas those in coarse wool are found to be flat. Due to this reason, fine wool is more prone to felting than coarse wool [3]. Felting is an irreversible process of entanglement of the wool fiber under mild alkaline/acidic conditions. It is quite often promoted by mechanical agitation. This unique property of wool is sometimes desirable but causes a nuisance in certain contexts. Most of the fiber modification techniques involve the flattening of the scale structure of wool. Once the sharp edges of the wool fiber got flattened, it loses its inherent property of felting. To enhance the physical attributes of wool fiber and to make it suitable for various applications, many attempts have been reported to be performed on the surface modification of wool fiber through physical, chemical, and biochemical treatments. The main objective of surface modification of wool fiber is to reduce the sharpness of the surface scales and, as a result, the fiber may show the hydrophobic or hydrophilic properties. Although wool has very high moisture management properties, the presence of scales inhibits the immediate penetration of water inside the wool fiber and the fiber is intrinsically hydrophobic.

Earlier, chemicals like sodium chlorite were employed to achieve anti felting properties to woolens. While this technique is highly effective in preventing the shrinkage of wool, but not eco-friendly. Recently eco-friendly processing techniques such as radiation, biopolymer, graft copolymerization, and enzyme treatments are employed to modify the surface. These treatments not only affect the surface characteristics of wool fiber but also have a direct impact on functional properties like dyeability, wettability, thermal conductivity, luster, electrical insulation, and moisture management. Additionally, it may also enhance the aesthetic appearance, wear comfort, odor control, hygiene, and market potential of the woolens. This chapter reviews the conventional and advanced surface modification treatments of wool. The effect of surface modification on the physico-mechanical properties and the possible applications of the surface-modified woolens are also covered in this chapter.

## 9.2 Wool Fiber Morphology and Structure

The major protein found in wool is keratin, which is made up of 18 amino acids such as glutamic acid, cysteine, and glycine, many of which include hydrophilic amino and carboxyl groups, allowing the wool fiber to retain moisture. The porous nature of the wool fiber acts as a thermal insulator, trapping air in the pores. The wool fiber's core cellular component is surrounded by an outer cellular component and an outer layer of cuticle cells. Cortical cells are made up of macrofibrils, which are composed of numerous microfibrils (Figure 9.1). These microfibrils are fabricated by aggregated keratin intermediate filament proteins encircled by keratin-associated matrix proteins with high sulfur and tyrosine levels. The disulfide bond within the cortex has a significant influence on wool's physical and mechanical properties [4].

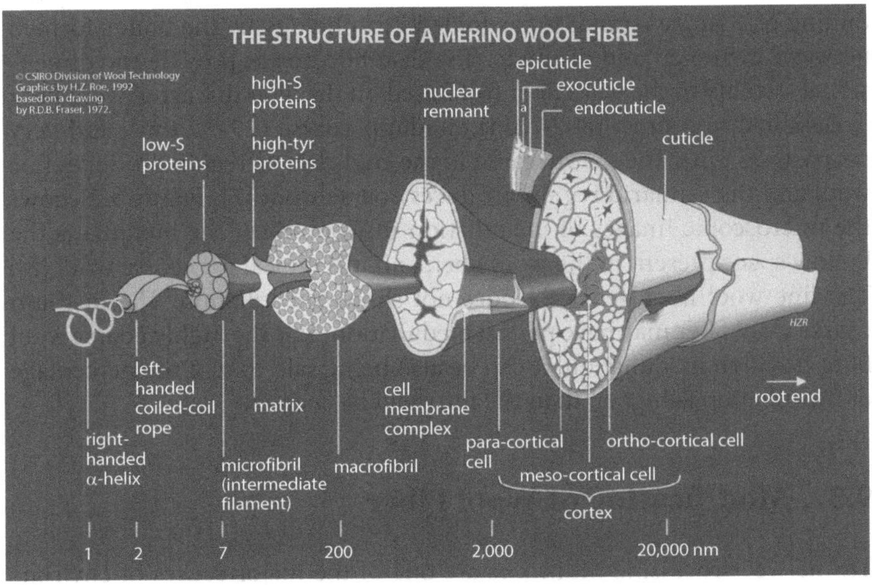


**Figure 9.1** Schematic diagram of the internal structure of the wool fiber. (http://www.scienceimage.csiro.au/image/2489) (with the permission from CSIRO Division of Wool Technology) [5].

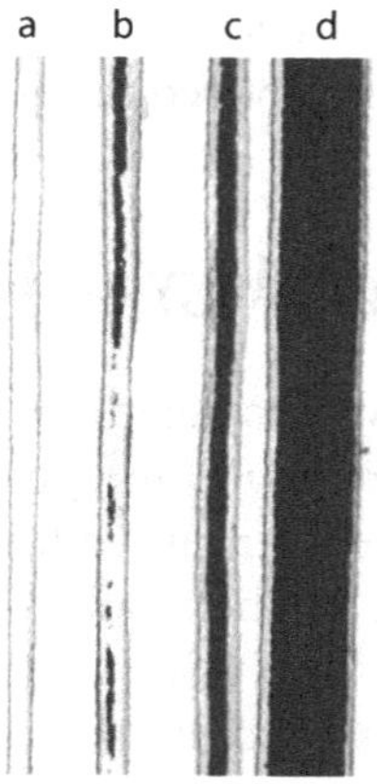


**Figure 9.2** Microscopic image showing medullation in fine, medium coarse and highly coarse wool. (a) fine, (b) medium, (c) coarse and (d) highly coarse wool.

The type and quantity of cross-links within the intermediate determine the stability of a wool fiber structure. The outer cuticle cell is made up of outer exo and inner endocuticle layers and has a laminar structure. Wool filament keratin and intermediate filament keratin are structurally similar proteins. The helix is the protein's secondary structure, created by the folding of a single strand. It has folded morphology by the bonds formed between hydrogen and disulfide. The disulfide bonds provide the dimensional stability to the wool fiber [6]. Based on the fiber diameter, wool fiber is classified into fine (18–24 μm), medium-coarse (25–32 μm), and very coarse (>35 μm). The sheep reared in the cold climate region produces fine wool and the semiarid region produces coarse wool [7]. Figure 9.2 shows the microscopic images of medullation in wool fiber. The medulla is the hollow space present in the wool fiber, and it is typical for coarse wool [8]. The fine wool fiber generally does not shows medullation. The medium coarse wool fiber may have partial medullation, and the highly coarse wool fiber has high medullation. It can be also being said that, if the percentage of medullation is higher, then the fiber will be coarser.

## 9.3 Modification of Wool Fiber

Textile wet processing industries are one of the major causes of environmental pollution. The effluents generated in the industries contain alkalis, acids, dyes, and other heavy metals. Due to the increase in the awareness of ecological and environmental concerns, most of the countries imposed stringent regulations on the effluent discharging norms on the industries. The industries are now converting conventional chemical processes to more eco-friendly processes. The presence of unsaturated fats and the cuticle scale make the external wool surface exceedingly hydrophobic. This is undesirable during wet processing because the dyes and chemicals are hydrophilic and having constraints in the penetration inside the fiber. Industries are using wetting and leveling agents to overcome these issues.

Most of the surface modifications performed over wool fiber aims to make the fiber more compatible for technical applications like composites, blends, conductive polymers, and anti-felt woolens in addition to enhance the aesthetic appearance and dye uptake. With the advancement of science, many attempts have been reported on the physical and chemical treatment of woolens. Some of them are commercialized and few are still in the developmental stage. The recent physical, chemical, and biological treatment methods reported on the modification of wool are listed below.

### 9.3.1 Physical Methods

#### *9.3.1.1 Plasma Treatment*

One of the emerging methods for the surface modification of natural fibers is plasma treatments [9]. The plasma treatment of natural fibers could be performed either at low-pressure or at atmospheric pressure. When a suitable gas (oxygen, helium, nitrogen, or argon) is exposed to the electromagnetic field, the plasma is generated. The topographical changes that happened on the wool fiber surface depend greatly on the plasma treatment conditions. The properties of plasma exposed fibers are determined by the nature of gas, energy, and duration of application. The advantages include short treatment time, low treatment temperature, and eco-friendliness [10]. The basic reactions attained by the surface modification are oxidation of the surface of materials when special gases are used [11].

The low-temperature plasma (LTP) treatment with oxygen is effective to produce the etching on wool fiber surfaces by producing cracks in the micro-level. The surface etching increases the tensile strength, fiber to fiber friction, water penetration, and dyeability of wool [12, 13]. The plasma treatment removed surface lipids and oxidized cysteine within the exocuticle, according to attenuated total reflectance fourier transform infrared spectroscopy (ATR-FTIR) examination. Plasma radiation, on the other hand, has reduced the crystallinity of wool fiber (Figure 9.3). The oxygen plasma treatment increased the average stress of wool fiber yarn by 34.6% for 5 minutes and 30.5% for 2.5 minutes of exposure. A comparative study of plasma

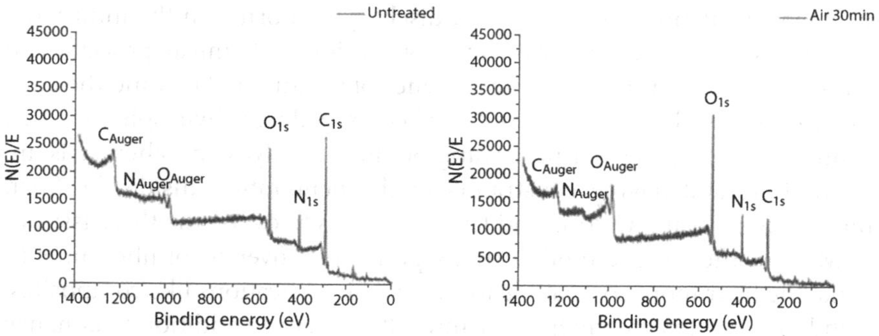


**Figure 9.3** XPS spectra of untreated wool and air plasma treated wool [15].

generated by various gases (oxygen, nitrogen, argon, or air) on the water absorption properties of wool yarn showed that the cysteine groups present in the wool were changed into cysteic acid groups, thus causing improvement in the wettability of the wool and the wicking ability increased with the increase in the duration of treatment. This perhaps happened because of the removal of the fatty acid layer on the cuticle layer of wool fiber. Among the four investigated gases, the oxygen treatment imparted the best wicking properties [14, 15]. Additionally, the plasma treatment could be adopted for increasing the dyeability of the textile fiber. After plasma treatment with helium gas, a notable increase in the hydrophilic properties was observed, which resulted in higher dye uptake of woolen fabric when dyed with acid dyes and natural dyes [16, 17]. A study reports the effect of air and argon plasma treatment on the properties of the wool knitted fabric in terms of variation of treatment duration and energy. Among various parameters studied, the thermal resistance and moisture management properties of the wool fabric were found to be enhanced after treatment, whereas the air permeability got substantially reduced [10].

#### *9.3.1.2 Corona Treatment*

Corona treatment is a kind of LTP treatment at atmospheric pressure. Corona treatment enhances the positive charge and zeta potential by bringing chemical changes on the fiber surface. The corona treatment involves exposing the wool surface to electrical discharges caused by electrical field of high frequency or voltage. Talebian *et al.* (2009) studied the corona treatment of wool fabrics at two different power levels (500 and 1,000 W) and different number of passages (3, 7, 11, 15, and 200). The SEM analysis revealed that the particles on the surface of the wool were related to the edge of wool scales that can be removed by corona treatment. The etching effect at 1,000

W and 20 passages completely removed the wool scales from the surface. The etching effect increased the tensile strength and Young's modulus [18].

Recent application of corona discharge (CD) treatment on wool fabrics in the natural dyeing of weld in acidic medium improved the color strength without change in the fastness properties when compared with the conventional method using tannic acid as mordant. It is due to the partial etching of cuticle scales by corona treatment [19]. The combined effect of physical and chemical surface modification improves the physical and mechanical properties of wool. The surface modified wool fiber after CD treatment comprises an oxygen functional group, with C=O groups being the main oxidation component. Because of the modest chemical alterations to the epicuticle, the color intensity of CD-treated wool increased only marginally as its wettability and capillary liquid transport capability improved. Chitosan coating on CD-treated wool cloth boosts color intensity in direct proportion to its concentration. The deposition of deacetylated chitosan increased the positive charge on the wool and provided more sites for the adsorption of negative acid dye ions [20]. The hydroxyl groups expose at the surface of the corona-treated wool fabric could be made available for the covalent bond formation with quaternary ammonium compound 2,3-epoxypropyltrimethylammonium chloride to improve the functional and mechanical properties. This combined effect improves the static propensity, hydrophilicity, and tensile strength than untreated and corona-treated fabrics [21].

### *9.3.1.3 Radiation Techniques*

Radiation is the process of moving energy in waves or particles, which can excite molecules and atoms of materials and change their properties. The ultraviolet, gamma, X-ray, infrared, and radio waves are reported for the physical modification of wool fiber [22]. The radiation interaction improves the dye uptake and photo-bleaching of wool. Unlike other chemical treatment methods, this technique is highly eco-friendly. Alternative use of these high-energy radiations is the conservation of historic textiles in museums. The moths of textile degrading pests will be killed by the direct exposure of high-energy radiations [23, 24].

### *9.3.1.4 UV Treatment*

The UV radiation (UV-A) mainly has been employed to produce shrink-resistant wool. While the textiles are exposed to UV radiation, it may cause excitation and/or dissociation and the process is known as the photosensitized oxidation [25]. In comparison with the X-rays and gamma rays,

UV radiations are having relatively low energy. Hence, it cannot penetrate inside the fiber. So the UV–wool fiber interaction is superficial. The studies reported that the exposure of wool to UV rays causes surface oxidation of disulfide bonds. This may help the dyer to color the wool fabric at relatively low temperatures with a better dye uptake [26]. The UV pretreatment is also employed for the shrink-resistant finishing of wool fabric.

Gamma radiation has the smallest wavelength and highest vitality of any wave in the electromagnetic spectrum. The surface modification using gamma radiation is one of the promising methodologies. The textiles grasp the electromagnetic radiation of varying wavelengths in the visible range of the spectrum. The advantages of this radiation treatment are low cost, improvement in dye uptake with minimum dye consumption. The literature about the effect of gamma irradiation on the properties of wool fabric is scanty. In comparison to the UV radiations, the gamma radiations are having high energy and can penetrate the wool fiber. This may cause potential damage to woolens, if not carefully handled [27]. Also, because gamma radiations are harmful to human, proper precautions must be taken by the researchers while handling the instruments.

A comparative study of the effect of UV and gamma radiation on woolen fabric has been reported [28]. Both the radiations changed the color of the woolen fabric. The UV radiation causes an initial color change from white to green and, later, to yellow, whereas the gamma radiation changed the color of the wool into pink-red. Interestingly, it is observed that the color change is greatly depending upon the energy of radiation. The color produced by UV radiation was removed by peroxide bleaching. However, in the case of gamma-radiated wool fabric, the color remained almost unaltered even after strong hydrogen peroxide bleaching. Another study reported that the coating of chitosan on wool fabrics followed by gamma radiation treatment increased the dye uptake [29].

#### *9.3.1.5 Microwave Irradiation*

The dielectric constant of wool (6.0–10) is higher than that of natural and synthetic fibers, and, hence, its polarization ability is higher in microwave field. The energy generated by microwaves is distributed uniformly inside the wool material at faster rate while comparing the conventional method. This treatment damages the scale-like structure of the wool fiber surface, and the concentration of cystine S-S bonds is lower than the untreated one. The crystallinity of treated wool varies with the humidity, and the concentration of alpha-helical conformation was changed in wool fiber [30].

#### 9.3.1.6 Xe excilamp

A Xe excilamp (lambda = 172 nm) is another tool for surface modification of wool, and the photo-irradiated substrate is rich in high concentration of oxygen, sulfur, and nitrogen along with hydrophilic functional groups. The resultant-treated fibers possess low contact angle and wet directional friction effect also capable to produce a deeper hue with acid dyes [31].

#### 9.3.1.7 Electron Beam Irradiation

Electron beam irradiation of scoured wool containing a balanced humidity without degreasing in the range of 0–350 kGy dose exhibited significant fluctuation in the quantity of reactive products depending upon both the absorbed dose and time. The formation of oxidized products, cysteic acid, was appeared to be faster due to the presence of humidity as an oxygen source when compared to the dry and degreased wool. The conformational pattern of the secondary structure of wool showed a large variation depending on absorbed dose as well as post exposure time [30].

#### 9.3.1.8 Laser Treatment

The smooth edges of the scales of untreated wool surface became modified by etching using $CO_2$ laser treatment. High energy level (1.67 mJ/cm$^2$) laser damages the fiber cuticle, and, as a result, the felting shrinkage as well as the wettability of the wool fabric got reduced. The FTIR spectral studies revealed the oxidation of S-O groups of wool scale surface to S-O groups of sulfonic acid. The peak at 1,640 cm$^{-1}$ corresponding to $NH_3^+$ of ATR spectra of laser-treated wool distinguished the effect of low- (0.21 mJ/cm$^2$), medium- (0.42mJ/cm$^2$), and high-energy–treated wool. By varying the power level and sweep speed of laser, the felting shrinkage resistance can be improved [32].

### 9.3.2 Chemical Methods

Wool is a natural fiber commonly used in the textile industry for suiting, shawl, carpet, and blankets. However, it is prone to felting shrinkage at favorable conditions. Research is going well since 1950 to solve the felt shrinkage through different techniques. Initially, the two primary goals of chemical modification of wool were to confer shrink resistance and moth proofing. Many compounds have been explored, but only a few have found practical usage. Surface treatment and other alterations to wool can provide opportunities for woolen manufacturers to create diverse woolen products by combining aesthetic features based on client demand, as well as improving fabric durability.

#### 9.3.2.1 *Steam Explosion*

Steam explosion is a hydrothermal process of treatment of natural fiber. This process is often performed at high temperatures for short time to several minutes, followed by explosive decompression [33]. The treatment temperature, duration, and pressure of the process greatly influence the process. Steam explosion is generally performed in autoclaves. In the lignocellulosic fiber, the steam explosion has been reported for the separation of individual fibers from the fiber bundle [34]. This method has also been adopted for the extraction of nanofibers [35]. The effect of steam explosion on wool fiber has been studied by a few researchers. The process brought slight damage to the fiber surface and also produced small grooves. With the increase in the steam pressure, the moisture content and tensile properties of the wool fiber got reduced, perhaps due to the breakage of disulfide linkages and hydrogen bonds under high temperature and humidity. This observation was further confirmed by FTIR analysis. The elevation of pressure beyond the critical point causes fiber degradation. This method is also adopted for the extraction of keratin from wool [33]. Wool is a protein fiber, sensitive to high temperature and pressure. The steam explosion treatment on wool has to be done with proper precaution because overcooking may result in keratin hydrolysis due to the partial or complete destruction of certain amino acids. This may ultimately lead to the destruction of the fibrous structure [36].

#### 9.3.2.2 *Grafting*

Grafting in the natural fiber is performed to overcome some of the natural drawbacks, aiming at specific applications. The majority of grafting treatments are used to improve the poor hydrophobic and thermal stability of natural fibers. There have been reports on grafting natural fibers with silane, chitosan, isocyanate, and acetic anhydride, and other materials [37, 38]. Wool is more susceptible to chemical reactions than other cellulosic fibers such as cotton because it contains two functional groups, *viz.*, amide and a carboxylic acid group. Wool with better functional properties can be utilized to create novel products such as wool-based adsorbents for metal ion recovery and functional textiles with improved antibacterial, electrical conductivity, and color uptake.

Chemical methods of grafting include oxidation and reduction, and they are preferred over radiation techniques to avoid accidental chances of fiber damage [39]. The introduction of functional polymers into the internal structure of wool could improve its properties like tensile strength, hydrophilicity, and thermal resistance [40]. Cyclodextrins are well established in textile industries for furnishing various functional finishing due to their peculiar properties of encapsulation. Grafting of β-cyclodextrin and β-chlorotriazinyl cyclodextrin on wool fabric in the presence of citric acid exhibited antimicrobial properties [41]. They proposed a mechanism for the reaction of citric acid with the hydroxyl groups of β-cyclodextrin and terminal amino groups of wool, leading to the formation of ester cross-linkages or ionic bonds. The grafting introduced microbial resistance to wool fabric against gram +ve and −ve bacteria in addition to the increase in the mechanical properties and dyeability. The influence of β-cyclodextrin grafting on wool on the dyeing and antimicrobial properties of the fabric has also been reported elsewhere [42]. Chitosan has been employed for grafting on wool by many researchers. Zhang *et al.* explained the chemistry of grafting wool with chitosan. The chitosan treatment leads to the reduction of disulfide bonds in the cuticle cells for the thiol group (Figure 9.4). The produced thiol then reacts with thiolated chitosan through the reconstruction of disulfide bonds, leading to grafting on wool. This grafting enhanced shrink and microbial resistance of wool fabric [38]. Similar observations were also reported while grafting chitosan on wool fabric using anhydrides [37]. The cross-linking ability of chitosan on the wool fabric was also exploited for furnishing aroma finishing on woolens [43].

In comparison with other natural fibers, wool is having better UV protection properties. However, prolonged exposure to solar radiation causes photo yellowing of wool. Attachment of para-aminobenzoic acid on wool fiber effectively resists photo yellowing and exhibited significant enhancement in UV protection factor values of wool fabric. The said chemical can bind with wool by the covalent bond formation and is fast to washing [44]. Curled human hair becomes straight when drenched in water and recovers its original shape after drying. This shape memory property is also applicable to wool fiber due to the presence of disulfide bonds and has been elevated by the addition of disulfide bonds in the wool fiber through chemical modification using cystamine. This property can be exploited for making smart and intelligent textiles [45]. Conductive textile is an emerging field of technology, where the non-conductive natural and synthetic fibers are converted into conductive through chemical treatments. For this, polypyrrole, polyaniline (PANI), and polythiophene were well-tried [46]. In such a study, the wool fabric was coated with chitosan and polyaniline to

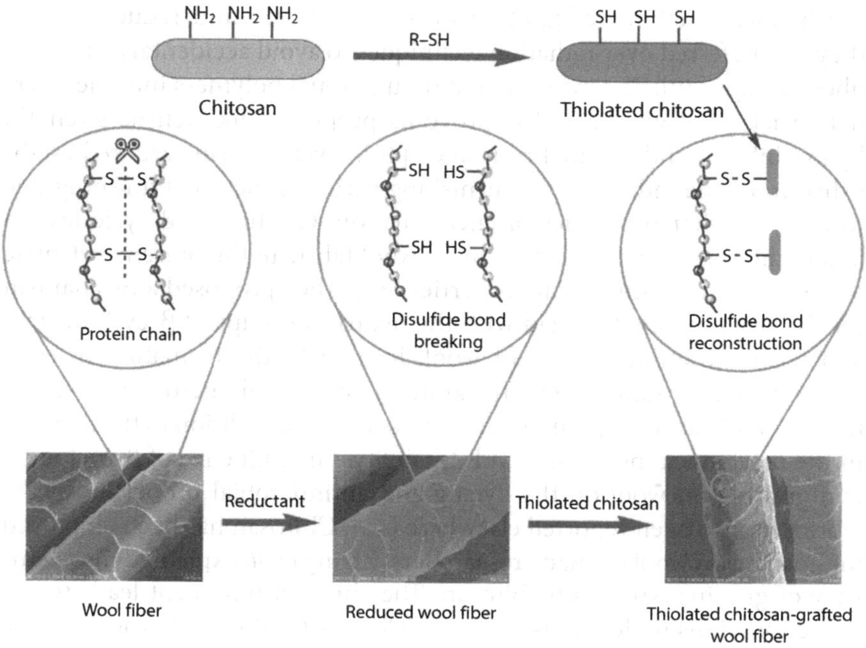


**Figure 9.4** Schematic illustration of the grafting reaction between chitosan and wool fiber [38].

make it conductive through *in-situ* polymerization. The pre-treatment of woolen fabric was performed by oxygen plasma treatment. This pre-treatment removes the scale structure and enhanced the affinity of wool with chitosan through hydrogen bonding. In addition to conductivity, the chitosan/PANI -coated wool fabric showed excellent antimicrobial properties against *E. coli* and *S. aureus* [47].

Sodium lignosulfonate (SLS) was employed to improve the surface adhesion between wool fabric and natural rubber latex. After the treatment with SLS, the wool fabric showed significant improvement in the diffusion of rubber latex as a result the modified wool-latex composite showed 32% increase in the areal density in comparison with untreated wool-latex composite [48]. In another study, the wool fiber was modified with tris(2-carboxyethyl) phosphine to produce thiol-active groups on wool fibers. The modified wool fiber was then treated with N-phenylmaleimide. The N-phenylmaleimide finished wool fabric showed excellent and durable bacterial resistance against *S. aureus* and *E. coli*. The finishing also imparted good anti-felting performance [49]. The reduction of disulfides in keratin with $NaBH_4$ to the corresponding thiols plays a key role in the modification and functionalization of wool fiber. Buchacher *et al.* established a method of potentiometric back titration method for thiol detection in wool fiber [50]. The suggested method could be employed for keratin characterization, process optimization, and technical quality control in chemical wool processing.

### 9.3.2.3 Chlorite Treatment

Wool chlorination, also known as "degradative treatment," is the process of stripping the fiber of its scales in order to reduce felting potential and allow the creation of machine-washable wool garments. Commercially, the most major degrading procedures have been chlorination and per mono sulfuric acid treatment, both of which are widely employed in the production of wool yarns, clothing, and textiles.

A study on improving the properties of wool by chemical treatment was discussed by Rippon and his coworkers [48]. They showed that chlorination is mainly for the treatment of machine washable garments. These treatments produce an accessible level of fiber damage that weakens the wool and leads to the formation of the harsh handle. The polymer used in the treatment is cationic and connected to the fiber surface by anionic cystic acid groups. The most commonly used polymers are polypeptide/epichlorohydrin condensate (Hercosett 125), which is cross-linked to the wool and reacts with the amino group. The Hercosett will significantly decrease the felting shrinkage of chlorinated wool by encapsulation and fiber-fiber bonds [51]. Hassan *et al.* [4] examined the sustainable method in imparting shrink resistance to wool fabric by using chlorine/Hercosett, Ozone/Hercosett, and Kroy/Hercosett. The results show that Chlorine/Hercosett treatment gives the most durable and best shrink to resist performance to wool and reduces the inherent stain resistance of wool fiber. The low cost, high production, and speed make the chlorine/Hercosett treatment more advantageous.

### 9.3.2.4 Zeolite Treatment

Zeolite treatment is a way of introducing hydrophilic character to wool. Introduction of micron-sized Zeolite molecular sieve 5A transforms the hydrophobic wool surfaces into super hydrophilic surfaces. As the chemical composition of Zeolite type A is 1:1 Si:Al ratio, the high level $Al^{3+}$ in the frame work is susceptible to acid treatment resulting in the collapse of the structure. The enhanced silanol (Si-OH) groups on zeolite surfaces undergo a condensation reaction with the silane bound to the wool, resulting in plasticization of the wool protein and low stiffness [52]. Chemical attachment of Linde Type A (LTA) nanozeolite (100 nm) on wool keratins in aqueous dispersion *via* binding agent silane led to good coverage of wool surface. Because LTA contains a high concentration of $Si^{4+}$, the nanozeolite's structural framework is chemically resistant. After surface modification, the nanozeolite structure maintained the final wool product's hydrophilicity, softness, and wear comfort. FTIR spectroscopy investigations showed a significant interaction of zeolite with wool proteins *via* silane as a linker. Chemical alteration leads to the oxidative breakdown of disulfide bonds, resulting in cysteic acid residues. In addition to $-SO_3H$ groups, carboxylic acid groups of aspartic and glutamic acid residues contribute 20–25 mol% of wool keratin, which aids in the binding of zeolites to wool keratin *via* silane coupling agents [53].

### *9.3.2.5 Nano-Finish*

Nanotechnology has brought a new sector of applications to the textile industry. Coating the surface of textiles with nanoparticles has become the usual approximation for the production of highly active surfaces to have antimicrobial, flame retardant, UV blocking, water repellent, and self-cleaning properties [54, 55]. The nano-finish creates carefree fabrics that reduce superior liquid repellence and furnish wrinkle resistance. These enhanced fabrics allow water and oil spills to easily bead and roll off the fabric without piercing the fibers and preserve in fabric the entire life of the garment.

For achieving multifunctional properties of wool fabric, many nano-formulations were attempted. Woolens are prone to the attack of moth and it is an age-old problem associated with wool. Plenty of efforts have been reported to prevent the moth attack; however, in the later stage, few of them were withdrawn from the market due to ecological and human toxicity. A couple of works about the use of nanoparticles for the prevention of moth has been reported in the literature. Nazari and his co-workers treated wool fabric with nano-$TiO_2$ in presence of citric acid (cross-linker) for imparting a mothproof finish to woolens [56]. The exhaust method of application of nanoparticles on wool fabric showed the antifeedant property to *A. verbasci*. Kaolinite is a natural mineral consisting of hydrous silicate with sheet-like structures stacked over one another [56]. Jose *et al.* used nano-kaolinite particles for mothproofing of woolens. They have reported that the kaolinite nanoparticles can be incorporated into the wool fabric using the room temperature

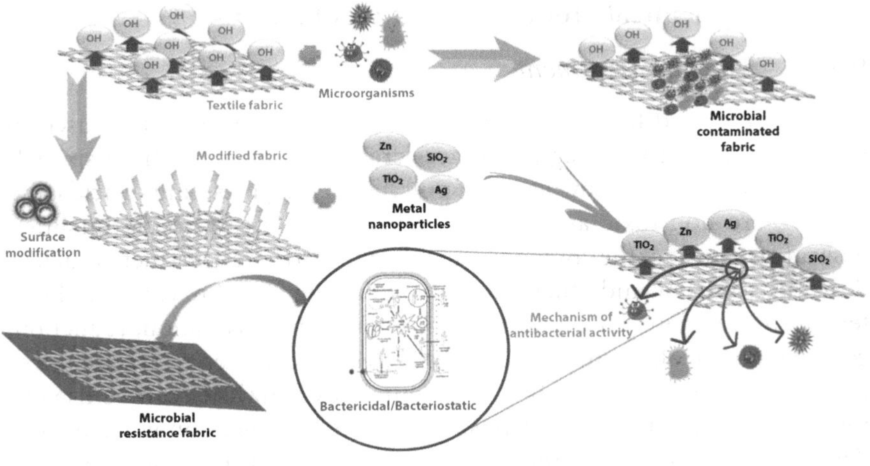


**Figure 9.5** Antimicrobial activity of metal nanoparticles on woolen textiles [58].

padding method followed by drying. The application of 1.0% nanoparticles restricts the moth attack by imparting anti-feeding property. The fabric damage was reported to be below 1.5% and the process is energy-saving and eco-friendly [57].

The application of various nanoparticles to cater to different functional finishing of woolens has been also described. The important metal and metal oxide nanoparticles that have been employed for achieving the antimicrobial properties on woolens have been reported elsewhere [58] (Figure 9.5). The treatment of $TiO_2$ nanoparticles on the plasma modified wool fabric was found to be effective to obtain self-cleaning properties [59]. The nanoclay was reported as an effective fire retardant formulation for lightweight wool fabric [55]. The plasma irradiated wool fibers were grafted with Ag-loaded $SiO_2$ and the resultant fibers showed good microbial resistance against *E. coli* and *S. aurous* [60]. The shrink-resistant properties of the wool fabric were introduced by the deposition of nano-polyurethane multilayer coating in presence of diallyldimethylammonium chloride. Interestingly, unlike conventional coatings, the formation of nanolayers did not alter the tensile strength, dyeability, and whiteness index of the wool fabric. Among 4, 8, and 16 layers of coating, the latter showed the highest shrink resistance properties [61]. In another work, the hydrothermal deposition of nano-$TiO_2$ on a wool fabric produced 10-nm-size nanoparticles on the wool fabric surface. While the dyeability of the wool got slightly diminished, the nano-$TiO_2$ coating resulted in higher thermal degradation and ultraviolet protection of wool [62].

### 9.3.3 Biochemical Modification of Wool Fiber

#### *9.3.3.1 Enzyme Treatment*

Enzymes are used for a variety of industrial processes. Enzymes used in textiles are primarily designed to accelerate reaction rates, have specialized activity, and emit low levels of pollutants. Enzymes, unlike acids and alkalies, are non-toxic and have no effect on fabric handling qualities. Nowadays, most textile companies use enzymes for pre-treatments such as desizing, scouring, and other functional finishing [63]. Enzymes modified textiles by a variety of chemical reactions, including oxidation, reduction, coagulation, breakdown, and hydrolysis [64]. Because enzymes are selective in their application, only a small number of enzymes can be used to modify wool. The primary purpose of enzyme treatment on woolens is to flatten the surface scale of the fiber. Depending on the enzymes, the treatment may improve physicomechanical characteristics, dye uptake, shine, aesthetic value, and comfort. Enzymes can also help to prevent wool felting [65]. Wool-specific enzymes preferentially attack disulfide bonds, providing hydrophilicity and shrink resistance to woolens. There are scientific investigations on the usage of lipase, laccase, protease, and transglutaminase enzymes on woolens [66].

As discussed previously, because the enzymes are highly selective, they operate at specific conditions of temperature, pH, and duration. When transglutaminase enzyme was applied on Rambouillet wool fibers, the physico-mechanical properties of wool get enhanced in terms of softness, moisture content, and luster. However, the higher concentrations of enzymes inversely affect the fiber strength and dye uptake [67]. It also made the wool slightly slippery [68]. Transglutaminase induces cross-linking in the wool protein [69].

Protease is a well-known protein degrading enzyme that is used on wool to reduce the scale of the fibers for shrink-resistant applications. Additionally, it may reduce the fabric's rigidity. Protease action is more localized to the scales' outer surface by increasing the protein's susceptibility to proteolytic destruction. However, while treating with protease, proper measures are to be taken to avoid excessive degradation of wool [70]. Similarly, the lipase enzyme treatment on woolens caused an increase in the water absorbency, pilling resistance, and imparted anti-shrinkage properties. However, it resulted in a marginal reduction in weight and tensile strength [71]. Laccase is a well-established biocatalyst for many industrial processes, extracted from plants, fungi, and microorganisms. The treatment of wool fabric with laccase showed improvement in pilling resistance, whiteness, dimensional stability, and dyeability of wool fabrics [72].

The degrading effect of protease on wool fabric could bring better dye uptake [73]. CD followed by enzymatic treatment resulted in the softening of wool, with a marginal reduction in the tensile strength [74]. The smooth fiber surface morphology plays an important role in the penetration of water and chemicals inside the fiber. When the wool fiber surface was modified by chlorination, it enhanced the diffusion and targeted activity of transglutaminase and tyrosinase enzymes [75]. Similar studies have also been reported elsewhere [76]. Lipase enzyme produced by *Bacillus aryabhattai* B8W22 was used for the bio scouring of wool. The researchers immobilized the enzyme on sericin-based discs to enhance its stability and to make it reusable. A good improvement in the dyeability of wool fiber against acid, basic, and reactive dyes was observed after scouring. The process aims to reduce consumption of water and energy during wet processing of wool [77].

#### 9.3.3.2 *Biopolymer Treatment*

Biopolymers are macromolecules created by living organisms or through chemical synthesis. In textiles, they are employed as thickening agents and dyeing auxiliaries. The advantages of synthesized biopolymers include the potential to create a sustainable industry as well as amplification in various properties of textiles such as tensile strength, flexibility, durability, clarity, and high gloss.

Chitosan is the most often used biopolymer in woolen products. It has several goals, including increased hydrophilicity, dye uptake, shrink resistance, and antibacterial finish [79, 80]. The treatment of wool fabric with various biopolymers such as wheat starch, gum arabic, and chitosan to improve shrink resistance has been reported [78]. Among all the above said treatments, the chitosan treatment improved wool's tensile and bending capabilities, although gum arabic and wheat starch coating had no significant effect on these qualities (Figure 9.6). The biopolymer treatment improved the dye uptake and decreased the area shrinkage of wool fabric. The schematic outline of biopolymer treatment on wool fiber is shown in Figure 9.7. Table 9.1 shows various physical and chemical treatments for achieving different functional properties on wool. Whey protein is a by-product of the dairy industry. Abdelghaffar and El-Bisib coated the wool fabric with soybean flour and whey protein concentrate [81]. The coating resulted in higher UV protection, antimicrobial activity, and dyeability with acid and reactive dyes. For the coated fabric, the dyeability was found to be better in acidic medium than alkaline. The coating is eco-friendly, and it ultimately resulted in the low chemical and dye consumption and high dye uptake during the dyeing of woolens.

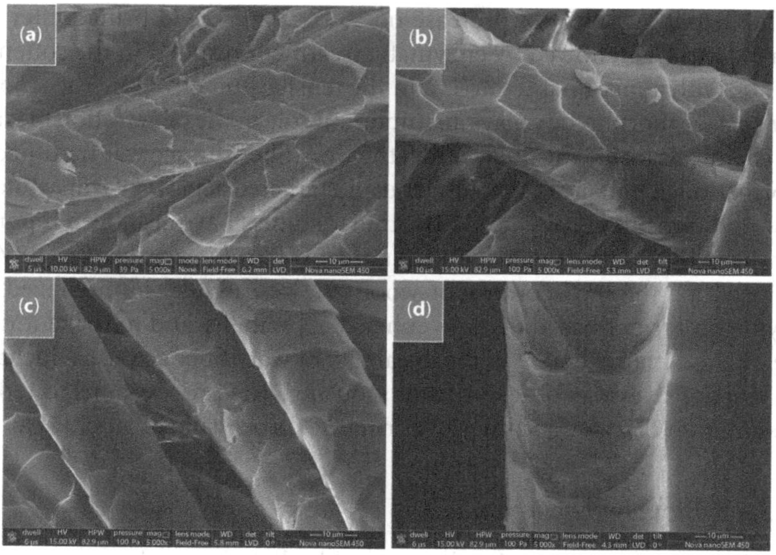


**Figure 9.6** FE-SEM images (a) untreated wool, (b) gum arabic, (c) chitosan, and (d) wheat starch biopolymer treated wool fiber samples [78].

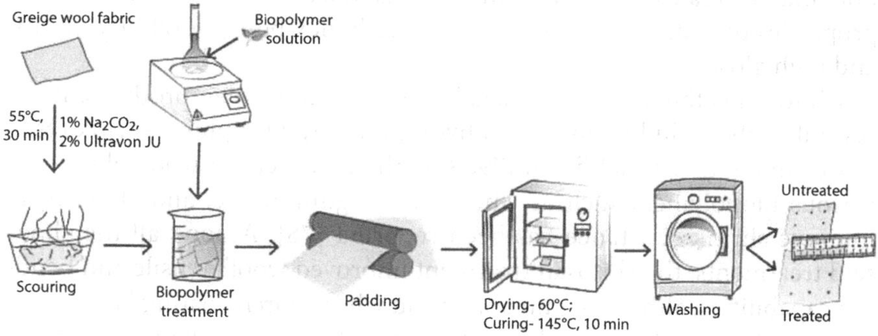


**Figure 9.7** Schematic outline of biopolymer treatment on wool fabric [78].

**Table 9.1** Effects of various surface modification treatments on woolens.

| Treatment | Properties |
|---|---|
| Plasma treatment | Hydrophilicity, dyeability, and wettability |
| Chlorite treatment | Shrink resistance |
| Biopolymer treatment | Shrink resistance |
| Enzyme treatment | Shrink resistance, dyeability, handling, and luster |

## 9.4 Future Scope and Challenges

Wool is widely employed as a warm clothing material in the textile industry, as well as technical textiles in various industrial areas, due to its unique and remarkable characteristics. Wool fiber can be easily chemically changed because it is composed of amino acids and contains amide and carboxylic acids. This certainly gives a great advantage over cellulosic and lignocellulosic fibers, which are having only one reaction site for modification. The chemical treatments like chlorination through highly efficient result in tragic environmental and economic issues. In this context, novel emerging technologies like CD [82], plasma discharge [83], microwaves, grafting, enzymes, and biopolymer could be used as an alternative in fiber processing [84, 85]. Because wool is a delicate fiber and sensitive to alkali, proper precautions are to be taken while modifying to avoid excessive damage of the fiber. The modified and functionalized wool could find potential applications in the development of new products to meet the market demand. In the deconstruction and reconstruction of disulfide bonds in fibers, chitosan can be further explored. The biochemical modification using enzymes and bio-polymers is another promising step. Recently, the nanotechnology is also launched in wool finishing. Many functional finishing on woolens using nanotechnology is commercialized today. The new and innovative technologies not only reduce the production cost but also save the environment.

## 9.5 Conclusion

Several treatment methods were developed for the modification of woolens for domestic and technical textiles. Many of the conventional and non–eco-friendly processes were replaced by sustainable methods from time to time. Chemical modification of wool could be employed as a powerful technique to enhance some inferior performance of wool fibers and to introduce novel functional features suited for technological implementation. Plasma and other radiation treatments are developing as an alternative solution for many treatments and are now adopted on the industrial scale. This may induce the low consumption of chemicals, water, cost, and time. Enzymes, as natural products, are fully biodegradable and perform their functions quietly without emitting contaminants. The breakthrough in the surface modification of wool fabrics will also aid in the production of superior textile goods with quality and economy while addressing the risks associated with the generation of highly polluted effluents from the traditional chemical process.

**AI disclosure statement:** The author acknowledges the use of AI-based tools for minor language editing or grammatical correction. However, all intellectual contributions, analyses, and conclusions presented in this chapter are entirely the author's own.

## References


1. Kadam, V., Jose, S., Kumar, A., Kumar, A., Diversified applications of coarse wool in various technical and industrial sectors: A review. *Indian J. Small Ruminants*, 30, 2, 218–232, 2024.
2. Jose, S., Boruah, S., Rani, S., Wool-Polymer Composites: A Comprehensive Review on Preparation, Properties, and Applications. *J. Nat. Fibers*, 22, 1, 2495921, 2025.
3. Rani, S., Kadam, V., Rose, N.M., Jose, S., Shakyawar, D.B., Yadav, S., Effect of enzyme treatment on wool fabric properties and dimensional stability. *Indian J. Fibre Text. Res.*, 46, 83–90, 2021.
4. Hassan, M.M. and Carr, C.M., A review of the sustainable methods in imparting shrink resistance to wool fabrics. *J. Adv. Res.*, *18*, 39–60, 2019.
5. https://www.scienceimage.csiro.au/tag/wool-and-woollen-products/i/2489/diagram-of-wool-fibre-structure/
6. Caven, B., Redl, B., Bechtold, T., An investigation into the possible antibacterial properties of wool fibers. *Text. Res. J.*, *89*, 4, 510–516, 2019.
7. Kumar, A., Prince, L.L., Jose, S., Sustainable wool production in India, in: *Sustainable fibres and textiles*, pp. 87–115, Woodhead Publishing, Cambridgeshire, United Kingdom, 2017.
8. Jose, S., Paduvilan, J.K., Prasad, A., Thomas, S., Kadam, V., George, S., Natural rubber latex coated coarse wool fabric composite—Preparation, characterization, and application. *Polym. Eng. Sci.*, 65, 5, 2582–2592, 2025.
9. Praveen, K.M., Thomas, S., Grohens, Y., Mozetič, M., Junkar, I., Primc, G., Gorjanc, M., Investigations of plasma induced effects on the surface properties of lignocellulosic natural coir fibres. *Appl. Surf. Sci.*, *368*, 146–156, 2016.
10. Karahan, H.A., Özdoğğan, E., Demir, A., Koçum, I.C., Öktem, T., Ayhan, H., Effects of atmospheric pressure plasma treatments on some physical properties of wool fibers. *Text. Res. J.*, *79*, 14, 1260–1265, 2009.
11. Höcker, H., Plasma treatment of textile fibers. *Pure Appl. Chem.*, *74*, 3, 423–427, 2002.
12. Barani, H. and Maleki, H., Plasma and ultrasonic process in dyeing of wool fibers with madder in presence of lecithin. *J. Dispersion Sci. Technol.*, *32*, 8, 1191–1199, 2011.

13. Barani, H. and Calvimontes, A., Effects of oxygen plasma treatment on the physical and chemical properties of wool fiber surface. *Plasma Chem. Plasma Process.*, *34*, 1291–1302, 2014.
14. Jeon, S.H., Hwang, K.H., Lee, J.S., Boo, J.H., Yun, S.H., Plasma treatments of wool fiber surface for microfluidic applications. *Mater. Res. Bull.*, *69*, 65–70, 2015.
15. Molina, R., Espinós, J.P., Yubero, F., Erra, P., Gonzalez-Elipe, A.R., XPS analysis of down stream plasma treated wool: Influence of the nature of the gas on the surface modification of wool. *Appl. Surf. Sci.*, *252*, 5, 1417–1429, 2005.
16. Haji, A., Ashraf, S., Nasiriboroumand, M., Lievens, C., Environmentally friendly surface treatment of wool fiber with plasma and chitosan for improved coloration with cochineal and safflower natural dyes. *Fibers Polym.*, *21*, 743–750, 2020.
17. Sajed, T., Haji, A., Mehrizi, M.K., Boroumand, M.N., Modification of wool protein fiber with plasma and dendrimer: Effects on dyeing with cochineal. *Int. J. Biol. Macromol.*, *107*, 642–653, 2018.
18. Talebian, A., Nourbakhsh, S., Ebadi, A.G., Valipour, P., Variations of Surface Properties of Corona Treated Wool. *Asian J. Chem.*, *21*, 1, 682, 2009.
19. Talebian, A., Habibi, S., Neshat, P., Green dyeing of weld on corona discharge treated wool fabric. *J. Text. Inst.*, *112*, 1, 144–151, 2021.
20. Ristić, N., Jovančić, P., Canal, C., Jocić, D., Influence of corona discharge and chitosan surface treatment on dyeing properties of wool. *J. Appl. Polym. Sci.*, *117*, 5, 2487–2496, 2010.
21. Hassan, M.M., Enhanced antistatic and mechanical properties of corona plasma treated wool fabrics treated with 2, 3-epoxypropyltrimethylammonium chloride. *Ind. Eng. Chem. Res.*, *53*, 27, 10954–10964, 2014.
22. Abdel-Latif, E. and Gaffer, H.E., Effect of gamma irradiation on the color properties of synthetic fabrics dyed by arylazo-pyrazolone disperse dyes. *J. Text. Inst.*, *108*, 5, 653–656, 2017.
23. Machnowski, W., Perkowski, J., Wrzosek, H., The influence of the ionizing radiation applied to the conservation of historical textiles on the natural fibers properties. *Note Konserw.*, *13*, 71–88, 2010.
24. Medha, K., Rajna, S., Devi, L.J., Samant, L., Jose, S., A comprehensive review on moth repellent finishing of woolen textiles. *J. Cult. Heritage*, *49*, 260–271, 2021.
25. Shahidi, S. and Wiener, J., Radiation effects in textile materials, in: *Radiation effects in materials*, pp. 309–328, Intech Open, London, United Kingdom, 2016.
26. Xin, J.H., Zhu, R., Hua, J., Shen, J., Surface modification and low temperature dyeing properties of wool treated by UV radiation. *Color. Technol.*, *118*, 4, 169–173, 2002.
27. Machnowski, W., Gutarowska, B., Perkowski, J., Wrzosek, H., Effects of gamma radiation on the mechanical properties of and susceptibility to biodegradation of natural fibers. *Text. Res. J.*, *83*, 1, 44–55, 2013.

28. Millington, K.R., Comparison of the effects of gamma and ultraviolet radiation on wool keratin. *Color. Technol.*, *116*, 9, 266–272, 2000.
29. Rana, M.S., Mamun, M., Biswas, S., Sourov, M.R.F., Surface modification of wool fabric with chitosan and gamma radiation. *Manuf. Sci. Technol.*, *4*, 1–10, 2017.
30. Xue, Z. and Jin-Xin, H., Effect of microwave irradiation on the physical properties and structures of wool fabric. *J. Appl. Polym. Sci.*, *119*, 2, 944–952, 2011.
31. Yang, M., Yan, K., Zhou, A., A study on Xe excilamp treatment of wool fibers. *J. Appl. Polym. Sci.*, *118*, 1, 241–246, 2010.
32. Nourbakhsh, S., Ebrahimi, I., Valipour, P., Laser treatment of the wool fabric for felting shrinkage control. *Fibers Polym.*, *12*, 521–527, 2011.
33. Tonin, C., Zoccola, M., Aluigi, A., Varesano, A., Montarsolo, A., Vineis, C., Zimbardi, F., Study on the conversion of wool keratin by steam explosion. *Biomacromolecules*, *7*, 12, 3499–3504, 2006.
34. Kukle, S., Grāvītis, J., Putniņa, A., Stikute, A., The effect of steam explosion treatment on technical hemp fibres, in: *Environment. Technologies. Resources. Proceedings of the international scientific and practical conference*, vol. 1, pp. 230–237, 2011.
35. Deepa, B., Abraham, E., Cherian, B.M., Bismarck, A., Blaker, J.J., Pothan, L.A., Kottaisamy, M., Structure, morphology and thermal characteristics of banana nano fibers obtained by steam explosion. *Bioresour. Technol.*, *102*, 2, 1988–1997, 2011.
36. Ningthoujam, D.S., Tamreihao, K., Mukherjee, S., Khunjamayum, R., Devi, L.J., Asem, R.S., Keratinaceous wastes and their valorization through keratinolytic microorganisms, in: *Keratin*, IntechOpen, London, United Kingdom, 2018.
37. Ranjbar-Mohammadi, M., Arami, M., Bahrami, H., Mazaheri, F., Mahmoodi, N.M., Grafting of chitosan as a biopolymer onto wool fabric using anhydride bridge and its antibacterial property. *Colloids Surf., B*, *76*, 2, 397–403, 2010.
38. Zhang, P., Zhang, N., Wang, Q., Wang, P., Yuan, J., Shen, J., Fan, X., Disulfide bond reconstruction: A novel approach for grafting of thiolated chitosan onto wool. *Carbohydr. Polym.*, *203*, 369–377, 2019.
39. Shavandi, A. and Ali, M.A., Graft polymerization onto wool fibre for improved functionality. *Prog. Org. Coat.*, *130*, 182–199, 2019.
40. Martinez-Hernandez, A.L., Velasco-Santos, C., Icaza, M.D., Castaño, V.M., Grafting of methyl methacrylate onto natural keratin. *e-Polymers*, *3*, 1, 016, 2003.
41. Gawish, S.M., Ramadan, A.M., Abo El-Ola, S.M., Abou El-Kheir, A.A., Citric acid used as a cross-linking agent for grafting β-cyclodextrin onto wool fabric. *Polym. Plast. Technol. Eng.*, *48*, 7, 701–710, 2009.
42. Molakarimi, M., Khajeh Mehrizi, M., Haji, A., Effect of plasma treatment and grafting of β-cyclodextrin on color properties of wool fabric dyed with Shrimp shell extract. *J. Text. Inst.*, *107*, 10, 1314–1321, 2016.

43. Gogoi, M., Kadam, V., Jose, S., Shakyawar, D.B., Kalita, B., Multifunctional finishing of woolens with lemongrass oil. *J. Nat. Fibers*, *19*, 4, 1353–1365, 2022.
44. Chen, N., Liu, C.K., Brown, E.M., Latona, N., Environment-friendly treatment to reduce photoyellowing and improve UV-blocking of wool. *Polym. Degrad. Stab.*, *181*, 109319, 2020.
45. Song, Y.X., Rong, M.Z., Zhang, M.Q., Improvement of multiple-responsive shape memory effects of wool through increasing the content of disulfide bonds. *Polymer*, *188*, 122130, 2020.
46. Wu, J., Zhou, D., Looney, M.G., Waters, P.J., Wallace, G.G., Too, C.O., A molecular template approach to integration of polyaniline into textiles. *Synth. Met.*, *159*, 12, 1135–1140, 2009.
47. Yu, J., Pang, Z., Zhang, J., Zhou, H., Wei, Q., Conductivity and antibacterial properties of wool fabrics finished by polyaniline/chitosan. *Colloids Surf., A*, *548*, 117–124, 2018.
48. Jose, S., Thomas, S., Jibin, K.P., Sisanth, K.S., Kadam, V., Shakyawar, D.B., Surface modification of wool fabric using sodium lignosulfonate and subsequent improvement in the interfacial adhesion of natural rubber latex in the wool/rubber composites. *Ind. Crops Prod.*, *177*, 114489, 2022.
49. Liu, G., Wang, W., Yu, D., Facile fabrication of durable antibacterial and anti-felting wool fabrics with enhanced comfort *via* novel N-phenylmaleimide finishing. *Bioprocess. Biosyst. Eng.*, *45*, 5, 921–929, 2022.
50. Buchacher, M., Bechtold, T., Pham, T., Characterisation of reduction state of cystine linkages on wool fibre surface under heterogeneous reaction conditions. *Polym. Test.*, *106*, 107438, 2022.
51. Rippon, J.A. and Evans, D.J., Improving the properties of natural fibres by chemical treatments, in: *Handbook of natural fibres*, pp. 245–321, Woodhead Publishing, Cambridgeshire, United Kingdom, 2020.
52. Carran, R.S., Ghosh, A., Dyer, J.M., The effects of zeolite molecular sieve based surface treatments on the properties of wool fabrics. *Appl. Surf. Sci.*, *287*, 467–472, 2013.
53. Carran, R.S., Ghosh, A., Dyer, J.M., Modification of surface properties of wool fabric with linde type a nano-zeolite. *J. Appl. Polym. Sci.*, *132*, 32, 2015.
54. Samanta, A.K., Bhattacharyya, R., Jose, S., Basu, G., Chowdhury, R., Fire retardant finish of jute fabric with nano zinc oxide. *Cellulose*, *24*, 1143–1157, 2017.
55. Jose, S., Shanmugam, N., Das, S., Kumar, A., Pandit, P., Coating of lightweight wool fabric with nano clay for fire retardancy. *J. Text. Inst.*, *110*, 5, 764–770, 2019.
56. Nazari, A., Montazer, M., Dehghani-Zahedani, M., Nano TiO2 as a new tool for mothproofing of wool: protection of wool against Anthrenus verbasci. *Ind. Eng. Chem. Res.*, *52*, 3, 1365–1371, 2013.
57. Jose, S., Nachimuthu, S., Das, S., Kumar, A., Moth proofing of wool fabric using nano kaolinite. *J. Text. Inst.*, *109*, 2, 225–231, 2018.

58. Bhandari, V., Jose, S., Badanayak, P., Sankaran, A., Anandan, V., Antimicrobial finishing of metals, metal oxides, and metal composites on textiles: a systematic review. *Ind. Eng. Chem. Res.*, *61*, 1, 86–101, 2022.
59. Haji, A., Mousavi Shoushtari, A., Mazaheri, F., Tabatabaeyan, S.E., RSM optimized self-cleaning nano-finishing on polyester/wool fabric pretreated with oxygen plasma. *J. Text. Inst.*, *107*, 8, 985–994, 2016.
60. Wang, S., Hou, W., Wei, L., Jia, H., Liu, X., Xu, B., Antibacterial activity of nano-SiO2 antibacterial agent grafted on wool surface. *Surf. Coat. Technol.*, *202*, 3, 460–465, 2007.
61. Uğur, S.S. and Sariişik, A.M., Nano polyurethane based surface modification on the anti-felting functionalization of wool fabrics. *Prog. Org. Coat.*, *77*, 8, 1249–1252, 2014.
62. Zhang, H. and Zhu, H., Modification of wool fabric treated with tetrabutyl titanate by hydrothermal method. *J. Text. Inst.*, *103*, 10, 1108–1115, 2012.
63. Samant, L., Jose, S., Rose, N.M., Shakyawar, D.B., Antimicrobial and UV protection properties of cotton fabric using enzymatic pretreatment and dyeing with Acacia catechu. *J. Nat. Fibers*, *19*, 6, 2243–2253, 2022.
64. Ammayappan, L., Application of enzyme on woolen products for its value addition: an overview. *J. Text. Appar. Technol. Manag.*, *8*, 3, 2013.
65. Prajapati, C.D., Smith, E., Kane, F., Shen, J., Selective enzymatic modification of wool/polyester blended fabrics for surface patterning. *J. Cleaner Prod.*, *211*, 909–921, 2019.
66. Fu, J., Su, J., Wang, P., Yu, Y., Wang, Q., Cavaco-Paulo, A., Enzymatic processing of protein-based fibers. *Appl. Microbiol. Biotechnol.*, *99*, 10387–10397, 2015.
67. Soun, B., Kaur, D., Jose, S., Effect of transglutaminase enzyme on physico-mechanical properties of rambouillet wool fiber. *J. Nat. Fibers*, *17*, 6, 793–801, 2020.
68. Kotlińska, A. and Lipp-Symonowicz, B., Research on the enzymatic treatment of wool fibres and changes in selected properties of wool. *Fibres Text. East. Eur*, *19*, 3, 86, 2011.
69. Zhang, R., Cai, Z., Zhang, H., Studies on the remedial effect of transglutaminase on protease anti-felting treated wool. *J. Text. Inst.*, *101*, 11, 1015–1021, 2010.
70. Lenting, H.B.M., Schröder, M., Gübitz, G.M., Cavaco-Paulo, A., Shen, J., New enzyme-based process direction to prevent wool shrinking without substantial tensile strength loss. *Biotechnol. Lett*, *28*, 711–716, 2006.
71. Mojsov, K., Janevski, A., Andronikov, D., Jordeva, S., Gaber, S., Ignjatov, I., Enzymatic treatment of wool fabrics with lipase in the improvement of some propertires of wool fabrics. *Tekst. Ind.*, 4–11, 2020.
72. Osma, J.F., Herrera, J.L.T., Couto, S.R., Banana skin: A novel waste for laccase production by Trametes pubescens under solid-state conditions. Application to synthetic dye decolouration. *Dyes Pigm.*, *75*, 1, 32–37, 2007.

73. Karanikas, E.K., Kosolia, C.T., Zarkogianni, M.C., Nikolaidis, N.F., Tsatsaroni, E.G., Effect of enzymatic treatment on the dyeing properties of protein wool fibers. *Fibers Polym.*, *14*, 223–229, 2013.
74. Park, M., Kim, H.Y., Jin, F.L., Lee, S.Y., Choi, H.S., Park, S.J., Combined effect of corona discharge and enzymatic treatment on the mechanical and surface properties of wool. *J. Ind. Eng. Chem.*, *20*, 1, 179–183, 2014.
75. Lantto, R., Ellis, J., Fatarella, E., Cortez, J., Influence of different pretreatments on the accessibility of transglutaminase and tyrosinase to wool fibre proteins. *J. Text. Inst.*, *103*, 1, 55–63, 2012.
76. Heck, T., Faccio, G., Richter, M., Thöny-Meyer, L., Enzyme-catalyzed protein crosslinking. *Appl. Microbiol. Biotechnol.*, *97*, 461–475, 2013.
77. Abou Taleb, M., Gomaa, S.K., Wahba, M.I., Zaki, R.A., El-Fiky, A.F., El-Refai, H.A., El-Sayed, H., Bioscouring of wool fibres using immobilized thermophilic lipase. *Int. J. Biol. Macromol.*, *194*, 800–810, 2022.
78. Rani, S., Kadam, V., Rose, N.M., Jose, S., Yadav, S., Shakyawar, D.B., Wheat starch, gum arabic and chitosan biopolymer treatment of wool fabric for improved shrink resistance finishing. *Int. J. Biol. Macromol.*, *163*, 1044–1052, 2020.
79. Vilchez, S., Manich, A.M., Jovancic, P., Erra, P., Chitosan contribution on wool treatments with enzyme. *Carbohydr. Polym.*, *71*, 4, 515–523, 2008.
80. Sadeghi-Kiakhani, M., Arami, M., Gharanjig, K., Application of a biopolymer chitosan-poly (propylene) imine dendrimer hybrid as an antimicrobial agent on the wool fabrics. *Iran. Polym. J.*, *22*, 931–940, 2013.
81. Abdelghaffar, R.A. and El-Bisi, M.K., Improvement of the Dyeability and Salt-Free Dyeing for Wool Fabrics with Anionic Dyes by Pretreatment with Whey and Soybean Proteins. *J. Nat. Fibers*, *19*, 15, 11003–11020, 2022.
82. Xu, W., Shen, X., Wang, X., Ke, G., Effective methods for further improving the wool properties treated by corona discharge. *Sen'i Gakkaishi*, *62*, 5, 111–114, 2006.
83. Kan, C.W., Chan, K., Yuen, C.W.M., Miao, M.H., Surface properties of low-temperature plasma treated wool fabrics. *J. Mater. Process. Technol.*, *83*, 1-3, 180–184, 1998.
84. Kadam, V., Rani, S., Jose, S., Shakyawar, D.B., Shanmugam, N., Biomaterial based shrink resist treatment of wool fabric: A sustainable technology. *Sustainable Mater.Technol.*, *29*, e00298, 2021.
85. Allafi, F.A.S., Hossain, M.S., Ab Kadir, M.O., Shaah, M.A.H., Lalung, J., Ahmad, M.I., Waterless processing of sheep wool fiber in textile industry with supercritical CO2: Potential and challenges. *J. Cleaner Prod.*, *285*, 124819, 2021.

# 10

# Advances in Nanofinishing for Textiles

**Çağlar Sivri**

*Department of Engineering Management, Faculty of Engineering and Natural Sciences, Bahçeşehir University, Beşiktaş, İstanbul, Türkiye*

***Abstract***

After development of the functionality criteria, nanofinishing is one of the most preferable subjects in the last 10 years in the surface modification of textiles, on account of several prominent benefits, including multifunctionality, effective retention, high penetrating efficiency of finishing agents, and enhanced performance. One useful interpretation to why the performance of nanofinished fabrics exceeds that of traditional finished fabrics, even when they are treated with the same textile material and the same equipment, is based among other factors in the nature of the nanotechnology itself, because nano-level materials behave considerably different comparing to the larger-scale material. Naturally, fibers and yarns as well as knitted and woven or nonwoven textile structures can all be treated, using the nanofinishing method. Nanocoating and lamination, sol-gel, synthesis, and integration of nanoparticles are the popular methods of nanofinishing of textiles nowadays. Water repellency, antimicrobial ability, UV protection, flame retardancy, self-cleaning, odor inhibition, coloration fastness, thermal comfort, enhanced comfort, etc., can be achieved by applying nanofinishing. The most recent advances in textile nanofinishing surface modification, including material, cost, applicability, and sustainability, and future development prospect are discussed and future work suggestions are given in this chapter.

***Keywords*:** Nanotechnology, nanofinishing, textile finishing, nanocoatings, nanoparticles

*Email*: caglar.sivri@bau.edu.tr

Aminoddin Haji and Çağlar Sivri (eds.) Surface Modification and Functionalization of Textiles, (229–252) © 2026 Scrivener Publishing LLC

## 10.1 Introduction

Nanotechnology has revolutionized nearly every industry by simultaneously enhancing material properties and advancing production processes. This widespread impact, in turn, has accelerated the adoption and dissemination of nanotechnology across various sectors. Among these, the textile industry stands out as one of the most significantly influenced, driven by evolving consumer demands for advanced performance, functionality, and superior product qualities [1].

While textile production and hand-applied finishing techniques have existed for thousands of years, systematic and modern finishing processes only began to emerge in the early 19th century, as noted by Mitchell [2]. The convergence of nanotechnology and textile finishing occurred much later, taking shape in the late 20th and early 21st centuries, when the practical potential of nanotechnology started to materialize [3].

Nanofinishing has brought about a transformative shift in textile surface properties, enabling functions like repellency, self-cleaning, conductivity, sensing, and UV protection, among others [4]. Just as materials behave differently at the nano-level, the surface of nanofinished textiles behaves differently, and, thus, different functions are achieved.

Novel versions of the ever-known materials at the nano-level, newly synthesized polymers and derivatives, sophisticated techniques, and advanced equipment have boosted the development of nanofinishing and its integration into textiles in the last decade. Conscious consumers are shifting expectations and demanding new features or more durable versions of the existing features relevant to the surfaces. Increasing R&D centers in the companies and staff holding PhD degrees having expertise in this area also contribute to this development, especially, the companies are pushing universities for collaboration to transform lab-scale findings into commercially sound products. Also, understanding the basics of traditional finishing methods and how they evolved to nanofinishing techniques that create extraordinary features can revolutionise the R&D efforts and the fruits of these efforts. Considering all these factors together, the attention on nanofinishing increases day by day, so the related works accordingly.

This chapter presents a detailed and most up-to-date overview related to nanofinishing in textiles. Firstly, nanofinishing techniques were discussed, and nanofinishing for different textile structures (fiber, woven, knitted, and nonwoven) was reviewed. The end uses of nanofinished textiles have been reviewed in the following section. Lastly, the outlook and future work advises part provides guiding remarks for the researchers and developers in this area.

## 10.2 Nanofinishing Techniques

Today, there are a variety of nanofinishing methods that can enhance textiles with multiple functional properties. However, it is important to note that not every technique delivers equally successful outcomes in terms of performance, durability, or cost. Achieving all these desired qualities at once with a single method is often challenging.

For instance, one approach might offer excellent performance and long-lasting durability but come at a high financial cost. Conversely, another technique may be more budget-friendly and still deliver good performance yet fall short in terms of durability. These trade-offs are an essential part of selecting an appropriate method for a given application.

The following sections explore these techniques in detail, evaluating each for its strengths, limitations, and suitability depending on the intended use.

### 10.2.1 Nanocoating

Nanocoating and nanolamination represent more advanced iterations of traditional coating and lamination techniques, offering greater precision and the ability to functionalize surfaces at a much smaller scale. While conventional coating methods often struggle with drawbacks like reduced material strength, weak adhesion, poor resistance to abrasion, and limited durability, nanocoating overcomes many of these limitations.

By applying an ultra-thin layer at the nanometer scale, nanocoating significantly improves the surface's reactivity, mechanical strength, and long-term durability, making it a superior choice for enhancing textile performance [5].

#### *10.2.1.1 Sol-Gel Technique*

The sol-gel method is a versatile nanocoating approach that involves applying nano-scale functional materials with benefits such as precision, eco-friendliness, and low processing temperatures. This technique allows for the development of individual nanomaterials or nanocomposites that exhibit multiple functionalities [6]. In practice, nano-sized components with targeted properties are dispersed within a polymeric matrix, and this mixture is then applied as a coating. Thanks to its advantageous characteristics, the sol-gel method has a broad range of applications, including self-cleaning surfaces, flame retardancy, antibacterial treatments, UV protection, and wrinkle resistance. Figure 10.1 demonstrates some of the most well-known and promising sol-gel applications in textile finishing [7].

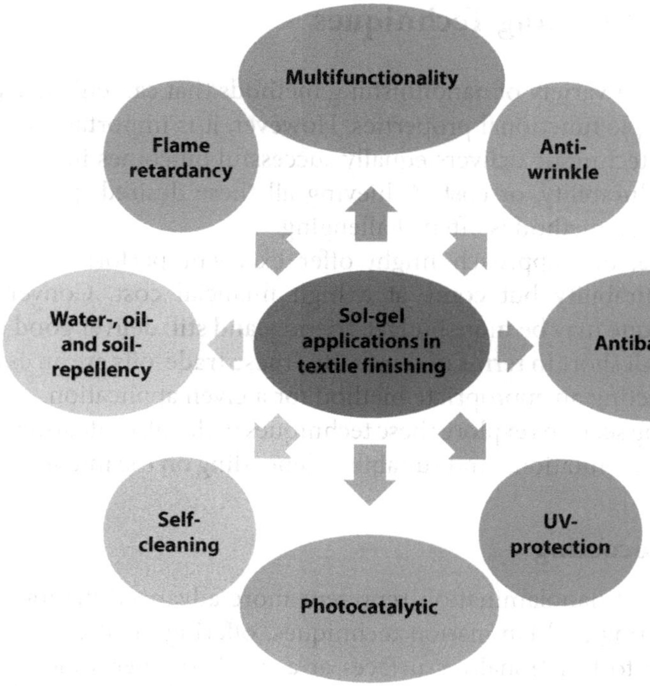


**Figure 10.1** Sol-gel applications in textile finishing [7].

Colleoni and colleagues developed a novel sol-gel multilayer method to enhance cotton fabrics using nanosol treatments. They used TEOS as the precursor and found that a higher concentration of the inorganic matrix on the fabric surface significantly improved the material's protective qualities [8]. Similarly, Abramova *et al.* designed a sol-gel process that applies nano-sized antibacterial particles onto textile surfaces with the help of ultrasonic waves. Their research showed that this method could be scaled up to a semi-industrial level using roll-to-roll techniques, allowing for even, consistent coatings and demonstrating the practical potential of the approach [9].

### *10.2.1.2 Layer-by-Layer (LbL) Technique*

The layer-by-layer (LbL) technique allows for the precise construction of nano-scale films on various substrates such as fibers, textiles, or nanofibers through electrostatic interactions. This method provides a more refined and controlled approach compared to traditional coating techniques. The LbL process typically involves a series of dipping and rinsing steps that drive ion exchange and the association of polyelectrolytes [10].

For instance, Ugur *et al.* applied a multilayer of nano-$Al_2O_3$ films onto cotton fabrics using LbL, resulting in enhanced fire resistance, tensile strength, and UV protection [11]. In another recent study, thin, multifunctional coatings incorporating polyaromatic hybrid nanoparticles were applied *via* LbL, yielding textiles with improved water and stain repellency for outdoor wear, enhanced hygiene for industrial clothing, and added comfort and UV protection for medical and athletic apparel [12].

#### 10.2.1.3 Plasma Polymerization

Plasma polymerization is a specialized process that modifies textile surfaces by breaking chemical bonds within fiber polymers, leading to changes in morphology, activation of the fiber surface, and increased surface area [13]. Gleissner *et al.* highlighted its value in improving surface properties of high-performance fibers, including wettability, interface strength, orientation, and thermal, mechanical, and chemical characteristics [14].

In another study, plasma polymerization was combined with silver nanoparticles through a co-sputtering method. This allowed for uniform distribution of the nanoparticles within the plasma layer, which, in turn, added antimicrobial functionality to the treated textile surfaces [15].

#### 10.2.1.4 Electroless Deposition

Electroless deposition is based on reducing metal ions in an aqueous solution to form thin, functional coatings without the need for an external electrical source. Key parameters like thickness, conductivity, elemental composition, and surface morphology are tailored to produce effective coatings for both laboratory and industrial applications [16].

Researchers used this method to apply copper coatings on woven cellulose lyocell fabrics, creating conductive layers suitable for temperature-sensing textiles that could monitor a patient's skin temperature when embedded in clothing [17]. Another innovative use involved developing wearable energy storage fabrics using stretchable textiles infused with nickel and cobalt bimetallic phosphide, showcasing impressive mechanical and electrochemical properties ideal for future wearable electronics [18].

#### 10.2.1.5 Vapor Deposition

Vapor deposition techniques are used for modifying surfaces of shielding materials, smart textiles (textronics), garments, and specialty accessories. These techniques are divided into physical vapor deposition (PVD) and chemical vapor deposition (CVD) [19]. PVD is a coating technique where the material first turns into vapor—one atom or molecule at a time—before settling onto the surface of the textile. This process allows for precise, uniform coatings without the need for liquid intermediates [20]. In CVD, vapor precursors react (usually through heating) to synthesize materials that are then deposited [21]. A hybrid system combining both techniques was developed by Mundra *et al.*, enabling substrates to move directly from the PVD to the CVD chamber. This system successfully supported graphene growth and demonstrated the advantages of integrating both deposition methods [22].

#### *10.2.1.6 Liquid-Phase Deposition (LPD)*

LPD is an environmentally friendly and economical method that uses controlled hydrolysis to produce oxide films. Parameters like pH, temperature, and surface characteristics are finely tuned to develop coatings with unique functionalities, particularly superhydrophilic properties [23].

For example, Tsuge *et al.* created transparent, highly adhesive titanium dioxide ($TiO_2$) films using LPD and self-assembly methods. These coatings exhibited outstanding superhydrophilic behavior [24]. LPD is also capable of producing optical films with strong photoluminescence, thanks to its ability to tailor phase composition and surface morphology [25].

#### *10.2.1.7 Langmuir–Blodgett (LB) Films*

The LB technique involves immersing a substrate into a water subphase to form a monolayer, which can then be layered to create thin, uniform coatings at the molecular scale. This method supports the creation of multilayered, multifunctional composites [26].

Rauf *et al.* applied this method to fabricate smart textile sensors for detecting humidity using metal–organic frameworks. These sensitive films were directly applied to fabrics with interdigitated electrodes [27]. LB films have also been developed at the thread level, enabling applications in toxin filtration, solar shielding, and gas sensing due to their conductive and insulating properties [28].

#### *10.2.1.8 Nanofiber Coating*

Nanofibers are valued in engineering for their fine diameter, high surface area, interconnected pores, and tunable structure [29]. In recent years, coating textile surfaces with nanofibers has become a key focus for adding advanced functions to traditional fabrics [30].

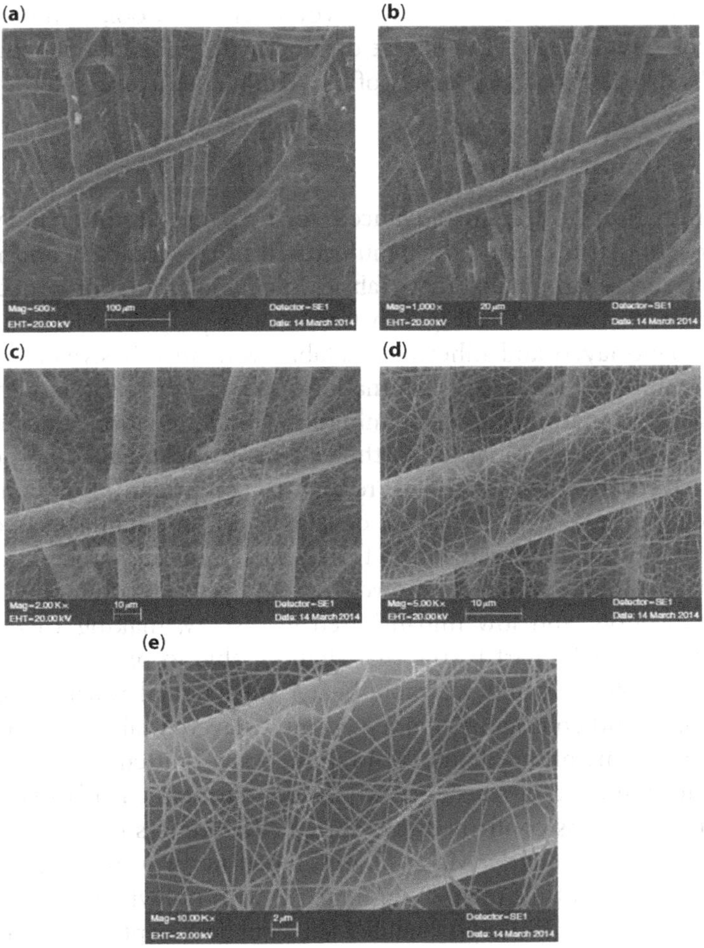


**Figure 10.2** Mask specimen covered with electrospun nanofibers using mixture of superabsorbent polymer and polyvinyl alcohol (a) 500×; (b) 1,000×; (c) 2,000×; (d) 5,000×; (e) 10,000× [31].

Sivri demonstrated this by applying electrospun nanofibers—made from a polyvinyl alcohol and superabsorbent polymer blend—onto polypropylene (PP) face masks (Figure 10.2). The coatings improved liquid absorption and provided significant viral and contaminant protection due to pore blockage while air permeability decreased. Scanning electron microscopy (SEM) and fourier transform infrared spectroscopy (FTIR) analyses confirmed uniform nanofiber distribution [31].

Other studies have shown that nanofiber coatings can also enhance surface roughness to achieve superhydrophobicity [32, 33]. For example, researchers used silicone nanofilaments on fibers like wool, cotton, polyester, viscose, and acetate. The degree of superhydrophobicity depended on the scale, distribution, and density of the nanofilament layers [34].

### 10.2.2 Nanolamination

Nanolamination is a more advanced form of the traditional lamination process that is created or manufactured microscopically and nano-/micro-layers are laminated on the fabric. In a nanolamination process, the process often involves creating a material (i.e., a polymer or fiber) in the form of nano-layers and adhering to a fabric with an adhesive in a process that resembles the traditional lamination process, with heat and pressure [35]. Flame laminating, hot melt lamination, film lamination, and adhesive system–based nanolamination methods can be applied for bonding the various layers together, even with respect to compound(s) of laminated materials used [36, 37]. Gao *et al.* developed MXene nanosheets on the surface of cationic modified cotton through a laminating process, and the fabricated nanocomposite has showed a great electrical conductivity, high solar absorption, and low mid-infrared emissivity, making this a kind of multifunctional smart textile that shall be able to provide both heat-insulating protection and wearable heater in personal warming aspects [38]. Knížek and colleagues engineered multi-layered fabrics by laminating nanofiber membranes onto woven and warp-knitted textiles, aiming to enhance the weather resistance of outdoor garments. Performance and comfort evaluations confirmed the fabric's effectiveness in harsh weather conditions [39]. Similarly, Ma *et al.* developed nonwoven composites by alternating layers of PP meltblown material with polyvinylidene fluoride nanofibers. This LbL design enabled efficient interception of particles, making the composite a promising candidate for air filtration applications [40]. In a separate study, researchers explored the acoustic performance of hybrid laminated structures combining needle-punched and meltblown nonwoven layers. These materials, incorporating fibers ranging from micro- to nanoscale, were arranged in various configurations to assess their sound absorption capabilities. The study also revealed a link between thermal conductivity and acoustic performance (Figure 10.3), indicating that improvements in thermal conductivity could enhance the material's ability to absorb sound [41].

Earlier research has demonstrated that applying nanolamination or nanofinishing techniques to textile layers can significantly enhance the overall durability of composite materials. Notably, these methods have been effective in reducing or even eliminating delamination issues commonly

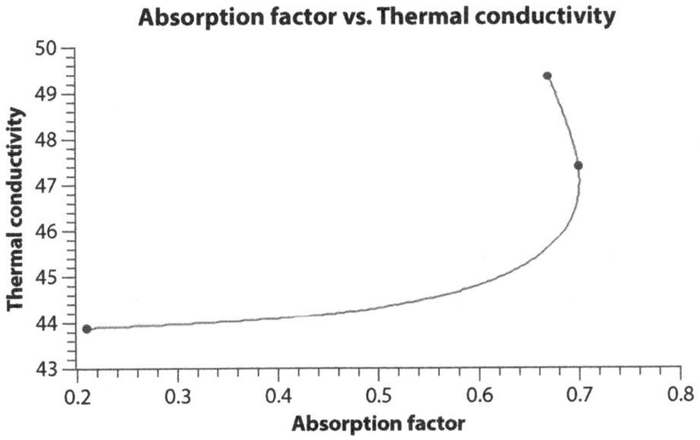


**Figure 10.3** The relationship between thermal conductivity and sound absorption coefficient of the specimens [41].

encountered during everyday use or washing. This improvement is largely attributed to the stronger bonding provided by nano-scale contact points and the increased surface roughness of the treated composites, both of which contribute to better adhesion between layers [42, 43]. Given these advantages, nanolamination is increasingly seen as a promising approach for surface functionalization in advanced textile applications.

### 10.2.3 Nanoparticle Synthesis and Textile Integration

A nanoparticle—also known as an ultrafine particle—is defined as a particle with a diameter ranging from 1 to 100 nanometers (nm) [44]. These particles exhibit exceptional mechanical strength, an increased surface-area-to-volume ratio, and enhanced chemical reactivity and stability due to their minute size. As such, nanoparticles hold significant promise for functionalizing various surfaces, including textiles [45]. Nanoparticles come in various forms, including metal-based, carbon-based, polymeric, ceramic, and lipid-based types [46].

Their synthesis methods are broadly categorized into top-down and bottom-up approaches [47]. Top-down techniques include sonication, laser ablation, ball milling, lithography, sputtering, etching, arc discharge, and electrospinning [48]. In contrast, bottom-up methods involve strategies like pyrolysis, spinning, sol-gel synthesis, green synthesis, and CVD [49]. Among these, metal-based and polymer-based nanoparticles are the most prevalent in textile finishing applications.

Metal nanoparticles are typically synthesized using specific metallic precursor salts as initiators. These salts are added to a solution, and the mixture is stirred in a dark environment at room temperature. Phytochemicals derived from natural extracts stabilize the solution before the required reactants are introduced to initiate nanoparticle formation. This process is often regarded as quick, cost-effective, and simple [50, 51]. Figure 10.4 offers a visual representation of the synthesis method.

Polymeric nanoparticles, on the other hand, are commonly produced through self-assembly or surfactant-guided methods [52]. Initially, a polymer is dissolved in an organic solvent and then emulsified into nanoscale droplets using high-speed homogenization. These droplets form an aqueous colloidal suspension, after which the solvent is evaporated, solidifying the particles. Finally, the nanoparticles are washed and collected by centrifugation [53].

Green synthesis has gained traction due to its eco-friendly attributes, low cost, and biocompatibility [54]. Especially popular are methods utilizing plant extracts, which are biodegradable, low in toxicity, and readily available from various plant parts such as stems, bark, roots, flowers, leaves, seeds, and fruits [55].

Nanoparticles can be applied on the surface or incorporated within textiles to impart functional properties [56, 57]. For example, Staneva *et al.* enhanced cotton fabrics with hydrogels containing barium hexaferrite nanoparticles, yielding antibacterial effects against *Escherichia coli* and *Pseudomonas aeruginosa* [58]. Literature consistently highlights the synergistic benefits of combining eco-friendly biopolymers with nanoparticles to improve textiles' barrier, antimicrobial, and antioxidant capabilities [59]. A notable case is the work of Shankar and Rhim, who developed chitosan films containing sulfur nanoparticles. These films showed increased hydrophobicity, mechanical strength, water vapor resistance, and antibacterial properties, making them ideal for outdoor clothing [60].

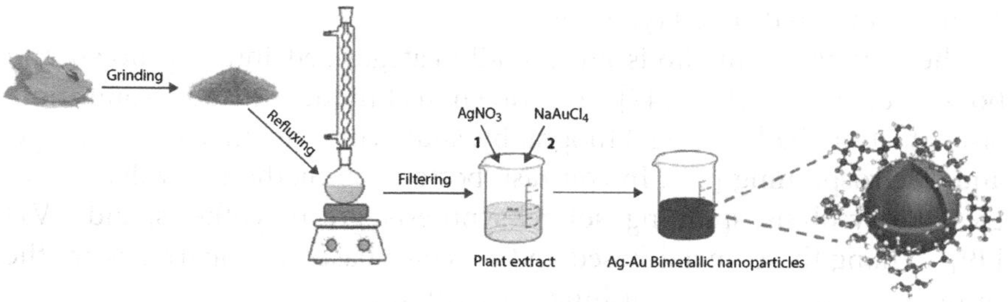


**Figure 10.4** The green synthesis method for Ag-Au bimetallic nanoparticles utilizing an aqueous extract of bay leaves (*Laurus nobilis* L.) [50].

Although relatively rare, clay-based nanoparticles also offer exciting possibilities for textile enhancement [61, 62]. Göcek applied nano-modified montmorillonite clay to cotton fabrics, improving flame retardancy, washing durability, and burn resistance [63]. In another instance, kaolinite-$TiO_2$ hybrid nanoparticles were used to functionalize cellulose fibers, leading to improved mechanical, thermal, and flame-resistant properties validated by advanced material characterization methods [64].

Electrospinning stands out as an accessible and scalable method for synthesizing and integrating nanoparticles into textiles [65]. It allows for not only the creation of nanoparticles but also their simultaneous encapsulation within nanofibers. The process creates a profiled fiber surface capable of retaining both micro- and nanoparticles in the polymer matrix [66, 67]. In Radacsi *et al.* pioneered a low-cost, single-step electrospinning method where nanoparticles were formed directly onto nanofibers, enhancing surface area and enabling their use in high-performance applications such as energy storage electrodes [68]. Moreover, electrospinning can be utilized to directly coat textiles with nanoparticles, thereby bestowing additional functional properties [69, 70]. In another innovative approach, Özkayalar and Alay Aksoy developed nonencapsulated phase change materials (PCMs) and bicomponent PCM nanofibers using co-axial electrospinning (Figure 10.5). The technique produced continuous, bicomponent, and uniform nanofibers capable of storing thermal energy, making them ideal for smart fabric applications [71]. Maleki *et al.* expanded on this by fabricating core-shell structured composite yarns containing PCMs and poly(l-lactic acid) using electrospinning. These yarns demonstrated superior heating and cooling behavior, making them highly suitable for thermoregulatory medical textiles involved in wound healing and drug delivery [72].

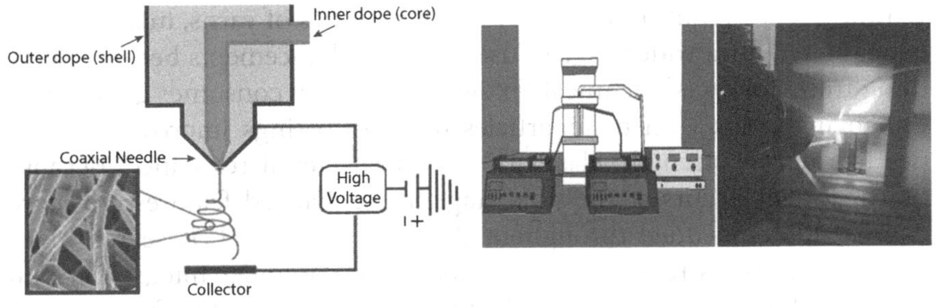


**Figure 10.5** Co-axial electrospinning setup [71].

## 10.3 Nanofinishing for Different Textile Structures

Nanofinishing is tailored to different textile structures—fibers, yarns, woven, knitted, and nonwoven fabrics—each responding distinctively to nano-scale modifications. Given the structural diversity across these categories, the effectiveness and type of nanofinishing technique must align accordingly.

### 10.3.1 Nanofinishing for Fibers

Applying nanotechnology directly to fibers enhances their surface characteristics before they are assembled into yarns or fabrics. By incorporating nanomaterials or utilizing nanoscale processes, nanofinishing enhances the structural, chemical, and biological characteristics of fibers [73]. Nanofinishing is carried out on individual fibers before they are woven or knitted into fabric. By treating each fiber separately, the process directly modifies its surface, leading to enhanced durability, consistent results, and specific functional benefits—such as self-cleaning capabilities or antiviral protection [74]. For instance, nanofinishing on cotton yields antimicrobial and UV protective traits [75]; on carbon fibers, it enhances conductivity and strength [76]. Similarly, wool gains shrink resistance and odor control [77], nylon improves in abrasion resistance and water repellency [78], silk exhibits antimicrobial qualities [79], and polyester achieves self-cleaning capabilities [80]. Although less common, fibers like hemp, bamboo, alpaca, and cashmere have also seen nanofinishing applications aimed at boosting thermal insulation, moisture management, and mechanical strength [81–84].

### 10.3.2 Nanofinishing for Yarns

Incorporating nanomaterials—like nanoparticles, nanotubes, or nanocoatings—can significantly improve the properties of yarns, making them better suited for a wide range of uses. These enhancements benefit everything from advanced technical textiles to everyday consumer goods [85]. Enhancing yarns with nanoparticles or nanocoatings improves properties like tensile strength, elasticity, and antibacterial resistance. Because the treatment occurs at the yarn stage, the enhanced features are more consistently embedded throughout the final fabric [86]. Research highlights applications ranging from antibacterial nanofiber integration into yarns to the fabrication of conductive silk yarns suitable for wearable electronics [87, 88].

### 10.3.3 Nanofinishing for Woven Fabric Structures

The tightly packed, stable weave of woven fabrics allows for even distribution and better retention of nanofinishing agents [89]. Currently, the literature does not point to a single, standardized method for applying nanofinishing to woven textiles. As a result, various finishing techniques can be employed depending on the desired outcome. Among the most common benefits achieved through nanofinishing are enhanced strength, water resistance, and self-cleaning properties [90]. For example, polysaccharide-based fabrics treated *via* dip-coating processes have demonstrated improved antimicrobial and UV protective properties [91]. Other studies introduced MXene coatings for conductive cotton, yielding high washability and shielding performance while preserving flexibility [92].

### 10.3.4 Nanofinishing for Knitted Fabric Structures

Due to their looped and elastic architecture, knitted fabrics pose unique challenges and opportunities for nanofinishing. While their open structure supports deeper nanoparticle penetration, it can complicate the uniformity of application [93]. Additionally, their stretchability may compromise finish durability. Zhu *et al.* introduced a novel method for improving the anti-pilling performance of knitted wool fabrics by utilizing engineered water nanostructures. At the nano scale, water molecules facilitated greater mobility among the entangled fibers, which, in turn, enhanced the fabric's resistance to pilling [94]. In recent years, the application of nano-sized metallic materials to finish knitted fabrics has gained significant traction, largely due to the superior conductive and thermal properties they offer. For example, one study demonstrated a metallization technique applied to weft-knitted fabrics to produce wearable biaxial strain sensors [95]. In another case, researchers applied a MXene coating to weft-knitted alginate/viscose fabrics, resulting in multifunctional wearable heaters with improved flame resistance [96].

### 10.3.5 Nanofinishing for Nonwoven Fabric Structures

Nonwovens, known for their high porosity and typically short life spans, require specially designed nanofinishing processes. Common methods include electrospinning, dipping, and spraying. These allow for more substantial application volumes and enhanced functionalities [97]. Gadkari *et al.* developed an antibacterial needle-punched nonwoven fabric by incorporating silver-loaded polyester nanocomposite fibers, achieved through surface etching using alkaline hydrolysis. This nanofinishing process significantly enhanced the functionality of the traditionally used nonwoven surface, making it well-suited for various medical applications such as therapeutic dressings, diapers, and surgical gowns [98]. In a separate study, cobalt carbonate hydroxide was grafted onto nonwoven fabric to introduce nanoscale surface roughness. The resulting nanocomposite material demonstrated effective performance as a separation medium for surfactant-stabilized organic solvent/water emulsions [99].

## 10.4 End Uses and Functions of Nanofinished Textiles

The wide array of functionalities enabled by nanofinishing techniques has greatly diversified the applications of textiles across numerous sectors, including apparel, healthcare, sportswear, home, and industrial uses. In clothing, features such as UV protection, liquid repellency, self-cleaning, breathability, moisture regulation, and antimicrobial activity are among the most frequently applied enhancements [12]. For instance, a process developed by Tarek *et al.* utilized silicone and ZnO nanoparticles in a pad-dry-cure technique, producing cotton and polyester fabrics with lasting self-cleaning and anti-wrinkle properties [100]. In another study, researchers applied nanofinishing techniques to alter the surface properties of clothing fabrics. As a result, the treated textiles exhibited super-hydrophobicity while remaining breathable and comfortable. Additionally, the finishing process imparted cooling functionality and UV protection [101].

Nanofinishing has also become vital in the field of medical textiles. Enhancements such as antimicrobial and antiviral activity, wound healing support, and resistance to fluids significantly improve hygiene and safety. One notable advancement involves antimicrobial coatings incorporating organo-selenium compounds, which demonstrated effectiveness even after repeated laundering [102]. Furthermore, nanofibers modified with MXene nanobelts have shown promising wound healing and drug delivery capabilities, offering both thermal responsiveness and biocompatibility [103].

In the domain of athletic wear, the pursuit of all-weather functionality remains a priority. Consumers demand garments that offer protection against wind, water, and cold while still maintaining breathability and moisture management. Nanofinishing addresses these needs effectively, enhancing performance textiles to ensure comfort and adaptability during physical activity [104].

Home textiles have, likewise, benefited from nanotechnology, particularly in enhancing anti-allergenic properties, stain resistance, and flame retardancy.

Treatments involving hydrophobic and oleophobic finishes—ranging from fluorocarbons to dendrimers and nano-enabled coatings—have proven effective in sustaining performance even after multiple laundering cycles [105]. Flame retardant nanofinishing, especially on upholstery fabrics, has become crucial in reducing household fire risks. Here, nanoparticles and nanotube-based coatings provide enhanced safety without compromising comfort [106].

Industrial and automotive textiles rely on nanofinishing for increased durability and specialized functionality. Fire resistance, high strength, chemical resilience, and sound absorption are key requirements met through advanced nanocoatings and nanofiber technologies. These modifications not only extend the life of the materials but also improve vehicle safety and comfort by enhancing thermal regulation and mechanical properties [107].

Emerging applications such as nanoyarns are reshaping future possibilities. These nanostructured yarns, capable of being woven or knitted, offer multifunctionality and scalability [108]. Among the promising techniques, bubble electrospinning shows potential to revolutionize nanoyarn production by simulating spinning mechanisms *via* controlled bubble formation [109]. Bubble electrospinning has been depicted in Figure 10.6.

Recent attention has also turned toward nano-enabled dyeing processes. Cotton fabrics dyed with nanoparticles exhibit improved durability and UV resistance with reduced toxicity, offering an environmentally safer alternative [110]. Likewise, nanocellulose-based dyeing has demonstrated success in reducing water contamination during textile processing, indicating a promising shift toward greener dyeing technologies [111].

Sustainability remains at the forefront of current research [112, 113]. Efforts to develop eco-friendly nanofinishing methods have led to innovations such as beetroot extract-based silver nanoparticle synthesis, yielding cost-effective antimicrobial cotton fabrics [114]. Other green approaches—including plasma-assisted treatments and bio-based nanocomposites—highlight the shift toward minimizing ecological footprints while maintaining high-performance standards [115].

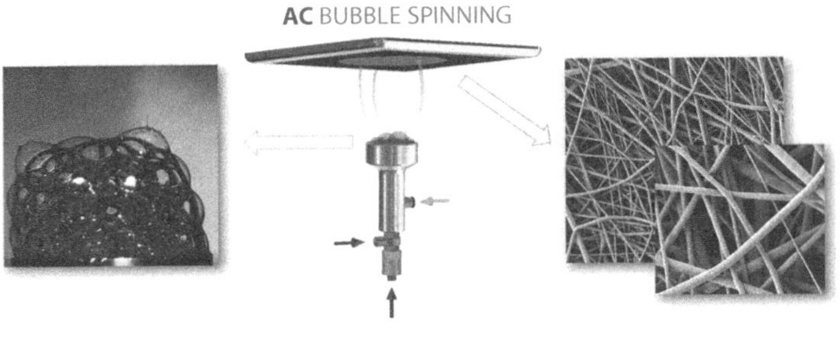


**Figure 10.6** Schematic representation of bubble electrospinning [109].

## 10.5 Outlook and Future Work Advises

This chapter has explored the latest innovations in nanofinishing, focusing on its applications in surface modification across various textile structures. It has outlined how nano-enabled techniques enable functions such as self-cleaning, UV shielding, chemical resistance, and all-weather protection in both technical and everyday textiles. What once prioritized performance and affordability has increasingly shifted toward sustainability in response to climate challenges and regulatory pressures.

The market expansion of technical textiles—from specialized sectors to consumer applications—has significantly accelerated the adoption of nanofinishing. Each textile form, due to its unique structural characteristics, benefits differently from these processes. While nanoparticles and nanofibers have already proven their versatility, thin-film applications still face issues like delamination, which future research must address to improve long-term performance.

Nanoyarns are emerging as a pivotal innovation, with potential applications growing rapidly, provided improvements in their tensile properties and scalability are realized. Additionally, integrating artificial intelligence into the design phase of nanofinishing could optimize material selection, functionality, and processing methods. AI-driven insights may soon guide researchers in matching textile substrates with the most suitable nano treatments before transitioning to production—paving the way for smarter, more efficient manufacturing in the textile industry.

**AI disclosure statement:** The author acknowledges the use of AI-based tools for minor language editing or grammatical correction. However, all intellectual contributions, analyses, and conclusions presented in this chapter are entirely the author's own.

## References

1. Malik, S., Muhammad, K., Waheed, Y., Nanotechnology: A Revolution in Modern Industry. *Molecules*, 28, 2, 2023.
2. Mitchell, B., *The Rise of Cotton Mills in the South*, University of South Carolina Press, Columbia, SC, 2001.
3. Rossbach, V., Analysis of Low Molecular Weight Homologues of Fiber-Forming Polycondensates. *Angew. Chem. Int. Ed. Engl.*, 20, 10, 831–840, 1981.
4. Atwah, A.A. and Khan, M.A., Influence of microscopic features on the self-cleaning ability of textile fabrics. *Text. Res. J.*, 93, 1–2, 450–467, 2023.

5. Montazer, M. and Harifi, T., Nanocoating and lamination, in: *Nanofinishing of Textile Materials*, pp. 95–107, Duxford, UK, Woodhead Publishing (Elsevier), 2018.
6. Sfameni, S., Rando, G., Plutino, M.R., Perspective Chapter: Functional Sol-Gel Based Coatings for Innovative and Sustainable Applications, in: *Sol-Gel Method-Recent Advances*, IntechOpen, London, UK, 2023.
7. Camlibel, N.O. and Arik, B., Sol-Gel Applications in Textile Finishing Processes, in: *Recent Applications in Sol-Gel Synthesis*, U. Chandra (Ed.), InTech, Rijeka, Croatia, 2017.
8. Colleoni, C., Donelli, I., Freddi, G., Guido, E., Migani, V., Rosace, G., A novel sol-gel multi-layer approach for cotton fabric finishing by tetraethoxysilane precursor. *Surf. Coat. Technol.*, 235, 192–203, 2013.
9. Abramova, A.V., Abramov, V.O., Bayazitov, V.M., Voitov, Y., Straumal, E.A., Lermontov, S.A., Cherdyntseva, T.A., Braeutigam, P., Weiße, M., Günther, K., A sol-gel method for applying nanosized antibacterial particles to the surface of textile materials in an ultrasonic field. *Ultrason. Sonochem.*, 60, 104788, 2020.
10. Lu, P. and Ding, B., Nano-modification of textile surfaces using layer-by-layer deposition methods, in: *Surface Modification of Textiles*, pp. 214–237, Elsevier, Cambridge, UK, 2009.
11. Uğur, SS., Sarıışık, M., Aktaş, A.H., Nano-Al2O3 multilayer film deposition on cotton fabrics by layer-by-layer deposition method. *Mater. Res. Bull.*, 46, 8, 1202–1206, 2011.
12. Babaeipour, S., Nousiainen, P., Kimiaei, E., Tienaho, J., Kohlhuber, N., Korpinen, R., Kaipanen, K., Österberg, M., Thin multifunctional coatings for textiles based on the layer-by-layer application of polyaromatic hybrid nanoparticles. *Mater. Adv.*, 5, 15, 6114–6131, 2024.
13. Radetić, M. and Marković, D., A review on the role of plasma technology in the nano-finishing of textile materials with metal and metal oxide nanoparticles. *Plasma Processes Polym.*, 19, 4, 2100197, 2022.
14. Gleissner, C., Landsiedel, J., Bechtold, T., Pham, T., Surface Activation of High Performance Polymer Fibers: A Review. *Polym. Rev.*, 62, 4, 757–788, 2022.
15. Hegemann, D., Hossain, M.M., Balazs, D.J., Nanostructured plasma coatings to obtain multifunctional textile surfaces. *Prog. Org. Coat.*, 58, 2–3, 237–240, 2007.
16. Petrova, M., Georgieva, M., Dobreva, E., Avdeev, G., Electroless deposition of nanodisperse metal coatings on fabrics. *Bulg. Chem. Commun.*, 44, 1, 92–98, 2012.
17. Root, W., Aguiló-Aguayo, N., Pham, T., Bechtold, T., Conductive layers through electroless deposition of copper on woven cellulose lyocell fabrics. *Surf. Coat. Technol.*, 348, 13–21, 2018.
18. Sun, P., Qiu, M., Li, M., Mai, W., Cui, G., Tong, Y., Stretchable Ni@NiCoP textile for wearable energy storage clothes. *Nano Energy*, 55, 506–515, 2019.
19. Miśkiewicz, P., Frydrych, I., Cichocka, A., Application of Physical Vapor Deposition in Textile Industry. *Autex Res. J.*, 22, 1, 42–54, 2022.

20. Kayaoğlu, B.K., Göcek, I., Kizil, H., Trabzon, L., Functional Nano and Micro-Scale Thin Film Deposition on Textiles: Emerging Technologies and Applications. *J. Text. Eng.*, 19, 88, 39–47, 2012.
21. Wilson, J.I.B., Textile surface functionalisation by chemical vapour deposition (CVD), in: *Surface Modification of Textiles*, pp. 126–138, Elsevier, Cambridge, UK, 2009.
22. Mundra, S.S., Pardeshi, S.S., Bhavikatti, S.S., Nagras, A., Development of an integrated physical vapour deposition and chemical vapour deposition system. *Mater. Today: Proc.*, 46, 1229–1234, 2021.
23. Yao, L. and He, J., Recent progress in antireflection and self-cleaning technology – From surface engineering to functional surfaces. *Prog. Mater Sci.*, 61, 94–143, 2014.
24. Tsuge, Y., Kim, J., Sone, Y., Kuwaki, O., Shiratori, S., Fabrication of transparent TiO2 film with high adhesion by using self-assembly methods: Application to super-hydrophilic film. *Thin Solid Films*, 516, 9, 2463–2468, 2008.
25. Yan-chao, H., Jian-feng, H., Xuan, Z., Li-yun, C., Li-Xiong, Y., Preparation of Sm2S3 thin films by liquid phase deposition method with self-assembled monolayers. *J. Alloys Compd.*, 536, 119–123, 2012.
26. Bodik, M., Jergel, M., Majkova, E., Siffalovic, P., Langmuir films of low-dimensional nanomaterials. *Adv. Colloid Interface Sci.*, 283, 102239, 2020.
27. Rauf, S., Vijjapu, M.T., Andrés, M.A., Gascón, I., Roubeau, O., Eddaoudi, M., Salama, K.N., Highly Selective Metal–Organic Framework Textile Humidity Sensor. *ACS Appl. Mater. Interfaces*, acsami.0c07532, 2020.
28. Rauf, S., Andrés, M.A., Roubeau, O., Gascón, I., Serre, C., Eddaoudi, M., Salama, K.N., Coating of Conducting and Insulating Threads with Porous MOF Particles through Langmuir-Blodgett Technique. *Nanomaterials*, 11, 1, 160, 2021.
29. Wang, P., Lv, H., Cao, X., Liu, Y., Yu, D.-G., Recent Progress of the Preparation and Application of Electrospun Porous Nanofibers. *Polymers*, 15, 4, 921, 2023.
30. Huang, Y., Yuan, Q., Lu, Z., Wang, W., Liu, K., Chen, Y., Wang, X., Wang, D., Qiu, Y., Facile preparation and characterization of a nanofiber-coated textile with durable and rechargeable antibacterial activity. *New J. Chem.*, 43, 43, 17116–17122, 2019.
31. Sivri, C., Improvement of protective and comfort properties of face masks using superabsorbent polymer containing nanofibers. *IJCST*, 30, 5, 668–686, 2018.
32. Sas, I., Gorga, R.E., Joines, J.A., Thoney, K.A., Literature review on superhydrophobic self-cleaning surfaces produced by electrospinning. *J. Polym. Sci. B Polym. Phys.*, 50, 12, 824–845, 2012.
33. Meng, L.-Y., Han, S., Jiang, N., Meng, W., Study on superhydrophobic surfaces of octanol grafted electrospun silica nanofibers. *Mater. Chem. Phys.*, 148, 3, 798–802, 2014.
34. Zimmermann, J., Reifler, F.A., Fortunato, G., Gerhardt, L., Seeger, S., A Simple, One-Step Approach to Durable and Robust Superhydrophobic Textiles. *Adv. Funct. Mater.*, 18, 22, 3662–3669, 2008.

35. Joshi, M. and Butola, B.S., Application technologies for coating, lamination and finishing of technical textiles, in: *Advances in the Dyeing and Finishing of Technical Textiles*, pp. 355–411, Elsevier, Oxford, UK, 2013.
36. Giraud, S., Bourbigot, S., Rochery, M., Vroman, I., Tighzert, L., Delobel, R., Poutch, F., Flame retarded polyurea with microencapsulated ammonium phosphate for textile coating. *Polym. Degrad. Stab.*, 88, 1, 106–113, 2005.
37. Singha, K., A Review on Coating & Lamination in Textiles: Processes and Applications. *AJPS*, 2, 3, 39–49, 2012.
38. Gao, Y., Yan, B., Zhou, M., Yu, Y., Wang, Q., Wang, P., MXene Nanosheets Assembled onto Cationic-Modified Cotton for Smart Multiprotection Textiles. *ACS Appl. Nano Mater.*, 7, 11, 13568–13578, 2024.
39. Knížek, R., Karhánková, D., Bajzík, V., Jirsák, O., Lamination of Nanofibre Layers for Clothing Applications. *Fibres Text. East. Eur*, 27, 1(133), 16–22, 2019.
40. Ma, W., Qi, H., Zhang, Y., Lin, M., Qiu, Y., Zhang, C., Fabrication of Laminated Micro/Nano Filter and Its Application for Inhalable PM Removal. *Polymers*, 15, 6, 1459, 2023.
41. Sivri, C. and Haji, A., Surface Coating of Needle-Punched Nonwovens with Meltblown Nonwovens to Improve Acoustic Properties. *Coatings*, 12, 8, 1092, 2022.
42. Rossini, M.B., Sadeghi, M.Z., Quadflieg, T., Schneiders, T.A., Gries, T., Schröder, K.-U., Delamination assessment of a composite laminate interleaved with electrospun TPU fibers under Mode-I and Mode-II loading. *Int. J. Adhes. Adhes.*, 117, 103007, 2022.
43. Armağan, O.G., Kayaoglu, B.K., Karakas, H.C., Guner, F.S., Adhesion strength behaviour of plasma pre-treated and laminated polypropylene nonwoven fabrics using acrylic and polyurethane-based adhesives. *J. Ind. Text.*, 43, 3, 396–414, 2014.
44. Altammar, K.A., A review on nanoparticles: characteristics, synthesis, applications, and challenges. *Front. Microbiol.*, 14, 1155622, 2023.
45. Khan, Y., Sadia, H., Ali Shah, S.Z., Khan, M.N., Shah, A.A., Ullah, N., Ullah, M.F., Bibi, H., Bafakeeh, O.T., Khedher, N.B., Eldin, S.M., Fadhl, B.M., Khan, M.I., Classification, Synthetic, and Characterization Approaches to Nanoparticles, and Their Applications in Various Fields of Nanotechnology: A Review. *Catalysts*, 12, 11, 1386, 2022.
46. Namakka, M., Rahman, Md.R., Said, K.A.M.B., Abdul Mannan, M., Patwary, A.M., A review of nanoparticle synthesis methods, classifications, applications, and characterization. *Environ. Nanotechnol. Monit. Manage.*, 20, 100900, 2023.
47. Harish, V., Ansari, M.M., Tewari, D., Gaur, M., Yadav, A.B., García-Betancourt, M.-L., Abdel-Haleem, F.M., Bechelany, M., Barhoum, A., Nanoparticle and Nanostructure Synthesis and Controlled Growth Methods. *Nanomaterials*, 12, 18, 3226, 2022.
48. Baig, N., Kammakakam, I., Falath, W., Nanomaterials: a review of synthesis methods, properties, recent progress, and challenges. *Mater. Adv.*, 2, 6, 1821–1871, 2021.

49. Meena, M., Zehra, A., Swapnil, P., Harish, Marwal, A., Yadav, G., Sonigra, P., Endophytic Nanotechnology: An Approach to Study Scope and Potential Applications. *Front. Chem.*, 9, 613343, 2021.
50. Hoda, N., Budama Akpolat, L., Mert SïVrï, F., Kurtuluş, D., Biosynthesis of Bimetallic Ag-Au (core-shell) Nanoparticles Using Aqueous Extract of Bay Leaves (Laurus nobilis L.). *J. Turk. Chem. Soc. Sect. A Chem.*, 8, 4, 1035–1044, 2021.
51. Mert Sivri, F., Akkoc, S., Önem, E., Uysal, E., Biosynthesis of Ag nanoparticles using *Laurus nobilis leaf* extract and biomedical applications. *Inorg. Nano-Metal Chem.*, 1–8, 2024.
52. Perumal, S., Polymer Nanoparticles: Synthesis and Applications. *Polymers*, 14, 24, 5449, 2022.
53. Zielińska, A., Carreiró, F., Oliveira, A.M., Neves, A., Pires, B., Venkatesh, D.N., Durazzo, A., Lucarini, M., Eder, P., Silva, A.M., Santini, A., Souto, E.B., Polymeric Nanoparticles: Production, Characterization, Toxicology and Ecotoxicology. *Molecules*, 25, 16, 3731, 2020.
54. Ying, S., Guan, Z., Ofoegbu, P.C., Clubb, P., Rico, C., He, F., Hong, J., Green synthesis of nanoparticles: Current developments and limitations. *Environ. Technol. Innovation*, 26, 102336, 2022.
55. Singh, H., Desimone, M.F., Pandya, S., Jasani, S., George, N., Adnan, M., Aldarhami, A., Bazaid, A.S., Alderhami, S.A., Revisiting the Green Synthesis of Nanoparticles: Uncovering Influences of Plant Extracts as Reducing Agents for Enhanced Synthesis Efficiency and Its Biomedical Applications. *IJN*, 18, 4727–4750, 2023.
56. Aguda, O.N. and Lateef, A., Recent advances in functionalization of nanotextiles: A strategy to combat harmful microorganisms and emerging pathogens in the 21st century. *Heliyon*, 8, 6, e09761, 2022.
57. Liu, Y., Yu, Y., Zhao, X., The influence of the ratio of graphite to silver-coated copper powders on the electromagnetic and mechanical properties of single-layer coated composites. *J. Text. Inst.*, 112, 11, 1709–1716, 2021.
58. Staneva, D., Koutzarova, T., Vertruyen, B., Vasileva-Tonkova, E., Grabchev, I., Synthesis, structural characterization and antibacterial activity of cotton fabric modified with a hydrogel containing barium hexaferrite nanoparticles. *J. Mol. Struct.*, 1127, 74–80, 2017.
59. Vrinceanu, N., Bucur, S., Rimbu, C.M., Neculai-Valeanu, S., Ferrandiz Bou, S., Suchea, M.P., Nanoparticle/biopolymer-based coatings for functionalization of textiles: recent developments (a minireview). *Text. Res. J.*, 92, 19–20, 3889–3902, 2022.
60. Shankar, S. and Rhim, J.-W., Preparation of sulfur nanoparticle-incorporated antimicrobial chitosan films. *Food Hydrocolloids*, 82, 116–123, 2018.
61. Assefi Pour, R., Bagheri, R., Naveed, T., Ali, N., Rehman, F., He, J., Surface functionalization of wool *via* microbial-transglutaminase and bentonite as bio-nano-mordant to achieve multi objective wool and improve dyeability with madder. *Heliyon*, 6, 9, e04911, 2020.

62. Abou El-Kheir, A. and El-Gabry, L., Potential Applications of Nanotechnology In Functionalization of Synthetic Fibres (A Review). *Egypt. J. Chem.*, 2022.
63. Göcek, I., Functionalization of Textile Materials with Nanoclay Incorporation for Improved Characteristics. *Politek. Derg.*, 22, 2, 509–522, 2019.
64. De Oliveira, C.R.S., Batistella, M.A., Guelli Ulson De Souza, S.M.D.A., Ulson De Souza, A.A., Functionalization of cellulosic fibers with a kaolinite-TiO2 nano-hybrid composite *via* a solvothermal process for flame retardant applications. *Carbohydr. Polym.*, 266, 118108, 2021.
65. Khalil, A., Jouiad, M., Khraisheh, M., Hashaikeh, R., Facile Synthesis of Copper Oxide Nanoparticles *via* Electrospinning. *J. Nanomater.*, 2014, 1, 438407, 2014.
66. Prabu, G.T.V. and Dhurai, B., A Novel Profiled Multi-Pin Electrospinning System for Nanofiber Production and Encapsulation of Nanoparticles into Nanofibers. *Sci. Rep.*, 10, 1, 4302, 2020.
67. Vargas-Molinero, H.Y., Serrano-Medina, A., Palomino-Vizcaino, K., López-Maldonado, E.A., Villarreal-Gómez, L.J., Pérez-González, G.L., Cornejo-Bravo, J.M., Hybrid Systems of Nanofibers and Polymeric Nanoparticles for Biological Application and Delivery Systems. *Micromachines*, 14, 1, 208, 2023.
68. Radacsi, N., Campos, F.D., Chisholm, C.R.I., Giapis, K.P., Spontaneous formation of nanoparticles on electrospun nanofibres. *Nat. Commun.*, 9, 1, 4740, 2018.
69. Stramarkou, M., Tzegiannakis, I., Christoforidi, E., Krokida, M., Use of Electrospinning for Sustainable Production of Nanofibers: A Comparative Assessment of Smart Textiles-Related Applications. *Polymers*, 16, 4, 514, 2024.
70. Ali, T., Najam-ul-Haq, M., Mohyuddin, A., Musharraf, S.G., Hussain, D., Next-generation functional nanotextiles—Prospects and challenges. *Sustainable Mater. Technol.*, 36, e00640, 2023.
71. Özkayalar, S. and Alay Aksoy, S., Production and Characterization of Nanoencapsulated Phase Change Materials (PCMs) and Bicomponent PCM Nanofibers. *Tekst. ve Konfeksiyon*, 31, 3, 156–170, 2021.
72. Maleki, H., Semnani Rahbar, R., Alay Aksoy, S., Yilmaz, D., Development of PCM-Loaded Composite Yarns for Enhanced Thermoregulation in Medical Textiles. *Fibers Polym.*, 2024.
73. Tarafder, N., Applications of nanotechnology for textile products: a review. *NR*, 1, 3, 15–22, 2018.
74. Caldas, A., Sousa, J., Carneiro, E., Carvalho, S., Functionalization of fabrics by Ag-TiO2 Nanoparticles deposition by sol-gel method. *JTEFT*, 10, 1, 49–53, 2024.
75. Silva, D.J., Barbosa, R.F.S., Souza, A.G., Ferreira, R.R., Camani, P.H., Rosa, D.S., Morphological, UV blocking, and antimicrobial features of multifunctional cotton fibers coated with ZnO/Cu *via* sonochemistry. *Mater. Chem. Phys.*, 286, 126210, 2022.

76. Chen, J., Xiao, P., Xiong, X., The mechanical properties and thermal conductivity of carbon/carbon composites with the fiber/matrix interface modified by silicon carbide nanofibers. *Mater. Des.*, 84, 285–290, 2015.
77. Hassan, M.M. and Carr, C.M., A review of the sustainable methods in imparting shrink resistance to wool fabrics. *J. Adv. Res.*, 18, 39–60, 2019.
78. Shabanian, S., Lahiri, S.K., Soltani, M., Golovin, K., Durable water- and oil-repellent textiles without long- or short-chain perfluoroalkylated substances. *Mater. Today Chem.*, 34, 101786, 2023.
79. Rehan, M., El-Naggar, M.E., Al-Enizi, A.M., Alothman, A.A., Nafady, A., Abdelhameed, R.M., Development of silk fibers decorated with the *in situ* synthesized silver and gold nanoparticles: antimicrobial activity and creatinine adsorption capacity. *J. Ind. Eng. Chem.*, 97, 584–596, 2021.
80. Jeong, E., Woo, H., Moon, Y., Lee, D.Y., Jung, M., Lee, Y.-S., Bae, J.-S., Self-Cleaning Polyester Fabric Prepared with TiOF2 and Hexadecyltrimethoxysilane. *Polymers*, 13, 3, 387, 2021.
81. Quan, H., Chen, J., Kong, X., Wu, J., Ni, L., Construction and Performance Evaluation of Functional Coating with Multi-Layered Structure on the Surface of Cashmere Fibers Based on Dispersed Graphene by Cationic Waterborne Polyurethane. *J. Nat. Fibers*, 20, 1, 2181910, 2023.
82. Sun, H., Li, X., Li, H., Hui, D., Gaff, M., Lorenzo, R., Nanotechnology application on bamboo materials: A review. *Nanotechnol. Rev.*, 11, 1, 1670–1695, 2022.
83. Altangerel, Z., Purev-Ochir, B., Ganzorig, A., Tsagaantsooj, T., Lkhamsuren, G., Choisuren, A., Chimed, G., Superhydrophobic modification and characterization of cashmere fiber surfaces by wet coating techniques of silica nanoparticles. *Surf. Interfaces*, 19, 100533, 2020.
84. Liang, Y., Pakdel, E., Zhang, M., Sun, L., Wang, X., Photoprotective properties of alpaca fiber melanin reinforced by rutile TiO2 nanoparticles: A study on wool fabric. *Polym. Degrad. Stab.*, 160, 80–88, 2019.
85. Singh, K.V., Sawhney, P.S., Sachinvala, N.D., Li, G., Pang, S.-S., Condon, B., Parachuru, R., Applications and future of nanotechnology in textiles. *National Cotton Council Beltwide Cotton Conference*, vol. 2, 2006.
86. Ghasemi, S., Tajvidi, M., Bousfield, D.W., Gardner, D.J., Reinforcement of natural fiber yarns by cellulose nanomaterials: A multi-scale study. *Ind. Crops Prod.*, 111, 471–481, 2018.
87. Qiu, Q., Chen, S., Li, Y., Yang, Y., Zhang, H., Quan, Z., Qin, X., Wang, R., Yu, J., Functional nanofibers embedded into textiles for durable antibacterial properties. *Chem. Eng. J.*, 384, 123241, 2020.
88. Hwang, B., Lund, A., Tian, Y., Darabi, S., Müller, C., Machine-Washable Conductive Silk Yarns with a Composite Coating of Ag Nanowires and PEDOT:PSS. *ACS Appl. Mater. Interfaces*, 12, 24, 27537–27544, 2020.
89. Rashid, A., Albargi, H.B., Irfan, M., Qadir, M.B., Ahmed, A., Ferri, A., Jalalah, M., Sonoprocesses for sustainable textile processing and nanofinishing: a review. *Text. Res. J.*, 00405175241252562, 2024.

90. Mohammadipour-Nodoushan, R., Shekarriz, S., Shariatinia, Z., Heydari, A., Montazer, M., Improved cotton fabrics properties using zinc oxide-based nanomaterials: A review. *Int. J. Biol. Macromol.*, 242, 124916, 2023.
91. Ramadan, M.A., Taha, G.M., EL-Mohr, W.Z.E.-A., Antimicrobial and UV protection finishing of Polysaccharide-Based Textiles using Biopolymer and AgNPs. *Egypt. J. Chem.*, 63, 7, 2707–2716, 2020.
92. Zhang, H., Ji, H., Chen, J., Wang, N., Xiao, H., A multi-scale MXene coating method for preparing washable conductive cotton yarn and fabric. *Ind. Crops Prod.*, 188, 115653, 2022.
93. Schwaiger, M., Bender, M., Schirmer, H., Taesler, J., Feuchter, M., Resch-Fauster, K., Effect of different weft-knitted structures on the mechanical performance of bio-based flexible composites. *Composites, Part C: Open Access*, 13, 100436, 2024.
94. Zhu, L., Ding, X., Wu, X., A novel method for improving the anti-pilling property of knitted wool fabric with engineered water nanostructures. *J. Mater. Res. Technol.*, 9, 3, 3649–3658, 2020.
95. Tai, C.-Y., Lin, C.-Y., Liu, T.-C., Jia, L.-C., Jones, T., Abdolvand, A., One-step metallization of weft-knitted fabrics for wearable biaxial strain sensors. *Sci. Rep.*, 12, 1, 20029, 2022.
96. He, G., Wang, L., Bao, X., Lei, Z., Ning, F., Li, M., Zhang, X., Qu, L., Synergistic flame retardant weft-knitted alginate/viscose fabrics with MXene coating for multifunctional wearable heaters. *Composites, Part B*, 232, 109618, 2022.
97. Wang, W., Feng, L., Song, B., Wang, L., Shao, R., Xia, Y., Liu, D., Li, T., Liu, S., Wang, L., Xu, Z., Deng, H., Fu, H., Fabrication and application of superhydrophobic nonwovens: a review. *Mater. Today Chem.*, 26, 101227, 2022.
98. Gadkari, R.R., Ali, W., Das, A., Alagirusamy, R., Configuration of a unique antibacterial needle-punched nonwoven fabric from silver impregnated polyester nanocomposite fibres. *J. Ind. Text.*, 51, 1_suppl, 1511S–1527S, 2022.
99. Cui, C., Wang, W., Lv, X., Jiao, S., Pang, G., Fabrication of superwetting non-woven fabric by grafting one-dimensional inorganic nanostructure for efficient separation of surfactant-stabilized organic solvent/water emulsions. *Colloids Surf., A*, 663, 131068, 2023.
100. Tarek, D., Mohmoud, A., Essam, Z., Sayed, R., Maher, A., Maamoun, D., Abdel Salam, S., Mohamed, H., Hassabo, A., Khattab, T., Development of Wrinkle Free and Self-Cleaning Finishing of Cotton and Polyester Fabrics. *J. Text. Color. Polym. Sci.*, 2023.
101. Aijaz, M.O., Alnaser, I.A., Farooq, I., Siddiqui, Md.I.H., Yang, S.B., Shakeel, F., Karim, M.R., Developing novel multifunctional protective clothes for disabled individuals using bio-based electrospun nanofibrous membranes. *Int. J. Biol. Macromol.*, 275, 133598, 2024.
102. Hoque, E., Tran, P., Jacobo, U., Bergfeld, N., Acharya, S., Shamshina, J.L., Reid, T.W., Abidi, N., Antimicrobial Coatings for Medical Textiles *via* Reactive Organo-Selenium Compounds. *Molecules*, 28, 17, 6381, 2023.

103. Jin, L., Guo, X., Gao, D., Wu, C., Hu, B., Tan, G., Du, N., Cai, X., Yang, Z., Zhang, X., NIR-responsive MXene nanobelts for wound healing. *NPG Asia Mater.*, 13, 1, 24, 2021.
104. Syduzzaman, Md.., Hassan, A., Anik, H.R., Akter, M., Islam, Md.R., Nanotechnology for High-Performance Textiles: A Promising Frontier for Innovation. *ChemNanoMat*, 9, 9, e202300205, 2023.
105. Pavlidou, S. and Paul, R., 3 - Soil repellency and stain resistance through hydrophobic and oleophobic treatments, in: *Waterproof and Water Repellent Textiles and Clothing*, J. Williams (Ed.), pp. 73–88, Woodhead Publishing, Duxford, UK, 2018.
106. Attia, N.F., Elashery, S.E.A., El-Sayed, F., Mohamed, M., Osama, R., Elmahdy, E., Abd-Ellah, M., El-Seedi, H.R., Hawash, H.B., Ameen, H., Recent advances in nanobased flame-retardant coatings for textile fabrics. *Nano-Struct. Nano-Objects*, 38, 101180, 2024.
107. Joshi, M. and Adak, B., 6. Nanotechnology-based Textiles: A Solution for the Emerging Automotive Sector, in: *Rubber Nanocomposites and Nanotextiles*, B. Banerjee (Ed.), pp. 179–230, De Gruyter, Berlin, Germany, 2019.
108. Sivri, C., Nanoyarns: Recent Advancements in Production Techniques, Applications, and Future Prospects, in: *Handbook of Nanofibers and Nanocomposites*, pp. 87–124, Jenny Stanford Publishing, Singapore, 2023.
109. Erben, J., Kalous, T., Chvojka, J., ac Bubble Electrospinning Technology for Preparation of Nanofibrous Mats. *ACS Omega*, 5, 14, 8268–8271, 2020.
110. Khandual, A., Rout, N., Verma, S.K., Patel, P., Pattanaik, P., Luximon, Y., Kumar, A., Nayak, R.K., Suar, M., Controlled nano-particle dyeing of cotton can ensure low cytotoxicity risk with multi-functional property enhancement. *Mater. Today Chem.*, 17, 100345, 2020.
111. Liyanapathiranage, A., Peña, M.J., Sharma, S., Minko, S., Nanocellulose-Based Sustainable Dyeing of Cotton Textiles with Minimized Water Pollution. *ACS Omega*, 5, 16, 9196–9203, 2020.
112. Periyasamy, A.P., Militky, J., Sachinandham, A., Duraisamy, G., Nanotechnology in Textile Finishing: Recent Developments, in: *Handbook of Nanomaterials and Nanocomposites for Energy and Environmental Applications*, O.V. Kharissova, L.M.T. Martínez, B.I. Kharisov (Eds.), pp. 1–31, Springer International Publishing, Cham, 2021.
113. Ledwani, L. and Sangwai, J.S. (Eds.), *Nanotechnology for Energy and Environmental Engineering*, Springer International Publishing, Cham, 2020.
114. Dos Santos, O.A.L., De Araujo, I., Dias Da Silva, F., Sales, M.N., Christoffolete, M.A., Backx, B.P., Surface modification of textiles by green nanotechnology against pathogenic microorganisms. *Curr. Res. Green Sustainable Chem.*, 4, 100206, 2021.
115. Naebe, M., Haque, A.N.M.A., Haji, A., Plasma-Assisted Antimicrobial Finishing of Textiles: A Review. *Engineering*, 12, 145–163, 2022.

# 11

# Microencapsulation in Surface Modification of Textiles

**Liliana Indrie[1*], Nor Dalila Nor Affandi[2†], Najua Tulos[2], Simona Tripa[1], Zlatin Zlatev[3] and Mohd Rozi Ahmad[2]**

[1]*Department of Textiles, Leather and Industrial Management, University of Oradea, Universitatii Str. No. 1, Oradea, Romania*
[2]*Textile Research Group, Faculty of Applied Sciences, Universiti Teknologi MARA, Shah Alam, Selangor, Malaysia*
[3]*Trakia University, Faculty of Technics and Technologies, Yambol, Bulgaria*

***Abstract***
The textile industry is adapting to consumer preferences by focusing on producing functional textile products with added value. Recent advancements in finishing techniques aim to enhance the functionality of textile substrates. Microencapsulation processes have emerged as a promising method for creating new textiles with various advantages over conventional products. This technique imparts desirable characteristics like antibacterial effects, fragrance, mosquito repellency, and UV protection. By encapsulating both volatile and non-volatile components within a thin polymeric coating, microencapsulation ensures a controlled release, resulting in sustained functional effects. Various active ingredients, including vitamins, enzymes, medications, insecticides, and essential oils, have been successfully enclosed in microcapsules made from different polymeric materials.

This chapter comprehensively explores the principles of microencapsulation, microencapsulation technology and the application of microcapsules in developing functional textiles. It also highlights significant research in this area and provides an overview of the primary essential oils utilized in microencapsulation, along with their associated functional benefits.

***Keywords*:** Microcapsules, functional textiles, essential oils

**Corresponding author*: lindrie@uoradea.ro
†*Corresponding author*: dalila@uitm.edu.my

Aminoddin Haji and Çağlar Sivri (eds.) Surface Modification and Functionalization of Textiles, (253–276) © 2026 Scrivener Publishing LLC

## 11.1 Introduction

A central market requirement currently revolves around the functionality of textile products. Functionalities can be integrated into textiles through the use of various encapsulated materials, which include both natural and synthetic options. Since the middle of the 20th century, microencapsulation has developed into the leading technology for refining textile materials. Industrialized nations are particularly interested in textiles with innovative properties and significant added value, such as functional textiles and medical textiles. Consequently, textile industries are proactively adopting advanced technologies such as microencapsulation to achieve desired surfaces with the required properties, a performance that may not be achievable using alternative methods.

Originally used in the paper industry to produce carbon-free copy paper [1], microencapsulation is now being explored in various sectors including food processing, pharmaceuticals, chemicals, cosmetics, and agriculture. There are a number of microcapsules available on the market that contain various useful materials. This approach promises significant improvements in both process and product performance. In particular, the exceptional stability of volatile components against oxidation, coupled with a controlled release mechanism, ensures outstanding effectiveness and longevity.

The microencapsulation process applies thin polymer coatings to tiny solid or liquid particles or microdroplets, effectively packaging ingredients on a microscale [2].

Microcapsules, also called artificial cells, are manufactured with a design that prevents their contents from leaking out, thereby preventing immunological reactions. However, they are designed in such a way that they allow their ingredients to interact unrestrictedly in biochemical reactions.

Microencapsulated materials have various applications in the textile industry as finishing agents for fibers, yarns, and fabrics using techniques such as spraying, screen printing, exhaustion, impregnation, and pad-dry curing.

Typically, microcapsules have a diameter of a few micrometers to a few millimeters. This technology has become increasingly popular in textile finishing.

The content enclosed within the microcapsule is called the core material, whereas the outer layer is called the shell, coating, or membrane [3].

Through the microencapsulation of the core material, substances can be given various special and functional properties. The core material can include any substance that is intended to fulfill a specific function of the substance.

Microencapsulation has been applied to a variety of materials, from polymers, fats, and waxes for the shell to drugs, agrochemicals, enzymes and fragrances for the core. This versatility allows the encapsulation of a wide range of substances and covers various applications in different industries [4].

Fluidized bed coating, spray cooling, spray drying, extrusion and coacervation are some of the methods used in microencapsulation. A specific technique can be selected on the basis of the characteristics of the shell, the core material, and the intended properties and shape of the capsules. There are usually three main steps in the encapsulation process:

1. Adding the active ingredients to the matrix or core of the microcapsule is the first step. This can be achieved by producing a liquid or dispersed solid system through the use of techniques such as suspension, emulsion, or solution. Methods such as grinding, sieving, drying, mixing, and dispersing can be used in this phase.
2. In the second step, an automated process is used. This includes dispersing the liquid through emulsification or microemulsification, as well as creating droplets in the air through techniques such as spraying or dripping. As the particles are stirred in a solid matrix, a solution is sprayed onto them using techniques such as fluidized bed, pan coating, or agglomeration. By following these procedures, the matrix material is applied to the particles and distributed evenly.
3. Fixing the droplets or coating solution is the final step. To achieve this, chemical reactions such as polymerization, physicochemical techniques such as gelation and coacervation, or physical processes such as drying or solidification can be used. Thanks to these techniques, the microcapsules remain stable and intact.

These steps are repeated iteratively until the desired final structure is achieved.

Microencapsulation is used to produce and improve textiles with novel and improved properties such as fragrance [5], thermoregulation [6], flame retardancy [7], UV protection [8], antibacterial effect [9], photochromic properties [10], antioxidant effect [11], or multifunctional properties by combining two or more functional elements.

### 11.1.1 Principles of Microencapsulation

Microencapsulation is a method to encapsulate one substance within another [12]. The encapsulated substance may be referred to as the core material, active ingredient, or agent, whereas the material utilized to encapsulate the core is referred to as the wall material, shell, membrane, or coating [13]. Figure 11.1 depicts the structure of microencapsulate particle containing wall and core materials. The performance of microencapsulated materials is influenced by the features of the wall materials [14]. As the concentration of the wall materials increases, it can enhance the microencapsulation efficiency [15]. In addition, the characteristics of the microencapsulated particles are influenced by the physico-chemical properties such as its solubility, molecular weight, glass or melt transition, crystallinity, diffusibility, and film-forming and emulsifying capabilities [15].

Wall materials can be made of either natural or synthetic polymers. Starch, cellulose, nylon, polyethylene, and polyester are some examples the polymers that can be used to encapsulate core materials and help to achieve the desired properties in the final product [15]. The core material, which refer to specific material to be coated, might have a liquid or solid composition [17]. This core material can be essential oil (EO), enzymes, dyes, and pigments as well as phase change materials [15]. The ability to modify the core material offers a certain degree of flexibility in creating and developing microcapsules with specific properties [17]. The uniqueness of the microencapsulated materials lies in their ability to be handled, separated, and stored in a micron size, allowing the active substances to be released under specific conditions [18]. According to Wangai *et al.* [19], a microencapsulation approach can maintain citronella repellency for up to 30 days on treated fabric stored at room temperature. This approach produces core-shell structures with diameters ranging from several hundred to thousands of microns. The procedure usually starts with the production of a protective film layer, which is usually made of a polymeric material [13, 20].

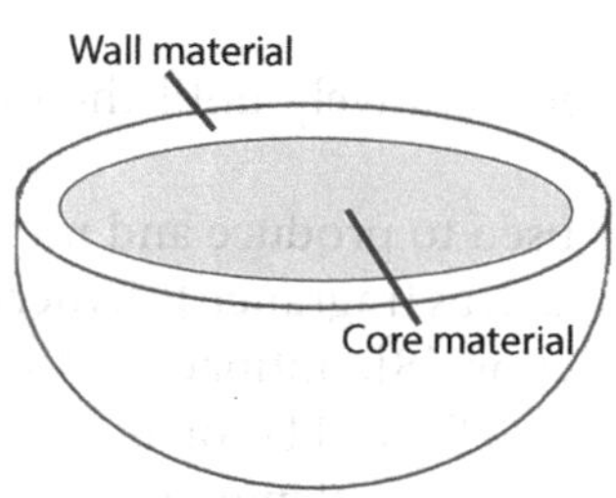


**Figure 11.1** The schematic cross section of microcapsule [16].

Microencapsulation can be in a form of spherical structure with a continuous wall surrounding the core (mononuclear) or unsymmetrically (irregular), where a matrix of several tiny droplets of core material are scattered throughout the capsule [13, 21].

The selection of microencapsulation technique is based on a number of factors, including particle size, the physical and chemical properties of the wall and core, microencapsulation, and controlled release mechanisms [22]. This technique is employed to modify the physical properties of a volatile chemical, making it more manageable and protecting it from various external factors such as sunlight, evaporation, humidity, alkalinity, or unintended friction [23]. In addition, the microencapsulation is used to improve the product's shelf life and provide controlled release of the encapsulated substance [12].

## 11.2 Microencapsulation Technology

The technique of microencapsulation makes possible the encapsulation of a compound within a small sphere, called a microsphere or microcapsule, with diameters ranging from as small as 1 mm to several hundred micrometers [24]. In textile applications, these microspheres or microcapsules can be formed using different microencapsulation methods. These methods are generally categorized into three categories,

1) Physical microencapsulation involves different preparation techniques, including spray drying, inclusion encapsulation, complex coagulation, and solidification [25].
2) Chemical microencapsulation encompasses processes such as *in situ* polymerization, interfacial polymerization and photopolymerization [26].
3) Physico-chemical microencapsulation is achieved through simple coacervation, complex coacervation, and molecular inclusion [26].

*In situ* polymerization involves the polymerization process forming a capsule shell in which a particular monomer or pre-polymer is suspended in a stable emulsion of the immiscible core material and a continuous solution [27]. Common shell materials for *in situ* polymerization are amine-aldehyde resins (aminoplasts), such as urea-formaldehyde, melamine-formaldehyde, urea-melamine-formaldehyde, or resorcinol-modified melamine-formaldehyde synthetic polymers [26].

Interfacial polymerization microencapsulation is one monomer that dissolves in the aqueous solution and the other in the organic lipophilic solvent of the emulsion during microencapsulation process. A polymer membrane, which is the microcapsule shell can be formed at the droplet interface by the reaction of both monomers [26].

Coacervation involves the phase separation of a solution into a coacervate phase and a supernatant phase, typically induced by alterations in temperature, pH, or ionic strength.

Spray drying involves atomizing a solution or suspension that contains the material to be encapsulated into fine droplets. These droplets are swiftly dried in a hot air chamber, resulting in a powder that contains encapsulated particles.

Each method has its own benefits and drawbacks. For instance, the chemical microencapsulation method, particularly aminoaldehyde microcapsules *via in situ* polymerization technique, offers outstanding technological features such as durability and resistance to chemical and physical agents. However, this method exhibits low environmental degradability and formaldehyde release [26]. In another study, Sumiga discovered that microcapsules created through *in situ* polymerization were smaller, featuring spherical, rigid, and impermeable walls made of pressure-sensitive synthetic polymers, whereas coacervate microcapsules were larger, with elastic walls made of natural polymers and partial permeability.

In the context of physico-chemical microencapsulation, this method often applies the environmentally friendly shell materials that are naturally sourced, safer for human skin, and less likely to degrade the textile after use. The drawback is that they are less durable and resistant to physical and chemical agents during the application of microcapsules and other functional textiles processes, as well as washing [26]. Zhang *et al.* [25] found that, although molecular entrapment with cyclodextrin using β-cylodextrin as the wall material has the advantage of being inexpensive, easy to perform, safe, and non-toxic, the material's less-than-ideal holes increase the microcapsules' long-term and high-temperature sustained-release characteristics. The complex coagulation method does not require high temperatures; therefore, the manufacturing conditions are moderate and the efficiency of microcapsule encapsulation is high, and the release effects are both well controlled and sustained.

Spray-drying as shown in Figure 11.2 is one of the most widely used physical microencapsulation methods for encapsulating EOs or volatile compounds, owing to its low cost and ease of usage when compared to other methods [28]. In addition, biodegradable compounds including Arabic gum, chitosan, ethyl cellulose, and polylactic acid can also be

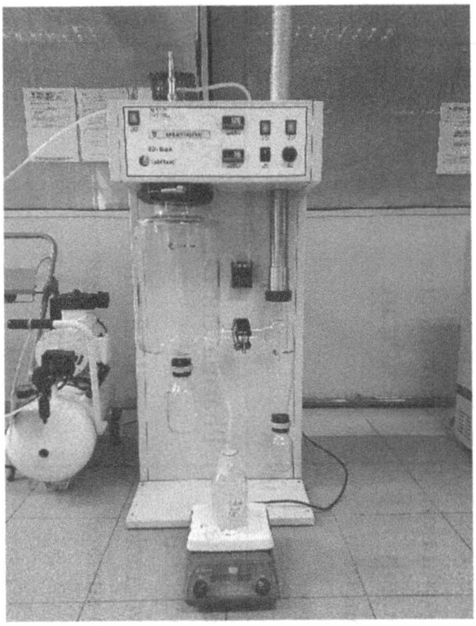

**Figure 11.2** Spray drying process setup [16].

used in spray drying [26]. Phenolic compounds has been shown to be an excellent approach for spray drying and microencapsulation [29]. In spray drying and other microencapsulation techniques, the core material, preparation method, and wall material all have an impact on the size of microcapsules, the concentration of EOs, which is a measure of encapsulation efficiency, the preparation process, as well as the sustained release performance [25]. The sustained release performance of microcapsules needs enhancement, particularly in applications where the core materials' functionality must be preserved over a long period [25].

Microcapsules may feature single, double, or multiple layers to effectively retain the volatile nature of active substance within the core [13]. Double-layered microcapsules, in particular, have been shown to enhance sustained release performance and offer superior protection for the core material, especially in the case of EOs [25]. These double-layered microcapsules specifically improve the overall protection and storage stability of the encapsulated Eos [25].

In a previous study, researchers prepared bilayer polyurethane/polyurea-formaldehyde microcapsules containing clove oil using *in situ* and interfacial polymerization reactions. They discovered that the double-layered microcapsules had a significant longer clove oil release period than the single-layered ones [25]. When compared to single-layer microencapsulation, double-embedding technique can improve encapsulation efficiency, thermal stability, and storage stability [25]. In the application

**Table 11.1** Methods to encapsulate essential oils (EOs) for textile applications [16].

| Coating technique | Essential oil (EOs) | Material used | Preparation technique | Formation | Application | References |
|---|---|---|---|---|---|---|
| Double layered (microcapsule shell, microcapsule core) | Lavender | B-cyclodextrin (first layer), chitosan and sodium alginate (second layer) | Complex coacervation | Smaller particle size distribution, roughly spherical | Textile, food science, biomedicine, healthcare. | Zhang *et al.*, [25] |
| Microcapsule multiple wall material | *Mentha spicata* | Inulin and Arabic gum | Spray dry | Various size, spherical, few cracks on surface | Pharmaceutical, food applications (flavor agent and chewing gum) | Mehran *et al.*, [31] |
| Single layered | Cinnamon | Sodium alginate | Spray dry | Average size, spherical with a homogenous surface | Antifungal treatment | Makimori *et al.*, [32] |

of microcapsule technology, this technology offers a lot of potentials specifically on textiles [30]. An overview of the methods for EO microencapsulation in textile applications is tabulated in Table 11.1.

Microencapsulation is a widely utilized technology in the food and pharmaceutical industries [33]. In food products, microencapsulation is primarily used to protect and isolate volatile and sensitive components [34]. Because of their size, shape, wall material, and active chemical release mechanism, microcapsules provide promising properties for textile processing techniques and the final goods that are produced [35]. In textile applications, microcapsules can be used for flame retardants, protection against atmospheric agents, and functional finishes [33]. With the development of functional fabrics with varied effects to tackle problems that cannot be addressed by traditional processing, the use of microencapsulated active chemicals in cotton and polyester fabrics has gotten a lot of attention [33]. Microencapsulation shows great potential for enhancing textile performance and represents a rapidly advancing technology that has recently gained considerable application in the textile industry [13].

## 11.3 Utilization of Microcapsules for Developing Functional Textiles

The functional properties of textiles can be modified through the incorporation of encapsulated materials, comprising both natural and synthetic substances. Recent studies have indicated a growing interest among contemporary consumers in the utilization of sustainable natural materials for the functionalization of textiles.

Microcapsules can be applied to textiles through various methods such as padding, coating, dipping, spraying, and printing [36, 37].

These techniques offer flexibility in integrating microcapsules into textile materials and enable the integration of functional properties such as fragrances, antimicrobial activity, or temperature regulation.

***a. Fragrant Textiles***

The enhancement of textiles with fragrances represents a significant commercial goal and represents a technical challenge. Fragrances from natural sources such as plants or EOs not only provide a pleasant aroma, but are also used for their antiseptic, anti-inflammatory, antimicrobial, and soothing properties. Aromatherapy textiles contain these EOs from plant sources and aim to increase both mental and physical well-being. As a result of wear and tear, the mechanical pressure on the microcapsules in these textiles causes them to burst and release the active ingredients.

The conclusion drawn from the literature summary is that factors such as polymerization method, mixing speed, fragrance type, composition of wall materials, mechanical properties, and size of capsules have significant impact on the ability of the microcapsules to release fragrances. Moreover, washing changes the number of microcapsules incorporated into textiles, which requires the use of cross-linking materials to increase stability by binding to the fiber surface.

Various studies have shown that scents such as lemon, orange, jasmine, lavender, and rose, among others, were often encapsulated using various methods described in the literature. These encapsulated fragrances were applied to textiles for various purposes.

Rodrigues *et al.* [38] developed microcapsules containing a masculine perfume using the interfacial polymerization technique. They manufactured PU-urea microcapsules with lemon oil as the core material and integrated them into textiles. Their research demonstrated precise control over the release of perfume from the textile during dry cleaning. Results showed a 24% decrease in lemon oil levels after one dry cleaning cycle, with almost all of the oil removed after five cycles. In addition, corrosion tests showed a reduction rate of 40% after 3,000 revolutions and 60% after 9,000 revolutions.

In a study conducted by Kumari *et al.* [39], geranium oil was added to cotton fabric after microencapsulation. Using the complex coacervation technique, acacia gum was selected as the wall material for encapsulation and geranium oil as the core material. The ideal process required the combination of oil, gum, and gelatin in a ratio of 1:4:4, a temperature of 50°C, and an initial and final pH of 4.5 and 9.0, respectively. The long-lasting scent finish on the textiles was achieved by effectively controlling the aroma release rate through the use of microencapsulation.

K. J. Sannapapamma *et al.* [40] developed vetiver oil microcapsules for application to organic cotton knitwear using interfacial polymerization. The fabric was coated with these microcapsules through the pad-dry curing method, which created a larger zone of inhibition compared to the pull-out method. Tests showed that the fabric treated with the pad-dry-cure method retained its antimicrobial activity and retained more fragrance after 20 washes, demonstrating superior durability and effectiveness.

***b. Antimicrobial Textiles***

The medical textile industry has increasingly adopted antimicrobial finishes to combat microorganisms. In the technical textile sector, these finishes are valued for their ability to prevent bio-deterioration, particularly in natural fibers that can be damaged by mold, mildew, and fungi when exposed to the elements. Additionally, there is concern about microbial growth and dust mites in floor coverings, as well as the need to protect consumers from insects and pests. With rising awareness of hygiene and consumer demand for hygienic products, there is also concern about the potential contact between antimicrobials and the human body [41].

Textiles are given antibacterial properties by a variety of antimicrobial agents such as triclosan, nanosilver, and quaternary ammonium salts. However, due to the limitations of these active ingredients—such as the synthetic nature of the manufacturer, high cost, high toxicity and limited availability—inexpensive, environmentally friendly antimicrobial chemicals are increasingly being used in textile finishing. The effectiveness and durability of natural extracts as antimicrobial finishes are widely studied, but these properties are not always easy to achieve. Most available EOs exhibit antimicrobial properties, making them suitable for the antimicrobial finishing of textiles. To address the limitations of EOs, particularly their volatility, encapsulating them in suitable polymer shells can be an effective solution.

The literature contains numerous articles on the application of EO containing microcapsules for antimicrobial finishing.

Chitosan is a naturally occurring polymer that exhibits good biocompatibility and minimal toxicity, based on investigations conducted thus far. Recent research has not only examined the antibacterial properties of this polysaccharide on textiles, but also how it promotes other benefits including shrink resistance and color absorption.

Indrie *et al.* [42] conducted antifungal tests on three antique maps (silk, canvas, and paper) using the AATCC 30-2004 method. After immersing the maps in a chitosan solution (10 g/L), they examined them with a field-emission scanning electron microscope (SEM). The results demonstrated that chitosan effectively removes external agents and microorganisms from fiber surfaces due to its antimicrobial properties. This simple and effective method suggests that chitosan could be a cost-effective antibacterial solution for preserving cultural artifacts.

In another study, the antimicrobial activity of a textile material treated with orange (*Citrus sinensis*) and rosemary (*Rosmarinus officinalis*) EOs was examined against pathogenic strains of *Aspergillus niger*, *Candida albicans*, *Trichoderma viride*, *Aspergillus flavus*, and *Epidermophyton floccosum*. The textile material consisted of a fiber mix with 56% cotton and 44% polyester. Biological tests revealed different percentage reduction rates depending on the strain and oil concentration tested. Treatment with rosemary EO applied to textiles contaminated with Aspergillus niger resulted in a contamination reduction of up to 22.12%, whereas treatment with orange EO resulted in a reduction of up to 51.45%. Both treatments applied to textiles contaminated with *Candida albicans* produced 100% reduction rates. In the case of contamination with *Trichoderma viride*, textiles treated with rosemary EO recorded a contamination reduction of up to 76.48%, and 100% with orange EO treatment. For textiles contaminated with *Aspergillus flavus*, the maximum efficiency of the rosemary oil treatment was 18.3%, and 60.57% with orange oil. In the case of contamination with the dermatophyte *Epidermophyton floccosum*, the maximum reduction rate with rosemary oil treatment was 56.99%, whereas orange oil treatment resulted in a reduction of up to 92.48% [43].

Positive results were also obtained by applying EOs like lavender, lemon, mint, tea tree, marjoram, and oregano to cotton, leather, and wool effectively inhibited six fungal species (*Alternaria*, *Aspergillus*, *Botrytis*, *Cladosporium*, *Mucor*, and *Penicillium*) and a yeast (*Candida guilliermondii*) [44].

In another study, the antimicrobial effects of *Moringa stenopetala* seed oil applied to cotton fabrics were analyzed. The results showed that, on treated fabrics, the growth of gram-positive bacteria (*S. aureus*) and gram-negative bacteria (*E. coli*) was reduced by 77.6%–100% before washing and by 45.8%–85.2% after washing, depending on the concentrations of the extract used [45].

Another solution to improve the antibacterial properties of cotton fabrics is the application of a mixture of olive oil and ethanol in different concentrations. Both before and after washing, fabrics treated in this manner show outstanding resistance to the bacteria *S. aureus* and *E. coli*. It was found that as the concentration of the oil increases, the antimicrobial efficiency also increases [46].

El-Shafei *et al.* [47] tested the antimicrobial effect of chamomile, sage, and green tea extracts applied to cotton fabrics. The authors showcased the antimicrobial activity of fabrics treated with chamomile, sage, and green tea against unicellular fungi (*Candida albicans*, *Candida tropicalis*, and *Candida pseudotropicalis*) as well as gram-positive and gram-negative bacteria (*Salmonella tyhimurium* and *Escherichia coli*). The main limitation of these treatments is their poor durability during washing. However, fabrics treated with these extracts can be used as disposable items, such as surgical bandages, which do not require washing.

L. Ammayyappan *et al.* [48] conducted a study on the antimicrobial activity of aloe vera, chitosan, and curcumin individually and, in various combinations, on cotton, wool, and rabbit hair. The cotton samples were pretreated, i.e., semi-bleached with hydrogen peroxide, whereas the wool and rabbit hair were treated with formic acid. Application of aloe vera alone showed lower antimicrobial activity than chitosan and curcumin; application of aloe vera along with chitosan and curcumin increased antimicrobial activity. All pretreated material samples retained their antimicrobial activity for up to 25 washing cycles.

D. Abramiuc *et al.* [49] investigated the antimicrobial effects of propolis, beeswax, and chitosan on knitwear consisting only of cotton of different densities. When evaluating comfort and antibacterial properties, the variant with the highest impregnation was taken into account. The results of the study demonstrated that the cotton knitwear that had undergone treatment exhibited notable antibacterial activity against a range of bacterial strains, including both gram-positive and gram-negative bacteria. These bacteria included β-hemolytic streptococci, as well as *Escherichia coli* and *Pseudomonas aeruginosa*. The treatment not only made the cotton substrates more hygroscopic and less permeable to air but also improved the wearing comfort of the knitted fabric.

Tea tree oil (TTO) was encapsulated by Beşen using gum Arabic (GA), polyvinyl alcohol (PVA), and β-cyclodextrin (β-CD). These capsules were then applied using the padding process at a concentration of 20% (w/v) to surfaces made entirely of viscose nonwoven fabric. The fabric samples treated with these capsules were analyzed using SEM and fourier transform infrared (FTIR) spectroscopy. Additionally, the antibacterial effectiveness of the samples against *E. coli* and *S. aureus* bacteria was assessed. The findings showed that capsules containing TTO encapsulated in β-CD, PVA, and GA could be effectively applied to textiles. The treated samples exhibited varying levels of antibacterial activity, depending on the specific wall material used for encapsulation [50].

The antioxidant and antibacterial activity of lime oil (*Citrus aurantifolia*) was demonstrated by Wijesirigunawardana and Perera who applied EO microcapsules onto a cotton fabric. Upon mechanical disruption of the microcapsules, the lime oil-infused cotton fabric exhibited notable antibacterial activity. Furthermore, even after a gentle washing step, the fabric retained both its microcapsules and antibacterial effectiveness [51]. The antibacterial properties of lime EO against typical skin bacteria such as *S. aureus*, *E. coli*, *K. pneumoniae*, and *S. epidermidis* have been demonstrated by E. Julaeha *et al.* [52].

EOs have been successfully applied to treat heritage objects.

A study on a handmade men's short coat made of hemp, cotton, and sheepskin from Valea Marei, Maramureş, Romania, found that EOs—lavender, lemon, peppermint, tea tree, marjoram, and oregano—strongly inhibited six fungal colonies and one present on the garment yeast (*Candida guilliermondii*). The treatments had no discernible effects on the physical properties of the fur [44].

The antifungal effects of lemon, lavender, and peppermint EOs were tested on a Romanian "ie" garment stored in a Beiuş museum, showing inhibition against *Botrytis* sp., *Cladosporium* sp., and *Rhodotorula mucilaginosa* fungi [53].

In another study, wild rosemary, *Vitex agnus-castus*, and lemongrass oils were tested against fungi from an Egyptian child mummy on flax fibers, with *Vitex agnus-castus* oil proving the most effective in inhibiting fungal growth [54].

Indrie *et al.* [55] studied the effects of Thyme and Salvia EOs on the tensile strength of hemp and cotton textiles. Salvia oil increased cotton fabric's strength by 20% in the warp and 39% in the weft directions, whereas Thyme oil reduced it by 29.9%. For hemp, both oils reduced strength, with Salvia decreasing it by 36% and Thyme by 40%. The study emphasizes the need to match EOs with specific materials.

The antimicrobial and antifungal effects of chitosan have been studied through its application on various textiles from Romania's cultural heritage. These treatments have also contributed to improving the shrink resistance of the materials, enhancing dye absorption, and improving whiteness due to the cleaning of existing stains on these products [9, 56].

In the study by Kadoglu *et al.* [57], grape seeds, known for their antibacterial properties, were applied to cotton fabric using a coating method to impart antibacterial properties. Color fastness tests (ISO 105-C06) showed that the coating did not negatively affect the wash fastness of the fabric. All coated fabrics achieved a value of 5, indicating high wash resistance. Friction resistance tests (TS EN ISO 105-X12) also confirmed that the coated fabrics were durable. Strength tests (TS EN ISO 13934-1) showed increased tensile strength and elongation at break of the coated materials. Antibacterial tests (AATCC 100) showed better effectiveness against *E. coli* with improved antibacterial properties with increasing coating thickness.

***c. Mosquito Repellent***

Mosquitoes are considered to be among the most harmful insects, due to their role in the transmission of arboviruses such as dengue fever, malaria and Zika [58]. Various studies have examined the mosquito-repellent properties of fabrics and highlighted this as a promising area of textile functionality. Mosquito-repellent textiles offer a revolutionary by providing protection against mosquito bites and reducing the risk of diseases such as malaria, dengue fever, and filariasis [59]. Both natural and synthetic repellents are used, but natural options such as EOs are gaining more interest due to their non-toxic and environmentally friendly properties. However, EOs have short-lived effectiveness due to their high volatility and the need for frequent reapplication.

A wide range of commercial insect repellent solutions are available on the market in spray, aerosol, and lotion forms. They could cause unwanted skin reactions as a result of their direct application to the skin. An alternative strategy to prevent overuse of pesticides is to encapsulate them. With micro- and nanoscale capsules in textile finishing, wash-resistant textile products can be produced that are effective even with less active ingredient application. N,N-diethyl-meta-toluamide (DEET) has a very good effect in protecting people from mosquito bites. According to Gryboski *et al.* [60], prolonged use of DEET has been linked to negative effects, including encephalopathies and skin reactions.

Because of these harmful effects, researchers have studied various plant species to develop natural alternatives to synthetic repellents. Successful repellents against mosquitoes have been found in plants such as *Mentha spicata* [61], *Magnolia grandiflora* [62], and *Citrus aurantifolia* [63]. In addition, certain EOs such as thymol and carvacrol have shown repellent properties against various species of mosquitoes.

The application of limonene to cotton fabrics has been demonstrated by Türkoglu *et al.* [64] as a method to prevent mosquito bites.

Citronella oil is known for its mosquito-repellent properties, but its use in topical products is limited because it evaporates quickly.

In their study, Solomon *et al.* [65] aimed to slow the evaporation of citronella oil through microencapsulation. They prepared microcapsules using the gelatin coacervation method with sodium sulfate as the coacervation agent and formaldehyde as the cross-linker. They examined the effects of stirring speed, oil loading, and cross-linker concentration on encapsulation efficiency. The optimal conditions resulted in an encapsulation efficiency of 60% and a sustained release profile, with 70% of the contents released after 10 hours. Topical formulations showed that microencapsulation reduced CO membrane permeation by at least 50%, with the PEG base providing the highest release. Microencapsulation stabilizes the oil and reduces CO permeation. A simple, inexpensive, scalable and repeatable way to obtain encapsulated EOs for textile applications is complex coacervation.

In the study by Specos *et al.* [66], microcapsules of citronella EO were created through a sophisticated coacervation process and then applied to cotton textiles to study the fabrics' repellent ability. The extractable content of the main components of citronella served as an indirect indicator of release from treated textiles. The repellent effectiveness of a human hand and arm covered with treated textiles was tested using *Aedes aegypti* mosquitoes. In comparison with fabrics that have been treated with an ethanol solution of the EO, those treated with microencapsulated citronella have been shown to provide a more robust and prolonged protection against insects, with a repellency rate of over 90% for a period of 3 weeks. Repellent textiles were produced by using a traditional pad-dry method to pad cotton fabrics with microcapsule slurries.

Many natural herbal products have the ability to repel mosquitoes. Mosquito-repellent properties have been found in extracts from a variety of plant species' roots, stems, leaves, flowers, fruits, and seeds. These extracts can be applied as microcapsules to enhance shelf life and enable controlled release, or used directly as raw agents for textile finishing. One of these plants with known therapeutic properties is lemongrass.

To find out whether lemongrass oil can keep mosquitoes away, the polyester fabric used in the study was treated with it [67]. When comparing the fabric treated with the pad-dry curing method with the microencapsulated polyester fabric, the results show that the latter had a lower mosquito repellent effect (92%). Compared to the control, FTIR analysis of the finished microcapsules encapsulated in the fabric also shows the addition of new functional groups to the fabric.

### *d. Flame Retardant*

In order to achieve long-term flame retardancy, organic or inorganic flame retardants were applied to textile substrates in microencapsulation. Microencapsulation has been applied to overcome the hydrophilicity of materials, prevent reactions with textile polymers, and/or prevent fire retardant chemicals from leaking or sublimating. In addition to textiles for home and automotive industries, the items include military and fire safety apparel.

Two common techniques for making fabrics flame-retardant are padding and coating. However, these techniques often cause the treated fabrics to become stiff and lose some of their color. A possible solution to these problems is microencapsulation. Microencapsulation, as described by Anal *et al.* [68], is the technology of packaging solid, liquid or gaseous materials into tiny sealed capsules that may (or may not) release their contents at a controlled rate under certain conditions. This process ensures that the active ingredients are kept away from the environment, protected from harmful substances and stored safely in the capsules. Microcapsules containing a flame retardant have reportedly been used in recent years to provide reliable fire protection. The substance most commonly used in such studies is di-ammonium hydrogen phosphate.

The goal of Golja *et al.* [69] was to produce fire-retardant textiles using microencapsulated flame retardants. Microcapsules with a melamine-formaldehyde wall and a triphenyl phosphate core were applied to polyester and cotton fabrics using screen printing and impregnation. SEM analysis, vertical burning tests, and thermal analysis thermogravimetric analysis (TGA) and differential scanning calorimetry (DSC) showed that these techniques could successfully apply the microcapsules. However, effective flame retardancy was only achieved at the highest concentration, also affected the mechanical properties of the materials.

## 11.4 Conclusions

Microencapsulation offers numerous possibilities to enhance properties or add new functions in the production of textiles, clothing, and accessories, thereby increasing the market value and usability of the products. When properties are improved or new functions are added through microencapsulation, functional textiles have a higher overall value and more possible uses.

Microencapsulation provides protection from external influences and enables the controlled release of active ingredients, extending the lifespan of treated textile products. There are various microencapsulation techniques including physical, chemical and physicochemical methods. Each method has its own advantages and disadvantages that affect the final properties of the microcapsules. The selection of microencapsulation method depends on the physical and chemical properties of the core and shell materials and the intended properties of the final product.

Microcapsules can be applied to textiles using various methods such as padding, coating, dipping, spraying, and printing. These techniques offer flexibility in integrating microcapsules into textile materials and enable the incorporation of functional properties such as fragrances, antimicrobial activity or temperature regulation. They are used in textile coatings to give them properties such as antimicrobial activity, flame retardancy, insect protection, and prolonged fragrance release. These new properties transform textile products into versatile and long-lasting solutions for consumers.

Interest in this area is growing, as evidenced by the increase in scientific publications and patents. Despite current advancements, there is a need to further explore the long-term applications, environmental sustainability, and performance optimization of microencapsulated materials in textiles.

**AI disclosure statement:** The author confirms that no artificial intelligence (AI)-generated content or AI-assisted writing tools were used in the preparation of this chapter (including summaries and the abstract). All text, analysis, and conclusions are the result of the author's original work.

## References


1. Green, B.K. and Schleicher, L., Oil-Containing Microscopic Capsules and Method of Making Them, U.S., 1957.
2. Hawkins, S., Wolf, M., Guyard, G., Greenberg, S., Dayan, N., Microcapsules as a Delivery System, in: *Delivery System Handbook for Personal Care and Cosmetic Products*, M.R. Rosen (Ed.), pp. 191–213, William Andrew Inc, Norwich, NY, USA, 2005, Link https://doi.org/10.1016/b978-081551504-3.50014-6 [Google Scholar].
3. Gutcho, M., *Microcapsules and Microencapsulation Techniques*, Noyes DataCorporation, Ridge Park, NJ, 1976.
4. Lu, Y. and Park, K., Microencapsulation: Methods and Pharmaceutical Applications, in: *Encyclopedia of Pharmaceutical Science and Technology*, 4th edn, Informa Healthcare, USA, 2012.
5. Rodrigues, S.N., Martins, I.M., Fernandes, I.P., Gomes, P.B., Mata, V.G., Barreiro, M.F., Rodrigues, A.E., Scentfashion (R):Microencapsulated perfumes for textile application. *Chem. Eng. J.*, 149, 463–472, 2009. [CrossRef].
6. Shin, Y. and Yoo D and Son, K., Development of thermoreg-ulating textile materials with microencapsulated phasechange materials (PCM). IV. Performance propertiesand hand of fabrics treated with PCM microcapsules. *J. Appl. Polym. Sci.*, 97, 910–915, 2005, https://doi.org/10.1002/app.21846.
7. Forte Tavčer, P., Magovac, E., Šumiga, B., Boh, B., Bischof Vukušić, S., Treatment of cotton fabric with encapsulated flame-retardant triphenyl phosphate. *Book of abstracts of the 42nd Symposium on Novelties in Textiles*, Ljubljana, p. 40, 2011.
8. Hong, S.A., Kim, I.Y., Kim, H.R., *et al.*, Manufacture of UV absorbers and UV protection fabrics using microcapsules. *Fibers Polym.*, 12, 491–498, 2011, https://doi.org/10.1007/s12221-011-0491-8.
9. Indrie, L., Bonet-Aracil, M., Ilieş, D.C., Albu, A.V., Ilieş, G., Herman, G.V., Baias, #0218., Costea, M., Heritage ethnographic objects – antimicrobial effects of chitosan treatment. *Ind. Textila*, 72, 3, 284–288, 2021, http://doi.org/10.35530/IT.072.03.1812.
10. He, Z., Bao, B., Fan, J., Wang, W., Yu, D., Photochromic Cotton Fabric Based on Microcapsule Technology with Anti-FoulingProperties. *Colloids Surf., A*, 594, 124661, 2020. [CrossRef].
11. Türkoğlu, G.C., Sariişik, A.M., Erkan, G., Kayalar, H., Kontart, O., Öztuna, S., Determination of antioxidantcapacity of capsule loaded textiles. *Indian J. Fibre Text. Res.*, 42, 189–95, 2017.

12. Mahdavi, S.A., Jafari, S.M., Assadpoor, E., Dehnad, D., Microencapsulation optimization of natural anthocyanins with maltodextrin, gum Arabic and gelatin. *Int. J. Biol. Macromol.*, 85, 379–385, 2016.
13. Chavan, P.P. and Pandit, P., Advanced insect repellent agents for protective textiles and clothing, in: *Advances in Functional and Protective Textiles*, pp. 335–357, Woodhead Publishing (an imprint of Elsevier), Oxford, United Kingdom, 2020.
14. Hu, Q., Li, X., Chen, F., Wan, R., Yu, C.W., Li, J., McClements, D.J., Deng, Z., Microencapsulation of an essential oil (cinnamon oil) by spray drying: Effects of wall materials and storage conditions on microcapsule properties. *J. Food Process. Preserv.*, 44, 11, 1–15, 2020.
15. Bakry, A.M., Abbas, S., Ali, B., Majeed, H., Abouelwafa, M.Y., Mousa, A., Liang, L., Microencapsulation of Oils: A Comprehensive Review of Benefits, Techniques, and Applications. *Compr. Rev. Food Sci. Food Saf.*, 15, 1, 143–182, 2016.
16. Anuar, S.N.S., *A* Study on Formation and Characterization of Microencapsulated Geraniol Essential Oil, MSc Dissertation, Universiti Teknologi MARA, Malaysia, 2022.
17. Rani, S. and Goel, A., Microencapsulation technology in textiles: A review study. *Pharma Innov. J.*, 10, 5, 660–663, 2021.
18. Thilagavathi, G., Bala, S.K., Kannaian, T., Microencapsulation of herbal extracts for microbial resistance in healthcare textiles. *Indian J. Fibre Text. Res.*, 32, 3, 351–354, 2007.
19. Wangai, N.L., Kamau, K.K., Nderu, M.G., Maina, D., Gitau, E., Murigi, W., Kamau, M., Njuguna, M.S., Gichuki, J., Otieno, F., Efficacy of Plant-based Repellents Against Anopheles Mosquitoes: A Systematic Review. *Biomed. Sci.*, 6, 3, 44, 2020.
20. De Araújo, J.S.F., De Souza, E.L., Oliveira, J.R., Gomes, A.C.A., Kotzebue, L.R.V., Da Silva Agostini, D.L., De Oliveira, D.L.V., Mazzetto, S.E., Da Silva, A.L., Cavalcanti, M.T., Microencapsulation of sweet orange essential oil (Citrus aurantium var. dulcis) by liophylization using maltodextrin and maltodextrin/gelatin mixtures: Preparation, characterization, antimicrobial and antioxidant activities. *Int. J. Biol. Macromol.*, 143, 991–999, 2020.
21. Peanparkdee, M., Iwamoto, S., Yamauchi, R., Microencapsulation: a review of applications in the food and pharmaceutical industries. *Rev. Agric. Sci.*, 4, 56–65, 2016.
22. Da Veiga, R.D.S., Da Silva-Buzanello, R.A., Corso, M.P., Canan, C., Essential oils microencapsulated obtained by spray drying: a review. *J. Essent. Oil Res.*, 31, 6, 457–473, 2019.
23. Atiénzar-Navarro, R., Bonet-Aracil, M., Gisbert-Payá, J., Rey, R.D., Picó, R., Sound absorption of textile fabrics doped with microcapsules. *Appl. Acoust.*, 164, 107285, 2020.

24. Dubey, R., Shami, T.C., Bhasker Rao, K.U., Yoon, H., Varadan, V.K., Synthesis of polyamide microcapsules and effect of critical point drying on physical aspect. *Smart Mater. Struct.*, 18, 2, 025021, 2009.
25. Zhang, T., Luo, Y., Wang, M., Chen, F., Liu, J., Meng, K., Zhao, H., Double-layered microcapsules significantly improve the long-term effectiveness of essential oil. *Polymers*, 12, 8, 2020.
26. Podgornik, B.B., Šandri, S., Kert, M., Microencapsulation for Functional Textile Coatings with emphasis on biodegradability—a systematic review. *Coatings*, 11, 11, 1371, 2021.
27. Singh, and Sheikh, Microencapsulation & Its Application In Production of Functional Textiles. *Indian J. Fibre Text. Res.*, 45, 495–509, 2020.
28. Khounvilay, K., Estevinho, B.N., Rocha, F.A., Oliveira, J.M., Vicente, A., Sittikijyothin, W., Microencapsulation of citronella oil with carboxymethylated tamarind gum. *J. Sci. Technol.*, 15, 7, 515–527, 2018.
29. Lu, T., Li, B., Sun, D., Hu, M., Ma, J., Sun, G., Advances in controlled release of microcapsules and promising applications in self-healing of asphalt materials. *J. Cleaner Prod.*, 294, 126270, 2021.
30. Sun, S., Gao, Y., Han, N., Zhang, X., Li, W., Reversible photochromic energy storage polyurea microcapsules *via in-situ* polymerization. *Energy*, 219, 119630, 2021.
31. Mehran, M., Masoum, S., Memarzadeh, M., Microencapsulation of Mentha spicata essential oil by spray drying: Optimization, characterization, release kinetics of essential oil from microcapsules in food models. *Ind. Crops Prod.*, 154, 112694, 2020.
32. Makimori, R.Y., Endo, E.H., Makimori, J.W., Zanqueta, E.B., Ueda-Nakamura, T., Leimann, F.V., Gonçalves, O.H., Dias Filho, B.P., Preparation, characterization and antidermatophytic activity of free- and microencapsulated cinnamon essential oil. *J. Mycol. Med.*, 30, 2, 100933, 2020.
33. Bezerra, F.M., Carmona, O.G., Carmona, C.G., Lis, M.J., de Moraes, F.F., Controlled release of microencapsulated citronella essential oil on cotton and polyester matrices. *Cellulose*, 23, 2, 1459–1470, 2016.
34. Botrel, D.A., de Barros Fernandes, R.V., Borges, S.V., Microencapsulation of Essential Oils Using Spray Drying Technology, in: *Microencapsulation and Microspheres for Food Applications*, pp. 235–251, Academic Press imprint (an imprint of Elsevier), Oxford, United Kingdom, 2015.
35. Voncina, B., Kreft, O., Kokol, V., Chen, W.T., Encapsulation of rosemary oil in ethylcellulose microcapsules. *Text. Polym. J.*, 1, 1, 13–19, 2009.
36. Bonet-Aracil, M., Monllor, P., Capablanca, L., Gisbert, J., Díaz, P., Montava, I., A comparison between padding and bath exhaustion to apply microcapsules onto cotton. *Cellulose*, 22, 2117–2127, 2015. [Google Scholar] [CrossRef] [Green Version].
37. Bonet-Aracil, M., Capablanca, L., Monllor, P., Díaz, P., Montava, I., Studying bath exhaustion as a method to apply microcapsules on fabrics. *J. Text. Inst.*, 103, 629–635, 2012. [Google Scholar] [CrossRef].

38. Rodrigues, S.N., *et al.*, Scentfashion®: microencapsulated perfumes for textile application. *Chem. Eng. J.*, 2009.
39. Kumari, P., Rose, N.M., Jeet Singh, S.S., Development of fragrant microcapsules for woven cotton fabric. *J. Appl. Nat. Sci.*, 9, 2, 1017–1021, 2017.
40. Sannapapamma, K., Lokanath, H.M., Naikwadi, S., Antimicrobial and Aroma Finishing of Organic Cotton Knits Using Vetiver Oil Microcapsules for Health Care Textiles. *Int. J. Mater. Text. Eng.*, 12, 2, 82–91, 2018.
41. Menezes, E., Antimicrobial finishing for specialty textiles. *Int. Dyer*, 187, 92, Dec 2002.
42. Indrie, L., Frances, L.C., Erden, E., Kilic, M., Ilies, D.C., Grama, V., Safarov, B., Zlatev, Z., Old maps and the effectiveness of chitosan as antimicrobial agent. *Ind. Textila*, 73, 4, 420–425, 2022, http://doi.org/10.35530/IT.073.04.202156.
43. Iordache, O., *et al.*, Antimicrobial activity of textiles treated with Rosemary and orange essential oils against a selection of pathogenic fungi. *Sci. Bull. Ser. F Biotechnol.*, 20, 362–369, 2006.
44. Ilieş, D.C., Hodor, N., Indrie, L., Dejeu, P., Ilieş, A., Albu, A., Caciora, T., Ilieş, M., Barbu-Tudoran, L., Grama, V., Investigations of the Surface of Heritage Objects and Green Bioremediation: Case Study of Artefacts from Maramureş, Romania. *Appl. Sci.*, 11, 14, 6643, 2021, https://doi.org/10.3390/app11146643.
45. Gebino, G., *et al.*, Study on efficacy of moringa stenopetala seed oil extract for antimicrobial activities on textile materials. *Res. J. Text. Apparel*, 2021.
46. Sadannavar, M.K., *et al.*, Antibacterial cotton functionalized with olive oil for developing medical textiles. *Indian J. Fibre Text. Res. (IJFTR)*, 46, 3, 303–310, 2021, http://op.niscpr.res.in/index.php/IJFTR/article/view/38246.
47. El-Shafei, A., *et al.*, Herbal extract as an ecofriendly antibacterial finishing of cotton fabric. *Egypt. J. Chem.*, 61, 2, 317–327, 2018, https://ejchem.journals.ekb.eg/article_5295_cd95c7a21fe8b226d61b859277fb7c11.pdf.
48. Ammayappan, L. and Jeyakodi Moses, J., Study of antimicrobial activity of aloevera, chitosan, and curcumin on cotton, wool, and rabbit hair. *Fibers Polym.*, 10, 2, 161–166, 2009, https://link.springer.com/article/10.1007/s12221-009-0161-2.
49. Abramiuc, D., *et al.*, Antibacterial finishing of cotton fabrics using biologically active natural compounds. *Fibers Polym.*, 14, 11, 1826–1833, 2013.
50. Beşen, B.S., Production of disposable antibacterial textiles *via* application of tea tree oil encapsulated into different wall materials. *Fibers Polym.*, 20, 2587–2593, 2019.
51. Wijesirigunawardana, P.B., Perera, B., Gayani, K., Development of a cotton smart textile with medicinal properties using lime oil microcapsules. *Acta Chim. Slov.*, 65, 1, 150–159, 2018.
52. Julaeha, E., *et al.*, Microencapsulation of lime (Citrus aurantifolia) oil for antibacterial finishing of cotton fabric. *RSC Adv.*, 11, 3, 1743–1749, 2021.
53. Wendt, J.A., Indrie, L., Dejeu, P., Albu, A., Ilieş, D.C., Costea, M., Caciora, T., Ilieş, G., Hodor, N., Josan, I., Berdenov, Z., Grama, V., Safarov, B., Natural

sources in preventive conservation of naturally aged textiles. *Fibres Text. East. Eur*, 2021.
54. Mansour, M.M.A., EL-Hefny, M., Salem, M.Z.M., Ali, H.M., The biofungicide activity of some plant essential oils for the cleaner production of model linen fibers similar to those used in ancient Egyptian mummification. *Processes*, 8, 1, 79, 2020, https://doi.org/10.3390/pr8010079.
55. Indrie, L., Affandi, N.D.N., Díaz-García, P., Haji, A., Ilies, D.C., Zlatev, Z., Taghiyari, H.R., Grama, V., Farima, D., Mechanical and Morphological Properties of Cellulosic Fabrics Treated with Microencapsulated Essential Oils. *Coatings*, 12, 20221958, https://doi.org/10.3390/coatings12121958.
56. Bou-Belda, E., Indrie, L., Ilieş, D.C., Hodor, N., Berdenov, Z., Herman, G., Caciora, T., Chitosan – a non-invasive approach for the preservation of historical textiles. *Ind. Textila*, 71, 6, 576–579, 2020, http://doi.org/10.35530/IT.071.06.1756.
57. Kodaloğlu, F.A. and Demiralay, H., Application of grape seed coating for antibacterial cotton fabric. *Ind. Textila*, 74, 3, 332–336, 2023, http://doi.org/10.35530/IT.074.03.202217.
58. Benelli, G. and Mehlhorn, H., *Mosquitoe-borne diseases*, Springer, Dusseldorf, 2018.
59. Prabha, R. and Vasugi, R.N., Filarial repellent finish using medicinal plants. *Indian J. Sci.*, *1*, 1, 74–76, 2012.
60. Gryboski, J., Weinstein, D., Ordway, N.K., Toxic encephalopathy is apparently related to the use of an insect repellent. *N. Engl. J. Med.*, 264, 289–291, 1961, doi: 10.1056/NEJM196102092640608. [PubMed] [CrossRef] [Google Scholar.
61. Azeem, M., Zaman, T., Tahir, M., Haris, A., Iqbal, Z., Binyameen, M., Nazir, A., Shad, S.A., Majeed, S., Mozūraitis, R., Chemical composition and repellent activity of native plants essential oils against dengue mosquito, *Aedes aegypti. Ind. Crops Prod.*, 140, 111609, 2019, doi: 10.1016/j.indcrop.2019.111609. [CrossRef] [Google Scholar.
62. Ali, A., Tabanca, N., Demirci, B., Raman, V., Budel, J.M., Baser, K., Khan, I.A., Insecticidal and biting deterrent activities of *Magnolia grandiflora* essential oils and selected pure compounds against *Aedes aegypti. Molecules*, 25, 1359, 2020, doi: 10.3390/molecules25061359. [PMC free article] [PubMed] [CrossRef] [Google Scholar.
63. Soonwera, M., Efficacy of essential oils from Citrus plants against mosquito vectors *Aedes aegypti* (Linn.) and *Culex quinquefasciatus* (Say). *J. Agric. Technol.*, 11, 669–681, 2015. [Google Scholar].
64. Türkoğlu, G.C., Sarıışık, A.M., Erkan, G., Deveci, O., Yıkılmaz, M.S., Kontart, O., Öztuna, S., Microencapsulated limonene loadedtextile and its insect repellent behavior, in: *14th National & 1stInternational recent developments, textile technology and chemistry symposium*, Turkey, 2013.
65. Solomon, B., Sahle, F.F., Gebre-Mariam, T., Asres, K., Neubert, R.H., Microencapsulation of citronella oil for mosquito-repellent application:

formulation and *in vitro* permeation studies. *Eur. J. Pharm. Biopharm.*, 80, 1, 61–6, 2012 Jan, doi: 10.1016/j.ejpb.2011.08.003. Epub 2011 Sep 8. PMID: 21924356.
66. Specos, M.M., García, J.J., Tornesello, J., Marino, P., Vecchia, M.D., Tesoriero, M.V., Hermida, L.G., Microencapsulated citronella oil for mosquito repellent finishing of cotton textiles. *Trans. R. Soc. Trop. Med. Hyg.*, 104, 10, 653–8, 2010 Oct, doi: 10.1016/j.trstmh.2010.06.004. Epub 2010 Jul 31. PMID: 20673937.
67. Anitha, R., *et al.*, Microencapsulation of lemon grass oil for mosquito repellent finishes in polyester textiles, 2011.
68. Anal, A.K., Stevens, W.F., Remuñán-López, C., Ionotropic cross-linked chitosan microspheres for controlled release of ampicillin. *Int. J. Pharm.*, 312, 166, 2006.
69. Golja, B., *et al.*, Application of flame retardant microcapsules to polyester. *Mater. Technol.*, 48, 1, 105–111, 2014.

# 12

# Chemical Vapor Deposition (CVD) and Physical Vapor Deposition (PVD) Techniques for Textile Surface Enhancement

**Habib Ali[1]*, Afreen Anjum[2] and Shafat Ahmad Khan[1]**

[1]*Department of Chemistry and Sustainability, School of Basic Sciences, Galgotias University, Greater Noida, Uttar Pradesh, India*
[2]*Department of Applied Science and Humanities, GL Bajaj Institute of Technology and Management, Greater Noida, Uttar Pradesh, India*

***Abstract***

This chapter explores the transformative potential of chemical vapor deposition (CVD) and physical vapor deposition (PVD) processing to revolutionize textile surface modification, as well as their new innovative approaches. Among the various techniques, CVD and PVD are two elemental techniques characterized by atomic-level precision, scalability, and versatility. This chapter thoroughly reviews how PVD and CVD processes improve their diverse properties on textile surfaces, including hydrophilicity, wettability, dyeability, and functionality. CVD may be defined as a reaction of volatile components at the surface. In contrast, PVD utilizes material that evaporates from a solid source onto textile surfaces. This vaporizing coating process involves an atom-by-atom or molecule-by-molecule transfer of material from the solid phase to the vapor phase, which is then deposited onto the surfaces of textiles. Updating the techniques used to optimize process parameters and obtaining the needed surface functionality are imperative. These processes are crucial in demonstrating aspects such as substrate compatibility and precursor selection, thereby revealing possibilities on how changes can occur while keeping textile integrity based on current studies and applications in practice. This text also deals with modern progress in textiles. Availability of CVD and

**Corresponding author*: habibali1989@gmail.com

Aminoddin Haji and Çağlar Sivri (eds.) Surface Modification and Functionalization of Textiles, (277–318) © 2026 Scrivener Publishing LLC

PVD technologies, along with sustainability and scalability issues, has mapped the journey for textiles.

***Keywords***: Chemical vapor deposition (CVD), physical vapor deposition (PVD), surface modification, precursor chemicals, thin-film coating, textile, surface enhancement, eco-friendly coating

## 12.1 Introduction

Fabric surface characteristics are very significant in that they define several textile properties, including the Hydrophilic nature, wettability, dyeability, durability, and functionality of a material [1–4]. These surface characteristics are helpful for the feel and look of textiles and help decide on the particular functionalities that may be required, such as water resistance, antibacterial properties, or conductive properties [5]. The alteration of textile fiber surfaces can potentially introduce a new or improved material with the prospect of introducing new application areas in technical textiles, intelligent textiles, advanced textiles, or even high-performance clothing [6, 7]. Therefore, the technology involved in processing involves techniques that allow one to control the behavior of a textile in its environment, including moisture, chemicals, light, or mechanical energy. Chemical vapor deposition (CVD) and physical vapor deposition (PVD) techniques are instrumental in the present context among the various surface modification techniques [8]. CVD and PVD provide similar deposition of thin films and coatings to textile materials and reasonably control the deposited layer in composition, thickness, and layering [3, 4]. These methods enable the deposition of material on surfaces to alter the nature of the substrate significantly, without harming textile material. This was made possible by CVD and PVD technologies that have enhanced the ability to gasify and deposit materials on the textile surface to achieve the desired characteristics for the garment's use, like high durability, improved barrier value, or a specific novel coat for health, electronics, or fashion consumer needs [9–11]. This chapter focuses on CVD and PVD to better understand how they are used in textile surface modification. It outlines the general concepts behind these processes, the procedure of every method, and the varieties of materials that can be deposited by vapor deposition. Also, the chapter focuses on how these techniques affect the physicochemical characteristics of textile fibers and enhance surface energy, bonding, and wear resistance. There are numerous opportunities to apply CVD and PVD in the textile industry, including providing textiles with hydrophobic features and increasing the mechanical properties of fibers in protective clothing.

Thus, this chapter identifies the application areas of CVD and PVD in modifying textiles and discusses the benefits and drawbacks of these technologies to enhance the reader's awareness of how these technologies are revolutionizing the textile industry. Deposition processes have been fine-tuned to enable developers to create textiles with specific surface characteristics to suit many new and potential applications.

## 12.2 Overview of CVD and PVD Techniques

CVD and PVD together fall under the physical type of surface modification [9]. In this type, CVD is the method by which thin films are laid down on the textile substrate CVD without the need for any wet chemical. CVD, on the other hand, is a process where there is a chemical reaction of the vapor phase precursors forming a thin film. At the same time, PVD is a coating technique where material is deposited from a vapor phase using either evaporation or sputtering.

## 12.3 Chemical Vapor Deposition (CVD)

### 12.3.1 Principles of Chemical Vapor Deposition (CVD)

CVD is a useful method of depositing oxidation layers and films on the substrate as a result of chemical reactions of gaseous reagents [12, 13]. Here, gaseous reagents are admitted into the reaction chamber, which may encompass the substrate, especially in surface treatment, and the substrate is most preferably a textile material. When they reach the surface of the substrate, these gases undergo other chemical reactions that form a solid film on the surface [13–15]. The solid films formed here include metal oxides, nitrides, carbides, or any other compound that settles down from the gaseous phase and is deposited on the substrate [15].

CVD provides a lot of flexibility for controlling the film's thickness, composition, and homogeneity, which makes its usage in industries where accuracy and material properties are significant, such as electronics, aerospace, and textile engineering [13, 15]. The process can be further refined to coat all materials, from metals to semiconductor polymers, textiles, and other fabrics. Thus, it enables textiles to be functionally hydro-phobic, antibacterial, electrically conductive, or thermally resistant [12, 14].

CVD can be classified as shown in Figure 12.1.

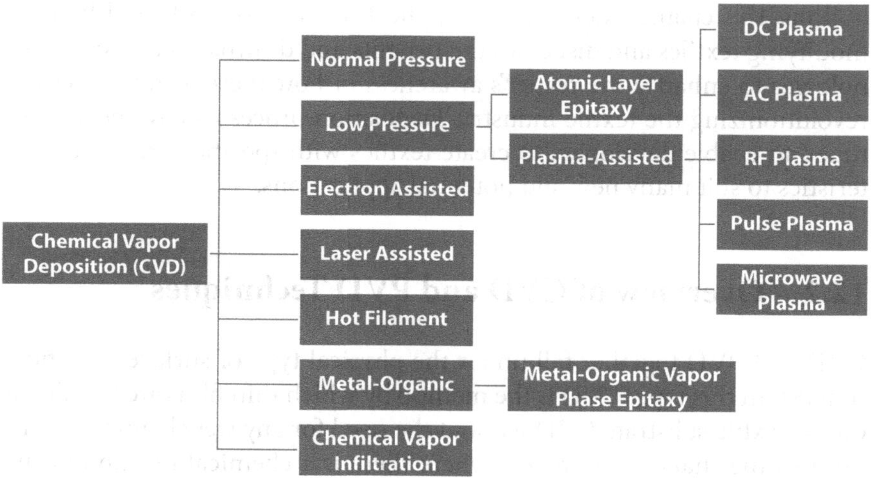


**Figure 12.1** Classification of CVD.

### *12.3.1.1 The CVD Process*

The CVD process involves several stages:

**Precursor Gas Introduction:** Either one or both of the precursor gases then enter the CVD reaction vessel, which contains the textile substrate in which the reaction takes place. These gases could be organometallic compounds or silicon, halide, or any reagent that could chemically interact at the substrate surface [16, 17].

**Chemical Reaction:** As indicated above, when the precursor gases are deposited on the substrate, they change chemically and can be stimulated by heat, plasma energy, or light. Gas and solid catalysts are used in this reaction to produce a solid material that is deposited on the surface of the textile substrate and removes volatile by-products from the reaction chamber [12, 16–19].

**Film Formation:** The deposition goes on by depositing a layer of solid material onto the surface, at the same time as the formation of a thin and homogeneous film. It has been seen that the properties of this film are related to the type of precursor used in reaction conditions like temperature, pressure, and substrate used in the formation of this film [20, 21]. As these parameters can be regulated, one can accurately fine-tune the film characteristics, including thickness, adhesion, and uniformity.

**By-Product Removal:** These gaseous by-products that do not take part in the formation of the films will be taken out of the chamber so that they do not affect the process of deposition on the substrate [14].

CVD is most useful for achieving conformal deposition, wherein the growing film will conform to the shape of the substrate and complex features of the substrate surface.

### *12.3.1.2 Common Variants of CVD*

Several CVD processes were developed to improve specific aspects of the deposition process or accommodate various substances and surfaces. The most commonly used variants in the textile industry are the following:

**Plasma-Enhanced Chemical Vapor Deposition (PECVD):** PECVD works by enhancing the chemical processes that are employed while using the CVD technique using plasma [22]. PECVD uses plasma with charged species of ions and electrons, reducing the required reaction temperature, and does not affect heat-sensitive materials such as textiles. PECVD reduces the chances of destroying the base textile fibers by lowering the reaction temperature and depositing a superior-quality film. This technique has higher application in preparing membranes or coatings that need to function in a certain way, including hydrophobic coatings, conducting coatings, or abrasion-resistant coatings. It has a vast application in the area of technical textiles.

This is a clear advantage offered by PECVD at lower deposition temperatures than thermal CVD. A plasma generated from several sources, such as direct current (DC), radio frequency (alternating current) (RF(AC)), and microwave, is made to react more efficiently between the precursors. The reacting gases in DC plasma get decomposed and ionized by a DC electrical discharge to form plasma, whereas, in AC plasma, the reacting gases are decomposed and ionized to generate time-varying plasma that is repeatedly initiated and extinguished shown in Figure 12.2. Nowadays, the RF PECVD is being widely used to deposit SiC films on the silicon substrate with the mixture of $SiCl_4$, $CH_4$, $H_2$, and Ar. Microwave PECVD qualifies the creation of plasma *via* microwaves. The major effect of microwaves is on electron oscillation, and electrons colliding with gas atoms and molecules produce a great fraction of ionization. Also, it provides good substrate-specific selective growth. Microwave (MW)-CVD is typically used in the fabrication of vertically aligned arrays of carbon nanotubes.

**Metal-Organic Chemical Vapor Deposition (MOCVD):** MOCVD, on the other hand, is a sub-process of CVD in which metal-organic compounds are

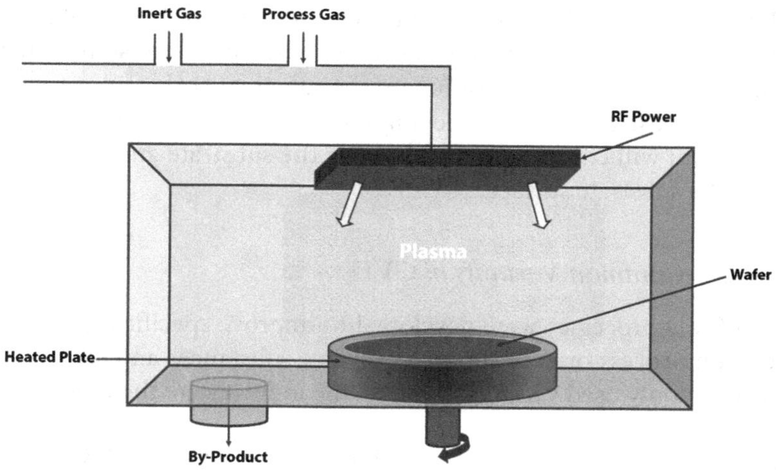


**Figure 12.2** Schematic of plasma-enhanced CVD.

employed as the precursors [15, 17, 21]. These precursors dissolve at the substrate's surface and form metal-containing films. MOCVD is used in other cases where metal oxides, nitrides, and carbides are employed. Concerning textiles, MOCVD can come in handy in laying down a thin layer that will enhance the conductivity, mechanical, or thermal characteristics of the fabric, respectively. This process is also useful in adopting layers with still bacteriostatic or fungistatic properties obtained by the incorporation of metals like silver, copper, or zinc into the surface of the textile material.

**Low-Pressure Chemical Vapor Deposition (LPCVD):** LPCVD is a type of CVD that occurs at low pressures as opposed to conventional CVD. Lower pressure has the advantages of better uniformity of the deposit and thickness, and fewer and less cumbersome effects from admixed reactions. LPCVD is most proficient for use in large-scale production of textile films where uniformity of the film across large areas of textiles is mandatory. Lower pressure also increases precursors' utilization efficiency; therefore, LPCVD can be an economical process for some coating systems.

#### *12.3.1.3 Key Factors Influencing the CVD Process*

Various considerations affect the process of CVD as well as the result of the process. These factors include the following:

**Substrate Temperature:** Because the growth of the film involves the reaction of the precursor gases and deposition on the substrate, the substrate temperature needs to be well-regulated. When the temperature is too low, the precursors will not react well, and, thus, the formation of the film will not be ideal. On the other hand, extreme temperatures over this range may result in undesirable reactions or even deterioration of the textile substrate [21].

**Precursor Composition:** They found that the type of film that developed on the textile surface directly depends on the precursor's choice. The precursor must be vaporized, capable of undergoing the required chemical change, and yet be inert enough not to form a complete layer while depositing on the substrate. By varying the chemistry of the precursors, it is also possible to place selected materials, for example, metal oxides, nitrides, or polymers, at the substrate surface, depending on the required surface characteristics.

**Reaction Chamber Pressure:** Another parameter that can control the deposition process is the pressure inside the reaction chamber, which influences the rate of the gases' interaction and the formation of the thin film on the substrate. The conventional method, however, entails using lower pressures because it is easier to produce uniform films by avoiding side reactions and contamination.

**Deposition Rate:** Hence, the rate at which the film is being deposited has to be well adjusted to the proper rate to allow the formation of a high-quality, non-defective film. High depositing rates of the coating material may lead to a rough and porous surface that may not have the desired properties. The rates also have to be efficient for sizable large-scale deductions, and low rates are, therefore, unsuitable for large-scale production.

**Gas Flow Rate:** Precursor gases required during the CVD process must be supplied at appropriate rates. Inadequate gas flow will cause variations in the pattern of deposition on the textile material, especially if it is large or the shape of the material is complicated.

#### *12.3.1.4 Applications of CVD in Textiles*

In the textile industry, CVD is used as a sensor to give fabrics unique performance characteristics. Applications include the following:

**Hydrophobic Coatings:** CVD can be applied in manufacturing waterproof fabrics by coating the layers of hydrophobic contaminants like fluorocarbons or silicone. These coatings improve the waterproofing of fabrics used in clothing without affecting the fabric's ability to vapor transfer and flexing.

**Antimicrobial Textiles:** Silver- or copper-based metal coats can also be deposited through CVD for medical purposes and used in sportswear and all hygienic products.
**Wear-Resistant Coatings:** Fragile and strong films formed by CVD can enhance the wear resistance of the fabrics to increase the durability of the fabrics in industries such as automobiles, marine, and outdoor uses.
**Barrier Coatings:** CVD is an outstanding technique for synthesizing gas or chemical barrier layers on textiles to protect clothing or filtration.

## 12.3.2 CVD Process for Textiles

The CVD process in textiles is a complex process for depositing a thin film on textile materials. Here, depositor material in the form of vapor is reacted on the textile substrate in a controlled atmosphere to form a thin layer. This involves several issues, such as the selection of the reactor system and the temperature control, and the handling of precursors. To develop high-performance materials, CVD in the textile industry is used to improve surface characteristics, including wear resistance, hydrophobic, and electrical conductance.

This section of the work has explained and illustrated the basic configuration of the CVD system, the process controls used, and how these parameters vary to improve the deposition process for textiles.

### *12.3.2.1 CVD Setup for Textiles*

The CVD setup for textiles typically consists of the following core components:

**Reaction Chamber (Reactor):** The reactor is the core or the central part of the CVD mechanical system where chemical reactions and deposition occur. In the case of textiles, the reactor design should be able to sufficiently support the flexible characteristics of the substrate and supplement the diffusion of the precursor gases to all the zones of the fabric. Depending on the application, reactors can vary in size and design but are generally classified into "hot-wall" and "cold-wall reactors," types of reactors, as well as other reactors used in solar energy production.
**Hot-Wall Reactor:** In hot-wall CVD, the chamber temperature is maintained at a constant value by heating the walls. This method becomes important when coating any complex or large textile surface (refer to Figure 12.3). The tubular type reactor is most commonly used for the CVD of tantalum. It usually consists of a high-temperature cylindrical quartz or

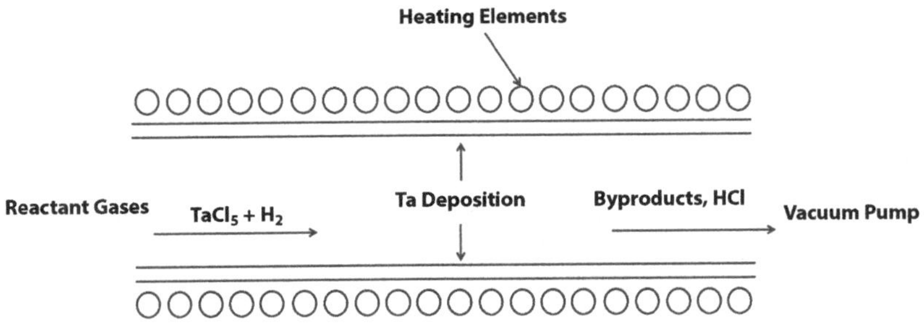


**Figure 12.3** Illustration of tantalum CVD in a tubular reactor.

graphite tube. The substrates are placed inside the tube, and the reactant gases, including the tantalum precursors ($TaCl_5$ or $TaBr_5$) plus the inert carrier gases, argon, or hydrogen, are introduced. The tantalum precursor is reduced to a solid tantalum film onto the substrate by the reducing gas, which is usually hydrogen. The chamber is kept either under vacuum or under reduced pressure to ensure an efficient deposition and to remove the by-products. The outline of the process is as follows:

**Reactor:** Heated quartz/graphite tube with substrate inside.
**Introduction of gases:** $TaCl_5$+$H_2$ and carrier gas.
**Deposition:** Gases flow over the heated substrate to reduce it and deposit tantalum films.
**Control:** Temperature, gas flow rates, and pressure are controlled.
Examples: In depositing tantalum carbide, $TaCl_5$ is used with a carbon source (e.g., $C_3H_6$) and hydrogen to deposit the carbide layer on the substrate.
**Cold-Wall Reactor:** Here, the wall is relatively cooler while the substrate is heated. This is especially helpful in cutting out side reactions with the reactor walls and funneling the energy to the textile substrate and is, therefore, ideal for heat-sensitive fabrics.

The fabric has GOT to be passed through what we call "textile handling mechanisms" such as holders, frames, or continuous feed rollers to maintain the steadiness of the fabric and enable it to be exposed to the precursor gases evenly during deposition.

**Precursor Handling System:** The CVD process requires using gases or vapors, also known as precursor gases, and their management should be precise. They are usually stored in containers interconnected with the reactor with the help of valves and pipes. In some cases, precursors could be in a gaseous or liquid state, depending on the nature of the precursors. In the case of the liquid precursors, the liquid is vaporized before it enters the reactor. Some techniques used in vaporizing liquid precursors are the "bubblers" or "flash vaporization systems" that offer a consistent and controlled stream of the precursor vapors.

The type of precursor materials has to be carefully selected because they have to decompose on the textile substrate, forming the required thin film. Some organic chemical reagents that can be used to introduce materials onto textiles during CVD include metal-organic compounds, silanes, and halides, whereby metals, oxides, nitrates, or carbides can be deposited onto the textile.

**Gas Delivery and Flow Control System:** When gaseous from the precursors, gases are channeled into the reaction chamber through tubes or nozzles. The requirement to provide an even distribution of the above gases to the face area of textiles necessitates proper flow regulation. This is acutely controlled using what is known as "mass flow controllers," which are used to control the rate at which the gases are introduced into the chamber. Further, "carrier gases" including hydrogen, nitrogen, or argon are also used to deliver the above-said precursor gases onto the substrate. The flow of these carrier gases is also regulated to achieve the optimum deposition method and to ensure that the precursor gases are distributed uniformly throughout the textile.

**Temperature Control System:** Temperature is one of the parameters that has an enormous impact on the CVD process as it affects the reaction of the gases used in depositing the film, the adhesion of the deposit, and the quality of the coating formed. This is achieved through a Temperature control system, normally known as "heaters" or "thermal elements," installed on the reactor to control the desired temperatures of the substrate and gases. In particular, in applications where the substrate is a textile material, the temperature needs to be chosen in such a way that it would not harm the textile material itself while depositing the thin films. Textiles are generally susceptible to heat; therefore, techniques such as PECVD in which plasma is used to initiate the reaction at a relatively low temperature.

**Vacuum System:** Another important aspect of CVD is the use of a vacuum system that is used to maintain low pressure within the reaction chamber, especially in the LPCVD. The vacuum system includes vacuum pumps that remove air and different impurities from the chamber; as a result, only precursors and carrier gases are inside the chamber. The use of a controlled vacuum is beneficial to discourage reactions with atmospheric gases and to produce a layer that is of high quality and has the same thickness all over the substrate.

#### *12.3.2.2 Typical Process Parameters in CVD for Textiles*

Several factors that characterize the film properties of textiles need to be regulated to meet the set standards during the CVD process. These include the following:

**Temperature:** Temperature may be considered one of the most important parameters that govern the CVD process because it determines the rate of chemical reactions and the properties of the deposited layer. Deposition temperature for textiles varies between "200°C and 600°C," depending on the precursor that is used and the type of textile that is being coated. Yet in some processes like PECVD, the deposition can take place at much lower temperatures to accommodate heat-sensitive fabrics such as synthetic polymers.

For instance, some metal oxides like titania or alumina can be covalently bonded at a temperature of 300°C–500°C, whereas some polymer coatings may be deposited at low temperatures of 100°C–300°C. They also found that the choice of temperature in the process not only affects the adhesion of the coating but also the thickness as well as the crystallinity of the said coating.

**Pressure:** Another variable is the pressure inside the reaction chamber. More than 90% of CVD operations are carried out at low pressure, referred to as LPCVD, to enhance the uniformity of the coating and suppress secondary reactions. Some of the textbook pressure ranges are as follows 0.1 and 10 torr. Working at low pressure makes the transportation of the precursor gases to all the faces of the textile, also making the deposition process controlled and completed in the shortest time possible.

**Gas Flow Rates:** There is a need to adjust the flow rate of the precursor gases and the carrier gases to provide a uniform deposit of the textile substrate. Gas flow rates are in the range of 10 to 500 cm (standard cubic centimeters per minute) based on the size of the reactor, the kind of precursor, and the requirement of the thickness of the film. Precise control of the flow rate becomes critical while synthesizing the thin films because it influences the amount of precursor used as well as the quality of the deposited film.

**Deposition Time:** The length of the CVD process is also among the important parameters that affect the thickness and the quality of the deposit. Increasing the deposition time leads to an increase in the coating thickness;

however, very long deposition times could incur wastage of film material and coating stress or affect the textile base or substrate. Deposition times are usually taken between minutes and some hours for textile coatings to be realized, depending on the type of thin-film material or thickness needed.

#### *12.3.2.3 CVD Optimization for Textile Applications*

Enhancement of the CVD process focused specifically on applications in textiles requires the fine-tuning of these parameters to achieve the necessary surface properties while not compromising the textile structure of the fibers [23]. Regarding the process parameters like temperature, pressure, and gas flow rates, they have to be optimized according to the textile material type, which is going to be treated, and the functional characteristics of the end product.

For instance, if fluoropolymers are being applied on outdoor fabrics to make them waterproof, then one has to select the right temperature and the precursor composition so that the fabric does not lose its breathability or its flexibility wholly to the coat to develop a long-lasting waterproof surface. On the other hand, in instances such as conductive layers for smart textiles, much emphasis would be placed on high electrical charge conveyance and binder properties with little impact on textile fibers.

### 12.3.3 Applications in Textile Surface Enhancement Using CVD

CVD has multiple chances of increasing the surface characteristics of the textile materials to perform functions that can be suitable for industrial, medical, and consumer uses. CVD essentially involves the application of thin, conformal coatings on textile fibers, which greatly enhance their performance to produce fabrics that not only possess mechanical durability but also possess inherent properties such as hydrophobicity, antimicrobial properties, UV protection, and better dye absorption [23, 24]. This section further explains different aspects of CVD, which are used to coat textile surfaces for hydrophobic and oleophobic; for the deposition of nanostructure materials for improving mechanical properties; and functionalized to apply UV protection, antimicrobial, and to increase dye affinity.

#### *12.3.3.1 Hydrophobic and Oleophobic Coatings*

Among all the applications of CVD, it is used predominantly for creating hydrophobic and oleophobic surfaces on textiles [20]. In weather-resistant textiles such as awning fabrics, athletic apparel, and healthcare clothing, hydrophobic coatings are vital because the fabrics need protection from water and any form of moisture or stain. Oleophobic coatings that can prevent the adhesion of oils and other organic compounds are widely employed in work wear, furnishing, and apparel to be used in food processing sections and hospitals [22].

The principles of these coatings involve the ability to lessen the surface energy of the fabric textures, particularly regarding water, oil, or any other liquid. CVD makes it possible to deposit thin films of fluoro-carbons, silicones, and siloxane materials that give these properties. The advantage of using CVD for these coatings is the uniformity and precision that it offers, ensuring that the textile fibers are completely covered, even in areas with complex geometries.

**Hydrophobic Coatings:** These coatings, however, are quite significant on fabrics that experience outdoor conditions such as tents, jackets, and tarps, among others [20]. This can be achieved through the use of CVD, where a water-repellent layer is applied on the textiles to make them waterproof while being able to "breathe." This is of paramount importance for fabrics to be used in apparel, for instance, the air permeability should not be hindered. Hydrophobic CVD coatings also extend the durability of textiles, as moisture accelerates the aging process and causes mold and mildew formation, which frequently occurs with fabrics that are used in zones with high humidity levels [20].

CVD can accommodate the deposition of some materials, such as silicon dioxide ($SiO_2$) or fluoropolymers, which are mostly used in hydrophobic coatings. Such materials create a nano-thin barrier layer that hinders droplets of water from spreading and wetting the surface of the fabric that they have as they tend and rather roll off [25].

**Oleophobic Coatings:** Antigrease finishes impart to textiles an ability not to be affected by oils and other organic compounds. Such functionality is crucial in applications where textiles are used in contact with oils, greases, or any other similar substances, for instance, kitchen textiles, uniforms, and upholstery textiles [23–25]. CVD can be utilized in the deposition of materials such as perfluoro octyl trichlorosilane or other fluorinated silanes to design oleophobic surfaces [24].

Besides the oil resistance, these coatings also factor in the easy-clean nature of textile materials and stains from food or industrial particles. CVD-derived oleophobic coatings are highly effective because they exclude chemical elements that bond chemically with the textile fibers so that the fabric remains oleophobic even when washed multiple times [25].

#### *12.3.3.2 Deposition of Nanostructured Materials for Enhanced Mechanical Strength*

CVD can also be applied for the synthesis of nanostructure materials on textiles, which has the potential to improve the mechanical characteristics of textiles such as tensile strength, abrasion resistance, and toughness [26, 27]. This is very helpful in applications where fabrics are exposed to immense mechanical pull, pressure, or tension, for example, protective wear, industrial and technical fabrics for aerospace and auto mobile industries, among others.

**Nanostructured Coatings for Mechanical Strength:** Application of nanostructures: Carbon, such as carbon nanotubes (CNTs), graphite or graphene, or metal oxides like titania or alumina, when coated on textiles through CVD, may enhance the mechanical properties of such fabrics. It appears that such nanostructured materials can enhance the tensile strength, abrasion, and tear strength of fabrics with less adding weight, making them suitable for high-end usage.

For example, the carbon nanotubes deposited on fabrics through CVD can be aligned with the fibers, to yield higher strength-to-weight composites. Such materials are preferably suited for industries such as aerospace because they require materials that are both light and strong. Also, the use of nanostructured coatings that have wear and friction-resistant properties for industrial fabrics is used in severe conditions.

**Flexible and Lightweight Textiles:** An obvious advantage of nanostructured materials is that they can be applied in very thin layers, which will not hamper the flexibility and softness of the textile. This is important in apparel, especially in wearable technology applications, because the mechanical properties of the fabric need to be improved while minimizing the impact on the comfort and flexibility of the garment. CVD makes it possible to deposit these nanostructured materials with the structure upgraded for enhanced mechanical performance while retaining most of the fabric properties.

#### *12.3.3.3 Functionalization for UV Protection, Antimicrobial Properties, and Increased Dye Affinity*

The CVD approach provides the opportunity to modify textiles with desired properties for various purposes. The various functional improvements include ultraviolet light barrier, antimicrobial action, and better acceptability of dyes. These functionalities are commonly used in such sectors as healthcare and fashion, among others, whereby performance, hygiene, and appearance are of paramount importance [25].

**UV Protection:** UV protection is the other functionality that may be incorporated in textiles that are used in outfits for outside activities, furnishing, and protective wear. UV exposure to fabrics harms them, reduces their colors, weakens fibers, and causes damage to the general fabric of the cloth. When the UV-blocking material is incorporated into CVD deposition, then materials such as ZnO or $TiO_2$, which can absorb and reflect UV, can be deposited.

These materials also act as barriers to UV radiation, thus protecting the textile surface from any harm. Further, anti-UV treatments can also extend the durability of colored fabrics by slowing down the bleaching action of the sun. This is particularly beneficial in fabrics used in outdoor settings, such as awnings, curtains, and outdoor apparel.

**Antimicrobial Properties:** Healthcare and other hygienic products, sportswear, and accessories are significant end uses for antimicrobial textiles, as any microbial growth causes severe infections in the human body [28]. CVD can covalently immobilize antimicrobial agents on textile substrates; these may include silver nanoparticles (AgNPs) or copper oxides (CuO). These materials have been used because of their non-specific mode of killing microbes, which is effective in inhibiting microbial growth on fabrics [29].

It is worth stating that the antimicrobial coatings realized with the help of CVD are very stable, thus preserving their activity after multiple washings. This makes them suitable to be used in medical textiles like hospital gowns, bed sheets, and face masks, to mention but a few, because hygiene is of great importance. Furthermore, the application of antimicrobial coatings in sportswear and activewear can help in preventing odor formation due to microbial activities on the fabric, thus making the fabric comfortable to wear [30].

**Increased Dye Affinity:** Self-organization of textile materials also applies to CVD, where control over the surface characteristics of the fabric provides an opportunity to improve the dyeing characteristics of the materials. CVD also alters the surface chemistry of textile fibers in a manner that enhances their affinity for specific dye molecules, and, hence, the resultant product is characterized by better color depth, color fastness, and color balance [31].

For instance, CVD can be applied to put down the thin films of those materials, which enhances the texture finish or chemical activity of textile fibers to have better adsorption of dyes. This is particularly helpful, especially when using synthetic fabrics such as polyester, which usually exhibit low dye uptake as compared to natural fibers like cotton. CVD-treated textiles have better dye uptakes due to which enables text to include dyeability can be increased to produce brighter as well as more durable colors, which are additional advantages for fashion as well as home textile uses.

Such high selectivity over the deposition process gives manufacturers the capability of producing fabrics that suit particular functions, from high-tech fabrics and apparel wear to even normal consumer use. Thus, being one of the efficient tools for surface modification and functionalization, CVD is expected to play a more important role in expanding the possibilities of application in the continuously developing field of advanced textile materials.

### 12.3.4 Advantages and Limitations of CVD for Textile Applications

The CVD technique in textile surface modification has certain advantages that make it a useful technique for the improvement of textile materials [8, 19, 21]. However, as with any technique, there are disadvantages of CVD, especially in its versatility when it comes to the type of textiles and conditions under which the process will be effective. CVD in textile applications: potential and limitations outline the benefits of CVD and also future risks to provide insight into the ability of CVD in efficiency improvement and the disadvantages involved in it.

#### *12.3.4.1 Advantages of CVD for Textile Applications*

**Uniform Coating:** Another major asset of CVD is that a planar substrate or an irregular surface forms the base of the component; even surfaces are achieved, even on small or large scales. This is especially significant in textiles, in which fibers are constructed into complicated arrangements such as weave or knit structures, thereby making regular surface coverages unattainable. CVD does not allow the presence of any uncoated area of the textile, even if it is the last inner crust and smallest fiber layer in the textile to be functionalized.

For example, a hydrophobic or antimicrobial coating deposited by CVD provides strong adhesion of the coating material to textile fabric, and the coating thickness is uniform throughout the fabric. This uniformity is required for applications in applications involving high-performance textiles, such as composite applications where reliability is paramount.

**Scalability:** Originally, CVD is a large-scale production technique, but recent developments have made it possible for its use in small production. Owing to its flexibility, CVD treatments can be applied to textiles in a roll-to-roll manner, therefore making it suitable for commercial applications where there is a need to treat large volumes of fabric in a relatively shorter time. The following are some of the advantages of CVD, which allows for vigorous production scaling that is convenient for industries such as apparel, automotive textiles, and protective clothing.

Furthermore, the process of CVD is extensible to deposit numerous types of materials, comprising metal oxides and polymers that can be adjusted for textile applications. Depending on the desired functional parameters, CVD is flexible enough to produce coatings with increased scales of durability, antimicrobial agents, or UV protective characteristics.

**Compatibility with Various Textile Substrates:** CVD can be used on a large number of textile materials; it is compatible with natural fibers such as cotton, wool, and silk, as well as synthetic fibers such as polyester, nylon, and polypropylene. Such a wide applicability of CVD makes it possible to be used across the diverse textile sectors involving fashion textiles and technical textiles.

Moreover, CVD is capable of depositing coats on fibers while at the same time not affecting the fundamental structure and mechanical characteristics of these fibers. This is important so that the fabric maintains its inherent qualities like flexibility, breathability, and softness, and new properties like water resistance or antibacterial properties are incorporated.

#### *12.3.4.2 Limitations of CVD for Textile Applications*

**High Process Temperatures:** One of the major limitations associated with CVD in textile applications is that the deposition reactions involved often demand high process temperatures. Conventional CVD processes work at a temperature ranging from about 200°C to about 600°C and/or associated gas phase temperatures, particularly the substrate temperature, generally unsuitable for heat-sensitive textile materials, such as certain synthetic or natural fibers that undergo thermal degradation at the CVD temperatures.

This relatively high level of heat treatment can also be very destructive to the textile substrate through the processes of thermal degradation, which, in most cases, manifests in color changes, strength loss, or texture alterations. Although there has been an effort to address this issue by developing low-temperature processes such as PECVD, not all materials can be deposited at low temperatures, making the range of CVD limited for textile uses.

**Potential Degradation of Heat-Sensitive Textiles:** Apart from the fact that high temperatures are a universal problem, the following fabrics—polyester, nylon, or fabrics produced from natural and synthetic blends—will degrade when used in high-temperature processes. This restricts the application of CVD for textiles that undergo low-temperature processing and have superior surface characteristics. Although techniques such as PECVD play a part in somewhat reducing this, the thermal considerations are still a problem for slender or easily combustible materials.

### 12.3.5 Case Studies and Recent Advances in CVD for Textiles

CVD technique in textile surface engineering has attracted considerable research interest and implementation due to the versatile functionalities delivered to textile fibers. In the recent past, there have been cases and many industrial examples that have shown that CVD can be effective in improving textiles' performance. These innovations not only prove the use of CVD in the development of the apparel industry and textiles but also show how CVD can alter the characteristics of fabrics required for a variety of end uses like tech textiles, fashion textiles, medical textiles, and more.

#### *12.3.5.1 Recent Research on CVD in Textiles*

Some prior/current research investigations have attempted to investigate the application of CVD in altering the surface chemistry of textiles, particularly for the creation of functional fabrics. For instance, a research work analyzed the application of CVD in the deposition of silicon dioxide ($SiO_2$) coatings on cotton fabrics to make them superhydrophobic. The treated textiles had a contact angle of more than 150°; thus, the result proved that the tested specimens had high water repellency. This was in sharp contrast with untreated cotton, which, as soon as it came into contact with water, would instantly absorb it. The study focused on the homogeneity of the coating imparted by CVD, whereby it offered a homogeneous, durable, waterproof, and highly flexible hydrophobic layer ideal for outdoor and sportswear use.

The other research work used titanium dioxide ($TiO_2$) nanocoatings on polyester textiles through CVD for enhancing functional characteristics, including UV protection and self-cleaning. This was found to be due to higher UV-blocking efficiency by the fabrics coated with $TiO_2$ than the uncoated fabrics, as well as photocatalytic behavior necessitating organic stains and light. Self-cleaning properties together with UV protection are especially important in the technical fabric used in outdoor applications and the healthcare sector.

Scientific literature review revealed that CVD is an efficient technique for coating silver nanoparticles (AgNPs) on both cotton and polyester fibers used in antimicrobial textiles. Silver-coated textiles in this study showed very high antimicrobial activity against bacterial species such as *E. coli* and *S. aureus* and are, therefore, suitable for medical textiles, sportswear, and apparel for sporting activities. The study also revealed that CVD facilitates the controlled and uniform deposition of silver nanoparticles, thus offering improved durability and wash-fastness as compared to the wet chemistry techniques.

#### *12.3.5.2 Industrial Applications of CVD in Textiles*

Today, several industries have entered the industrial-scale application of CVD for enhancing the surfaces of textile materials. For instance, the outdoor apparel industry has applied CVD to develop fabrics that are high-performance, waterproof, and stain-proof. Manufacturers of technical outerwear have applied CVD in applying fluoropolymer coatings on synthetic material fabrics, resulting in weather-resistant clothing that also fits the breathability factor.

In the automotive industry, CVD-treated textiles find application in cars where they are used in interior parts and specifically for UV protection, antimicrobial activities, and stain resistance. These functionalized textiles improve the comfort, durability, and safety of car interiors, and at the same time deny the aspect of beauty.

#### *12.3.5.3 Comparative Analysis of CVD-Treated versus Untreated Textiles*

CVD effectiveness in the treatment of textiles is apparent when fabrics that have been treated with CVD are compared with fabrics that have not been treated. Whenever untreated fabrics are compared with those treated with CVD, the latter perform better in terms of functional and non-technical attributes. For instance, when testing the water resistance, it took only a few seconds for normal cotton fabric to get soaked while the one treated with CVD is water repellent and still so even after it has been washed severally times. Likewise, protective textiles that were coated with antimicrobial material through CVD exhibit a significantly lower bacterial count compared to fabrics that may otherwise develop bacteria and smell.

In summary, the advantages of CVD-treated textiles over untreated ones are clear: they are enhanced water repellence, increased UV protection, enhanced capability of preventing bacterial growth, and increased durability. Both of the above-mentioned case studies and industrial applications endorse the emerging significance of CVD for the enhancement of textile surfaces and new possibilities for development across various domains.

## 12.4 Physical Vapor Deposition (PVD)

PVD is a well-developed and popular technique for coating, and it is specifically popular in the textile industry where requirements such as durability, wear resistance, and improved visuals are getting higher [32]. PVD is an entirely distinct method of coating textile substrates, where the material is converted into vapor and then recollected on the fabric surface in a vacuum [33, 34]. They produce thin and strong films that lead to the change of the textile's physical and chemical states.

### 12.4.1 Principles of PVD

PVD is diametrically opposite to CVD in the aspect that while the former uses physical processes to deposit materials onto a substrate, the latter depends on chemical reactions [35]. First, a solid or liquid material is vaporized and transferred in the form of vapor to the textile substrate located in a vacuum chamber. The vapor, on the other hand, then solidifies on the surface of the textile, resulting in the formation of a thin and uniform layer on the textile.

Something that can be used to describe PVD is that it operates in a vacuum, which is useful in making the deposited material free from contaminants and providing uniform adhesion to the surface of the textile. A schematic of the PVD process for tungsten (W) coating is shown in Figure 12.4. This system also precludes a certain level of control over the deposition process, through which structurally sound coatings with particular desired characteristics can be produced [36].

There are several techniques of PVD, and each method has its strengths, taking into account the textile substrate and the end use of the finished product. They include evaporation, sputtering, and ion plating. Evaporation is a process of deposition in which the target material is heated so that it evaporates and forms a thin film on the substrate [33, 35].

PVD techniques can be classified as illustrated in Figure 12.5.

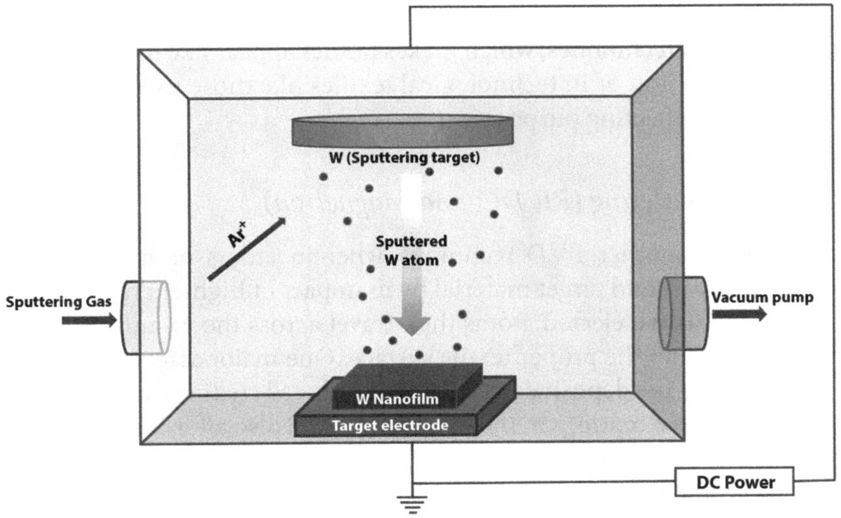


**Figure 12.4** Schematic of physical vapor deposition for W coating.

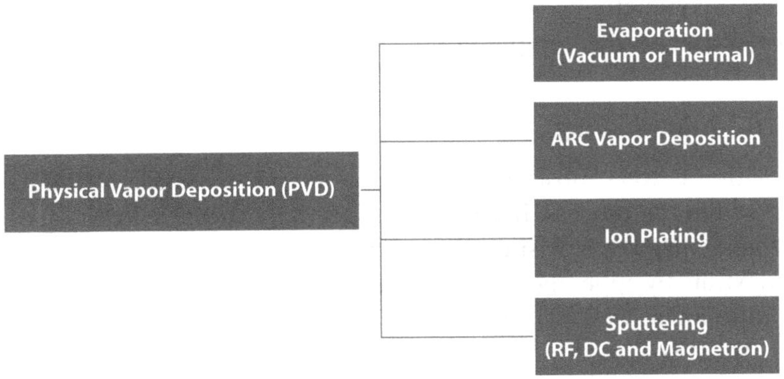


**Figure 12.5** Classification of PVD.

### *12.4.1.1 Evaporation (Vacuum or Thermal)*

In the evaporation method, the material that is to be deposited is subjected to heat until the material attains the vaporization stage. This vaporized material is then transported through the vacuum chamber, and, on the substrate of the textile material, it forms a very thin layer of film [35]. The benefit of evaporation is its non-complexity and versatility of laying down metal, ceramic, polymer, and other types of films. This method is particularly used to apply ornamental deposits, mirror layers, or protective layers on the fabrics.

For instance, deposits such as metal layers can be deposited on textiles by evaporation techniques, which makes fabrics appear like metals and can be used for fashion or in technological textiles like those used in automobiles for heat reflecting purposes.

### *12.4.1.2 Sputtering (RF, DC, and Magnetron)*

Sputtering is another PVD technique wherein atoms or molecules are expelled from a solid target material by in impact of high-energy particles, usually ions. These ejected atoms then travel across the vacuum chamber then deposit onto the proper textile surface. One major asset of sputtering is the feature of the deposit density of the produced coatings and the strong adhesion of these coatings to the substrate [33]. It also affords more control over the robustness and constituency of the layer that is to be coated as opposed to evaporation.

This article points out that sputtering is often applied for making functional coatings of textiles, like; antimicrobial, conductivity, or wear-resistant textiles. In technical textiles, such functions as UV-protection, electrical conductivity, or chemical resistance of webs containing metal oxides can be created by sputtering.

### *12.4.1.3 Ion Plating*

Ion plating follows most of the evaporation and sputtering techniques as discussed in the process. In this technique, the material is first evaporated and then ionized by a plasma. The ionized vapor is directed toward the textile substrate by applying an electric field, which assists in achieving better film adhesion as well as uniformity [37]. Ion plating is most recommended where there is a need for a strong bond between the applied coating and textile substrate, and this makes it good for use where there is likely to be a lot of wear and tear [38, 39].

For instance, ion-plated coatings are utilized for high-performance textural applications in aerospace and military industries where fabrics are expected to blend abrasion, heat, and corrosive chemical environments [40].

### *12.4.1.4 ARC Vapor Deposition*

Arc vapor deposition is a form of PVD where a high-current, low-voltage electrical arc vaporizes a metal target (cathode) inside a vacuum. The ionized metal vapor then condenses to form a cold, hard coating on the substrate. The technique provides high deposition rates and coating adherence and acts as a deposition method for functional and wear-resistant coatings on various surfaces like textiles.

Thus, the PVD method is one of the most effective and highly controlled coating processes best suited for modifying the surface of textiles. Whereas vacuum technology allows precise vaporization and deposition of matter, PVD prepares sleek, even, and tenacious films on textiles. Such methods as evaporation, sputtering, and ion plating give a good variety of opportunities to give textile products some additional properties such as wear-resistant layers, UV protection, antimicrobial effect, and even aesthetic looks. Due to its effectiveness in working with several types of materials and substrates, PVD plays an important role in the further improvements of properties and designs of textile materials.

## 12.4.2 PVD Process for Textiles

PVD involves using a physical process where a material is converted into vapor and then deposited onto a textile substrate in a restricted environment [35]. This technique can only be done by using specific equipment to avoid any disparity in the coating methods. In this section, we will present the main elements of the PVD setup: the vacuum chambers, target materials, and substrates, and many details concerning the setup and the coating process parameters that have a decisive impact on the result of the coating [33].

### *12.4.2.1 PVD Setup and Components*

**Vacuum Chambers:** The central component of the PVD process is the vacuum chamber, also known as the deposition chamber, that forms a closed system. It plays a robust role in conserving the low-pressure conditions that are required to vaporize the target material and to avoid its contamination by air. For textile applications, the chamber size and shape may change according to the production scale, but the major challenge is to obtain a stable and clean physical vacuum for the deposition process [18].

Any PVD vacuum chamber can work at low pressure, which is achieved with the help of pumps installed in the chamber. This also assists in minimizing the chances that the metal or mixture being used will undergo oxidation or another chemical reaction that may be detrimental to the deposited film quality. Such a vacuum condition also enables the easy movement of the vaporized material from the source to the textile substrate for the right coating.

**Target Materials:** In the case of the PVD process, the target material is the material that will be in the form of vapor before depositing on the textile substrate. This can include metals such as titanium, gold, and silver; metal oxides such as titanium dioxide or zinc oxide; or some other functional material that must endow the textile with certain characteristics. One must know the type of finishing to be given to the textile product and the resultant properties, like antimicrobial, UV stable, conductive, or aesthetic, that are to be achieved in the end product [39].

Regardless of the technique used, in the vacuum chamber, the target material is heated, for example, using heat in evaporation or ion beams in sputtering to cause the target material to vaporize. After it is in the vapor state, the material moves across the chamber and then deposits itself on the textile substrate in a thin, even layer.

**Substrate Handling:** The coverage of the textile substrate during the PVD process is very sensitive and, therefore, needs to be done carefully to produce good results. Textiles are usually flexible, permeable, and thermal and, thus, must be handled cautiously when they are being deposited. The substrate often has a cradle or holder that turns inside the vacuum chamber to let the vaporized material expose the substrate uniformly from all sides [39]. This rotation also makes it possible to apply a uniform layer of the covering material on the working surface of the textile material, regardless of the complexity of the structure of the textile material being used.

For temperature-sensitive textiles, means for cooling the substrate holder may also be provided during the deposition process to avoid the textiles from getting hot. Because the PVD processes can use high-energy ions and heat, the substrate's temperature needs to be regulated so as not to compromise the textile material by thermal degradation.

### *12.4.2.2 Key Parameters in the PVD Process*

Several parameters influence the success and quality of the PVD process, particularly when working with textiles:

**Deposition Rate:** The deposition rate means the rate at which the vaporized material is deposited on the textile substrate. This is because the rate of deposition influences the thickness and additional uniformity of the resulting thin film. A fast deposition rate is also undesirable because it may lead to poor adhesion between the substrate and the coating or the formation of rough surfaces and inconsistent coatings. On the other hand, if the rate of change is low, then it turns out that the process of production is not very efficient. Maintenance of deposition parameters guarantees full coverage and performance of the coat without affecting the properties of the textile product.

**Chamber Pressure:** Pressure exerted in the chamber of PVD plays a very decisive role in the PVD process. The lower pressures are preferred in general because they aid in developing an efficient interface for vapor transferring while eliminating the dangers of contamination. However, the pressure must be carefully regulated to achieve the vaporization of the target material and the condensation of the material onto the textile substrate. Practical pressure in the PVD ranges while the material and the method of deposition are being deposited *via* evaporation, sputtering, etc.

**Substrate Temperature:** The temperature of the textile substrate during deposition is also an important parameter. In certain PVD techniques, such as sputtering or ion plating, high-energy particles cause heating of the substrate and, consequently, alter adhesion as well as the quality of the coating. This is particularly important for fabrics, especially those made from heat-sensitive materials; the proper control of Ag's substrate temperature is one of the prime requirements to avoid damage. Nonetheless, in specific cases associated with the target material and properties of the obtained coating, cooling systems can be used to maintain the temperature of the substrate safely.

## 12.4.3 Applications of PVD in Textile Surface Enhancement

PVD technology has exhibited a very strong application in modifying the surface characteristics of the textiles and therefore creating many functional uses, protective uses, and even aesthetic uses on the textiles. In the context of PVD, textiles can be deposited with thin layers that have boosted the nature of durability, abrasion, aesthetics, and functionality, such as conductivity and heat sink. The versatility of PVD makes it an ideal technique to be used in many sectors such as fashion, technical textiles, electronics, and health.

### *12.4.3.1 Enhancing Textile Durability and Abrasion Resistance*

Among the major employment areas of PVD in textiles, one of the primary aims is to enhance its manifestation to wear and abrasion. As conducted on the surface of a textile material, PVD can create hard and protective layers that could easily increase the working lifespan of the coated materials.

For instance, technical textiles employed in high-performance sportswear, workwear, and outdoor cladding are advantageously coated with PVD for protection against abrasion from frequent use or exposure to unfriendly physical climates. Titanium nitride (TiN) and chromium-based coatings are often applied during PVD to make fabrics wear-resistant. These coatings not only improve the overall mechanical characteristics of the textile but also retain gentle and comfortable wearing characteristics. Moreover, PVD coatings can be coated onto textiles used in the car and aircraft manufacturing industry because fabrics employed in those sectors endure rigorous rubbing abrasion, besides enduring probable high temperatures. These coatings help to keep the characteristics of the textiles at the intended functional and durability level, through the intended functionality.

#### *12.4.3.2 Decorative Coatings and Reflective Finishes*

PVD is also best suited for applying pretty-looking coatings and reflective layers for clothing fabric and other related textiles. For sparkle, reflective surfaces—aluminum, gold, silver, or titanium—can be applied on materials to make them shiny like fabrics used in fashion and interior design. Some of the coatings can produce a reflection to mirror-like luster, whereas others can make the garment surfaces look metallic and matte, and this enables the designers to create diverse looks on garments and decorative textile products.

Mirrored coatings produced by PVD are particularly useful in safety and performance wear that requires visibility. For instance, PVD coatings can be incorporated into textiles worn in reflective safety aprons, sportswear, or cycling wear to make the wearer conspicuous at night. The thin profile of the coating is such that the reflective quality is boosted while at the same time not compromising the feel or furnishings of the wearer when in the form of a garment.

In addition to fashion and safety, reflective coatings have applications in technical textiles as a kind of heat barrier blanket or thermal insulation products, where the coating moderates temperature by reflecting heat in the form of radiation.

#### *12.4.3.3 Electrical Conductivity and Thermal Management as Functional Layer*

PVD takes the textile through a process through which it can acquire properties of the functional layers, such as electrical conductivity and thermal control. These functional coatings are critically needed in the development of smart materials, which involves the incorporation of electronics into the fabric for wearables, sensors, or conductive fabrics.

Metallic coatings that include copper, silver, or gold are most preferred during the PVD process formulation, mostly to develop conductive layers on fabrics. Given that these coatings can be thin enough to maintain the flexibility of the textile, they also offer the necessary electrical paths required for several electronic uses. Conductive textiles find their application in wearable devices, medical electrodes, and flexible circuits in textile electronics garments.

Another commonly practiced application of PVD in textiles is thermal management. Other coatings that help control heat transfer are added to fabrics; these include aluminum oxide ($Al_2O_3$) or titanium dioxide ($TiO_2$), which enhance fabrics' thermal insulation or reflectivity to radiation heat. These materials can be used in textiles for heating garments, warm covers, and safety apparel in extreme temperature conditions, like during fire brigades or baking industries.

### 12.4.4 Advantages and Limitations of PVD for Textile Applications

PVD is particularly preferred for the treatment of textiles mainly because of the ease and ability to treat in environments. It has many features: Low process temperature will increase the quality of coated fabrics, and, on the other side, a wide range of coating materials means that it can cover a wide range of textile materials to solve many problems of application challenges. Indeed, PVD, like all technologies, has its drawbacks, especially when working in a vacuum and when coating complex geometries of textiles. In this section of the article, the possibilities for improvement of the textile surface modification by PVD will be discussed alongside the limitations.

#### *12.4.4.1 Advantages of PVD*

**Lower Process Temperatures Compared to CVD:** The benefit of PVD over CVD is that it is easier because it works at comparatively low process temperatures. PVD can be performed at temperatures that are effectively non-destructive to heat-sensitive textile base materials such as synthetic fabrics made from polyester, nylon, or acrylic. This feature makes PVD particularly suitable for the application to fabrics that cannot be treated at high temperatures, as in the case of CVD, therefore widening at the same time the range of fabrics that can use more advanced surface treatments.

Lower-temperature processing not only minimizes damage and shrinkages in fabrics but also reduces energy costs in large-scale production and processing.

**Versatility in Coating Materials:** PVD is excellent in its flexibility regarding the selection of coated materials. Depending on the nature of materials, members of the PVD family can be used to coat textiles with metals, metal oxides, nitrides, and many other compounds. This versatility makes it possible for the developer to manipulate the characteristics of the textile to satisfy certain utility or beauty needs.

For instance, a silver coating provides antimicrobial properties, or a copper, titanium dioxide, or zinc oxide coating provides the additional function of UV protection. By using PVD, designers and marketers of printed textiles can develop a variety of products such as decorative metallic finishes, reflective coatings, and even transparent functional layers that fit the fashion, technical as well and industrial textiles.

Also, PVD coatings are normally thin, uniform, tough, and able to withstand environmental stress, to improve the performance and durability of coated textiles.

**Environmentally Friendly Process:** PVD is known to be a green technology surface engineering process because it does not use hazardous solvents or other chemicals. In contrast to methods such as chemical bath deposition, printing techniques using intricate chemistry that poses a threat of contamination to the environment or require the use of complex procedures to remove chemical residues from the process, and PVD is a clean and organized process carried out in a vacuum condition with no negative effects to the environment.

Awaited, there is no effluent and volatile organic compound emission in the PVD process; thus, it is environmentally friendly as a surface enhancement technique. This environmentally friendly characteristic of PVD complements the increasing consumer consciousness of the need to promote sustainable materials usage in the production of textile-based products for fashion as well as technical textile purposes.

#### *12.4.4.2 Limitations of PVD*

**Requirement for Vacuum Conditions:** Perhaps the greatest drawback of PVD is the requirement for vacuum conditions during the deposition of the coatings. A vacuum is needed in the setup to minimize contamination of the deposited materials and to cause the target materials to evaporate or sputter.

Although vacuum technology is very efficient in depositing high-purity coatings, it is considered to complicate the deposition process and is expensive.

The need for a vacuum means that PVD cannot scale to some large, continuous production operations, as well as some wet chemical methods that can fit most mass textile manufacturing lines easily.

**Challenges in Coating Complex Textile Geometries:** Fabrics are usually compliant and permeable and possess a complex structure, so application of uniform PVD layering turns out to be very difficult, especially if the fabrics are of irregular geometry or possess high surface roughness. Compared to proximal PVD techniques for depositing thin films on flat substrates or semi-flat surfaces, coating fibers within a fabric or areas far and deep within the fabric structures poses a problem.

In some cases, this deposition is not uniform across the textile base; therefore, surface characteristics or performance may be lower in some regions of the used textile. This limitation means that the substrate has to be handled carefully and may require several layers to be applied to ensure that the coating is spread uniformly, thus adding to processing time and money.

#### 12.4.4.3 *Case Studies and Recent Advances*

Being an aesthetic technology, PVD has gradually worked its way into several emerging application areas due to flexible electronics and green clothes. PVD has created new possibilities in applications where normal surface treatment techniques cannot work because PVD deposits very thin layers of functional coatings to fabrics. In this section, the focus turns to some emerging fields and applications of PVD, for example, smart clothing and fashionable PVD coatings for sustainable apparel. We will also state our analysis of PVD's applicability at a large scale and cost considerations for industries.

#### 12.4.4.4 *PVD for Conductive Textiles*

Undoubtedly, the most captivating and revolutionary of PVD utility is that in fabricating conductive textiles. Such fabrics contain deposited thin metal or conductivity layers, thereby enabling such fabrics to perform electronic roles. Conductive textiles are intrinsic to wearable technology; fabrics containing sensors, circuits, and other electronic devices. PVD methods can coat metals like silver, copper, or gold used for conductivity, see that PVDs do not make the fabric stiff, and remain soft [41].

Applications of conductive textiles include the following:

**Wearable Electronics:** Clothing with built-in monitors and indicators of physiological and physical indicators, for example, respiration rate, temperature, and activity. These garments contain conductive fabrics, and data can be sent in real-time while wearing them, and the form is both comfortable and functional while preserving movement.
**Smart Fabrics for Energy Storage:** When fabrics are covered with thin films of conductive materials, they can be used as flexible batteries or capacitors, which can be built into the very fabric.
**Electrostatic Discharge (ESD) Protection:** In areas such as electronics manufacturing, the PVD coatings on conductive textiles enable the removal of static charges to guard electronic parts from harm.

This is particularly important for moldable fabrics that do not need the added physical properties that PVD could potentially introduce with its layers of amorphous films, which is a requirement for wearable and smart textiles.

### *12.4.4.5 PVD in Eco-Friendly Fashion*

Sustainability has only recently started to become one of the leading determinants of fashion, and PVD is ready to help this process in fashion through sustainable coatings. The dyeing and finishing process usually involves the use of chemicals and results in high pressure on water and facilities, thereby making PVD to be friendly to the environment. Unlike a wet chemical process, PVD is also a dry process that takes place in a vacuum, which avoids the use of water or other toxic chemicals, and, therefore, has an immense advantage over the environment.

In an eco-friendly fashion, PVD can be used to create the following:

**Decorative Coatings:** Other metallic finishes include thin layers such as titanium, aluminum, or zinc oxide that can be applied to textiles without the use of heavy metal dyes or toxic procedures.
**Durable, Non-Toxic Finishes:** PVD-coated garments are long-lasting, and wearing garments with PVD coatings means that people avoid constantly replacing their clothes, which helps prevent wastage.
**Functional Coatings:** Besides aesthetics, PVD can give textiles improved UV protection, antimicrobial features, or upgraded stain-resistant properties that would not only help in the longevity of garments but also, therefore, diminish the life cycle yet again in the long run.

These applications are in congruence with the relatively recent sustainable fashion, in which designers and producers employ sustainable solutions in production to reduce their effects on the environment.

#### *12.4.4.6 Scalability and Cost-Effectiveness in Industrial Applications*

But as we have seen, there are many benefits given to PVD that may be able to support it if the two key factors that would allow the textile industry to take off are achieved; these are scalability and cost. PVD processes may also be more difficult and costly than wet methods of finishing because the use of vacuum conditions and specific equipment is necessary. But, as has been seen, high volume vacuum deposition systems make it much more feasible for large production runs.

There are current efforts to enhance throughput, for example, roll-to-roll processing, where textiles are passed through a PVD chamber in an unending roll to give continuous fabric rolls that are coated. This method increases the scalability and reduces the unit cost significantly, making the PVD more suitable for competing with the traditional methods.

The self-etching characteristic of PVD makes it cost-efficient because the procedure rarely needs repeated application. PVD coatings are comparatively long-lasting, and the treated textiles prove to need replacement far less often than if they had been using conventional methods; as a result, the initial cost of PVD technology is recouped by cost savings necessary in the case of the conventional technologies. In high-performance segments, including technical textiles and smart wearables, the added value of PVD, which is the function enhancement and durability, can offset the higher initial investment of the technology; thus, it can be considered suitable for mass production.

In conclusion, the fact that PVD is used for new-generation applications such as conductive textiles and environmentally friendly fashion is evidence of the paramount importance of PVD for enhancing textile innovations. Due to growing developments in terms of scalability and pricing, PVD stands ready to transform into a core segment of textile surface improvement for performance attire, as well as eco-friendly apparel lines.

## 12.5 Comparative Analysis of CVD and PVD

Of the two processes, CVD and PVD are two techniques used in enhancing the surface of the textiles; however, there is a difference between CVD and PVD with regard to some process parameters, coating deposited onto the fabrics surface, and the performance characteristics. This section will draw a contrast between the two methods primarily based on their advantages and limitations. Also, the general guideline on how to decide which method is most suitable for which type of textile material, its properties, and application will be given.

### 12.5.1 Process Parameters

CVD happens at high temperatures where gaseous precursors react and produce a thin film present on the substrate. The process typically consumes 300°C–1000°C, and such temperatures are difficult to handle for delicate fabrics like synthetic fibers. CVD operates well on substrates that can withstand heat or substrates that have been pretreated to minimize the effects of CVD. Moreover, CVD is typically conducted under low pressure to drive the vapor-phase reaction, hardware is used for maintaining the pressure and temperature of the system.

On the other hand, **PVD** is a process where the material is evaporated and transported in a physical state as a vapor phase under vacuum from a solid or liquid phase source and then deposited onto a textile substrate. PVD works at lower temperatures than CVD exists in a range of 100°C–300°C and is therefore more appropriate for finishing textiles that are sensitive to high temperatures. Another common type of PVD also needs vacuum conditions to maintain purity in the deposition process; however, lower deposition temperatures and lack of chemistry make the system even simpler than CVD.

### 12.5.2 Coating Characteristics

Finally, CVD coatings have relatively good thickness uniformity and high interfacial adhesion compared to the reference substrate material. The details of film characteristics that can be controlled include the film thickness and the film deposition as metal, oxide, nitride, and diamond-like carbon. CVD can deposit high-quality and thick layers with improved mechanical characteristics, such as wear, chemical, and thermal stability. However, it is used on selective textiles due to its high temperatures and also due to the involvement of some reactive chemicals.

While PVD coats are generally less thick than layers produced by CVD coatings, their layers offer a smoother profile over the substrate and higher adhesion. PVD is suitable for aesthetic coatings, protective coatings including Ultraviolet protection, and deposition of nano coatings. Because the materials that are targeted for PVD coatings, including metals, alloys, and ceramics, can vary, it is used in applications in which appearance, electrical conductivity, or optical transmission are significant. Moreover, other PVD processes, including sputtering or evaporation, enable the deposition of a desired layer thickness that is useful for functional or aesthetic uses.

### 12.5.3 Performance Outcomes

There is information that the use of the CVD results in such mechanical characteristics as wear resistance, water repellency, and heat-shielding properties of the textile increasing. The chemical bonds that are created during the CVD make the coatings resistant, and they are long-wearing even when exposed to harsh environments. CVD is most appropriate in niches that demand high-performance fabrics, including protective clothing, filter media, or technical fabrics for industrial and aerospace customers.

On the other hand, textiles treated by PVD show exceptional results in applications where there is a need for a lightweight material with flexibility that has to feature electrical conductivity, UV, or decorative properties. As mentioned above, the characteristics of the PVD technique that can deposit thin, uniform layers of the coating while maintaining the process fabric flexibility make it effective for wearable electronics, smart clothing, and fashion.

### 12.5.4 Selection Criteria

The choice between CVD and PVD depends on several factors:

**Textile Type:** PVD is indeed used for easily fusible fabrics like polyester and nylon because it works at comparatively lower temperatures than other methods. On the other hand, CVD is beneficial for materials conducive to high temperatures, such as fabrics.

**Desired Properties:** CVD is most suitable for the deposition of advanced coatings with better mechanical and chemical characteristics for use in protective or technical textiles. PVD is used for decoration and coatings that need electrical conductivity, UV, and lightweight properties.

**Application Requirements:** For functional textiles such as wearable or smart fabrics, PVD appears to provide greater freedom and versatility. CVD, which forms more compact coatings and is hard-wearing, is a better option in applications that require more vigor and lasting power.

Thus, it is possible to state that each of the methods mentioned has its benefits, and the choice of the method to be used depends on the characteristics of the textile and requirements that are to be met by the coatings, as well as by the performance characteristics of the final product.

## 12.6 Future Trends and Innovations in CVD and PVD for Textiles

Due to increasing demand for functional and high-performance textiles, CVD and PVD technologies are gradually developing to fit the needs of the textile market. These techniques are on the frontier of innovations that are taking place in textile surface modification, and therefore, the capability to take fabrics to the next level in terms of functionality and performance enhancement is becoming rapidly achievable. This section provides novel development of CVD and PVD, and how these methods can be merged with other surface modification techniques for the development and advancement of smart textiles and wearable technologies.

### 12.6.1 Advancements in CVD and PVD Technologies for Textiles

Textile colonization, diamond-like carbon (DLC) coating, CVD, and PVD for textiles. They also found out that the future of CVD and PVD in the textile industry depends on the capacity of the two techniques to meet specific textile needs. Novelties are aimed at increasing the efficiency of such processes and are environmentally friendly as well as expanding the applicability of technologies to textiles with heat-sensitive base materials. For CVD, innovation includes low-temperature CVD processes that enable the deposition of high-quality coatings on polymer materials such as polyester, nylon, and cotton fabric, but not through thermal degradation. At the same time, new technologies appear, especially PECVD, that allow for obtaining high-quality coatings using less energy and at lower temperatures. This advancement presents possibilities for processing components that cannot withstand the usual treatments applied to the textile, especially where the intended functional or protective outcome is preferred. In PVD, enhancements are directed at scaling up the technology to allow it to be commercially applied to the industry. Another revolutionizing characteristic has been the use of roll-to-roll PVD systems, in which the textile does not need to be removed and replaced all the time, but it simply rolls into the deposition chamber, making for much higher throughput and reduced costs. Such dual process of PVD is also referred as orally as hybrid PVD techniques. As for the evaporation, sputtering, and ion plating techniques of PVD, there is a main and important focus on hybrid techniques, which can form the custom coatings both in precise functional use and appearance.

### 12.6.2 Integration with Other Surface Modification Methods

With rising tendencies toward increasing the attractiveness of the textiles of multi-functional usage in the sphere of the textile industry, the task of coalescing the CVD and PVD with the other methods of surface modification becomes one of the most paramount. Together with plasma treatment, UV curing, sol-gel coatings, and layer-by-layer assembly, the opportunity to achieve multifunctional textiles with performances that cannot be reached by each technique separately is given.

For instance, CVD or PVD coatings may be made together with the surface treated by plasma to increase the adhesion of the coating as well as to provide a uniform covering of intricate textile geometries. Similarly to that, sol-gel coatings can be offered as first barrier coatings, and other coatings can be deposited by CVD or PVD methods as second protective or functional layers. This integration will enable the generation of new fabrics that can assume the task of wear resistance alongside other functional properties such as hydrophobicity, self-cleaning, UV stability, and anti-microbial properties. As such, advances are of immense importance in such fields are healthcare and sports and, especially, in industries that deal with outdoor clothing.

### 12.6.3 Potential for Smart Textiles and Wearable Technologies

As to the smart textiles and wearable technologies, it is possible to name incredibly promising development perspectives for CVD and PVD technologies. The opportunity to apply thin layers of conductive materials on fabrics creates countless avenues for incorporating various electronics, sensors, and energy storage devices into fabrics-involved garments and textiles. This integration makes textiles perform as smart clothing for measuring body temperature, sending information, and charging devices at the same time.

CVD and PVD coatings are also applied to make conductive textiles for wearable electronics, where a combination of thin films of metals such as silver, copper, or gold can be applied on fabrics to create conductive paths. Some fabrics can be woven into clothing to be used in the detection of heart rate, rate of sweating, and level of physical activity in the most natural way possible for the user [42, 43].

In the future, nanostructured CVD and PVD coatings may be improved to manufacture textiles that are power storage devices that could power the wearables. For instance, clothes made from materials that can incorporate nano-sized carbon structures could serve as integrated flexible batteries or supercapacitors for small gadgets or sensors woven into the textiles. The above potential can stretch to healthcare, military, and civil sports for relevance to real-time data presentation. In addition, smart fabrics produced using CVD and PVD processes enable controlled modification of fabric properties such as color and texture, and can respond dynamically to external stimuli including light, temperature, moisture, and other environmental factors [44]. These innovations will revolutionize the fashion industry through the development of functional clothing with added wearer/environment feedback functionality. However, more technological improvements to CVD and PVD technologies will further influence the textile industry. The combination of the presented methods with other surface modification approaches, as well as their potential to generate smart clothing and wearable devices, is the start of a brand-new epoch in the functionality of fabrics. Applying CVD and PVD coating techniques, the textile industry is gradually approaching a new type of fabric that reacts to the environment and can provide the wearer with relevant information about the environment with which the wearer interacts. The developing trends will maintain the current pace of innovation in enhancing the surface of textiles for CVD and PVD.

## 12.7 Conclusion

This chapter has primarily focused on elaborating the role of CVD and PVD in the surface engineering of textiles. These vapor deposition methods provide specific advantages for creating new, higher-value-added textiles, including improvements in strength and surface treatment characteristics such as durability, mechanical strength, hydrophobicity, UV protection, and electrical conductivity.

Despite these precursors, CVD and PVD have their principles of operation while offering specific methods for thin-film deposition on many kinds of fabrics. CVD is based on the chemical reaction of gaseous precursors and the formation of solid films, and provides good control over the coating's thickness and composition. We are referring to its high-performance soundness in application, array, and tight-knit ability to adhere to a desirable density ideal for protective textiles and technical fabrics. Nonetheless, such processes as CVD demand high temperatures and can hardly be applied to heat-sensitive textiles—this is one of the main issues threatening the technique.

On the other hand, PVD is a technique that is exclusively a physical process where the material is vaporized and deposited through a vacuum process. The lower treatment temperatures of PVD suggest that it is ideal for textiles that cannot be heat-treated, such as synthetic fabrics, and the quality of the coating in terms of adhesion, cosmetic appeal, and functional improvements. However, there are some disadvantages to this technique as PVD makes use of vacuum environments, and it is difficult to coat a complex geometrical textile structure.

In this chapter, we have also reviewed details of textile surface enhancement about CVD and PVD throughout this chapter. These techniques include hydrophobic coatings and nanostructured materials to improve mechanical properties, functional layers for UV protection, antimicrobial activity, and electrical conductivity. These techniques are in a position to address advanced textile application demands. It has been shown in the aforementioned case studies with woolen, silk, cotton, acrylic, and polyester fabrics that the introduction of CVD and PVD has led to enhanced efficiency, strength, durability, texture, and utility of the treated textiles compared to untreated fabrics.

From this perspective, both CVD and PVD can dramatically transform the process of surface modification of textiles. Newer trends seem to indicate that there is a switch toward low-temperature CVD processes and large-area PVD systems that are best suited for the textile sector, based on some of the problems facing this sector, such as the temperature of the process and the cost of production. Furthermore, the combination of these types of vapor deposition methods with other surface engineering methods like plasma processing and sol-gel approaches allows for the creation of multipurpose textiles. This integration shall enable improvement in the performance of the textile material, thus providing properties such as self-cleaning, antimicrobial functionality, and energy storage, which are important for creating smart textiles and wearable technology.

Therefore, based on all the presented information, it is possible to state that both CVD and PVD are crucial in the modern textile industry and can create the base for numerous further developments. Their further development is expected to broaden the fields where they might find use to create environmentally friendly, high-performing fabrics that can fit various sectors fashion, sports, medical, and aviation. The continuously enhancing vapor deposition technologies will remain on the brink of defining the aspects of the future of textiles: green, functional, and smart textiles.

**AI disclosure statement:** The author confirms that no artificial intelligence (AI)-generated content or AI-assisted writing tools were used in the preparation of this chapter (including summaries and the abstract). All text, analysis, and conclusions are the result of the author's original work.

## References


1. Hedayet Ullah, M., Akther, H., Mahbubur Rahman, M., Foisal, A.B.M., Mahmud Hasan, M., Amir-Al Zumahi, S.M., Amri, A., Surface modification and improvements of wicking properties and dyeability of grey jute-cotton blended fabrics using low-pressure glow discharge air plasma. *Heliyon*, 7, 8, e07893, 2021, doi: 10.1016/j.heliyon.2021.e07893.
2. Mather, R.R., Surface modification of textiles by plasma treatments. Elsevier Inc., in: *Surface Modification of Textiles*, pp. 296–317, Heriot-Watt University, UK, 2009, doi: 10.1533/9781845696689.296.
3. Wei, Q., Xu, Y., Wang, Y., Textile surface functionalization by physical vapor deposition (PVD). Elsevier Inc., in: *Surface Modification of Textiles*, pp. 58–90, Heriot-Watt University, UK, 2009, doi: 10.1533/9781845696689.58.
4. Wilson, J.I.B., Textile surface functionalisation by chemical vapour deposition (CVD). Elsevier Inc., in: *Surface Modification of Textiles*, pp. 126–138, Heriot-Watt University, UK, 2009, doi: 10.1533/9781845696689.126.
5. Islam, M.R., Golovin, K., Dolez, P.I., Clothing Thermophysiological Comfort: A Textile Science Perspective. *Textiles*, 3, 4, 353–407, 2023, doi: 10.3390/textiles3040024.
6. Shah, M.A., Pirzada, B.M., Price, G., Shibiru, A.L., Qurashi, A., Applications of nanotechnology in smart textile industry: A critical review. *J. Adv. Res.*, 38, 55–75, 2022, doi: 10.1016/j.jare.2022.01.008.
7. Che, J. and Yang, X., A recent (2009–2021) perspective on sustainable color and textile coloration using natural plant resources. *Heliyon*, 8, 10, e10979, 2022, doi: 10.1016/j.heliyon.2022.e10979.
8. Habig, K.H., Chemical vapor deposition and physical vapor deposition coatings: Properties, tribological behavior, and applications. *J. Vac. Sci. Technol., A*, 4, 6, 2832–2843, 2019, doi: 10.1116/1.573687.
9. Abegunde, O.O., Akinlabi, E.T., Oladijo, O.P., Akinlabi, S., Ude, A.U., Overview of thin film deposition techniques. *AIMS Press*, 6, 2, 174–199, 2019, doi: 10.3934/MATERSCI.2019.2.174.
10. Tien, C.L., Special Issue 'Advanced Coating Technology by Physical Vapor Deposition and Applications. *Coatings*, 13, 2, 467, 2023, doi: 10.3390/coatings13020467.
11. Miskiewicz, P., Frydrych, I., Cichocka, A., Application of Physical Vapor Deposition in Textile Industry. *Autex Res. J.*, 22, 1, 42–54, 2022, doi: 10.2478/aut-2020-0004.
12. Saeed, M., Alshammari, Y., Majeed, S.A., Al-Nasrallah, E., Chemical Vapour Deposition of Graphene—Synthesis, Characterisation, and Applications: A Review. *Molecules*, 25, 17, 3856, 2020, doi: 10.3390/molecules25173856.
13. Escorcia-Diaz, D., García-Mora, S., Rendón-Castrillón, L., Ramírez-Carmona, M., Ocampo-López, C., Advancements in Nanoparticle Deposition Techniques for Diverse Substrates: A Review. *Nanomaterials*, 13, 18, 2586, 2023, doi: 10.3390/nano13182586.

14. Choy, K., Chemical vapour deposition of coatings. *Prog. Mater Sci.*, 48, 57–170, 2003. www.elsevier.com/locate/pmatsc, Uk.
15. Goncharov, O.Y., Treshchev, S.Y., Lad'yanov, V.I., Faizullin, R.R., Guskov, V.N., Baldaev, L.K., Tantalum chemical vapor deposition on substrates from various materials. *Inorg. Mater.*, 53, 10, 1064–1068, 2017, doi: 10.1134/S0020168517100089.
16. Sun, L., Yuan, G., Gao, L., Yang, J., Chhowalla, M., Gharahcheshmeh, M.H., Gleason, K.K., Choi, Y.S., Hong, B.H., Liu, Z., Chemical vapour deposition. *Nat. Rev. Methods Primers*, 5, 1, 1–20, 2021, doi: 10.1038/s43586-020-00005-y.
17. Lamm, B.W. and Mitchell, D.J., Chemical Vapor Deposition of Zirconium Compounds: A Review. *Coatings*, 13, 2, 266, 2023, doi: 10.3390/coatings13020266.
18. Otitoju, T.A., Ahmad, A.L., Ooi, B.S., Superhydrophilic (superwetting) surfaces: A review on fabrication and application. *J. Ind. Eng. Chem.*, 47, 19–40, 2017, doi: 10.1016/j.jiec.2016.12.016.
19. Pedersen, H. and Elliott, S.D., Studying chemical vapor deposition processes with theoretical chemistry. *Theor. Chem. Acc.*, 133, 5, 1–10, 2014, doi: 10.1007/s00214-014-1476-7.
20. Babul Reddy, A., Siva Mohan Reddy, G., Sivanjineyulu, V., Jayaramudu, J., Varaprasad, K., Sadiku, E.R. Hydrophobic/Hydrophilic Nanostructured Polymer Blends. *Design and Applications of Nanostructured Polymer Blends and Nanocomposite Systems,* 385-411, 2015, DOI: 10.1016/B978-0-323-39408-6.00016-9.
21. Mugabi, J.A., Eriksen, S., Petrushina, I., Christensen, E., Bjerrum, N.J., Kinetic Study of the Chemical Vapor Deposition of Tantalum in Long Narrow Channels. *Adv. Mater. Interfaces*, 3, 14, 1500795, 2016, doi: 10.1002/admi.201500795.
22. Franz, G., Plasma enhanced chemical vapor deposition of organic polymers. *Processes*, 9, 6, 980, 2021, doi: 10.3390/pr9060980.
23. Khan, S.A., Ahmad, A., Khan, M.I., Yusuf, M., Shahid, M., Manzoor, N., Mohammad, F., Antimicrobial activity of wool yarn dyed with Rheum emodi L.(Indian Rhubarb). *Dyes Pigm.*, 95, 2, 206–214, 2012.
24. Verbič, A., Golja, B., Stres, B., Likozar, B., Novak, U., Unveiling PFAS-free Solutions for Hydrophobic and Oleophobic Textile Coatings, University of Ljubljana, pp. 148–156, 2024, doi: 10.55295/psl.2024.i19.
25. Mahltig, B., Hydrophobic and oleophobic finishes for textiles, in: *Functional Finishes for Textiles: Improving Comfort, Performance and Protection*, pp. 387–428, Elsevier Inc., 2015, doi: 10.1533/9780857098450.2.387.
26. Kapil, N., Mayani, S.V., Gopal Bhattacharyya, K., Environmental implications of nanoceramic applications. *Results Chem.*, 5, 100724, 2023, doi: 10.1016/j.rechem.2022.100724.
27. Patra, J. and Gouda, S., Application of nano technology in textile engineering: An overview. *J. Eng. Technol. Res.*, 5, 5, 104–111, 2013, doi: 10.5897/JETR2013-0309.

28. Schneider, G., Bim, F.L., de Sousa, A.F.L., Watanabe, E., de Andrade, D., Fronteira, I., The use of antimicrobial-impregnated fabrics in health services: an integrative review. *Rev. Lat. Am. Enfermagem*, 1, 29, e3416, 2021, doi: 10.1590/1518-8345.4668.3416.
29. Tanasa, F., Teaca, C.A., Nechifor, M., Ignat, M., Duceac, I.A., Ignat, L., Highly Specialized Textiles with Antimicrobial Functionality—Advances and Challenges. *Textiles*, 3, 2, 219–245, 2023, doi: 10.3390/textiles3020015.
30. Gulati, R., Sharma, S., Sharma, R.K., Antimicrobial textile: recent developments and functional perspective. *Polym. Bull.*, 79, 5747–5771, 2022, doi: 10.1007/s00289-021-03826-3.
31. Alsukaibi, A.K.D., Various Approaches for the Detoxification of Toxic Dyes in Wastewater. *Processes*. 10, 10, 2022, doi: 10.3390/pr10101968.
32. Shahidi, S., Moazzenchi, B., Ghoranneviss, M., A review-application of physical vapor deposition (PVD) and related methods in the textile industry. *Eur. Phys. J. Appl. Phys.*, 71, 3, 31302, 2015.
33. Baptista, A., Silva, F., Porteiro, J., Míguez, J., Pinto, G., Sputtering physical vapour deposition (PVD) coatings: A critical review on process improvement and market trend demands. *Coatings*, 8, 11, 402, 2018, doi: 10.3390/COATINGS8110402.
34. Vorobyova, M., Biffoli, F., Giurlani, W., Mauro Martinuzzi, S., Linser, M., Caneschi, A., Innocenti, M., PVD for Decorative Applications: A Review. *Materials*, 16, 14, 4919, 2023, doi: 10.3390/ma16144919.
35. Savale, P.A., Physical Vapor Deposition (PVD) Methods for Synthesis of Thin Films: A Comparative Study. *Mater. Sci. Eng. Phys., Arch. Appl. Sci. Res.*, 8, 5, 1–8, 2016, http://scholarsresearchlibrary.com/archive.html.
36. Mostafavi, A.H., Mishra, A.K., Gallucci, F., Kim, J.H., Ulbricht, M., Coclite, A.M., Hosseini, S.S., Advances in surface modification and functionalization for tailoring the characteristics of thin films and membranes *via* chemical vapor deposition techniques. *J. Appl. Polym. Sci.*, 140, 15, e53720, 2023, doi: 10.1002/app.53720.
37. Baklanov, M.R., Gismatulin, A.A., Naumov, S., Perevalov, T.V., Gritsenko, V.A., Vishnevskiy, A.S., Rakhimova, T.V., Vorotilov, K.A., Comprehensive Review on the Impact of Chemical Composition, Plasma Treatment, and Vacuum Ultraviolet (VUV) Irradiation on the Electrical Properties of Organosilicate Films. *Polymers*, 16, 15, 2230, 2024, doi: 10.3390/polym16152230.
38. Fancey, K.S. and Matthews, A., Evaporative ion plating: Process mechanisms and optimization. *IEEE Trans. Plasma Sci.*, 18, 6, 869–877, 1990, doi: 10.1109/27.61497.
39. Seesaard, T. and Wongchoosuk, C., Flexible and Stretchable Pressure Sensors: From Basic Principles to State-of-the-Art Applications. *Micromachines*, 14, 8, 1638, 2023, doi: 10.3390/mi14081638.
40. Mattox, D.M., Ion plating-past, present and future. *Surf. Coat. Technol.*, 133–134, 517–521, 2000, Doi: 10.1016/S0257-8972(00)00922-1.

41. Pawlak, R., Korzeniewska, E., Koneczny, C., Hałgas, B., Properties of thin metal layers deposited on textile composites by using the PVD method for textronic applications. *Autex Res. J.*, 17, 3, 229–237, 2017.
42. Mokhtari, F., Cheng, Z., Raad, R., Xi, J., Foroughi, J., Piezofibers to smart textiles: A review on recent advances and future outlook for wearable technology. *J. Mater. Chem. A*, 8, 19, 9496–9522, 2020.
43. Ismar, E., Bahadir, S.K., Kalaoglu, F., Koncar., V., Futuristic clothes: electronic textiles and wearable technologies. *Global Challenges*, 4, 7, 1900092, 2020.
44. Hossain, Md.R., Ahmed, Md.R., Alam, Md.S., Smart textiles. *Text. Prog.*, 55, 2, 47–108, 2023.

# 13

# Modification and Functionalization of Textiles Integrating Electroless Deposition

**Md. Reazuddin Repon[1]* and Arnob Dhar Pranta[2,3]**

[1]*Department of Bioproducts and Biosystems, School of Chemical Engineering, Aalto University, Espoo, Finland*
[2]*ZR Research Institute for Advanced Materials, Sherpur, Bangladesh*
[3]*Department of Textile Engineering, Mawlana Bhashani Science and Technology University, Santosh, Tangail, Bangladesh*

***Abstract***
Textile modification, due to the integration of electroless deposition, opens a wide range of applications in different areas of textile application to improve conductivity for smart textiles with the addition of antibacterial qualities. To maintain the flexibility and comfort of the fabric, additional features can be added with customized solutions for a wide range of applications. The production of multifunctional textiles with improved and better traits is the main objective of electroless deposition technology. This book chapter aims to discuss the features, mechanisms, methods, challenges, and prospects of electroless deposition ranging from medical applications to smart wearable electronics. The use of electroless deposition to manufacture advanced textiles can add a new dimension to the textile application by the fabrication of conductive fabrics for smart wearables with antimicrobial properties.

***Keywords***: Electroless deposition, smart textiles, protective finishing, functional textiles, metal precursors

## 13.1 Introduction

Over the past few decades, there have been immense efforts observed in the development of textile materials that provide advanced features. As a

**Corresponding author*: reazmbstu.te@gmail.com

Aminoddin Haji and Çağlar Sivri (eds.) Surface Modification and Functionalization of Textiles, (319–336) © 2026 Scrivener Publishing LLC

result, the purpose of textiles and apparel expanded far beyond conventional usage to consumer demand. Different techniques have been introduced for manufacturing textiles that impart electroconductive features [1]. These include electrodeposition, chemical coating, metallization electroless deposition, electroplating, carbon and metallic compound inclusion, and electroconductive composites [2].

Electroless deposition is the process of depositing metals onto the surfaces of the substrate without using any kind of external power. To provide aesthetic appeal, to extend functionality, and to make the textiles more durable are the prime objectives of this approach with the surface modification of textiles. A wide range of metals such as nickel, cobalt, copper, and gold can be deposited using this method [3]. Coatings can be made to ensure specific requirements in terms of electrical conductivity, thermal insulation, or even aesthetics. Surface modifications of textiles by electroless deposition can enhance specific properties such as conductivity, antimicrobial activity, or even visual acceptance [4]. For example, conductive finishing can add the feature of making electronic functions in the finished apparel, which is a step to manufacture smart textiles. With the addition of a layer of conductive materials to the fabric, it is possible to determine electrical impulses from the human body [5]. Antibacterial coatings to textiles through electroless deposition help to improve hygiene and also extend the apparel's lifetime as it prevents degradation [6]. Compared to traditional electroplating techniques, electroless deposition technology provides many advantages such as a more uniform coating of complex structures and surfaces and the elimination of electrical contacts [7, 8]. Furthermore, textiles with electroless deposition open up numerous possibilities for customization and functionality enhancement to fulfill consumer demand.

This chapter will explore the detailed mechanism of the electroless deposition process. It will also discuss the methods, i.e., chemical deposition, chemical vapor deposition (CVD), spray coating, and inkjet printing. Furthermore, the chapter delves into the features imparted by electroless deposition including conductivity, antibacterial activity, and increased mechanical properties. This chapter also emphasizes on the application of textiles in which surfaces are modified with electroless deposition.

## 13.2 Mechanism of Electroless Deposition

Electroless deposition is commonly termed as chemical or auto-catalytic deposition, which has applications in various industries [9]. Preparation of the substrate surface for metal ion deposition begins with chemical treatments that activate it. After that, a metal ion or a complex catalytic agent is adsorbed onto the substrate to aid in the reduction of metal ions in solution [10]. There are nuclei formed on the surface of the substrate by these reduced metal ions [11]. An auto-catalytic cycle is maintained throughout this process, with the deposited metal layer serving as a source of additional catalytic sites for subsequent reduction and deposition [12]. When deposition is finished, the substrate is rinsed to remove any remaining chemicals.

The standard components of a plating solution include a metal salt, a reducing agent, complexing agents that improve the autocatalytic process, buffers, and stabilizers [13]. A typical five-step electroless plating process is intended to improve the deposition of copper and nickel particles onto polyester fabric.

**I. Pre-Treatment:** Each fabric sample is pre-treated by being washed in a 2% non-ionic detergent at pH 7 and 40°C for 20 minutes to eliminate any surface oils or contaminants that may have been introduced during production. Deionized water is used to rinse the fabric samples for approximately 5 minutes.

**II. Sensitization:** The surface sensitizer is applied to the pre-treated samples by slowly agitating them for 10 minutes at 25°C and pH 1. The mixture consisted of 5 g/L stannous chloride and 5 mL/L hydrochloric acid, with a concentration of 37%. Deionized water is subsequently used to rinse the sensitized textiles.

**III. Activation:** Surface activation is achieved by immersing the fabric samples in an activator that consists of 0.3 g/L palladium (II) chloride, 0.5 mL/L hydrochloric acid (concentration 37%), and 20 g/L boric acid. The activator is mixed at pH 2 and 25°C for 5 minutes. After activation, deionized water is used to rinse the fabrics.

**IV. Electroless Plating:** The electroless copper and nickel plating of polyester fabric was carried out under specific operating conditions, and the bath composition is detailed in Table 13.1. The electroless copper plating process uses tartrate as a complexing agent and formaldehyde as a reducing agent. Electroless nickel plating requires a reducing agent such as sodium hypophosphite. After being treated with a metallizing agent, polyester fabric was rinsed with deionized water at a temperature of 40°C for approximately 20 minutes.

**V. After Treatment:** After washing, the metallized polyester fabrics are steam-cured for 1 minute at 150°C. Before being evaluated, fabrics are subjected to a conditioning period of at least 24 hours under standard atmospheric pressure, relative humidity of 65 ± 2%, and temperature of 21 ± 1°C.

**Table 13.1** Composition and condition of electroless plating of copper and nickel [14].

| Electroless plating | Composition | Conditions |
|---|---|---|
| Copper | 10 g/L copper (II) sulfate 5-hydrate | 28 ± 1°C, pH 11, 20 min |
| | 45 g/L potassium sodium tartrate | |
| | 10 g/L sodium hydroxide | |
| | 20 ml/L formaldehyde solution | |
| Nickel | 15 g/L nickel (II) sulfate 7-hydrate | 40 ± 1°C, pH 9 20 min |
| | 8 g/L tri-sodium citrate | |
| | 18 g/L ammonium chloride | |
| | A few drops of sodium hydroxide | |
| | 15 g/L sodium hypophosphite monohydrate | |

## 13.3 Electroless Deposition Methods

A chemical reaction occurred in the electroless deposition process to deposit the desired metal ions or other precursors onto the substrate surface from a solution. This process can be applied to textiles in a variety of ways. Methods of electroless deposition are stated in Figure 13.1.

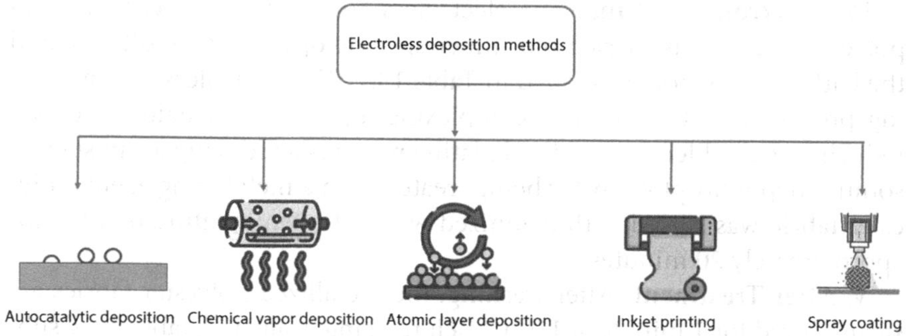


**Figure 13.1** Methods of electroless deposition.

### 13.3.1 Autocatalytic Deposition

Depositing a material onto a surface at an increasing rate throughout the deposition process is known as autocatalytic deposition. This happens because the deposited material aids in the deposition process by acting as a catalyst. More of the material will eventually settle to the surface, increasing the number of potential deposition sites and speeding up the process of growth. Autocatalytic plating is one of the electroless plating techniques that has been used the most frequently for coating metals on substrates [15]. In the process of autocatalytic plating, metallic elements are deposited on the surface through redox reactions that take place in the presence of a reducing agent [16]. When compared to immersion, autocatalytic plating produces layers that are of higher quality, larger in thickness, and more uniform. The rate of deposition typically ranges from 12.5 to 25 micrometers (0.0005 to 0.001 inches). Due to the slow deposition rate, the coating is typically less than 50 μm (0.002 in) in practice, even though it has been done up to 650 μm (0.026 inch) in thickness [17]. When compared to electroplating, plating tends to produce a uniform thickness. This is because there are no electric fields present, which eliminates the challenges associated with making the plating process uniform. Nickel and copper are the two most common elements utilized in electroless platings. As for nickel, the deposits are dense, relatively hard which ranges from 43 to 55 Hardness Rockwell C(HRC), increasing to approximately 65 HRC after 2 hours at 343°C (650°F) [18].

### 13.3.2 Chemical Vapor Deposition (CVD)

CVD involves a chemical reaction taking place on heated substrate surface to deposit a solid material from a vapor. A layer of powder, a single crystal, or even a thin film might be the end product of this solid process. Experimental conditions such as substrate material, substrate temperature, reaction gas mixture composition, and total pressure gas flows can be varied to grow materials with a broad range of physical, tribological, and chemical properties.

Even on complexly shaped substrates, the CVD method can produce coatings of consistent thickness and characteristics with low porosity due to its exceptional throwing power [19]. One more noteworthy quality is the ability to selectively deposit material onto pre-patterned substrates. Thin films made by CVD and related processes are used for a wide variety of purposes, such as coatings for heat resistance, oxidation barriers, conductive oxides, tribological and corrosion resistance, and microelectronic epitaxial layers [20]. Further uses for CVD include making solar cells, high-temperature fiber composites, particles with clearly defined sizes, and preparing materials with high melting points such as tungsten and ceramics.

Different variations of CVD techniques such as atmospheric pressure CVD, low-pressure CVD, and plasma-enhanced CVD are commonly practiced [21]. Every technique has its own benefits that have been engineered to meet certain requirements.

### 13.3.3 Atomic Layer Deposition (ALD)

ALD is another technique of deposition of a wide variety of thin materials from the vapor phase. This method has potential in the field of energy conservation. When it is compared to other methods, it has been seen that it has major advantages over other methods for this reason the demand for ALD made structures is on the rise as it is applicable for smaller and space efficient areas. ALD techniques provide a remarkable control over the deposition thickness and the composition of the final material can be achieved as expected before production. Furthermore, desirable traits such as cyclical and self-saturating nature can be characterized by the ALD process.

At the atomic level, this process can deposit a predetermined level of thickness onto the surface of the substrate in a very accurate way with exact composition [22]. By repeating cycles, it is quite possible to ensure atomic-scale control and uniform deposition over vast areas. ZnO and $Al_2O_3$ deposition at low temperatures on natural fibers increases fabric's antimicrobial activity performance. The thickness, uniformity, configuration, and adherence to the flexible substrate are the determinants of the nanostructured layer's efficiency [23].

### 13.3.4 Inkjet Printing

Inkjet printing is another promising method for the application of conductive material onto textiles. Unlike conventional printing methods, this specific method does not require any kind of toxic chemicals. Conductive ink onto the textile surface is implemented through this method without any requirement for a printing plate or screen [24]. This ink is a suspension consisting of metal ions such as silver or copper. These metal ions undergo reduction when the reducing agent is introduced and allow it to transform to its metallic state as a final result a uniform layer of conductive material forms on the textile. Electroless deposition inkjet printing has the potential to be scaled up for mass production, making it very suitable for industrial applications including smart textiles, wearable electronics, and functional clothes. At room temperature, it is possible to produce highly conductive textiles using polyaniline and polypyrrole formation [25].

### 13.3.5 Spray Coating

Spray coating is a versatile way to add metallic layers to achieve metallic properties. In this process, a spray solution that contains metal ions is subjected onto the surface of the textiles with a thin layer. After the pretreatment processes, the fabric is subjected to the spray that contains specific metal ions that particular properties need to be developed in the fabric surface. Metal spray coating deposition on textiles has numerous advantages. Metals with useful characteristics particularly conductivity or antimicrobial effects can be used, for example, silver, copper, or gold. In particular, the spray coating method affects the dispersal of $TiO_2$ particles used to make antimicrobial clothing [26]. Smart textiles, electromagnetic shielding, and even ornamentation can all benefit from these characteristics.

## 13.4 Factors Influencing of Electroless Deposition

### 13.4.1 Fiber Type

Electroless deposition in textiles is impacted by fiber type in various ways. Fiber compatibility with the deposition process is highly dependent on the surface characteristics and chemical composition. The deposited material tends to adhere better to fibers that have reactive functional groups, such as hydroxyl or amine groups. Coating coverage and adhesion strength are both improved by fibers with higher surface energies, which, in turn, affects wetting and adhesion. In comparison to natural fibers like wool or cotton, the surface energy of synthetics like polyester or nylon is usually lower. Furthermore, the amount and the rate of deposition are affected by the porosity and surface area of the fibers. Fibers with a higher porosity or surface area provide more nucleation sites for deposition. Coated fabric's performance and longevity are affected by the fiber's mechanical qualities including tensile strength and pliability.

In polyester fabric, the coating's copper content drops and its nickel content rise marginally with an increase in the concentration of $NiSO_4$. However, as pH rises, the coating's copper content increases and its nickel and phosphorus content decreases [27]. Study on electroless nickel-phosphorous (Ni-P) plating on polyester fiber with sodium hypophosphite as a reducing agent has shown that, as the concentrations of nickel sulfate, sodium hypophosphite, pH, and bath temperature rise, the deposition rates also rise. The deposition rates are found to decrease as the sodium citrate concentration increases [28]. On the other hand, electroless deposition coated cotton fabrics with ZnO crystallites. The contact angle measurements reveal that adding ZnO prisms to the fiber surface turns the cotton fabric from hydrophilic to superhydrophobic [29].

### 13.4.2 Deposition Layers

The performance of electroless deposition in textiles is greatly affected by deposition layers. Electroless deposition layer nucleation and growth can be influenced by the deposition layer's thickness, composition, and morphology [30]. A thin and consistent layer of deposition can create an ideal environment for electroless deposition, resulting in the controlled and uniform growth of the target material. However, when the deposition layer is too thick and uneven it will cause poor adhesion. As a result, this will hamper the process of electroless deposition layer from nucleating and growing. Multiple layers of deposition on the substrate allow for enhanced durability and eliminate delamination [31]. On top of that, it acts as a barrier against chemicals and moisture. Uniform coatings change the surface of textiles nature drastically with the addition of layers [32].

### 13.4.3 Method of Deposition

The method of electroless deposition in textiles plays a significant role in case of ensuring efficiency, adhesion, and uniformity. For instance, in autocatalytic deposition, a continuous and uniform coating is produced on textile material due to catalyzation. This reaction spreads out to other areas of the material. On the other hand, using CVD allows a solid coating application to textile substrate that is quite different from other methods. CVD is well suited for precise applications as it offers a high degree of controllability over the composition of films and the thickness of deposition. However, ALD allows for atomic-level precision in film thickness control by controlling the rate of sequential, self-limiting surface reactions. ALD is suitable for advanced-level applications as it confirms the uniform and conformal coating on textiles. In the spray coating process, a liquid precursor is atomized and later deposited onto the textile material. This method is great for coating complex shapes and large areas, but it cannot be controlled well, so the coating might not be uniform. In contrast, inkjet printing allows for the incredibly accurate and waste-free application of functional inks to fabrics. On the other hand, it might be difficult to get even coverage and stickiness, especially with some ink mixes. When it comes to electroless deposition in textiles, each deposition method has its own set of pros and cons, providing a range of alternatives for specific uses.

## 13.5 Properties and Performance Evaluation

### 13.5.1 Electric Conductivity

With electroless deposition, a substrate can be coated with metal or alloy without the use of an external electrical power source. Improvements in the material's electric conductivity are a major use of electroless deposition. Increased conductivity is achieved by covering a substrate with a conductive metal layer. Studies indicate that the presence of a silver layer on aramid fiber results in both excellent conductivity and strong adhesion to the metal layer. The aramid fiber that was prepared exhibited low electrical resistance of 0.38 Ω/cm, indicating excellent conductivity [33]. This is especially helpful in fields like smart conductive textiles where highly conductive materials are important for proper operation. Carbon-based nanoparticles coated with noble metals have the potential to provide exceptional electrical conductivity while avoiding the drawbacks of high weight and cost typically associated with metals like silver and gold [34]. Compared to untreated materials, treated ones have better electrical conductivity because the thickness and composition of the metal layer that is deposited can be precisely controlled.

### 13.5.2 Antimicrobial Activity

Antimicrobial activity of fabric is influenced by the treatment with electroless deposition. Because the surface of the textile is modified by the deposition of metal ions through the electroless deposition process, it allows the enhancement of the antimicrobial properties in the textile substrate. Besides, it limits and eliminates the inhibition of bacteria, fungi and other microorganisms by enhancing the textile substrates antimicrobial properties. Metals such as silver, copper, zinc, and others that possess antimicrobial traits add a major advantage when deposited on textile substrates. Due to the inherent antimicrobial properties of these metals, the growth of bacteria, fungi, and viruses is effectively resisted through coated surfaces. In addition, the density and coating consistency provide antimicrobial coverage to the entire surface that prevents microorganisms from developing in any vulnerable location. In the case of silk, electroless deposition surface modification was reported as very effective. The effectiveness of silver-plated silk fabric as an antibacterial agent against *E. coli* is 99.98%, whereas against *S. aureus* it is 99.83% [35]. Similarly, Cu nanoparticles were able to inhibit the growth of both *E. coli* and *S. aureus*. A higher antimicrobial index was observed as a result of *S. aureus*'s decreased resistance to Cu nanoparticles [36].

### 13.5.3 Hydrophobicity

Electroless deposition is a widely implemented method in surface engineering to enhance the hydrophobicity of textile materials. For example, the hydrophobic metal layer can be formed on textile fibers through electroless deposition on textiles. This coating can enhance the textile's ability to repel water by creating a barrier that prevents the blast of water. In addition, the metal layer can be precisely controlled in thickness and uniformity using electroless deposition that guarantees that the textile surface will have consistent hydrophobicity. One important feature is that the cotton fabrics functionalized with ZnO exhibit super hydrophobicity, meaning that water contact angles are greater than 150° [29]. Similarly, with a water contact angle of up to 112.8° and excellent self-cleaning properties, the Ni-plated cotton modified with (3-mercaptopropyl) trimethoxysilane demonstrated remarkable hydrophobicity [37]. Furthermore, textile electrodes made of highly conductive polyester fabric with improved multi-layer silver deposition exhibit excellent mechanical characteristics, thermal stability, and hydrophobicity [38].

### 13.5.4 Mechanical Properties

Electroless deposition is a method for improving the fabric's mechanical qualities. The fabric's tensile strength and durability are both enhanced by this process. Also, the deposited metal layer acts as a protective coating, making it more durable and less susceptible to wear and abrasion. It has been found that physical and mechanical properties are enhanced by physically entrapping inert particles in a Ni-P layer that is grown on top of the substrate [39]. Similarly, the mechanical properties of the composite materials made of nickel and titanium dioxide were improved when compared to those of pure nickel electrodeposition [40]. According to the results of the performance tests, the gold-coated para-aramid yarns exhibited favorable mechanical properties and satisfactory resistance to washing characteristics [41]. Even though electroless deposition makes the fabric stronger, it can keep the fabric's pliability and drape so that it will be comfortable to wear. In addition, it helps with dimensional stability, which means that it does not stretch or shrink as much over time.

## 13.6 Current Challenges

### 13.6.1 Technical Challenges

Electroless deposition is a chemically driven and inherently complex process that involves the reduction of metal ions from a metal salt solution onto a substrate surface without the use of external electric current. This process requires precise interaction between a reducing agent and the metal salt, and its success heavily depends on various substrate-related parameters. In the case of textiles, challenges arise due to their diverse structures, textures, compositions, and surface characteristics. These inherent variations can significantly affect the uniformity and adhesion of the deposited metal layer. For instance, surface roughness and chemical composition are critical in determining the quality of metal-substrate bonding. Textiles often exhibit inconsistent porosity and variable chemical make-up, which further complicate the deposition process and make it difficult to achieve consistent coating. Moreover, the compatibility between the textile substrate and the metal layer is particularly difficult to manage, as different fabrics respond differently to the same treatment conditions. Ensuring uniform, adherent metal deposition across diverse textile substrates requires rigorous control over process parameters such as temperature, pH, concentration of reactants, and treatment time. The inherent variability of textile materials makes it challenging to establish a standardized protocol, highlighting the necessity for substrate-specific optimization and precise control of process variables to achieve reproducible and homogeneous coatings.

### 13.6.2 Environmental Challenges

Electroless deposition presents significant environmental challenges, primarily due to its complex nature and the variability of textile substrates, which complicates process control and optimization. One of the key concerns is the difficulty in minimizing chemical usage, as precise control over reactant concentrations is often hindered by inconsistencies in textile material properties. Although the process itself does not require electrical energy, it typically demands elevated temperatures to facilitate metal ion reduction, thereby leading to high thermal energy consumption. This substantial energy input contributes to increased production costs and a larger environmental footprint. Additionally, electroless deposition generates chemically intensive wastewater containing heavy metals, organic compounds, and other hazardous pollutants, posing serious challenges for effective wastewater management and environmental compliance.

The combination of high energy requirements and potential environmental contamination underscores the need for more sustainable practices and advanced treatment solutions in electroless metallization of textiles.

### 13.6.3 Economic Challenges

Electroless deposition faces economic challenges, particularly in terms of material and equipment costs. The process relies on specialized chemicals, catalysts, and high-purity substrates, all of which contribute to elevated material expenses. Additionally, the infrastructure required for electroless deposition—including deposition chambers, agitation systems, and precise monitoring instruments—represents a significant capital investment. The setup often necessitates custom-built apparatus and specialized configurations to accommodate different textile substrates, further driving up installation and operational costs. These economic constraints can hinder large-scale implementation and limit the accessibility of electroless deposition technologies in cost-sensitive markets.

## 13.7 Applications and Future Prospects

Electroless deposition technology opens up vast areas of potential uses. Various applications of electroless deposited textile materials are stated in Figure 13.2.

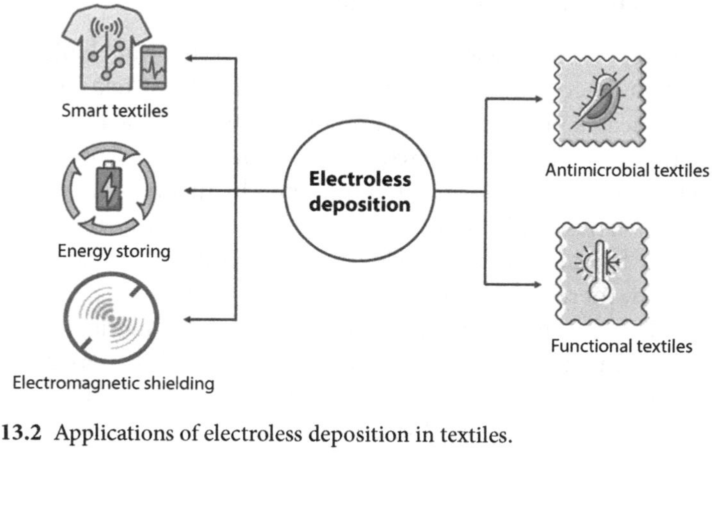


**Figure 13.2** Applications of electroless deposition in textiles.

### 13.7.1 Smart Textiles

Electroless deposition technology has opened new possibilities in a variety of fields through surface modification in the areas of smart textiles. Physicians and nurses are notified with real time data recorded from the patient as it has the trait to track and regulate specific parameters to detect any irregularities found in the patient's body. The host can be actively monitored and regulated through sensors and actuators integrated into the electroless deposited finished textiles. Additionally, the trajectory of safety and security will enhance with the passage of time as it enables the features to detect any possible hazards in protective garments. The sports and fitness sector can benefit from the application of smart textiles as the technology allows it to track athletes' progress and potential injuries.

### 13.7.2 Energy Storage

Storage of energy can be an interesting application of electroless deposited textile materials. Textiles that is manufactured with electroless deposition having conductive polymers or metal ions as nanoparticles allows to the transformation of thermal energy and sunlight into electrical power. Consequently, this feature can lead wearable electronics to be self-powered without depending on energy sources. When the electroless deposited textiles face any kind of stress, it induces a change in the deposited material and it becomes active which leads to electricity generation through tribo-electric effect. The most interesting part is that the generated electricity can be stored in capacitors and batteries for prospective future usage.

### 13.7.3 Electromagnetic Shielding

The area of electromagnetic shielding has a potential for future usage through electroless deposition finishing. Conductive materials that have been deposited through electroless deposition can protect the textiles from electromagnetic interference (EMI). Improving the shielding performance requires the selection of proper materials and correct deposition techniques to confirm superior protection against EMI across a wider frequency range in the future. Textiles with electroless deposition finishing have the potential to become integral parts of several electronic devices and systems as these devices cause EMI shielding.

### 13.7.4 Antimicrobial Textiles

The antimicrobial properties through electroless deposition is an exciting feature that is quite interesting and helps to move forward another step towards the improvement of medical textiles. Electroless deposition creates a coating that acts as a major barrier to microbial development and improves resistance to bacterial culture. Thus, the lifetime of the textiles increases with an effective solution especially required for medical purposes. As the hygiene issue is directly related to medical textiles, this special fabrication can lift this sector one step further. However, the medical sector is not only the potential place for the application of these specially fabricated textiles but also it can be implemented in outdoor apparel, active sportswear, and many more.

### 13.7.5 Technical and Functional Textiles

An electroless deposition allows the textile to impart special functional improvements with a wide array of traits including antibacterial, water repellency temperature regulation, and many more. This paves the way for the creation of adaptive clothing that can better meet the demands of the wearer in terms of comfort, performance, and safety. The potential of functional clothing is anticipated to be enhanced in the future by developments in material science and fabrication techniques, allowing for the incorporation of even more complex functionalities.

## 13.8 Conclusion

Electroless deposition is an important way to impart additional features to any textiles through surface modification. It helps to improve the functionality and aesthetic appeal of the fabric. Hence, it is a novel approach to manufacture special textiles with customized traits that have environmental advantages. Multi-functional features like improved conductivity, antibacterial activity, and aesthetic appeal can be achieved with a single treatment of electroless deposition. Moreover, it helps to provide better performance of textiles in varied contexts. These specific features with customized solutions can lead to consumer satisfaction to the next level. Innovative developments in fields as diverse as healthcare and fashion are fueling the rapid pace of technological advancement that promises even more complex features in the near future, such as self-cleaning fabrics and the ability to gather energy.

**AI disclosure statement:** The author acknowledges the use of AI-based tools for minor language editing or grammatical correction. However, all intellectual contributions, analyses, and conclusions presented in this chapter are entirely the author's own.

## References

1. Chatterjee, K., Tabor, J., Ghosh, T.K., Electrically conductive coatings for fiber-based E-Textiles. *Fibers*, 7, 6, MDPI, Jun. 01, 2019, doi: 10.3390/fib7060051.
2. Fallah, P., Rajagopalan, S., McDonald, A., Yue, S., Development of hybrid metallic coatings on carbon fiber-reinforced polymers (CFRPs) by cold spray deposition of copper-assisted copper electroplating process. *Surf. Coat. Technol.*, 400, 126231, Oct. 2020, doi: 10.1016/J.SURFCOAT.2020.126231.
3. Qiu, T. and Chu, P.K., Self-selective electroless plating: An approach for fabrication of functional 1D nanomaterials. *Mater. Sci. Eng.: R: Rep.*, 61, 1–6, 59–77, May 2008, doi: 10.1016/J.MSER.2008.03.001.
4. Silva, A.C.Q., Silvestre, A.J.D., Freire, C.S.R., Vilela, C., Modification of textiles for functional applications, in: *Fundamentals of Natural Fibres and Textiles*, pp. 303–365, Woodhead Publishing, Duxford, United Kingdom, Jan. 2021, doi: 10.1016/B978-0-12-821483-1.00010-3.
5. Shi, J., *et al.*, Smart Textile-Integrated Microelectronic Systems for Wearable Applications. *Adv. Mater.*, 32, 5, Wiley-VCH Verlag, Feb. 01, 2020, doi: 10.1002/adma.201901958.
6. Burnett-Boothroyd, S.C. and McCarthy, B.J., Antimicrobial treatments of textiles for hygiene and infection control applications: an industrial perspective, in: *Textiles for Hygiene and Infection Control*, pp. 196–209, Woodhead Publishing, Cambridge, United Kingdom, Jan. 2011, doi: 10.1533/9780857093707.3.196.
7. Khoperia, T.N., Electroless deposition in nanotechnology and ULSI. *Microelectron. Eng.*, 69, 2–4, 384–390, Sep. 2003, doi: 10.1016/S0167-9317(03)00325-3.
8. Ali, A., Baheti, V., Vik, M., Militky, J., Copper electroless plating of cotton fabrics after surface activation with deposition of silver and copper nanoparticles. *J. Phys. Chem. Solids*, 137, Feb. 2020, doi: 10.1016/j.jpcs.2019.109181.
9. Barshilia, H.C., Surface Modification Technologies for Aerospace and Engineering Applications: Current Trends, Challenges and Future Prospects. *Trans. Indian Natl. Acad. Eng.*, 6, 2, 173–188, Jun. 2021, doi: 10.1007/s41403-021-00208-z.
10. Blesa, M.A., Weisz, A.D., Morando, P.J., Salfity, J.A., Magaz, G.E., Regazzoni, A.E., The interaction of metal oxide surfaces with complexing agents

dissolved in water. *Coord. Chem. Rev.*, 196, 1, 31–63, Jan. 2000, doi: 10.1016/S0010-8545(99)00005-3.
11. Greene, J.E., Thin Film Nucleation, Growth, and Microstructural Evolution: An Atomic Scale View, in: *Handbook of Deposition Technologies for Films and Coatings: Science, Applications and Technology*, pp. 554–620, William Andrew, Oxford, United Kingdom, Jan. 2010, doi: 10.1016/B978-0-8155-2031-3.00012-0.
12. Levi, J., *et al.*, Comparing methods to deposit Pd-In catalysts on hydrogen-permeable hollow-fiber membranes for nitrate reduction. *Water Res.*, 235, 119877, May 2023, doi: 10.1016/J.WATRES.2023.119877.
13. Sudagar, J., Lian, J., Sha, W., Electroless nickel, alloy, composite and nano coatings – A critical review. *J. Alloys Compd.*, 571, 183–204, Sep. 2013, doi: 10.1016/J.JALLCOM.2013.03.107.
14. Jiang, S.Q. and Guo, R.H., Modification of textile surfaces using electroless deposition, in: *Surface Modification of Textiles*, pp. 108–125, Woodhead Publishing, Cambridge, United Kingdom ,2009.
15. Delaunois, F., Petitjean, J.P., Lienard, P., Jacob-Duliere, M., Autocatalytic electroless nickel-boron plating on light alloys. *Surf. Coat. Technol.*, 124, 2–3, 201–209, Feb. 2000, doi: 10.1016/S0257-8972(99)00621-0.
16. Muench, F., Electroless Plating of Metal Nanomaterials. *ChemElectroChem*, 8, 16, John Wiley and Sons Inc, 2993–3012, Aug. 13, 2021, doi: 10.1002/celc.202100285.
17. Mayrhofer, P.H., Mitterer, C., Hultman, L., Clemens, H., Microstructural design of hard coatings. *Prog. Mater Sci.*, 51, 8, 1032–1114, Nov. 2006, doi: 10.1016/J.PMATSCI.2006.02.002.
18. Groysman, A., Corrosion problems and solutions in oil refining and petrochemical industry. p. 356, Springer International Publishing, Gewerbestrasse, Switzerland, 2017.
19. Choy, K.L., Chemical vapour deposition of coatings. *Prog. Mater Sci.*, 48, 2, 57–170, Jan. 2003, doi: 10.1016/S0079-6425(01)00009-3.
20. Schalk, N., Tkadletz, M., Mitterer, C., Hard coatings for cutting applications: Physical vs. chemical vapor deposition and future challenges for the coatings community. *Surf. Coat. Technol.*, 429, 127949, Jan. 2022, doi: 10.1016/J.SURFCOAT.2021.127949.
21. Fraga, M. and Pessoa, R., Progresses in synthesis and application of sic films: From CVD to ALD and from MEMS to NEMS. *Micromachines*, 11, 9, MDPI AG, Sep. 01, 2020, doi: 10.3390/MI11090799.
22. Sneh, O., Clark-Phelps, R.B., Londergan, A.R., Winkler, J., Seidel, T.E., Thin film atomic layer deposition equipment for semiconductor processing. *Thin Solid Films*, 402, 1–2, 248–261, Jan. 2002, doi: 10.1016/S0040-6090(01)01678-9.
23. Popescu, M.C., *et al.*, Antibacterial efficiency of cellulose-based fibers covered with ZnO and Al2O3 by Atomic Layer Deposition. *Appl. Surf. Sci.*, 481, 1287–1298, Jul. 2019, doi: 10.1016/J.APSUSC.2019.03.268.

24. Gysling, H.J., Nanoinks in inkjet metallization — Evolution of simple additive-type metal patterning. *Curr. Opin. Colloid Interface Sci.*, 19, 2, 155–162, Apr. 2014, doi: 10.1016/J.COCIS.2014.03.013.
25. Stempien, Z., Rybicki, T., Rybicki, E., Kozanecki, M., Szynkowska, M.I., *In-situ* deposition of polyaniline and polypyrrole electroconductive layers on textile surfaces by the reactive ink-jet printing technique. *Synth. Met.*, 202, 49–62, Apr. 2015, doi: 10.1016/J.SYNTHMET.2015.01.027.
26. Ortelli, S., *et al.*, Influence of spray-coating process parameters on the release of TiO2 particles for the production of antibacterial textile. *NanoImpact*, 19, 100245, Jul. 2020, doi: 10.1016/J.IMPACT.2020.100245.
27. Guo, R.H., *et al.*, Effects of deposition parameters of electroless copper plating on polyester fabric. *Fibers Polym.*, 14, 5, 752–758, 2013, doi: 10.1007/s12221-013-0752-9.
28. Guo, R.H., Jiang, S.X., Yuen, C.W.M., Ng, M.C.F., Lan, J.W., Zheng, G.H., Influence of deposition parameters and kinetics of electroless Ni-P plating on polyester fiber. *Fibers Polym.*, 13, 8, 1037–1043, 2012, doi: 10.1007/s12221-012-1037-4.
29. Preda, N., *et al.*, Superhydrophobic properties of cotton fabrics functionalized with ZnO by electroless deposition. *Mater. Chem. Phys.*, 138, 1, 253–261, Feb. 2013, doi: 10.1016/j.matchemphys.2012.11.054.
30. Genova, V., Paglia, L., Marra, F., Bartuli, C., Pulci, G., Pure thick nickel coating obtained by electroless plating: Surface characterization and wetting properties. *Surf. Coat. Technol.*, 357, 595–603, Jan. 2019, doi: 10.1016/J.SURFCOAT.2018.10.049.
31. Sahasrabudhe, H., Harrison, R., Carpenter, C., Bandyopadhyay, A., Stainless steel to titanium bimetallic structure using LENS™. *Addit. Manuf.*, 5, 1–8, Jan. 2015, doi: 10.1016/J.ADDMA.2014.10.002.
32. Edwards, G.A., Bergren, A.J., Porter, M.D., Chemically Modified Electrodes, in: *Handbook of Electrochemistry*, pp. 295–327, Elsevier Science, Amsterdam, Netherlands, Jan. 2007, doi: 10.1016/B978-044451958-0.50021-5.
33. Yu, D., Mu, S., Liu, L., Wang, W., Preparation of electroless silver plating on aramid fiber with good conductivity and adhesion strength. *Colloids Surf. A Physicochem. Eng. Asp.*, 483, 53–59, Oct. 2015, doi: 10.1016/J.COLSURFA.2015.07.021.
34. Cauchy, X., Klemberg-Sapieha, J.E., Therriault, D., Synthesis of Highly Conductive, Uniformly Silver-Coated Carbon Nanofibers by Electroless Deposition. *ACS Appl. Mater. Interfaces*, 9, 34, 29010–29020, Aug. 2017, doi: 10.1021/acsami.7b06526.
35. Yu, D., Kang, G., Tian, W., Lin, L., Wang, W., Preparation of conductive silk fabric with antibacterial properties by electroless silver plating. *Appl. Surf. Sci.*, 357, 1157–1162, Dec. 2015, doi: 10.1016/j.apsusc.2015.09.074.
36. Balela, M.D.L. and Amores, K.L.S., Electroless deposition of copper nanoparticle for antimicrobial coating. *Mater. Chem. Phys.*, 225, 393–398, Mar. 2019, doi: 10.1016/j.matchemphys.2018.12.086.

37. Liu, W., Sun, D., Ma, H., Chen, Z., Gui, C., Huang, J., A Facile Process Combined with Electroless Deposition and Hydrophobic Treatment to Fabricate Self-Cleaning Radiation Protection Suits for Pregnant Woman. *Fibers Polym.*, 23, 5, 1309–1317, May 2022, doi: 10.1007/s12221-022-4773-0.
38. Qian, L., He, D., Cao, X., Huang, J., Li, J., Robust conductive polyester fabric with enhanced multi-layer silver deposition for textile electrodes. *Colloids Surf. A Physicochem. Eng. Asp.*, 644, 128857, Jul. 2022, doi: 10.1016/J.COLSURFA.2022.128857.
39. Selvakumar, A., Perumalraj, R., Jeevananthan, P.N.R., Archana, S., Sudagar, J., Electroless NiP-MWCNT composite coating for textile industry application. *Surf. Eng.*, 32, 5, 338–343, 2016, doi: 10.1080/02670844.2015.1104102.
40. Chiu, W.T., Chen, C.Y., Chang, T.F.M., Hashimoto, T., Kurosu, H., Sone, M., Ni–P and TiO2 codeposition on silk textile *via* supercritical CO2 promoted electroless plating for flexible and wearable photocatalytic devices. *Electrochim. Acta*, 294, 68–75, Jan. 2019, doi: 10.1016/J.ELECTACTA.2018.10.076.
41. Schwarz, A., *et al.*, Gold coated para-aramid yarns through electroless deposition. *Surf. Coat. Technol.*, 204, 9–10, 1412–1418, Jan. 2010, doi: 10.1016/j.surfcoat.2009.09.038.

# 14

# Cleaner Production Approaches in Textile Finishing: Enzyme Treatments and Immobilization

**Seyda Eyupoglu[1]* and Nigar Merdan[2]**

[1]*Department of Textile, Clothing, Footwear and Leather, Vocational School of Technical Sciences, Istanbul University-Cerrahpasa, Istanbul, Türkiye*
[2]*Department of Fashion and Textile Design, Architecture and Design Faculty, Istanbul Commerce University, Istanbul, Türkiye*

***Abstract***
In recent years, the use of environmentally friendly production approaches in textile finishing has become an important issue. The enzymes used in these processes are biocatalysts consisting of metabolic products of living organisms derived from bacteria. Researchers have turned to environmentally friendly manufacturing approaches in textile wet processes and have conducted many research studies on the realization of environmentally friendly production with the use of enzymes. In this purpose, amylase in desizing, catalase in the removal of hydrogen peroxide after bleaching, laccase in the removal of indigo dyes in denim fabrics, peroxidase in the oxidation of non-covalently bound dyestuff in reactive dyes, lipase desizing in the presence of oils in spinning, pectinase in the bio-cleaning of raw cotton or flax, cellulase in the enzymatic stone washing of jeans and denim garments and in the removal of free fibril ends from the fiber surface by mechanical action. In this context, this chapter aims to inform about history of enzymes and enzymatic treatments, enzyme immobilization, structure of enzymes, enzymatic treatments of textile finishing, and ecological approaches of enzymatic treatments are used. Enzymatic treatments can be accepted as an eco-friendly and sustainable technology to alternative the traditional textile finishing process and promising for cleaner production.

***Keywords*:** Enzymatic treatments, cleaner production, textile finishing, amylase, lipase, pectinase, sustainability

**Corresponding author*: seyda.eyupoglu@iuc.edu.tr

Aminoddin Haji and Çağlar Sivri (eds.) Surface Modification and Functionalization of Textiles, (337–370) © 2026 Scrivener Publishing LLC

## 14.1 Introduction

Recently, the demand of bio-catalytic enzyme technology has increased for textile industry because it is environmentally friendly. When enzyme technology was first popularized, amylase was used to desizing of starch, and protease was utilized to anti-felting finishing of wool. In 1980s and 1990s, as a result of the rapid increase in ecological problems, enzyme technology has become more widely used. Through the application of enzymes in textile processes, both environmental impact and product quality have been greatly benefited. Of the 7,000 known enzymes, only about 75 are widely utilized in textile processes. The main enzymes used in the textile industry are oxidoreductases and hydrolases. The group of hydrolases involves amylases, cellulases, cellulases, pectinases, proteases, and lipases/esterases.

In finishing processes achieved with enzymes, the obtained effects are permanent even after many washing steps. In conventional process, textile chemicals are not removed completely on fabrics and fabric image deteriorates. In this context, enzymes are an alternative competitor to a variety of textile chemicals in modern content that are completely biodegradable and pollution-free.

Textile sector is one of the earliest sectors subsisting from many centennials. Major relevance of the sector is to design, clothing, distribute, and produce yarn. Different chemical and physical operations are needed in order to perform these. Environmental considerations of textile industry have stood out all over the world because it is producing huge number of effluents including many chemicals and toxic substances. Enzymes are accepted as the best alternating to substitute or decrease in these toxic and hazardous chemicals. Many enzymes have been used in textile industry such as cellulase, amylase, pectinase, catalase, protease, lipase, and laccase. The use of enzymatic processes is particularly promising because they have low impact on the environment, yield high-quality goods, and save energy, water, and time [1]. Because of the development of textile processing, which is environmentally friendly, different enzymes are being utilized in textile sector for minimization of impurities, bleaching, fabric fuzzy, etc. To perceive the impact of enzymes, it is needed to understand some procedures of traditional textile processing [1].

Desizing: The existence of elements like carboxymethyl cellulose, starch, and polyvinyl alcohol on yarn induces some problems, and it must be removed. The removal of substrate is called as desizing. Traditional desizing is performed with sodium hydroxide and hydrogen peroxide.

Scouring: Scouring is a process that improves the water absorbency and whiteness of fabrics by moving away non-cellulosic substances such as proteins, waxes, and pectin from the fibers. Traditionally, scouring of cellulosic fibers is carried out with detergents, soap solutions, and alkali.

Bleaching: Bleaching indicates the elimination operations of inherent coloring in the fiber. It is performed to suggest a white appearance and to ease the dyeing process. Traditionally, hydrogen peroxide and sodium hypochlorite are used in fabric bleaching to remove the raw material's natural color and impurities.

Mercerizing: The fabric is handled with a caustic soda solution throughout the mercerizing process, and, to neutralize the chemicals utilized, a final acid wash is used. Its goal is to cause the fibers to expand in order to enhance their luster, strength, ability to absorb moisture, affinity for dyes, and reactivity. Usually, it comes after the bleaching process. Deeper hues of cotton can be achieved by mercerization, which alters the internal light scattering.

These pre-finishing processes have been applied to textile materials to increase attractiveness of textile materials and gain the suitable features for its uses. Finishing and pre-finishing process causes to increase water, energy, and chemical consumption. For these reasons, ecological production methods have been investigated, and enzymatic treatments take the lead in ecological process. This chapter emphasizes the importance of various enzymes used in the textile industry at different stages in addition to traditional textile processing.

## 14.2 Basic Information About Enzymes

The term "enzyme" springs from "enzymos," a Greek word, which means that "ferments or in the cell." These biological macromolecules have high molecular weight, catalyzing chemical reaction of process, and naturally available. Enzymes are organic soluble substance, manufactured by living biological organism whose diversified amino acids combined to form lengthy polypeptide chains (-HN-R-COHN-R-CO-). The presence of ionic groups and the distinctive three-dimensional form of enzymes contribute to complexity of structure.

During the last 30 years, enzyme industry has been rapidly enhanced. Enzymatic application has been great installed in many industries that are textiles, detergent, food, animal feeds, medicines, stuffs, etc. The significant reaction of enzymes can be abstracted with five parts. i. Due to the fact that enzymes can function in extremely mild circumstances, processes involving their destruction of fiber are safer. ii. Enzymes are biological macromolecules that are biodegrade harmful chemical and reduce waste load.

iii. Because enzyme activities are specialized, quality reduction is unaffected. iv. Enzymatic applications are eco-friendly as they use low steam energy and abstain using of costly auxiliaries and chemicals. v. As a catalyst, enzymes allow a reaction to occur and then return to its original state for reuse.

Enzymes are defined as biocatalysts consisting of metabolic products of living organisms derived from bacteria. Catalysts are substances participating in biochemical or chemical reactions that remain unaltered after the reaction. All enzymes are protein form or involve a protein part. Enzymes are named by adding the suffix "-ase" at the end of the effected substate or type of catalyzed reaction. The compound that the enzyme affects is addressed as "substrate," and the number of substrate molecules per second is called "turnover number of an enzyme." Substrate and enzyme work as a team; the end product of one enzyme becomes the substrate of the next enzyme. For example, amylase enzyme transforms starch into two-chain maltose; maltase enzyme also decomposes maltose into one-chain glucose.

Enzymes have been utilized for many years in medicals, some analysis, food chemistry, textile industry, beverage industry, and household detergents. Enzymes are present in 80% of household detergents. Modern gene technology and enzyme technology will, in the future, include new types with constant production and potentially wide applications. Widespread impact of enzymes can be listed as follows [2, 3]:

- Enzymes can reduce the activation energy of biological reactions.
- Enzymes always perform one type of reaction.
- Enzymes can perform the same type of reaction over and over again without deterioration.
- Enzymes ensure that the reaction to be realized reaches equilibrium quickly.
- Enzymes work in inanimate environment.
- Enzymes always start their reactions from the outer surface of the substance that they affect.

Significant enzyme properties were given in Figure 14.1.

Enzymes have some important properties. They have the ability to initiate and accelerate chemical reactions. These properties are used to convert molecules called substrates into a specific product. Enzymes work by recognizing and interacting with specific substrates. This enables enzymes to catalyze specific reactions. Enzymes are consumed in the reactions and only improve the reaction rate. Therefore, an enzyme can catalyze the same reaction many times. All enzymes have an optimum process condition and

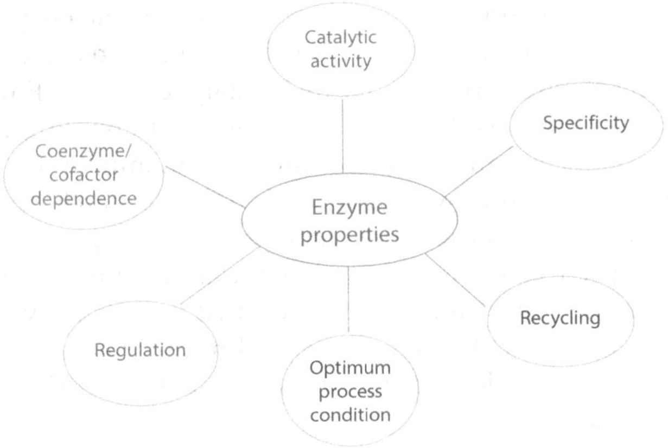


**Figure 14.1** Significant enzyme properties.

**Table 14.1** Categories of significant enzymes and their applications.

| Enzyme | Origin | Substrate | Reaction |
|---|---|---|---|
| Cellulase | *Aspergillus niger* | Cellulose | Cellulose hydrolyses |
| Amylase | *Bacillus subtilis* | Starch and amylose | Desizing |
| Pectinase | *Aspergillus niger* | Pectin | Pectin degradation |
| Protease | *Bacillius subtilis* | Protein | Degumming silk and scoring animal-based fibers |
| Lipase | Gastric juice and plant | Waxes and fats | Remove of waxes and fats |
| Laccase | *Aspergillus niger* and *penicilium* | Indigo dye | Denim finishing |
| Hemicellulose | Oruaze and Rhizopous | Hemicellulose | Hemicellulose hydrolysis |
| Catalase | *Aspergillus niger* and *Penicilium* | Peroxides | Decomposition of *in-situ* peroxide |

work most effectively under optimum conditions. Some enzymes require cofactors or coenzymes, which are small organic molecules that work together with cofactors, to carry out their catalytic activity. Enzymes have a specific binding site to their substrates. Substrate binding triggers the composition of the enzyme-substrate complex and initiates the reaction.

Around 1857, enzymes were first used to divide starch by amylase in order to desize cotton fabrics. The employment of enzymes in the textile finishing and pre-treatment process of natural fibers has become popularize. New enzymes have recently been created for use in a variety of wet processing methods for different fibers, and novel research is being conducted in this area. Table 14.1 lists the significant categories of enzymes and their applications.

## 14.3 Structure of Enzymes

Proteins of enzymes are coded by genes. Therefore, amino acid sequence of enzymes is one of a kind. Some enzymes (pepsin or urease) are formed from a protein; however, several enzymes comprise of two different parts. These parts are given in below:

- Protein Part (Apoenzyme part of the enzyme): This part specifies which substance the enzyme will act on.
- Coenzyme Part: It is an organic or inorganic molecule, often composed of phosphate, with a much smaller molecule than the protein part. The organic molecules that the enzyme needs for its activity are called "Coenzyme." This is the part that functions in the enzyme and does the main work. Some enzymes are active only when certain ions are added to the medium.

## 14.4 Operating Mechanism of Enzymes

Enzymes involve in two parts: apo-enzyme and coenzyme. The apo-enzyme part determines with which substrate the enzyme will work. There is a relationship between the apo-enzyme and the substrate. This relationship was deemed as induced-fit model by German chemist Emil Fischer. The coenzyme part functions in close proximity to the chemical bond, for example, by breaking ester bonds. It is believed that the apo-enzyme part of the enzyme binds or attaches to the substrate molecule at one or more sites (active sites)

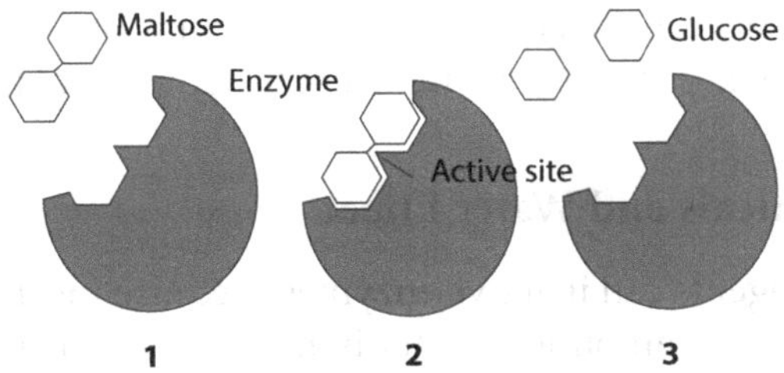


**Figure 14.2** Induced-fit model of enzyme and substrate.

(enzyme-substrate complex), whereas the coenzyme part degrades the substrate by actually combining or binding with the bonds on the substrate. Figure 14.2 indicates the induced-fit model of enzyme and substrate.

## 14.5 Effecting Factors of Enzyme Functioning

### 14.5.1 Temperature

Enzymes cannot work at high temperatures. Reaction rate of enzymes increases with reaction temperature. For example, the reaction rate of enzyme rises two times more when the temperature rises by 10°C; however, it reduces and finishes after a certain temperature. The temperature at which enzymes work with the highest efficiency is called optimum temperature.

### 14.5.2 pH

Enzymes are so sensitive to pH changes. They are commonly ineffective in very alkaline and acidic conditions. In some cases, enzymes show the highest activity at a certain pH called as optimum pH. For instance, the acidic pH 2 of the stomach is ideal for the breakdown of protein by the enzyme pepsin. Depending on the pH, various electrical charges on the protein molecule and its outer surface shape (tertiary structure) occur and provide substrate-enzyme compatibility. Perhaps this electrical charge increases the attraction between enzyme and substrate. Furthermore, strong acids and bases can coagulate enzymes.

### 14.5.3 Enzyme/Substrate Concentration

In a constant temperature and pH, a reaction rate is observed depending on the ratio between enzyme and substrate concentration. An excess of substrate or enzyme can affect this rate in different ways. Adding enzyme to a medium with plenty of substrate will raise the amount of the end product.

### 14.5.4 Chemicals and Water Effect

Many chemical agents can inactive enzymes. Because most of the enzyme's function in water, the amount of water is also a factor in enzyme function. Generally, enzymes do not function in environments containing less than 15% water.

## 14.6 Classification of Enzymes

Enzymes are generally categorized with four-digit number. For instance, in 3.6.1.3.3. "ATP phosphohydrolase," the first number gives the class, the second number the subclass, the third number the group, and the fourth number the unique sequence number.

1. Oxidoreductase: This group enzymes catalyzes redox reactions.
   a. Dehydrogenases: These enzymes affect electron gaining reactions.
   b. Oxidases: This group affects the reactions that lose electrons.
   c. Reductases: These enzymes reduce the substrate through a reductase.
   d. Transhydrogenases: These groups reduce one molecule by moving hydrogen from one molecule to another.
   e. Hydroxylases: These enzymes add a hydroxyl or water molecule to their substrate, for example, phenylalanine hydroxylase adds a hydroxyl group to phenylalanine to convert it to tyrosine.
2. Transferase Enzymes: These enzymes transfer an atom or group of atoms (methyl, carboxyl, glycosyl, amino, and phosphate groups) other than hydrogen from one molecule to another.
3. Hydrolase Enzymes: Enzymes that break down molecules by introducing a molecule of water or through a molecule of water. They act on ester, peptide, acid anhydride, and glycosidic bonds.

   a. Esterases: Enzymes that break down the ester bond (lipase, ribonuclease, phosphatase, pyrophosphatase, and glycosidase).
   b. Proteases: These enzymes degrade peptide bonds.
4. Liases: Enzymes that break molecules without removing water molecules.
5. Isomerases: These enzymes modify the molecule, changing its arrangement in space.
6. Ligases (= Synthetases): It is a group of enzymes that enable the binding of substrate molecules to each other using energy. For instance, they enable the activation of fatty and amino acids.

## 14.7 Cellulase Enzyme and Applications

Cellulase enzyme degrades glycosidic bonds in β-1 compounds into small sugars. In nature, a variety of microorganism can produce cellulase, and the term "cellulase" is used to refer to a mixture, not a single enzyme. Because the cellulase enzymes from these microorganisms are often composed of more than a dozen different cellulases. Cellulases, for short, I.U.B. (International Union of Biochemistry): 3.2.1.4 or 1,4-(1, 3; 1, 4)-β-DGlucan-4-glucanohydrolase, is the name given to the group of enzymes that act together to hydrolyze cellulose. Cellulases involve in complex mixture of "Endo-Gluconases," "Exo-Gluconases," "Sellobiohydrolases," and "β-Glucosidades." All these enzyme systems work synergetically and degrade cellulose. Endo-Gluconases hydrolyze β-glucosidic bonds, i.e., long-chain polymers, by means of static bridging bonds. This results in the formation of cellooligosaccharides, which assist the enzyme as redegradable substances. Exo-Gluconases provide to leave of non-reducing glucose unit in the last of cellulose chain. B-Glucosidades supply hydrolyzation of cellobiose to glycose. Cellulase enzyme has been investigated deeply for a long time for textile finishing process. It is clearly cognized that cellulases can hydrolyze jute, cotton, ramie, linen, and other cellulose-based fiber and guarantee a clean biopolished textile material. Furthermore, cellulases have been also used for stone and worn washed effect in denim fabrics. Cellulases are active in the temperature range from 30°C to 60°C. They are classified in accordance with pH sensitivity as acid stable (pH 4.5–5.5), neutral (pH 6.6–7), or alkaline stable (pH 9–10).

Vajpayee *et al.* extracted cellulosic fiber from lotus plant stem and produced lotus fabric from fibers. The fabric samples were treated with cellulose enzyme to enhance the fabric surface features. The optimum condition cellulase enzyme treatment was found 2% enzyme concentration, 1:20 flotte ratio, 5.5 pH, 55°C, and 60 min. Cellulase enzyme treatment causes to remove non-cellulosic impurities, increase in wettability, and change in fabric surface roughness and morphology. Due to the cellulase enzyme treatment, lotus fabric surface qualities have improved, which has improved natural coloring as well as the natural product absorption for an antibacterial effect [4]. The cellulase enzyme used in the biopolishing process is an increasingly popular process due to its applicability especially in knitted products instead of scorching. Sankarraj and Nattalhambi treated cotton fabric with immobilized cellulase. Immobilized cellulase enzyme treatment illustrated better tensile strength, lower weight loss, and high whiteness index [5]. Kumar *et al.* immobilized cellulase enzyme with epoxy resin and treated cotton fabric for biopolishing. The epoxy resin immobilized cellulase enzyme indicated maximum biopolishing activity instead of standard cellulase enzyme [6]. Saravanan *et al.* investigated the effect of cellulase enzyme on cotton fabric in the using of biopolishing process. Cellulase enzyme was applied to cotton fabric samples and weight loss was determined. Crystallinity index was not affected from cellulase enzyme treatment. Fourier transform infrared analysis shows an increased in $CH_2$ in-plane bending, -OH bending, and C-H vibrations of cellulose chains [7]. Uddin treated bleached cotton fabric with acid cellulase and neutral cellulase. Physical properties changes such as strength loss, pilling resistance, weight loss, abrasion resistance, and bending length of the cellulase enzyme treated fabrics were determined. According to the results, neutral cellulase enzyme treatment indicated better results at 55°C for 50 min. Another study is reported about biopolishing of viscose fabrics with cellulase enzyme in optimum conditions. The results show that the pilling of enzymes treated with different trademarks is the same. The most optimum treatment time was determined as 30 min. Increasing the enzyme concentration affects the effectiveness of the biopolishing process [8].

Denim stone-wash enzymes have been accepted in "denim stone wash," carried out with pumice stone, due to ecological and health concerns. Furthermore, enzymatic denim stone-wash reduces all damages compared with using pumice stone. Traditional stone washing of denim fabrics consists of amylases-mediated removal of starch sizing and abrasion of denim with pumice stone in huge washing machines. Cellulase enzyme transforms the cellulose from the denim fabric into sugar. The benefits of using cellulase instead of pumice stone involve rising productivity of the machines, less work-intensive, less damage of fiber, and environmentally safe. Biopolishing is defined as a biological process in which cellulase enzyme interacts on the surface of fabric. The enzyme molecule is too big to fit inside a cotton fiber because it is more than a thousand times bigger than a water molecule. The aim of biopolishing is removing of micro-fibrils of cellulose-based fabrics with the action of cellulase enzyme. Biopolishing provides a number of advantages, which are improved softness and water absorbency of fibers, reduced tendency for pilling formation, and cleaner surface structure. Biopolishing application of wool fabrics contains two steps. In order to whiten the wool fibers and eliminate the lipid layer that protects them, activated peroxide bleach is first applied at lower temperatures and in half the time as conventional procedures. Additionally, the peroxide treatment made the fibers more dye-receptive. The actuality of machine-washable wool is achieved in the second phase. The enzyme molecule is too big to fit inside a cotton fiber because it is more than a thousand times bigger than a water molecule [9, 10]. Cellulase enzyme has been generally used to cellulose-based fibers for biopolishing. Furthermore, the traditional stone-washing method with pumice stone has been replaced by the use of cellulase enzyme in recent years.

## 14.8 Pectinase Enzyme Treatment

The term "pectinases" is generally used for enzymes acting on the pectin main chain. Pectinases are divided into two subgroups, as hydrolases and lyases, according to their mechanism of action. In recent years, the use of various food-agricultural industry wastes in pectinase production has increased. In the production of pectinases, citrus peel, orange peel extract, apple pulp and corn flour mixture, wheat bran, pumpkin oil cake, and other agricultural where waste is used as a carbon source can be seen. *Erwinia*, *Bacillus*, *Saccharomyces*, *Kluyveromyces*, *Rhizopus*, *Aspergillus*, and *Penicillium* are the most common used microorganisms in microbial pectinase enzyme production [11].

Pectinases are a kind of enzymes that decompose pectin known as a plant cell-wall component. They involve rhamnogalacturonases, polygalacturonases, pectin lyases, and pectate lyases, which cleave different bonds [12]. Depending on how they interact with the polygalacturonate structure, pectinases can take on many forms. Methylation degree is the effective factor of the specific and enzymatic activity of pectinase enzyme. It was discovered that the chrysanthemum family was the original pectinase structure. It is the structural topology of the β-helix pectinase enzyme that differs from the β-strand turns. The active sites of pectinase enzyme are Asp201 and His223, which deal as a nucleophile and proton donator.

His223 induces proton transfer, breakage of glycosides bone, and covalent bond generation. Contemporaneously, by identifying the $H_2O$ molecule and saving the second product, the opposing residue restores the active site of the enzyme. Some of microorganisms consisted of pectinase fabrication are *Xanthomonas*, *Actinomycetes*, *Streptomycetes*, and *Pseudomonas* and fungal resources that are *Aspergillus niger*, *Fusarium oxysporum*, *Aspergillus versicolor*, *Rhizopus stolonifer*, *Aspergillus flavus*, *Penicillium jenseni*, *Mucor racemous*, *Penicillium citrinum*, *Trichoderm aviride*, and *Mucor hiemalis*. The practicableness of microbes causes to non-toxic effluents, decline of fiber damage, and energy consumption. Banik *et al.* assert the investigation about pectinase enzyme effect on the pilling of jute fibers and ability to improve jute yarn quality. Pectinases have a remarkable aptitude for degumming, retting, and breaking down complicated compounds in fiber crops [13–15].

Another study is related with pectinase enzyme treatment of hemp fiber. Pectinase enzyme treatment improves tensile strength, flexural strength, and Young's module of hemp fiber [16]. Li *et al.* investigated the effect on pectinase enzyme treatment of bamboo and eucalyptus wood fibers. The results proved that pectinase enzyme decreased the amount of pectin in fibers, resulting in development of mechanical behaviors [17]. Alix *et al.* researched the effect of pectinase enzyme on flax fiber. According to the results, pectinase enzyme treatment causes to mime retting and enhance the mechanical behaviors of flax fiber [18]. Other study reports that bio-scouring of desized cotton fabric by pectinase. After the enzymatic treatment reduces weight loss and sugar amount of cotton fabric. Furthermore, bio-scouring by pectinase enzyme application increases whiteness index, tearness, and tensile strength [19]. Colombi *et al.* searched the effect of pectinase enzyme on weight loss, pectin content, water absorption, functional groups, and whiteness index of knitted cotton fabrics. The results show that pectinase enzyme treatment induces highly cleaning of knitted cotton fabric [20]. Another study discusses the influence of pectinase enzyme pre-treatment on bamboo and coconut coir fiber reinforced composites. The fibers were treated with pectinase enzyme with ultrasonic-assisted method. Pectinase enzyme treatment improves the tensile strength of composites. Furthermore, pectinase enzyme treatment increases in interaction between matrix and fiber, moisture resistance, and bending strength [21]. Jana *et al.* investigated of sequential enzymatic treatment with xylanase, pectinase, and laccase of banana fiber. The sequential enzymatic application indicated the degumming efficiency as 22.6% weight reduction and 16.83% moisture sorption [22]. Abidin *et al.* research degumming process of ramie with pectinase enzyme treatment. Pectinase enzyme treatment improves whiteness index, tenacity, and fineness of ramie fibers [23]. Ibrahim *et al.* isolated locally acid pectinase enzyme manufacturing fungal strain from local agriculture soil and used acid pectinase enzymes in bio-scouring of gray cotton fabrics. Experimental results indicated that using of pectinase enzyme in bio-scouring of cotton fabrics in different structures develops wettability and dyeing properties and removes non-cellulosic impurities [24]. The non-cellulosic gummy materials on plant-based fibers are described as degumming, and removing of these materials is necessary before industrial applications. Pectinase enzyme treatment results in any damage to plant-based fibers with environmentally friendly and energy conservative properties [25]. Another study was reported that pectinase enzyme was used in retting process of *Mitsumata* bast fiber, classified as plant-based fiber [26]. Retting of flax, ramie, and jute fibers with pectinase enzyme causes any damages compared to cellulase enzyme treatment [27]. Bio-scouring refers the removal of undesirable non-cellulosic impurities such as fats, pectin, waxes, protein, natural colorants, water, and minerals that are soluble compounds present in the cellulosic matrix of main cell wall. Compared with alkali scouring, pectinase enzyme scouring decreases chemical and biological oxygen demand [28]. Pectinase enzyme is reported as cost effective for the processing of cotton fibers. Good quality cotton and blend fabrics can now be prepared in one step with a recently established, repeatable procedure for mill trials. Pectinase enzyme has been commonly used in scouring of cotton fiber-based fabrics without cellulose destruction. Cotton is a basic textile material, and bio-scouring is necessary, thus increasing dyeing properties, soft feel, and water absorbency of cotton [29]. Morozova *et al.* investigated the effect of pectinase enzyme treatment on wettability of cotton fabrics. After the pectinase enzyme treatment, wettability and dye uptake of cotton fabrics improved [30]. In another study, cotton fabric was scoured with acidic pectinase at different temperature, pH, and concentration. Two essential factors for achieving the best enzyme activity and bio-scouring outcomes are controlling pH and employing a non-inonic surfactant. At high liquor ratios, the efficiency of the enzymatic reaction increases. Lower enzyme concentrations and lower liquor ratios can be employed at ideal temperature conditions [31]. Alkali and thermostable pectinase application was investigated in textile materials for bio-scouring and desizing of cotton and micropoly fabrics. The optimum condition of desizing is determined as pH 9.5, 120 min, and 65°C. In addition to exhibiting improved physical qualities, fabrics using desizing and bio-scouring procedures demonstrated improved penetration and wetting properties. The use of alkaline chemicals in fabric scouring could be reduced or eliminated by enzyme scouring, which would facilitate the bleaching process and improve dye absorption. The approach presented in this paper has been demonstrated to be ecologically benign and convincingly replicable, making it a viable option for the textile sector to implement [19]. Ecological scouring of ramie fibers was carried out with pectinase enzyme. The effects of various bio-scouring factors such as buffer molarity, pH, and material-to-fluid proportion on brightness, yellowness, and whiteness were investigated. Pectinase enzyme treatment made the scouring process more eco-friendly and economically than chemical scouring [32]. In another study, it is reported that pectinase enzyme treatment is used as an ecological and sustainable process in bio-scouring of natural fibers, degumming and retting of plant-based fibers, and waste water treatments [33]. Singh *et al.* treated sisal fibers with pectinase enzyme for bio-scouring. Enzymatic treatment of sisal fibers increased the whiteness and brightness of sisal fibers and decrease in yellowness [34]. Other study was reported pectinase enzyme treatment in bio-scouring of jute fibers. An efficient reduction of impurities and development of hydrophobicity, without decreasing the tensile strength, was determined [35]. Optimum enzyme concentration is differing from pectinase enzyme treatment, but pectinase enzyme is general effective in low concentrations at 0.05%–2% range. Pectinase enzyme is also very effective in pH 5–9. The initial drawback of bio-scouring was longer treatment times; however, recent pectinase enzyme discoveries and additive use have made it viable for industrial use.

## 14.9 Amylase Enzyme Treatment

Amylases are most used enzyme in enzymatic treatments. Amylase, which is a carbonhydrase, is the first enzyme used commercially. Starch is depolymerized *via* breaking α-glycosidic bonds by amylase enzyme and water-soluble dextrin is formed. Amylase enzyme has commonly used in textile and paper industry, liquefaction of starch, bread, glucose and fructose production, and alcohol fermentation. Since the beginning of the last century, heavy irritating processes for desizing in the textile industry have been replaced by enzymatic treatment with amylase, which is known to be environmentally friendly. The characteristics of the process are suitable for the existing machine park and production speed. In addition, the process has the advantages of being economical and the availability of high temperature resistant enzymes. In textile industry, amylase enzyme has generally been used to remove sizing of cotton yarns. Starch is commonly preferred as a sizing agent in textile industry because of its natural structure and it is inexpensive and accessible.

Vigneswaran *et al.* estimated of the optimum operation parameters to perform ecological desizing of organic cotton fabrics with amylase enzyme. The results indicated that amylase enzymes are better at the temperature of 50°C–55°C and time of 30–40 min to degrade of polyvinyl alcohol starch [36]. In another study, desizing of cotton fabric was carried out with $Ca^+$ ion-independent α-amylase. The optimum desizing conditions and minimum weight loss were determined at 60°C, pH 5–6 for 30 min and 4.5 IU of amylase enzyme presence [37]. Pandey extracted flax fiber with amylase enzyme. The results show that enzymatic extraction method increases fiber extraction efficiency compared with chemical extraction methods [38]. Saha *et al.* used amylase enzyme in desizing of cotton, silk, and polystyrene fabrics. Amylase enzyme removed the starch on cotton (70%), silk (75%), and polystyrene (85%) fabric samples accomplishedly [39]. Vakar *et al.* researched the impact of lipase, amylase, and protease on the dyeing characteristics of natural dye on cotton fabric samples. It was obtained that the dye absorption capacity of samples improved [40, 41]. In another study, it has been reported that cotton and wool fabrics were treated with amylase enzyme before the natural dyeing process. Amylase enzyme treatment resulted in an improvement of dye uptake in all conditions compared with untreated samples [42]. Amylase enzyme has commonly used in textile process for desizing and surface modification of cellulosic fibers.

## 14.10 Lipase Enzyme Treatment

Lipases (triacylglycerol/triglyceride hydrolase, E.C. 3.1.1.3) are hydrolase group enzymes that catalyze the hydrolysis of glycerides to glycerol and free fatty acids. Besides their hydrolytic activity on triglycerides, lipases have the ability to catalyze reactions such as esterification, interesterification, acidolysis, alcoholysis, and aminolysis. In nature, lipases are founded commonly, and these group enzymes can be produced by a variety of plant, animal, and microorganism. Bacteria, fungi, ferment, and microbial-based lipases are preferred in organic biotechnological and chemistry applications. The talent of lipases to archived very specific chemical transformations makes them increasingly popular in the detergent, food, chemical, and pharmaceutical industries. In textile industry, lipases have been commonly used to remove sizing in cotton fibers. Furthermore, lipase enzyme has been generally used in improvement of wet process of polyester fibers. Lipase enzyme was treated polyester fabric samples to improve moisture regain. Lipase enzyme treatment improves moisture regains of polyester fabric by two times compared with untreated polyester fabric [43]. Kim and Song aimed to determine optimum conditions of increasing the hydrophilicity of polyester fabrics. The lipase hydrolytic activity, wettability, and moisture regain of PET fabrics were examined at different temperature, pH, concentration, and reaction time. Because lipase hydrolysis produced hydrophilic groups on the surface of polyester fabrics, lipase treatment proved an effective way to increase the moisture recovery and wettability of polyester fabrics [43]. Lipase enzyme hydrolyzes ester bonds in polyesters fibers and generates polar carboxylic and hydroxyl groups. Kumar and Kumar purposed to investigate and identify the effect of lipase enzyme treatment on surface modification and weight loss of polyester fabrics. Studies are conducted on the functional groups that exist both pre- and post-treatment, as well as the impact of enzyme treatment on the enhancement of dye absorption. The results indicates that enzymatic treatment forms less weight loss, surface damage, and improved dye uptake, shear properties, and moisture regain [44]. In another study, lipase enzyme was used to remove oil dirty from linen fabrics. Lipase enzyme provided to remove oil dirty in linen fabrics without any fabric damages compared with alkali and organic solvents [45]. Tegegne and Haile treated cotton/polyester blend fabric with lipase enzyme to improve wettability and water regain capacity of samples. The ideal lipase enzyme treatment conditions were obtained to be 30°C temperature, 50 min of exposure time, and 14% lipase concentration [46]. In other study, the effect of lipase enzyme treatment on wool fabric dyeability with reactive dyes is researched. Lipase enzyme treatment was improved the dyeability capacity of wool fabrics [47].

## 14.11 Protease Enzyme Treatment

Protease enzymes are hydrolase class enzymes that catalyze certain peptide bonds in the protein molecule. Protease enzymes are used in the finishing of wool fabrics to perform partial hydrolysis on the fiber surface and reduce the difference in frictional resistance in the root-end direction. Furthermore, protease enzymes have been used to remove sericin of silk fabrics. Parvinzadeh treated wool yarn with protease enzyme to improve some mechanical and physical behaviors such as dyeing affinity, smoothness, drapeability, and water absorbency. Wool yarn was treated with protease enzymes for 60 min at the concentration of 1%, 2%, 4%, and 6%. After the enzymatic treatment, the dyeing process was performed with 50% on weight of fabric (o.w.f.) madder. Protease enzyme treatment causes to improve light fastness of samples [48]. In another study, protease enzyme immobilized and treated wool tops and fabrics to improve whiteness, felting, and dyeability. Protease enzyme treatment causes to enhance dye-uptake of wool fibers [49]. Pooja and Fatima treated wool fabric with different concentration rate pectinase enzyme and observed the differences such as hand of fabric, surface structure, weight loss, absorbency, tensile strength, pilling, dyeability, color fastness, and fabric drape. After the enzymatic treatment, wool fabrics showed slight weight loss, improvement in softness, improved absorbency, smoother surface scale, and drapability. Protease enzyme treatment provided development in the colorfastness and physical properties of wool fabric [50]. Another study reported alkaline and acid enzyme treatment to wool fabric to enhance shrink resistance and handle. The acid protease enzyme treated wool fabrics indicates higher softness, lower weight, and strength losses than alkali protease enzyme. Furthermore, peroxide bleaching before acid protease enzyme treatment causes to improve whiteness index and shrinkage resistance [51]. In literature, it is reported that enzymes are used to develop eco-friendly process by decreasing in the concentration of chemical agents, energy, and water consumption. Majsov treated wool fabric with protease enzyme to observe the changes on softness, pilling resistance, absorbency, tensile strength loss, water retention, weight loss, felting shrinkage, alkali, and urea-bisulfite solubility. Protease-treated wool fabrics induced improvement in absorbency, softness, felting, and resistance shrinkage and a slight increase in tensile strength loss, weight loss, alkali, and urea-bisulfite solubility [52]. Protease enzyme treatment causes to increase 11.1% dye uptake at low dyeing temperature. After protease enzyme treatment, the dyeing effect of one-bath method dyed wool fabric was much the same that with the two-bath method and high boiling dyeing process. The results remark that protease treatment recommends and substitute of a one-bath dyeing operations, replacing conventional dyeing process and energy- and water-consuming two-bath method [53]. Recycling of wool/polyester blended fabric was investigated with protease enzyme treatment. Protease enzyme decomposed wool fibers in wool/polyester fibers blend, whereas polyester fibers do not effect. Recovery of polyester fibers from wool/polyester blend was obtained with pectinase enzyme treatment [54]. Another study aimed to reduce dyeing temperatures of polyamide 6/elastane knitted fabrics with protease enzyme treatment. Protease enzyme hydrolyzes amide bonds and causes to create new carboxyl and amine groups in fiber macromolecules that result in increasing functional group content that anionic dyes can attach [55]. Protease enzyme has commonly used to develop dye uptake, shrinkage, drapability, absorbency, pilling resistance, and felting of wool fiber-based materials.

## 14.12 Catalase Enzyme Treatment

Catalase [E.C. 1.11.1.6] is an oxidoreductase enzyme that catalyzes the reaction that converts hydrogen peroxide into molecular water and oxygen. Catalase enzyme has been purified and characterized from many plants, which are pumpkin, lentil, cucumber, and cotton. Because catalase enzyme is an enzyme that catalyzes the decomposed of hydrogen peroxide, it is used in all processes where hydrogen peroxide is used and its excess is required to be removed from the environment, especially in areas such as textile and paper industry. In textile industry, catalase enzyme is commonly used to neutralize after the peroxide bleaching. The use of catalase enzyme shortens the process time in peroxide bleaching and saves water and energy due to not required rinsing. A catalase enzyme was isolated from *Bacillus* SF to investigation of potential usage. Results revealed that catalase enzyme was stable to high temperatures and pH values; however, it is very sensitive to hydrogen peroxide. Catalase enzyme treatment causes to decrease in color differences in dyeing samples [56]. Another study reported that catalyze enzyme provides to reuse of the residual water from washing process of peroxide bleached fabrics [57]. Tasmili *et al.* used catalase enzyme to remove dyes and pigments from the waste water of textile dyeing process. The experimental results showed that catalase enzyme can be apply successfully of textile wastewater including reactive dyes [58]. Catalase enzyme can be used to decompose of hydrogen peroxide [59]. After bleaching process, any peroxide residual on the fabric can prevent the dye from binding to the fabric. For this reason, peroxide removal is required. In literature, catalase enzyme has been reported as remover and decomposer of hydrogen peroxide after bleaching [60].

## 14.13 Laccase Enzyme Treatment

The enzyme laccase, also known as benzenediol:oxygen oxidoreductase, EC 1.10.3.2, is a copper-containing enzyme that catalyzes the oxidation of various substances. Laccase enzyme catalyzes the reaction of hydrogen atom removal from hydroxyl groups by removing one electron from mono- and poly-phenolic substances and aromatic amines, and free radicals are formed as a result of this reaction. Many industrial uses have been proposed for laccase enzyme treatment. Besides laccase enzyme is commonly preferred in decolorization of textile wastewater, bleaching of textile materials, denim washing, and synthesis of dyes [61]. Avinc *et al.* investigated the effect of ultrasound on laccase enzyme treatment of cotton fabrics.

**Table 14.2** Some other enzymatic treatments.

| Fiber | Enzyme | Trading name | Process conditions | Effected fiber features | References |
|---|---|---|---|---|---|
| Flax | Laccase | - | 60°C, pH 5 | - Fiber chemical composition<br>- Functional groups | [69] |
| Hemp | Laccase | - | 60°C, pH 5 | - Fiber chemical composition<br>- Functional groups | [69] |
| Banana | Pectinase | - | 2 hours, 100:100 enzyme ratio | - Moisture content<br>- Mechanical properties<br>- Chemical characterization<br>- Surface morphology | [70] |
| Banana | Laccase | - | 2 hours, 100:100 enzyme ratio | - Moisture content<br>- Mechanical properties<br>- Chemical characterization<br>- Surface morphology | [70] |
| Banana | Pectinase + Laccase | - | 2 hours, 25:75 enzyme ratio | - Moisture content<br>- Mechanical properties<br>- Chemical characterization<br>- Surface morphology | [70] |
| Banana | Pectinase + Laccase | - | 2 hours, 50:50 enzyme ratio | - Moisture content<br>- Mechanical properties<br>- Chemical characterization<br>- Surface morphology | [70] |

*(Continued)*

**Table 14.2** Some other enzymatic treatments. (*Continued*)

| Fiber | Enzyme | Trading name | Process conditions | Effected fiber features | References |
|---|---|---|---|---|---|
| Agava americana L. | Laccase | Denolite II S (Novozymes) | 55°C, pH 5.5, 1 hour, 1:50 flotte ratio | -Hydrophilic properties<br>-Whiteness index<br>-Tensile strength | [71] |
| Flax | Cellulase | | 45°C, pH 4.5, 3 hour, 1:15 flotte ratio | - Color strength<br>- Spectrophotometric properties | [72] |
| Palm | Xylanase | -Talaromyces thermophilus<br>-Bichwood | pH 7, 50°C | - Surface characteristics<br>- Mechanical properties<br>- Reducing sugar analysis<br>- Moisture absorption<br>- Fiber chemical composition | [73] |
| Palm | Pectinase | Bichwood | pH 5, 50°C | - Surface characteristics<br>- Mechanical properties<br>- Reducing sugar analysis<br>- Moisture absorption<br>- Fiber chemical composition | [73] |
| Palm | Xylanase + pectinase | - | pH 5, 50°C | - Surface characteristics<br>- Mechanical properties<br>- Reducing sugar analysis<br>- Moisture absorption<br>- Fiber chemical composition | [73] |

(*Continued*)

**Table 14.2** Some other enzymatic treatments. (*Continued*)

| Fiber | Enzyme | Trading name | Process conditions | Effected fiber features | References |
|---|---|---|---|---|---|
| Hemp | Laccase | - | 40°C, 20–40 min (conventional method), 1–3 min (microwave method) | - Tensile strength | [74] |
| Flax | 1,4-β-xylanase | Feedlyve AXC 1500 L | pH 5, 50°C | - Wetting properties<br>- Tensile properties<br>- Functional groups<br>- Thermogravimetric analysis<br>- Surface characterization | [75] |
| Flax | Pectinase | Peclyve EXG | pH 5.5, 50°C | - Wetting properties<br>- Tensile properties<br>- Functional groups<br>- Thermogravimetric analysis<br>- Surface characterization | [75] |
| Flax | Xyloglucanase | - | pH 5, 50°C | - Wetting properties<br>- Tensile properties<br>- Functional groups<br>- Thermogravimetric analysis<br>- Surface characterization | [75] |

(*Continued*)

**Table 14.2** Some other enzymatic treatments. (*Continued*)

| Fiber | Enzyme | Trading name | Process conditions | Effected fiber features | References |
|---|---|---|---|---|---|
| Flax | Laccase | | pH 6.5, 45°C, 12 hours | - Chemical composition<br>- Physical properties<br>- Functional groups<br>- Crystallinity index | [76] |
| Flax | Laccase | EC 1.10.3.2 | - | - Color strength<br>- Color resistance<br>- Functional groups | [78] |
| Poly(ethyleneterephthalate) | Lipase, cutinase | - | 90°C, 10 min, 1:30 flotte ratio | - Moisture regain<br>- Surface wettability<br>Tensile strength | [79] |
| Banana | Cellulase | - | 55°C, 60 min, 1:20 flotte ratio, pH 5.5 | - Surface characteristic<br>- Functional groups<br>- Wet out time<br>- Bulk properties<br>- Water contact angle<br>- Hydrophilicity<br>- Tensile strength | [80] |
| *Sambucus ebulus* L. | Cellulase | Cellusoft L – Novozyme | 55°C, pH 5.5, 30 min | - Surface properties<br>- Elemental component<br>- Functional groups<br>- Thermal behaviors<br>- Crystallinity index<br>- Tensile strength | [81] |

(*Continued*)

**Table 14.2** Some other enzymatic treatments. (*Continued*)

| Fiber | Enzyme | Trading name | Process conditions | Effected fiber features | References |
|---|---|---|---|---|---|
| *Sambucus ebulus* L. | Pectinase | Scourzym L – Novozyme | 55°C, pH 5.5, 30 min | - Surface properties<br>- Elemental component<br>- Functional groups<br>- Thermal behaviors<br>- Crystallinity index<br>- Tensile strength | [81] |
| Cotton | Laccase | EC 1.10.3.2 | 50°C, pH 5 | - Reducing sugar<br>-Color strength | [82] |
| Lyocell | Cellulose | EC number 3.2.1.4 | 60°C, 60 min | - Physical and mechanical properties<br>- Crystallinity index<br>- Surface morphology and wettability | [33] |
| Cotton | Cellulose | EC number 3.2.1.4 | 60°C, 60 min | - Physical and mechanical properties<br>- Crystallinity index<br>- Surface morphology and wettability | [33] |

Enzymatic treatment was performed to cotton fabrics with ultrasonic energy. Whiteness, weight loss, tensile strength, fabric thickness, surface characteristic, and functional groups of fabric samples were investigated. The results illustrated that laccase enzyme treatment had no important damage of the cotton fiber [62]. In industry, laccases are known as responsible of biological degradation of lignin and drew much attention among scientists in last decade, owing to their wide-spread stability and applicability [63]. Laccase enzyme can use to degrade a variety of compounds involving phenolic and non-phenolic compounds [64]. Kalme *et al.* examined degradation of three different textile dyes [65]. In another study, laccase enzyme has been used as an alternative of harmful mordants in fixation, which are morin, rutin, and quercetin, in printing of cotton fabrics [66]. Laccase enzyme demonstrated considerable detoxification in tests involving the germination of rice seeds and could actually remove dyes from textile effluent. In conclusion, laccase may prove helpful in the future for wastewater decolorization and detoxification [67]. In other study, laccase-catalyzed coloration was carried out wool and polyethyleneimine. It is known that the development of polymeric color during laccase catalyzation of gallic acid and catechin is influenced by amino groups found in wool and polyethyleneimine. When wool is enzymatically colored, shrink-resistance and color depth can be greatly enhanced by adding a crosslinking agent, like glycerol diglycidyl ether, and a multi-primary amine molecule, like polyethyleneimine [68]. Some other enzymatic treatments were given in Table 14.2.

## 14.14 Enzyme Immobilization

Enzyme immobilization is the operation to localize or confine the enzyme onto the carrier while retaining its catalytic activity. Immobilization was used in the first time to produce of pure amino acids and hydrolyze of penicillin G. After, a variety of investigation has been made for the immobilization of distinctive enzymes. When compared to soluble enzymes, immobilized enzymes are used for a number of reasons, including their easier operation and improved operational stability. The immobilized enzymes have been used in different aims, which are treatments of pollutants, continuous analysis, and production of useful compounds. Enzyme immobilization history can be examined in three phases. The initial phase dates back 19th century when microorganisms that had been immobilized were utilized in industry to produce vinegar and create trickling filters that would clarify waste water. The second phase—when the foundation of immobilization technology was developed and the article of enzyme immobilization was greatly enhanced—occurred in 1960s. Until the 1970s, only single enzyme immobilization system was utilized. After 1970s, more complicate systems were promoted involving two-enzyme reaction with living cells and co-factor regeneration [33].

Enzymes are protein molecules that accelerate chemical reaction in living cells. Enzymes do not modify after reactions. Because of that, identical enzyme can be utilized a bit more; however, the existence of the enzyme in solution with reactants and/or products makes it difficult to separate the enzyme from the solution. In such cases, the products are removed from the environment, and the enzyme becomes re-usable by attaching the enzyme to a solid. Enzyme immobilization is a technique that prevents enzyme movements. Enzymes that are attached to a solid and enable the conversion of substrates passing over it into products are called immobilized enzymes. Enzyme immobilization is a process in which the movement of the enzyme is significantly restricted while its catalytic activity continues.

## 14.15 Advantages of Enzyme Immobilization

The enzyme attached to a solid by immobilization has many advantages over an enzyme in solution:

- The enzyme can be utilized many times and this reduces the cost.
- As a result of the removal of the enzyme from the environment, the reaction can be stopped quickly.
- There are any enzyme residues in the resulting products.
- The stability of the enzyme increases.
- Separation of the enzyme from the environment is facilitated.
- It can be operated in continuous system.
- It provides products simply.
- Waste liquid quantity diminished.
- Enzymatic activation can increase in some cases.
- Half-life of enzyme can extend [33].

Enzyme immobilization has many advantages; however, some reasons limit the using of enzyme immobilization in industrial applications. These disadvantages can be listed in below.

- Lower cost of soluble enzymes
- Unwillingness to change traditional methods

- The need for new investments to replace conventional processes
- Cost of the carrier to be used in the immobilization process
- Performance of the system

## 14.16 Enzyme Immobilization Methods

Enzyme immobilization is different for every enzyme. There is no general method recognized for each enzyme individually. When selecting enzyme immobilization method, chemical structure of enzyme, enzyme ingredients, properties of substrates and products, and usage areas of products should be considered. In addition, in the immobilization process, it is very significant to choose a method that will not change the active groups and chemical structure of the enzyme in the binding site and will not lead loss of activity in the enzyme. For this reason, reaction with the groups in the binding site of the enzyme should be avoided. In the immobilization process, the active site of the enzyme is protected by protective groups during binding. After binding, the protective groups are removed without causing loss of activity in the enzyme. In some cases, this protective effect can be provided by substrate or competing type inhibitor.

Immobilization process can be classified as follows;

1. Adsorption
2. Covalent binding
3. Affinity immobilization
4. Entrapment

Adsorption Method: Adsorption method is the most using enzyme immobilization technique. The mechanism of adsorption is based on weak intermolecular forces such as hydrophilic interactions, electrostatic interactions, and Van der Waal's forces. The enzyme is dissolved in a solution, and under ideal circumstances, the solid support is in contact with the enzyme solution for a predetermined amount of time to maintain the enzyme activity. The un-absorbed enzymes are taken away by washing from the surface. Because there is no functionalization of the support involved, immobilization by adsorption is a straightforward and affordable procedure that requires no reagents, is inexpensive, and typically has no negative effects on the activity of the enzyme [83].

Covalent Bonding Enzyme Immobilization: Covalent bonding enzyme immobilization is one of the commonly used techniques. This method is based on covalent binding of the enzyme to a water-insoluble carrier.

Usually, the binding takes place between the nucleophilic group of the enzyme and the functional group of the carrier. Functional groups involved in attachment groups: amino group ($NH_2$), carboxyl group ($CO_2H$), sulfhydryl group (SH), hydroxyl group (OH), imidazole group, phenolic group, thiol group, threonine group, and indole group [84].

Affinity Immobilization: Affinity immobilization takes advantage of an enzyme selectivity to sustain it in many physiological contexts. There are two methods to accomplish this: either the target enzyme affinity ligand is pre-coupled to the matrix, or the enzyme is conjugated to an entity that becomes affine to the matrix. Enzyme purification has also been accomplished simultaneously with affinity adsorbents. Complex affinity supports such as agarose-linked multilayered concanavalin and alkali stable chitosan-coated porous silica beads. Higher enzyme concentrations in a harbor contribute to greater stability and effectiveness) [84].

Entrapment Immobilization Method: In immobilization by entrapment method, the enzyme is held in a polymer matrix or membrane acting as a cage. This entrapment occurs in such a way that the substrate is allowed to enter while the protein is retained inside. The difference of this method from covalent binding is that the enzyme is free in solution, but its movement is restricted by being held in the gel matrix or membrane. For this reason, the entrapment method has gained wide applicability [84].

**Table 14.3** Some enzyme immobilization materials.

| **Immobilization supporter materials** | | |
|---|---|---|
| **Natural polymers** | **Synthetic polymers** | **Inorganic polymers** |
| - Alginate | - Amberlite | - Zeolites |
| - Chitin and chitosan | - DEAE cellulose | - Ceramics |
| - Collagen | - Glutaraldehyde | - Celite |
| - Carrageenan | - Polyethylene glycol | - Silica |
| - Gelatin | | - Glass |
| - Cellulose | | - Activated carbon |
| - Starch | | - Charcoal |
| - Pectin | | |
| - Sepharose | | |

In enzyme immobilization, many materials have been used for supports of immobilization process. These materials were given in Table 14.3 [84].

## 14.17 Conclusion

The present chapter discussed significant brief information about the use of enzymes in textile wet process to struggle an enormous variety of chemical-based industrial pollutants. Enzymes have an important operational inclination to efficiency of bioremediate. During the past many years, a variety of major efforts and steps have been done in that direction, although using several methodologies. In this respect, the improvement of enzyme systems, which are clean in nature, is offered as a potential instrument for many biotechnological and industrial applications for textile materials. The development and application of immobilized green catalysts, health concerns about hazardous compounds, non-toxicity and mild reaction conditions, and the long-term pursuit of environmentally friendly processes will probably continue to be the focus of intensive research in the future. In textile wet process, concerning the environmental pollution, it is hope that enzymes replace conventional chemicals. Enzymes have gained particular recognition in the textile processing industry for their capacity to substitute harsh traditional chemicals, save water and energy usage, and enhance aesthetic quality.

**AI disclosure statement:** The author confirms that no artificial intelligence (AI)-generated content or AI-assisted writing tools were used in the preparation of this chapter (including summaries and the abstract). All text, analysis, and conclusions are the result of the author's original work.

## References

1. Kumar, D., Bhardwaj, R., Jassal, S., Goyal, T., Khullar, A., Gupta, N., Application of enzymes for an eco-friendly approach to textile processing. *E.S.P.R.*, 1–11, 2021.
2. Stöhr, R., Enzymes-biocatalysts in textile finishing. Melliand Textilberichte International Textile Reports, 76, 1010–1015, 1995.
3. Hemmpel, W.H., The surface modification of woven and knitted cellulose fiber fabrics by enzymatic degradation. *Int. Text. Bull. Dye. Print. Finish.*, 37, 3, 5–11, 1992.

4. Vajpayee, M., Dave, H., Singh, M., Ledwani, L., Cellulase Enzyme Based Wet-Pretreatment of Lotus Fabric to Improve Antimicrobial Finishing with A. indica Extract and Enhance Natural Dyeing: Sustainable Approach for Textile Finishing. *ChemistrySelect*, 7, 25, e202200382, 2022.
5. Sankarraj, N. and Nallathambi, G., Enzymatic biopolishing of cotton fabric with free/immobilized cellulase. *Carbohydr. Polym.*, 191, 95–102, 2018.
6. Kumar, V.S., Meenakshisundaram, S., Selvakumar, N., Conservation of cellulase enzyme in biopolishing application of cotton fabrics. *J. Text. Inst.*, 99, 4, 339–346, 2008.
7. Saravanan, D., Lakshmi, S.S., Raja, K.S., Vasanthi, N.S., Biopolishing of cotton fabric with fungal cellulase and its effect on the morphology of cotton fibres. *Indian J. Fibre Text. Res.*, 38, 2, 156–160, 2013.
8. Bahtiyari, M.I., Duran, K., Körlü, A.E., Perincek, S., Usage of commercial cellulases in biopolishing of viscose fabrics. *Text. Apparel*, 20, 1, 57–64, 2010.
9. Sukumaran, R.K., Singhania, R.R., Pandey, A., Microbial cellulases-production, applications and challenges. *JSIR*, 64, 832–844, 2005.
10. Mojsov, K., Biopolishing enzymes and their applications in textiles: A review. *Tekst. Ind.*, 61, 2, 20–24, 2014.
11. Uzuner, S. and Çekmecelioğlu, D., Gıda atiklarinin pektinaz enzimi üretiminde kullanımı. *Gıda*, 41, 4, 259–266, 2016.
12. Rehman, H., Baloch, A.H., Nawaz, M.A., Pectinase: immobilization and applications. A review. *Trends Pept. Protein Sci.*, 6, 1–16, 2021.
13. Chares Subash, M. and Muthiah, P., Eco-friendly degumming of natural fibers for textile applications: a comprehensive review. *Clean. Eng. Technol.*, 5, 100304, 2021.
14. Banik, S. and Ghosh, S.N., Pectinolytic activity of microorganisms in piling of jute. *Indian J. Fibre Text. Res.*, 33, 151–156, 2008.
15. Chiliveri, S.R., Koti, S., Linga, V.R., Retting and degumming of natural fibers by pectinolytic enzymes produced from Bacillus tequilensis SV11-UV37 using solid state fermentation. *Springerplus*, 5, 1–17, 2016.
16. Saleem, Z., Rennebaum, H., Pudel, F., Grimm, E., Treating bast fibres with pectinase improves mechanical characteristics of reinforced thermoplastic composites. *Compos. Sci. Technol.*, 68, 2, 471–476, 2008.
17. Li, X., Xiao, R., Morrell, J.J., Wu, Z., Du, G., Wang, S., Zou, C., Cappellazzi, J., Improving the performance of bamboo and eucalyptus wood fiber/polypropylene composites using pectinase pre-treatments. *J. Wood Chem. Technol.*, 38, 1, 44–50, 2018.
18. Alix, S., Lebrun, L., Marais, S., Philippe, E., Bourmaud, A., Baley, C., Morvan, C., Pectinase treatments on technical fibres of flax: effects on water sorption and mechanical properties. *Carbohydr. Polym.*, 87, 1, 177–185, 2012.
19. Ahlawat, S., Dhiman, S.S., Battan, B., Mandhan, R.P., Sharma, J., Pectinase production by Bacillus subtilis and its potential application in biopreparation of cotton and micropoly fabric. *Process Biochem.*, 44, 5, 521–526, 2009.

20. Colombi, B.L., Valle, R., Valle, J., Andreaus, J., Advances in sustainable enzymatic scouring of cotton textiles: evaluation of different post-treatments to improve fabric wettability. *Cleaner Eng. Technol.*, 4, 100160, 2021.
21. Yang, X., Wang, M., Li, Y., Wang, W., Liu, Y., Improvement of mechanical and thermal expansion properties of pectinase treatment plant fiber reinforced starch/PBAT composites. *Polym. Compos.*, 2024.
22. Jana, T., Mondal, S., Pal, K., Mondal, K., Patra, S., Halder, S.K., Mondal, K.C., Quality improvement of banana fiber through sequential enzymatic treatment. *SMBA*, 1–14, 2024.
23. Abidin, K.Y., Nurhayati, N., Nandyawati, D., Sabbathini, G.C., Enzymatic degumming using xylanase and pectinase to improve brightness and fineness quality of ramie fiber (Boehmeria nivea l.) as textile raw material. *JBBI*, 10, 1, 76–85, 2023.
24. Ibrahim, N.A., Eid, B.M., Abdel Aziz, M.S., Hamdy, S.M., AbdAllah, S.E., Environmentally benign scouring of cotton knits using locally produced acid pectinase enzyme. *Fibers Polym.*, 20, 787–793, 2019.
25. Mojsov, K., Biotechnological applications of pectinases in textile processing and bioscouring of cotton fibers, in: *II International Conference Industrial Engineering and Environmental Protection 2012 (IIZS 2012), Proceedings*, Zrenjanin, Serbia, 31 October, 2012, pp. 314–322, 2012.
26. Yamamoto, R., Buschle-Diller, G., Takagishi, T., Eco-friendly processing of cotton: application to industrial manufacturing, in: *Proceedings of the American Chemical Society National Meeting*, 2001.
27. Patra, A.K., Chakraborty, J.N., Madhu, A., Studies on enzymatic pretreatment of linen. *I.J.F.T.R.*, 35, 4, 337–341, 2010.
28. Hebeish, A., Ramadan, M.A., Hashem, M., Sadek, B., Abdel-Hady, M., New development for combined bioscouring and bleaching of cotton-based fabrics. *Res. J. Text. Apparel*, 17, 1, 94–103, 2013.
29. Kohli, P. and Gupta, R., Alkaline pectinases: a review. *I.S.B.A.B.*, 4, 3, 279–285, 2015.
30. Morozova, V.V., Semenova, M.V., Salanovich, T.N., Okunev, O.N., Koshelev, A.V., Bubnova, T.V., Krichevskii, G.E., Timatkov, A.G., Barysheva, N.V., Sinitsyn, A.P., Application of neutral-alkaline oectate lyases to cotton fabric boil off. *Appl. Biochem. Microbiol.*, 42, 603–608, 2006.
31. Calafell, M. and Garriga, P., Effect of some process parameters in the enzymatic scouring of cotton using an acid pectinase. *Enzyme Microb. Technol.*, 34, 3–4, 326–331, 2004.
32. Singh, A., Varghese, L.M., Battan, B., Patra, A.K., Mandhan, R.P., Mahajan, R., Eco-friendly scouring of ramie fibers using crude xylano-pectinolytic enzymes for textile purpose. *ESPR*, 27, 6701–6710, 2020.
33. Rahman, M.A., Yun, C., Park, C.H., Development of a superhydrophobic cellulose fabric *via* enzyme treatment and surface hydrophobization. *Text. Res. J.*, 91, 1–2, 40–50, 2021.

34. Singh, A., Varghese, L.M., Battan, B., Patra, A.K., Mandhan, R.P., Mahajan, R., Environmental pollution reducing strategy for scouring of undegummed sisal fibers using xylanase and pectinase enzymes. *Bioprocess. Biosyst. Eng.*, 44, 607–615, 2021.
35. Stanescu, M.D., Applications of enzymes in processing cellulosic textiles—A review of the latest developments. *Cellul. Chem. Technol.*, 57, 1–15, 2023.
36. Vigneswaran, C., Anbumani, N., Ananthasubramanian, M., Rajendran, R., Prediction of optimum process parameters to achieve eco-friendly desizing of organic cotton fabrics with indigenously produced alpha-amylase and their enzyme kinetics. *J. Text. Inst.*, 103, 4, 422–433, 2012.
37. Chand, N., Nateri, A.S., Sajedi, R.H., Mahdavi, A., Rassa, M., Enzymatic desizing of cotton fabric using a Ca2+-independent α-amylase with acidic pH profile. *J. Mol. Catal. B-Enzym.*, 83, 46–50, 2012.
38. Pandey, R., Fiber extraction from dual-purpose flax. *J. Nat. Fibers*, 13, 5, 565–577, 2016.
39. Saha, P., Khan, M.F., Patra, S., Truncated α-amylase: an improved candidate for textile processing. *Prep. Biochem. Biotechnol.*, 48, 7, 635–645, 2018.
40. Vankar, P.S., Shanker, R., Mahanta, D., Tiwari, S.C., Ecofriendly sonicator dyeing of cotton with Rubia cordifolia Linn. using biomordant. *Dyes Pigm.*, 76, 1, 207–212, 2008.
41. Zhang, Y., Rather, L.J., Li, Q., Recent advances in the surface modification strategies to improve functional finishing of cotton with natural colourants-A review. *J. Cleaner Prod.*, 335, 130313, 2022.
42. Tsatsaroni, E. and Liakopoulou-Kyriakides, M., Effect of enzymatic treatment on the dyeing of cotton and wool fibres with natural dyes. *Dyes Pigm.*, 29, 3, 203–209, 1995.
43. Kim, H.R. and Song, W.S., Lipase treatment of polyester fabrics. *Fibers Polym.*, 7, 339–343, 2006.
44. Kumar, J.A. and Senthil Kumar, M., A study on improving dyeability of polyester fabric using lipase enzyme. *A. U. T.*, 20, 3, 243–249, 2020.
45. Ahmed, H.E., Gremos, S.S., Kolisis, F.N., Enzymatic removal of the oily dirt from a Coptic tunic using the enzyme lipase. *J. Text. Appar. Technol. Manag.*, 6, 3, 2010.
46. Tegegne, W. and Haile, A., Improving hydrophilicity and comfort characteristics of polyester/cotton blend fabric through lipase enzyme treatment. *Clean Technol. Environ. Policy*, 1–14, 2024.
47. Kantouch, A., Raslan, W.M., El-Sayed, H., Effect of lipase pretreatment on the dyeability of wool fabric. *J. Nat. Fibers*, 2, 2, 35–48, 2005.
48. Parvinzadeh, M., Effect of proteolytic enzyme on dyeing of wool with madder. *Enzyme Microb. Technol.*, 40, 7, 1719–1722, 2007.
49. Schumacher, K., Heine, E., Höcker, H., Extremozymes for improving wool properties. *J. Biotechnol.*, 89, 2–3, 281–288, 2001.

50. Pooja, E.S. and Fatima, N., Quality improvement of wool fabric using protease enzyme. *Environ. Ecol. Res.*, 2, 8, 301–310, 2014.
51. Raja, A.S.M. and Thilagavathi, G., Comparative study on the effect of acid and alkaline protease enzyme treatments on wool for improving handle and shrink resistance. *J. Text. Inst.*, 101, 9, 823–834, 2010.
52. Mojsov, K., Enzymatic treatment of wool fabrics-opportunity of the improvement on some physical and chemical properties of the fabrics. *J. Text. Inst.*, 108, 7, 1136–1143, 2017.
53. Cui, L., Yu, Y., Fan, X., Wang, P., Wang, Q., Effect of protease treatment on dyeing properties of wool fabrics for single bath. *Eng. Life Sci.*, 9, 2, 135–139, 2009.
54. Navone, L., Moffitt, K., Hansen, K.A., Blinco, J., Payne, A., Speight, R., Closing the textile loop: Enzymatic fibre separation and recycling of wool/polyester fabric blends. *Waste Manage. Res.*, 102, 149–160, 2020.
55. Atav, R., Soysal, A.B., Çağman, F.N., Karadağ, N.G., Determining the effect of enzymatic pretreatment on the low temperature dyeability of polyamide fabrics. *Color. Technol.*, 139, 1, 109–116, 2023.
56. Fruhwirth, G., Paar, A., Gudelj, M., Cavaco-Paulo, A., Robra, K.H., Gübitz, G., An immobilised catalase peroxidase from the alkalothermophilic Bacillus SF for the treatment of textile-bleaching effluents. *Appl. Microbiol. Biotechnol.*, 60, 313–319, 2002.
57. Costa, S.A., Tzanov, T., Carneiro, F., Gübitz, G.M., Cavaco-Paulo, A., Recycling of textile bleaching effluents for dyeing using immobilized catalase. *Biotechnol. Lett*, 24, 173–176, 2002.
58. Taslimi, P., Nadaroglu, H., Adıguzel, A., Removal of some textile dyes from aqueous solution by using a catalase-peroxidase from Aeribacillus pallidus (P26). *JPAM*, 7, 4, 2629–2640, 2013.
59. Miłek, J., Wójcik, M., Verschelde, W., Thermal stability for the effective use of commercial catalase. *Pol. J. Chem. Technol.*, 16, 4, 75–79, 2014.
60. Choudhury, A.K.R., Enzyme applications in textile chemical processing, in: *Sustainable Technologies for Fashion and Textiles*, pp. 91–115, Woodhead Publishing, 2020.
61. Usluoğlu, A., Bitkilerden elde edilecek bazı enzimlerin tekstil proseslerinde kullanılması, pHd Thesis, Sakarya Universitesi, Türkiye, 2016.
62. Avinc, O., Erişmiş, B., Eren, H.A., Eren, S., Treatment of cotton with a laccase enzyme and ultrasound. *De Redactie*, 55, 2016.
63. Singh, D. and Gupta, N., Microbial Laccase: a robust enzyme and its industrial applications. *Biol.*, 75, 8, 1183–1193, 2020.
64. Imran, M., Asad, M.J., Hadri, S.H., Mehmood, S., Production and industrial applications of laccase enzyme. *J. Cell Mol. Biol.*, 10, 1, 1, 2012.
65. Kalme, S., Jadhav, S., Jadhav, M., Govindwar, S., Textile dye degrading laccase from Pseudomonas desmolyticum NCIM 2112. *Enzyme Microb. Technol.*, 44, 2, 65–71, 2009.

66. El-Hennawi, H.M., Ahmed, K.A., Abd El-Thalouth, I., A novel bio-technique using laccase enzyme in textile printing to fix natural dyes. *Indian J. Fibre Text. Res.*, 37, 3, 245, 2012.
67. Motamedi, E., Kavousi, K., Motahar, S.F.S., Ghaffari, M.R., Mamaghani, A.S.A., Salekdeh, G.H., Ariaeenejad, S., Efficient removal of various textile dyes from wastewater by novel thermo-halotolerant laccase. *Bioresour. Technol.*, 337, 125468, 2021.
68. Yuan, M., Wang, Q., Shen, J., Smith, E., Bai, R., Fan, X., Enzymatic coloration and finishing of wool with laccase and polyethylenimine. *Text. Res. J.*, 88, 16, 1834–1846, 2018.
69. Morin, S., Lecart, B., Lang, M., Richel, A., Lignocellulosic fibres surface interactions in enzymatic reaction using data-mining. *Carbohydr. Polym.*, 254, 117412, 2021.
70. Paramasivam, S.K., Panneerselvam, D., Sundaram, D., Shiva, K.N., Subbaraya, U., Extraction, characterization and enzymatic degumming of banana fiber. *J. Nat. Fibers*, 19, 4, 1333–1342, 2022.
71. Eyupoglu, S., Merdan, N., Dayioglu, H., Demir, B., Hydrophilicity procedure of Agava Americana L. fibers with ecologic methods. *International Conference on Advance Technology & Sciences (ICAT'16)*, Konya, Türkiye, 01-03 September, 2016.
72. Zhao, Q., Feng, H., Wang, L., Dyeing properties and color fastness of cellulase-treated flax fabric with extractives from chestnut shell. *J. Cleaner Prod.*, 80, 197–203, 2014.
73. Chaari, R., Khlif, M., Mallek, H., Bradai, C., Lacoste, C., Belguith, H., Dony, P., Enzymatic treatments effect on the poly (butylene succinate)/date palm fibers properties for bio-composite applications. *Ind. Crops Prod.*, 148, 112270, 2020.
74. Merdan, N., Kocak, D., Mıstık, S.I., Yüksek, M., Akalın, M., Dyeing of Enzymatic Treated Hemp Fibre by Microwave Method. *The 4th RMUTP International Conference:Textile & Fashion*, Banghok, Thailand, 2012.
75. Seghini, M.C., Tirillò, J., Bracciale, M.P., Touchard, F., Chocinski-Arnault, L., Zuorro, A., Sarasini, F., Surface modification of flax yarns by enzymatic treatment and their interfacial adhesion with thermoset matrices. *Appl. Sci.*, 10, 8, 2910, 2020.
76. Samant, L., Goel, A., Mathew, J., Jose, S., Thomas, S., Effect of surface treatment on flax fiber reinforced natural rubber green composite. *J. Appl. Polym. Sci.*, 140, 12, e53651, 2023.
78. Kim, S., Lopez, C., Güebitz, G., Cavaco-Paulo, A., Biological coloration of flax fabrics with flavonoids using laccase from Trametes hirsuta. *Eng. Life Sci.*, 8, 3, 324–330, 2008.
79. Lee, S.H. and Song, W.S., Surface modification of polyester fabrics by enzyme treatment. *Fibers Polym.*, 11, 54–59, 2010.
80. Vajpayee, M., Singh, M., Dave, H., Ledwani, L., Enzymatic surface modification of banana fabric with cellulase: Characterization and antimicrobial finishing with plant extracts. *Ind. Crops Prod.*, 191, 115895, 2023.

81. Eyupoglu, S., Eyupoglu, C., Merdan, N., Investigation of the effect of enzymatic and alkali treatments on the physico-chemical properties of Sambucus ebulus L. plant fiber. *Int. J. Biol. Macromol.*, 266, 130968, 2024.
82. Kim, S.Y., Zille, A., Murkovic, M., Güebitz, G., Cavaco-Paulo, A., Enzymatic polymerization on the surface of functionalized cellulose fibers. *Enzyme Microb. Technol.*, 40, 7, 1782–1787, 2007.
83. Nguyen, H.H. and Kim, M., An overview of techniques in enzyme immobilization. *Appl. Sci. Converg. Technol.*, 26, 6, 157–163, 2017.
84. Datta, S., Christena, L.R., Rajaram, Y.R.S., Enzyme immobilization: an overview on techniques and support materials. *3 Biotech*, 3, 1–9, 2013.

15

# Surface Modification and Functionalization of Textiles by Ozone Application

**İdil Yiğit[1]*, Aliye Akarsu Özenç[2], Semiha Eren[2] and Hüseyin Aksel Eren[2]**

[1]*Bursa Uludag University, Textile, Clothing, Footwear and Leather Department, Vocational School of Orhaneli, Bursa, Turkey,*
[2]*Bursa Uludag University, Textile Engineering Department, Bursa, Turkey*

***Abstract***
Various methods are applied to textile products to give them new properties and make them functional. Ozonation is an innovative environmentally friendly surface modification method. Ozone gas creates physical and chemical changes on the fiber surface thanks to its strong oxidant properties. While changing the properties of fiber surfaces such as dye uptake, wettability, and adhesion, it is used in applications such as desizing, impurity removal, bleaching, color removal, patterning, denim industry, and decolorization in wastewater. In this study, the effects of ozone treatments applied to textile materials for different purposes are discussed, and it is emphasized that ozone can be an alternative to traditional chemical methods thanks to its environmentally friendly structure.

***Keywords***: Ozone, surface modification, textile application

## 15.1 Introduction

The performance properties of textile fibers are largely determined by surface interactions [1]. In recent years, surface modification techniques have emerged as effective methods for improving the functional attributes of

**Corresponding author*: idilyigit@uludag.edu.tr

Aminoddin Haji and Çağlar Sivri (eds.) Surface Modification and Functionalization of Textiles, (371–394) © 2026 Scrivener Publishing LLC

textile fibers [2, 3]. It has been demonstrated that the desired characteristics of textile materials can be achieved through appropriate pre-treatment and finishing processes [4]. Surface modifications can lead to substantial changes in the properties of fibers, significantly affecting parameters such as wettability [5–7], dyeability [8], flexibility [9], adhesion [10–12], flame retardancy [13], electrical conductivity [14], antibacterial properties [15], oil repellency [16], and thermal stability [17].

Surface modifications applied to textile materials are classified as physical, chemical, or a combination of both, depending on the technique used. These modifications can be achieved through various processes designed to enhance the performance and functionality of fibers [18].

Various techniques have been employed by researchers to enhance the surface properties of textile fibers. These methods include aqueous treatments such as alkali treatment [19], plasma treatment [20], laser treatment [21], sol-gel processes [22], surface grafting [23], enzymatic treatments [24], layer-by-layer deposition [25], coating techniques [26], UV radiation [27], and ozone surface modification [28]. Each technique offers distinct advantages and limitations. Recently, there has been an increasing emphasis on sustainable methods due to rising environmental concerns. Ozone gas treatments, in particular, have gained prominence in the textile industry due to their ability to minimize chemical usage and reduce water consumption, offering significant benefits for sustainable textile production. This section discusses the application and effects of ozone gas, an environmentally friendly and sustainable method, in surface treatments.

## 15.2 Oxidation Methods and Ozone

The term "advanced oxidation processes" (AOPs) refers to a group of techniques that include Fenton and Fenton-like processes, electrochemical oxidation, photocatalysis, and ozonation. Despite the fact that these processes employ various reacting systems, they are all distinguished by the same chemical characteristic the use of extremely reactive oxidizing agents, such as hydroxyl radicals (HO·) with redox potentials of 2.80 eV [29, 30]. The capacity of AOP to remove organic materials that are not biodegradable and to prevent the removal of residual deposits gives it a distinct edge over traditional alternative treatments [31]. Among the various AOPs, ozonation stands out especially for its high oxidation capacity (2.07 V) and environmentally friendly properties.

## 15.3 Overview of Ozone

Schönbein was the first to identify ozone as a separate chemical species in 1839, but Soret was the first to propose its formula in 1865, based on three oxygen atoms. Von Siemens was the first to establish a dependable method of synthesizing high-concentration ozone using an electrical discharge [32].

Ozone is easily produced *in situ* from oxygen or dry air, and, because of its short half-life, any remaining ozone in the system breaks down quickly into $O_2$. Thus, the entire ozone treatment process has the benefit of being readily available, nontoxic, and without requiring any more separation stages or concerns about environmental persistence [33]. Ozone is extensively used for treating natural and industrial water sources, polluted streams, pools, contaminated air, and for medical applications, owing to its oxidizing, antibacterial, and antiviral properties [34].

With a bond length of 1.26 Å between oxygen atoms, ozone has a cyclic structure and exists in several mesomeric forms that are in a state of dynamic equilibrium [35]. The reaction for ozone formation is as follows [36, 37]:

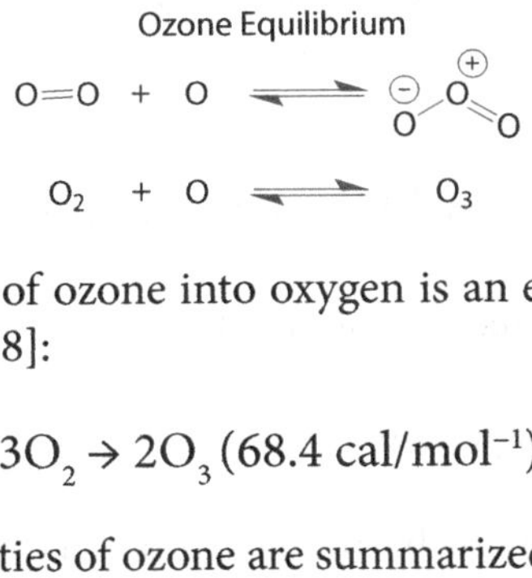


The decomposition of ozone into oxygen is an exothermic reaction, as illustrated below [37, 38]:

$$3O_2 \rightarrow 2O_3\,(68.4\ \text{cal/mol}^{-1})$$

The physical properties of ozone are summarized in Table 15.1 [39].

**Table 15.1** Some of the physical properties of ozone.

| | |
|---|---|
| Molecular weight (g/mol) | 48 |
| Oxidation potential (V) | 2.07 |
| Critical pressure (atm) | 65 |
| Critical density (g/ml) | 0.537 |
| Melting point (°C) | –251 |
| Boiling point (°C) | –112.3 |

## 15.4 Production of Ozone

Electrochemical, ultraviolet radiation, and electrical discharge methods are widely used for commercial ozone production [40].

Ozone production by electrochemical method has been reported to occur by electrochemical reaction of oxygen gas or water in an electrolysis cell [41, 42]. It was also emphasized that this method does not require high voltages, oxygen or pure air supply [43, 44].

Ozone production using UV-C lamps involves the formation of ozone ($O_3$) molecules through the interaction of UV-C radiation with oxygen molecules. This process relies on the absorption of molecular oxygen ($O_2$) at wavelengths shorter than 240 nm, with the most efficient ozone production occurring at around 160 nm [45].

In contrast, ozone production through electrical discharge involves breaking down oxygen molecules in a high-energy electric field, resulting in the formation of free oxygen atoms (O). These free atoms subsequently react with other oxygen molecules to produce ozone ($O_3$) [46]. This method emphasizes the importance of parameters such as air humidity, gas flow rate, and oxygen concentration in optimizing ozone production [47].

## 15.5 Advantages and Disadvantages of Ozone

While ozone provides effective solutions across various applications due to its strong oxidation capabilities, it can pose risks to human health and the environment if not managed properly.

- No environmentally harmful by-products and waste [48, 49].
- Ozone gas is ecological and economic [50].
- In drinking water treatment, ozone is used to disinfect and remove iron, manganese, scent, color, and smell [51].
- Extremely high oxidation power, allowing contaminants in wastewater to be oxidized in a matter of minutes with only a comparatively short contact time [52].
- Ozone is known to be unstable and cannot be stored, as it rapidly decomposes into oxygen [53].
- The half-life of ozone refers to the time required for its initial concentration to decrease by half. A shorter half-life may lead to less effective treatments during exposure. Relative humidity, airflow rate, and temperature are influential factors in determining the half-life of ozone [34, 54, 55].

Ozone stability is strongly influenced by temperature and phase. At low temperatures, gaseous ozone exhibits a relatively long half-life. For example, it can persist for up to three months at −50°C. However, its stability decreases significantly with increasing temperature, with the half-life reducing to eight days at −20°C and further to only three days at 20°C. This behavior indicates that temperature control is a critical parameter in ozone-based treatment processes.

Aqueous ozone, on the other hand, has a substantially shorter half-life. Its half-life is only 30 minutes at 15°C and 15 minutes at 25°C. At 35°C, the half-life of this rapid deterioration is only 8 minutes, and it continues to increase with temperature [56].

## 15.6 Applications of Ozone in Textile Industry

Ozone is increasingly recognized as an innovative and environmentally friendly agent in the textile industry. Due to its strong oxidative properties, ozone offers significant advantages in various textile applications, presenting a lower environmental impact and reduced energy consumption compared to traditional methods. It use in processes such as bleaching, disinfection, pre-dyeing treatments, and wastewater management enhances both sustainability and operational efficiency within the industry. This book chapter explores the role of ozone in the textile industry, highlighting its benefits and examining the future potential of this technology.

### 15.6.1 Textile Wastewater Color Removal

It was estimated that the textile industry significantly contributes to wastewater pollution, with dyeing processes alone generating approximately 70 billion tons of wastewater annually, accounting for a substantial portion of overall sewage discharge [57]. Currently, the textile industry utilizes aromatic and heterocyclic dyes. The complex and stable structures of these dyes make them particularly challenging to degrade, not only in textile wastewater but also in other complex matrices [58, 59]. Ozonation is one of several methods employed for the treatment of textile wastewater. Ozone effectively decomposes the conjugated double bonds in dye chromophores as well as other functional groups, including complex aromatic rings. This process results in the formation of smaller, non-chromophoric molecules, leading to a reduction in the color of the effluents [60, 61]. Ozone effectively oxidizes both organic and inorganic substances; however, in dyeing wastes containing suspended solids, it may not fully convert organic compounds into carbon dioxide and water, even at high doses [62]. Key factors influencing the effectiveness of ozonation in wastewater treatment include ozone dose, ozonation time, ambient pH, and dye concentration [63–66]. Numerous studies have reported successful textile wastewater decolorization using ozonation [67–69]. Research focusing on the reuse of decolorized textile wastewater after ozonation significantly contributes to sustainability efforts in the industry [70–73].

### 15.6.2 Bleaching

The goal of bleaching is to achieve lighter fabrics by removing the colorants using bleaching agents while minimizing damage to the fibers. These agents work by either oxidizing or reducing the colorants, which are then washed away, resulting in a permanent whiteness [74]. Ozonation is known for its successful application in bleaching cellulosic fibers. As a potent oxidizer, ozone converts hydroxyl groups in cellulosic structures into carbonyl and carboxyl groups [75] and attacks the bonds between carbonyl and alkoxy groups in the saturated wax structure of the fibers [76–78]. Research indicates that factors such as ozone phase, flow rate, ozonation time, fiber humidification, ambient pH, and temperature significantly influence the whiteness values of cellulosic fibers during bleaching [79–82]. Ozonation has been reported for bleaching various fibers, including cotton [83, 84], nettle fiber [85], flax [33], and jute [86]. In addition to cellulosic fibers, ozone has also been successfully used to bleach PLA fibers [87], soybean fibers [88], and Angora rabbit fibers [89].

### 15.6.3 Desizing

Desizing is a crucial step in the pretreatment of cotton fabrics before dyeing and printing, as it involves removing the sizing agents applied to the fabric [90]. Ozone has emerged as a promising and eco-friendly alternative for desizing, offering a sustainable method that reduces environmental impact. Below, some studies demonstrating its effectiveness in this area are presented. It was reported that both the whiteness values and water absorption capacities of fabrics improved during the sizing process [91]. Additionally, chemical oxygen demand (COD) analyses indicated lower values compared to conventional methods [92].

### 15.6.4 Color Removal and Patterning

The desire for garments with faded or vintage looks has grown in recent years, which has led to the development of numerous technologies and ways to produce these effects. However, a lot of conventional color fading techniques require a lot on a lot of water and chemicals, which pollute the environment [93]. Ozone treatment offers a more environmentally responsible option for color fading because it uses less water and fewer or any toxic chemicals to achieve the required faded effects in textiles.

Studies on the decolorization efficiency of dyed and printed textiles using reactive dyes with ozone gas have shown that ozone treatment can achieve higher decolorization percentages for certain dyestuff groups and colors compared to conventional methods. Specifically, ozone treatment has been found to be more effective for anthraquinone and phthalocyanine chromophore groups. This is likely because ozone targets the –N=N– or –C=C– double bonds in the azo chromophore groups that link aromatic rings, leading to effective decolorization. Additionally, it was observed that among the primary colors—yellow, red, and blue, yellow is the most resistant to fading, whereas blue is the least resistant. This indicates that different dyes respond variably to ozone exposure. Aromatic compounds with electron-donating groups (e.g., -OH and -NH2) interact more strongly with ozone due to their higher electronic density, whereas those with electron-acceptor groups (e.g., -COOH and -$NO_2$) exhibit weaker interactions.

Beyond its effectiveness in decolorization, ozone treatments also offer significant advantages in terms of cost and time efficiency. They can achieve up to a 98% reduction in COD values, highlighting their environmental and operational benefits over conventional methods [94–98].

Zulfiqar *et al.* [99] investigated the efficiency of ozone on reactive dyes with different chromophore groups and found that Reactive Blue 21, from the phthalocyanine group, achieved the highest color removal. Their results indicated that ozone's color removal efficiency varies depending on the type of reactive dye, with vinyl sulfone anchor groups being more susceptible to ozone compared to monochlorotriazine groups. Another study has suggested that ozone's effectiveness in dye removal follows the order: formazan > anthraquinone > phthalocyanine > monoazo chromophores [100]. This supports the observation that ozone's effectiveness is influenced by the chromophore type in dyes. Yigit *et al.* [77] reported that ozone application enhanced color removal rates in printing processes on reactive-dyed cotton fabrics, although it slightly reduced contour sharpness. Various parameters influencing the ozonization process have been studied in the literature, including duration and concentration. It has been noted that pH also plays a crucial role in the process. Specifically, ozone degradation of cellulosic materials is pH-dependent, with increased degradation at lower pH due to more direct ozone reactions. At low pH, ozone directly attacks azo dye groups, whereas, at higher pH, indirect reactions with free radicals contribute to enhanced discoloration [101].

### 15.6.5 Denim Industry

In denim garment production, various techniques such as whiskering, laser treatment, stone washing, spraying, and bleaching are employed to achieve the desired aged appearance. These methods are typically categorized as either dry or wet processes [102]. Conventional methods of washing denim are known for using a lot of water and adding a lot of waste to the environment. The use of sodium hypochlorite in denim bleaching raises serious environmental issues because it forms adsorbable halogenated organic molecules. Furthermore, there are concerns associated with stone-washing procedures for the environment and public health. Studies show that the finishing process for a single denim piece uses about 70 liters of water. An estimated 5 billion jeans are produced worldwide each year, which means that 350 million cubic meters of water are used in the denim manufacturing process [103, 104]. Ozone washing has many benefits over conventional methods of denim washing and is considered an environmentally friendly substitute [105, 106]. The denim industry was the first in the textile sector to adopt ozone for garment processing. Ozone efficiently breaks down indigo and other dyes. In addition to its use in denim washing, it is frequently applied to treat other textiles such as T-shirts, shirts, chinos, and casual wear [107].

Various studies have examined the advantages and effects of ozone use in denim processing. Sarker *et al.* [108] highlighted that ozone treatment for denim finishing requires less water and chemicals compared to traditional methods, reduces processing time, and generates less wastewater. Similarly, Nayak *et al.* [105] emphasized the potential of ozone technology to reduce environmental impacts in denim washing. Ben Hmida *et al.* [109] investigated the effects of ozone on denim fabrics and observed that ozone concentration, moisture level, and exposure time caused significant changes in the fabric's appearance and physical properties. These findings were corroborated by Ben Fraj *et al.* [104], who reported that ozone treatment increased the whiteness of the fabric while leading to changes in some physical and mechanical properties.

The authors explained that the mechanism of ozone decolorization of indigo dyes involves oxygen-centered radicals breaking the C=C bonds in the dye molecules. They noted that increasing the ozone feed rate leads to a higher concentration of these radicals, which enhances the fading effect. Furthermore, they observed that, while moisture content up to 50%

increases the fading effect due to enhanced ozone solubility and bleaching ability, a moisture content of 100% reduces this effect, likely because excess water depletes the ozone [98].

### 15.6.6 Oxidative Cleaning

Reduction clearing typically involves using a strong reducing bath, such as one containing sodium dithionite and caustic soda, to treat dyed fibers. However, alternative reducing agents can offer a more environmentally friendly approach, reduce water consumption, and facilitate easier and more precise control of the process [110]. Additionally, ozone has emerged as a viable option for polyester clearing, providing a sustainable alternative to traditional methods by minimizing chemical use and enhancing overall efficiency.

Research on the effectiveness of ozone in cleaning various fabrics yields significant insights. Eren and Aniş [111] demonstrated that ozone cleaning of polyester fibers after dyeing offers superior durability, energy efficiency, and time savings compared to conventional reduction cleaning. They found that cyclic trimer removal rates were comparable to those of traditional methods at 1 minute but improved with increased ozone exposure, reaching higher rates at 3 minutes. The authors highlighted ozone's strong oxidation potential and its ability to react with organic substances, such as trimers in polyester fibers, as key factors in its effectiveness. Avinç *et al.* [112] reported that ozonation achieved up to a 33% reduction in COD when cleaning disperse-dyed PLA fabrics, with lower ozone doses effectively addressing color difference issues. Additionally, Eren *et al.* [113] found that ozone treatment for cleaning disperse-dyed polyester fibers yielded similar color and wash durability results as conventional reduction cleaning.

## 15.7 Surface Modification and Functionalization Applications of Ozone in Textile Industry

### 15.7.1 Fiber Surface

Due to its many benefits, including affordability, plentiful supply, environmental control, biodegradability, sustainability, and advantageous physical qualities, natural and synthetic fibers are being utilized more and more in a variety of industrial applications [114]. The surface modification made to these fibers often determines their application's efficiency. Ozonation is one such modification technique that is utilized to improve their performance and mechanical characteristics. Some studies in this field are presented below.

Michael *et al.* [115] investigated the impact of UV/ozone surface modification on polymer fabrics, examining changes in surface properties over treatment times ranging from 1 to 20 minutes. Their study revealed significant alterations in the physical and mechanical properties of the fabrics, including changes in crystallinity and crystal size. They explain that UV/ozone treatments damage amorphous areas, leading to a rearrangement of polymer chains that enhances the crystallinity/amorphousity ratio. It was noted that, as treatment time increases, the tensile strength, elongation, and Young's modulus of the fabric decrease, reflecting a reduction in elasticity. This change is attributed to cross-linking reactions induced by UV/ozone exposure, which stiffen molecular chains and increase the crystalline structure, making the fabric stiffer and more brittle. Conversely, an increase in amorphousness would have the opposite effect. Similarly, Osbeck *et al.* [116] explored the effects of ultraviolet-generated ozone ($UV/O_3$) on the surface and near-surface functionality of carbon fibers. Their findings indicated that fibers subjected to ozone surface modification exhibited a high level of oxygen functionality within 5 minutes (O/C ratio of $0.11 \pm 0.03$), with oxygen content reaching saturation after approximately 40 minutes. X-ray photoelectron spectroscopy analysis identified the main functional groups as hydroxyl, alkoxide, carbonyl, and carboxyl. Additionally, Wang *et al.* [117] investigated the application of ozone surface modification on aramid fibers to enhance their surface properties. Their study demonstrated that ozone treatment slightly improved the wicking effect of aramid fabrics. Fattahi *et al.* [118] investigated ultraviolet/ozone treatment for the surface modification of polyester fibers. Their study involved applying ozone treatment to fibers in various states, including dry fibers and those wetted with water, hydrogen peroxide, and hydrogen peroxide/sodium silicate solutions. The results indicated that ozone treatment enhanced the oxygen-containing functional groups on the fiber surfaces and increased surface roughness at the nanoscale. Similarly, Yiğit *et al.* [119] explored the effects of ozone treatment on hemp fibers. They applied ozone at different gas flow rates and assessed the impact on hydrophilicity and tensile strength of the treated hemp fabrics. The authors suggest that the increase in hydrophilicity observed in the treated fibers is due to the chemical reactions between cellulose and ozone, which break glycosidic bonds and generate carboxyl and carbonyl groups. These changes are believed to enhance the formation of hydrogen bonds on the fiber surface, thus improving hydrophilicity. Furthermore, the increase in tensile strength of the samples is attributed to the combined effects of ozonation and moisturizing processes. Water enhances hydrogen bonding and elasticity, contributing to stronger fibers. Additionally, ozonation helps to remove lignin from the amorphous regions of the fibers and increases their crystallinity, further enhancing tensile strength.

In another study focused on enhancing and functionalizing surface properties, fluorocarbon treatment following UV-ozone surface modification was evaluated as an effective method for increasing the hydrophobicity of polyester fabrics [120]. The results have indicated that the UV-Ozone treatment increases surface topography, which enhances the penetration and effectiveness of the subsequent fluorocarbon treatment. This results in a more hydrophobic surface, demonstrating that the combination of UV-ozone irradiation and fluorocarbon treatment leads to superior water repellency and improved surface performance.

### 15.7.2 Fiber-Matrix

The characteristics of composites are enhanced by good fiber-matrix adhesion, which improves performance and durability [121]. Various methods are employed to achieve this, including ozone surface modification.

Fu *et al.* [122] highlighted that ozone surface modification enhanced the oxygen concentration on carbon fibers' surfaces, converting oxygen bonds from C–O to C=O. This modification reduced the water contact angle of the fibers to zero, improved the bond strength between the fibers and cement mortar, and enhanced the tensile strength, modulus and elasticity of carbon fiber-reinforced cement mortar. Similarly, Dinç *et al.* [11] reported that ozone treatment increased the interfacial strength of fibers by converting epoxy groups on the surface of carbon fibers into hydrogen and carboxyl groups. After 60 minutes of ozone treatment, the carboxyl groups increased from 1.27% to 8.12%, and hydrogen groups rose by 28.37%. In another study with carbon fiber, Downey *et al.* [123] employed UV-ozone surface treatment combined with aromatic and aliphatic epoxy fiber coatings to enhance the toughness of carbon fiber-reinforced epoxy composites. They reported that the UV-ozone treatment increased the concentration of oxygen-containing groups on the fibers, which improved fiber/matrix adhesion and consequently enhanced the Mode I crack toughness of the composite by 23%. Similarly, Wang *et al.* [124] focused on the increase in oxygen-containing groups in their study of ultra-high molecular weight polyethylene fibers. The fibers were modified using ozone followed by UV grafting with glycidyl methacrylate. This treatment resulted in a 79% increase in fiber-matrix adhesion.

Ozone surface modification has also been explored in studies involving natural fiber-reinforced composites. Rajendran Royan *et al.* [125] investigated the impact of ultraviolet-ozone ($UV/O_3$) surface modification on rice bran fibers used in high-density polyethylene (rHDPE) composites. They found that $UV/O_3$ treatment increased the roughness of the fiber surface, which improved adhesion to the rHDPE matrix and resulted in a 5% increase in tensile strength. Similarly, Sanchez *et al.* [126] examined the effects of different surface modifications, including ozone, on the physical and mechanical properties of bamboo fibers in biocomposites. Their study compared ozone treatment with other methods such as mercerization and plasma modification. The results explain that ozone treatment effectively removes impurities from the fiber surface and increases the surface roughness of bamboo fibers. Their findings confirm that ozone treatment enhances the exfoliation of fiber surfaces. This, which showed that higher concentrations of treatment further improve impurity removal and surface roughness. Surface modified fibers were found to increase dimensional stability and mechanical strength by more than 50%.

## 15.8 Ozone Treatment on Dye Uptake

One of the main goals is to use less dye while maintaining the qualities of the fabric and particularly its color. Reactive regions can be created on the fiber by surface modification techniques, which improve dye adsorption and thereby lower the quantity of contaminants in wastewater. Ozone has become a viable substitute in the quest for better dyeing processes [127].

In a study by Kim and Jang [128], meta-aramid films were subjected to UV/ozone treatment, which led to a raise in surface oxygen content. This modification decreased the water contact angle and increased the surface energy of the films. The formation of anionic and dipolar dyeing spots on the films due to ozone treatment is observed, which enhanced their dyeability with cationic dyes. Similarly, Dong and Jang [8] applied UV/ozone surface modification to aramid-based fabrics. Their study found that this treatment increased the oxygen content on the fabric surfaces, resulting in the formation of carbonyl and hydroxyl groups. This modification also decreased the surface zeta potential and increased hydrophilicity. Consequently, the modified aramid fabrics exhibited a higher affinity for cationic dyes, although the crystal structure of the fabrics remained unchanged.

In the study by Elnagar *et al.* [129], the effectiveness of UV/ozone pretreatment for dyeing polyester and polyamide synthetic fabrics with natural dyes was examined as a more environmentally friendly alternative.

The research demonstrated that pretreated polyester and polyamide fabrics showed an increased affinity for natural dyes, specifically curcumin and saffron, when subjected to UV/ozone pretreatment for various durations. The authors emphasized that the increase in dye uptake in UV/ozone-treated fabrics can be attributed to the interaction of UV light with the polymeric materials in the presence of oxygen, leading to the formation of free radicals (P•, POO•, HOO•, H•, and HO•). They also noted that the generation of new functional groups (C–O, C=O, and COOH) during the treatment enhances the affinity of dye molecules for the fabric, contributing to the increased dye uptake. Gabardo *et al.* [130] investigated the modification of polyester fabrics using ozone followed by dyeing with C.I. Disperse Yellow 211. The study found that ozone treatment led to both molecular and topographic changes in the fabric. Raman spectroscopy revealed an increase in hydrophilic bonds post-oxidation. Kinetic analysis indicated that dye adsorption adhered to a three-stage intraparticle diffusion model, with a 10.74% increase in dye uptake compared to non-ozone treated samples. In the study by Eren *et al.* [131], polyester fabrics were subjected to ozonation for varying durations and gas flow rates before dyeing with supercritical carbon dioxide ($scCO_2$). Results demonstrated that ozone surface modification enhanced color strength (K/S) and provided improved durability test outcomes. The results were attributed that the increase in dye uptake to the formation of carboxylate groups during ozone surface treatments, which enhance the interaction between the fiber and dye. Additionally, they observed more surface peeling and breakage in scanning electron microscope (SEM) images of fabric treated for 45 minutes at a 10 L/min ozone gas flow rate, contributing to the increased dye penetration.

In the literature, there are also studies on increasing dye uptake by ozone surface modification applied to natural fibers. In one of these studies, ozonation method was used to improve the dyeability of mohair fibers. In the study, a certain ozone concentration and process parameters such as humidification rate, pH level, and process time were investigated. The results that the increase in dye uptake of mohair fibers after ozonation is primarily due to the oxidative cleavage of disulfide linkages and the subsequent rise in amino groups, which enhance the binding of anionic dyes through electrostatic and covalent bonds. Additionally, the damage to the cuticle layer of the mohair fibers, reducing its barrier effect, further contributes to improved dyeability [132]. The study that is conducted with the effectiveness of ozone-based dyeing processes for cotton fabrics was evaluated, and ozone enhances dye interaction with the fabric by improving surface functionalization and facilitating the formation of new functional groups. The study reported that ozone was effective at various stages of the dyeing process, including during dyeing, dry ozonation, pre-wet ozonation, and post-dyeing also. Notably, the K/S values of ozone-based dyeing processes were 33.5% higher than those of traditional methods. It was also noted that, while ozone increases color intensity at lower doses, this effect tends to diminish at higher concentrations [133].

## 15.9 Impurity Removal in Lignocellulosic Fibers

Studies have highlighted that ozone surface modification can effectively remove impurities such as lignin, hemicellulose, and wax from lignocellulosic fibers. For example, Maqsood *et al.* [86] found that ozone treatment of jute fibers led to significant changes in both physical and chemical properties over time. The process caused the decomposition of lignin and partial dissolution of hemicellulose while enhancing fiber morphology through increased crystallinity and crystallite size. Similarly, Bhattacharyya and Baheti [134] demonstrated that ozone pretreatment effectively removed non-cellulosic components from jute, corn cob, and sisal fibers. Fourier transform infrared spectroscopy (FTIR) analysis confirmed a reduction in lignin, hemicellulose, and other non-cellulosic substances following ozone treatment.

Flax fibers were treated with ozone to remove surface impurities, with the study identifying optimal conditions as 130 mg $L^{-1}$ $min^{-1}$ ozone flow, 20 minutes of treatment, pH 2.0, and 70% moisture content. The results showed a decrease in the degree of polymerization after ozone treatment, as assessed by tests for lignin content, strength, elongation, thermal stability, and polymer chain length. Additionally, the copper number, which indicates oxidation and the presence of aldehyde groups, increased. The ozone treatment also led to enhanced crystallinity, whiteness, and surface roughness of the flax fibers [33].

In a study investigating the interaction of hemp fibers, which are rich in lignin, with ozone gas, it was found that ozone pretreatment significantly increased the whiteness of hemp fibers by 72%. The study highlighted that lignin removal was evidenced by a decrease in the FTIR peak at 1,732 $cm^{-1}$, associated with acetyl and aldehyde groups present in lignin and hemicellulose. Additionally, a reduction in the 1,616 $cm^{-1}$ peak, related to C=C stretching in aromatic rings of lignin, was observed in the ozonated samples [78].

## 15.10 Sericin Removal

Literature on the use of ozone for removing sericin from silk fabrics is limited. Notable studies investigated the effects of ozone on silk fibers.

Researches highlighted that ozone pre-treatment led to the formation of chromophoric compounds containing carbonyl groups, resulting in an increased yellowing index. Results were attributed the weight loss observed during the process to the removal of gaseous by-products such as ammonia, carbon dioxide, aldehydes, ketones, and nitrates [135]. In a separate study, the effects of ozone and low-frequency oxygen plasma treatments on the physical properties of raw and degummed silk fabrics were explored [136]. The authors noted that ozone surface modification significantly reduced the whiteness values of silk fabrics compared to plasma treatment and increased their yellowness values. They found ozone to be more effective in removing sericin and reported no significant changes in the mechanical properties of the fabric during this process. Furthermore, the SEM images at 2,000× magnification revealed that, while raw silk fabric had a smooth surface, ozone treatment led to noticeable roughness, including fibrillation and peeling. The results were expressed that the combined plasma and ozone treatment resulted in the most significant surface modifications, showing extensive fibrillation, and fiber breakage.

## 15.11 Flame Reterdancy

One of the innovative applications of ozone gas is in enhancing the flame retardancy of fibers. In a study conducted by Eren *et al.* [137], the impact of ozone on flame retardancy was evaluated during the finishing processes of polyester fabrics. The study found that flame retardancy diminished because of residual foreign substances and dye wastes on the fibers after finishing. However, the fibers regained their flame-retardant properties following ozonation. FTIR analysis revealed that ozonation led to the disappearance of C–H bonds in the 3,000–2,840 $cm^{-1}$ range, effectively removing dye residues from the fiber surface and restoring flame retardancy.

## 15.12 Conclusion

This book section explores various facets of ozone utilization in the textile industry. It covers the chemical and physical properties of ozone, its production methods, as well as its advantages and disadvantages. The discussion extends to current applications of ozone, including wastewater treatment, bleaching, desizing, color removal, patterning, and its specific uses in the denim sector. The section then focuses on the surface modification and functionalization of textiles using ozone, highlighting studies on fiber surface changes, fiber-matrix interactions, dye uptake, impurity removal in lignocellulosic fibers, sericin removal, and flame retardancy applications. The examination of ozone's effects on textile surfaces demonstrates that the resulting data can be applied not only within the textile industry but also in various related industrial fields, enhancing functional properties of textiles exposed to ozone.

**AI disclosure statement:** The author confirms that no artificial intelligence (AI)-generated content or AI-assisted writing tools were used in the preparation of this chapter (including summaries and the abstract). All text, analysis, and conclusions are the result of the author's original work.

## References

1. Al Kashouty, M., Elsayad, H., Twaffiek, S., Salem, T., Elhadad, S., An overview: textile surface modification by using sol-gel technology. *Egypt. J. Chem.*, 63, 9, 3301–3311, 2020.
2. Jocic, D., Jovancic, P., Petrovic, Z., Bertan, E., Navarro, A., Julia, M.R., Erra, P., The influence of surface modification on wool functional and dyeing properties, in: *Proceedings of the World Textile Conference 2nd AUTEX Conference" Textile engineering at the dawn of a new millennium: an exciting challenge"*, Bruges, Belgium, pp. 1–3, 2002, July.
3. Nemani, S.K., Annavarapu, R.K., Mohammadian, B., Raiyan, A., Heil, J., Haque, M.A., Sojoudi, H., Surface modification of polymers: methods and applications. *Adv. Mater. Interfaces*, 5, 24, 1801247, 2018.
4. Karahan, H.A., Demir, A., Özdoğan, E., Öktem, T., Seventekin, N., Some methods used for the surface modification of textiles. *Text. Apparel*, 17, 4, 248–255, 2007.
5. Li, M., Deng, T., Liu, S., Zhang, F., Zhang, G., Superhydrophilic surface modification of fabric *via* coating with nano-TiO2 by UV and alkaline treatment. *Appl. Surf. Sci.*, 297, 147–152, 2014.
6. Zhang, Q., Jiang, Y., Yao, L., Jiang, Q., Qiu, Y., Hydrophobic surface modification of ramie fibers by plasma-induced addition polymerization of propylene. *J. Adhes. Sci. Technol.*, 29, 8, 691–704, 2015.
7. Iqbal, M., Dinh, D.K., Abbas, Q., Imran, M., Sattar, H., Ul Ahmad, A., Controlled surface wettability by plasma polymer surface modification. *Surfaces*, 2, 2, 349–371, 2019.
8. Dong, Y. and Jang, J., The enhanced cationic dyeability of ultraviolet/ozone-treated meta-aramid fabrics. *Color. Technol.*, 127, 3, 173–178, 2011.
9. Kumar, N. and Das, D., Alkali treatment on nettle fibers: Part I: investigation of chemical, structural, physical, and mechanical characteristics of alkali-treated nettle fibers. *J. Text. Inst.*, 108, 8, 1461–1467, 2017.

10. Orue, A., Jauregi, A., Peña-Rodriguez, C., Labidi, J., Eceiza, A., Arbelaiz, A., The effect of surface modifications on sisal fiber properties and sisal/poly (lactic acid) interface adhesion. *Composites, Part B*, 73, 132–138, 2015.
11. Saka Dinç, Z., Eren, S., Öz, Y., Eren, H.A., Influence of surface modifications on interfacial and mechanical properties in fibre-reinforced composites. *J. Reinf. Plast. Compos.*, 07316844241259870, 2024.
12. Şahin, H.G., Akarsu Özenç, A., Saka Dinç, Z., Mardani, A., Eren, S., Investigation of fresh and hardened properties of 3D printable concrete containing ozone-modified carbon fiber. *J. Sustainable Cem.-Based Mater.*, 14, 3, 534–548, 2025.
13. Liu, Y., Ding, D., Lu, Y., Chen, Y., Liao, Y., Zhang, G., Zhang, F., Efficient and durable cotton fabric surface modification *via* flame retardant treatment. *Colloids Surf., A*, 648, 129005, 2022.
14. Nayab-Ul-Hossain, A.K.M., Sela, S.K., Hassan, M.N., Sarkar, A., Surface modification of ligno-cellulosic fiber (jute) to increase electrical conductivity. *Mater. Lett.: X*, 5, 100036, 2020.
15. Zheng, M., Yang, Y., Yang, H., Zhang, H., Zhang, L., Zheng, W., Wang, Z., Surface modification and antibacterial functionalization of polyester fabrics using deep eutectic solvent and its composite solution. *Fibers Polym.*, 25, 2, 557–564, 2024.
16. Vasiljević, J., Gorjanc, M., Tomšič, B., Orel, B., Jerman, I., Mozetič, M., Simončič, B., The surface modification of cellulose fibres to create super-hydrophobic, oleophobic and self-cleaning properties. *Cellulose*, 20, 277–289, 2013.
17. Gañan, P., Garbizu, S., Llano-Ponte, R., Mondragon, I., Surface modification of sisal fibers: effects on the mechanical and thermal properties of their epoxy composites. *Polym. Compos.*, 26, 2, 121–127, 2005.
18. Ferrero, F. and Periolatto, M., Modification of surface energy and wetting of textile fibers, in: *Wetting and wettability*, pp. 139–168, InTech, Rijeka, Croatia, 2015.
19. Gassan, J. and Bledzki, A.K., Alkali treatment of jute fibers: relationship between structure and mechanical properties. *J. Appl. Polym. Sci.*, 71, 4, 623–629, 1999.
20. Xie, J., Xin, D., Cao, H., Wang, C., Zhao, Y., Yao, L., Qiu, Y., Improving carbon fiber adhesion to polyimide with atmospheric pressure plasma treatment. *Surf. Coat. Technol.*, 206, 2-3, 191–201, 2011.
21. Genna, S., Leone, C., Ucciardello, N., Giuliani, M., Increasing adhesive bonding of carbon fiber reinforced thermoplastic matrix by laser surface treatment. *Polym. Eng. Sci.*, 57, 7, 685–692, 2017.
22. Sun, Z. and Mingming, W., Effects of sol-gel modification on the interfacial and mechanical properties of sisal fiber reinforced polypropylene composites. *Ind. Crops Prod.*, 137, 89–97, 2019.
23. Mori, M., Uyama, Y., Ikada, Y., Surface modification of polyethylene fiber by graft polymerization. *J. Polym. Sci., Part A: Polym. Chem.*, 32, 9, 1683–1690, 1994.

24. Wu, J., Cai, G., Liu, J., Ge, H., Wang, J., Eco-friendly surface modification on polyester fabrics by esterase treatment. *Appl. Surf. Sci.*, 295, 150–157, 2014.
25. Li, H. and Peng, L., Antimicrobial and antioxidant surface modification of cellulose fibers using layer-by-layer deposition of chitosan and lignosulfonates. *Carbohydr. Polym.*, 124, 35–42, 2015.
26. Hu, C., Zhou, Y., Zhang, T., Jiang, T., Zeng, G., Effect of fiber modified by alkali/polyvinyl alcohol coating treatment on properties of sisal fiber plastic composites. *J. Reinf. Plast. Compos.*, 39, 23-24, 880–889, 2020.
27. Eren, S. and Ozenc, A., Investigation of Bleaching of Cotton Fabrics with UV-TiO2. *Tekstilec*, 66, 2, 2023.
28. Pujokaroni, A.S., Ohtani, Y., Ichiura, H., Ozone treatment for improving the solubility of cellulose extracted from palm fiber. *J. Appl. Polym. Sci.*, 138, 1, 49610, 2021.
29. Krishnakumar, B. and Swaminathan, M., Influence of operational parameters on photocatalytic degradation of a genotoxic azo dye Acid Violet 7 in aqueous ZnO suspensions. *Spectrochim. Acta, Part A*, 81, 1, 739–744, 2011.
30. Asghar, A., Raman, A.A.A., Daud, W.M.A.W., Advanced oxidation processes for *in-situ* production of hydrogen peroxide/hydroxyl radical for textile wastewater treatment: a review. *J. Cleaner Prod.*, 87, 826–838, 2015.
31. Hutagalung, S.S., Muchlis, I., Khotimah, K., Textile wastewater treatment using advanced oxidation process (AOP). *IOP Conf. Ser.: Mater. Sci. Eng.*, 722, 1, 012032, 2020, IOP Publishing.
32. Oyama, S.T., Chemical and catalytic properties of ozone. *Catal. Rev.*, 42, 3, 279–322, 2000.
33. Zhang, L., Meng, C., Fu, J., Lou, J., Zhang, X., Gao, W., Fan, X., Effect of ozone treatment on the chemical and mechanical properties of flax fibers. *Ind. Crops Prod.*, 189, 115694, 2022.
34. Batakliev, T., Georgiev, V., Anachkov, M., Rakovsky, S., Ozone decomposition. *Interdiscip. Toxicol.*, 7, 2, 47–59, 2014.
35. Bocci, V., Ozone, pp. 75–84, Springer, Dordrecht, The Netherlands, 2011.
36. Magara, Y., Itoh, M., Morioka, T., Application of ozone to water treatment and power consumption of ozone generating systems. *Prog. Nucl. Energy, 29*, 175–182, 1995.
37. Eren, H.A., Yiğit, I., Eren, S., Avinc, O., Ozone: an alternative oxidant for textile applications, in: *Sustainability in the Textile and Apparel Industries: Production Process Sustainability*, pp. 81–98, Springer Nature Switzerland AG, Cham, Switzerland, 2020.
38. Ritchie, M., The thermal decomposition of ozone. *Proc. R. Soc. Lond. Ser. A, Contain. Pap. Math. Phys. Character*, 146, 859, 848–866, 1934.
39. Rubin, M.B., The history of ozone. IV. The isolation of pure ozone and determination of its physical properties. *Bull. Hist. Chem.*, 29, 2, 99–106, 2004.
40. Pekárek, S., DC corona discharge ozone production enhanced by magnetic field. *Eur. Phys. J. D*, 56, 91–98, 2010.
41. Silva, L.M.D., Santana, M.H., Boodts, J.F., Electrochemistry and green chemical processes: electrochemical ozone production. *Quim. Nova*, 26, 880–888, 2003.

42. Franco, D.V., Jardim, W.F., Boodts, J.F., Da Silva, L.M., Electrochemical ozone production as an environmentally friendly technology for water treatment. *CLEAN - Soil Air Water*, *36*, 1, 34–44, 2008.
43. De Sousa, L.G., Franco, D.V., Da Silva, L.M., Electrochemical ozone production using electrolyte-free water for environmental applications. *J. Environ. Chem. Eng.*, *4*, 1, 418–427, 2016.
44. Rodríguez-Peña, M., Pérez, J.A.B., Llanos, J., Sáez, C., Rodrigo, M.A., Barrera-Díaz, C.E., New insights about the electrochemical production of ozone. *Curr. Opin. Electrochem.*, 27, 100697, 2021.
45. Claus, H., Ozone generation by ultraviolet lamps. *Photochem. Photobiol.*, 97, 3, 471–476, 2021.
46. Meher, P., Deshmukh, N., Mashalkar, A., Kumar, D., Ozone (O3) generation and its applications: A review. *AIP Conf. Proc.*, 2764, 1, 2023, September, AIP Publishing.
47. Cuong, L.C., Nghi, N.H., Dieu, T.V., Oanh, D.T.Y., Vuong, D.D., Influence of oxygen concentration, feed gas flow rate and air humidity on the output of ozone produced by corona discharge. *Vietnam J. Chem.*, 57, 5, 604–608, 2019.
48. Eren, S. and Yetişir, I., Ozone bleaching of woven cotton fabric. *Pamukkale Univ. Muh. Bilim. Derg.*, 24, 7, 1245–1248, 2018.
49. Eren, S. and Özenç, A.A., Ozonlama İşleminin Ham İpek Kumaşların Fiziksel ve Mekanik Özelliklerine Etkisinin İncelenmesi. *Uluslararası Bilim Teknoloji ve Tasarım Dergisi*, 4, 2, 66–75, 2023.
50. Remondino, M. and Valdenassi, L., Different uses of ozone: environmental and corporate sustainability. Literature review and case study. *Sustainability*, 10, 12, 4783, 2018.
51. Leitzke, O., Applications of Ozone in Water Treatment: Advantages, Disadvantages and Developments, in: *The Modern Problems of Electrostatics with Applications in Environment Protection*, pp. 265–303, Springer Netherlands, Dordrecht, 1999.
52. Helmy, Q., Suryawan, I.W.K., Notodarmojo, S., Ozone-based processes in dye removal, in: *Advanced Oxidation Processes in Dye-Containing Wastewater*, vol. 1, pp. 91–128, Springer Nature Singapore, Singapore, 2022.
53. Novak, J.S. and Yuan, J.T., The ozonation concept: Advantages of ozone treatment and commercial developments, in: *Advances in Thermal and Non-Thermal Food Preservation*, pp. 185–193, Blackwell Publishing, Oxford, UK, 2007.
54. Taran, V.S., Krasnyj, V.V., Lozina, A.S., Chechelnitskyi, O.G., Schebetun, A.V., Investigation of ozone decay half-life in dependence of temperature and humidity as well as $H_2S$ and $NH_3$ oxidation mechanism. *Вопросы атомной науки и техники*, 2017.
55. McClurkin, J.D. and Maier, D.E., Half-life time of ozone as a function of air conditions and movement. *Julius-Kühn-Archiv.*, 425, 381, 2010.

56. Mrozek-Szetela, A., Rejda, P., Wińska, K., A Review of Hygienization Methods of Herbal Raw Materials. *Appl. Sci.*, 10, 2020. 10.3390/app10228268.
57. Siddique, K., Rizwan, M., Shahid, M.J., Ali, S., Ahmad, R., Rizvi, H., Textile wastewater treatment options: a critical review, in: *Enhancing cleanup of environmental pollutants: Volume 2: Non-biological approaches*, pp. 183–207, Springer, Cham, Switzerland, 2017.
58. Ding, S., Li, Z., Wang, R., Overview of dyeing wastewater treatment technology. *Water Resour. Prot.*, *26*, 73–78, 2010.
59. Holkar, C.R., Jadhav, A.J., Pinjari, D.V., Mahamuni, N.M., Pandit, A.B., A critical review on textile wastewater treatments: possible approaches. *J. Environ. Manage.*, *182*, 351–366, 2016.
60. Tehrani-Bagha, A.R., Mahmoodi, N.M., Menger, F.M., Degradation of a persistent organic dye from colored textile wastewater by ozonation. *Desalination*, *260*, 1-3, 34–38, 2010.
61. Navin, P.K., Kumar, S., Mathur, M., Textile wastewater treatment: a critical review. *Int. J. Eng. Res. Technol.*, *6*, 11, 1–7, 2018.
62. Tzitzi, M., Vayenas, D.V., Lyberatos, G., Pretreatment of textile industry wastewaters with ozone. *Water Sci. Technol.*, 29, 9, 151–160, 1994.
63. Sevimli, M.F. and Sarikaya, H.Z., Ozone treatment of textile effluents and dyes: effect of applied ozone dose, pH and dye concentration. *J. Chem. Technol. Biotechnol.: International Research Process, Environmental & Clean Technology*, 77, 7, 842–850, 2002.
64. Sevimli, M.F., Sarikaya, H.Z., Yazgan, M.S., A new approach to determine the practical ozone dose for color removal from textile wastewater. *Ozone Sci. Eng.*, 25, 2, 137–143, 2003.
65. Bilińska, L., Blus, K., Bilińska, M., Gmurek, M., Industrial textile wastewater ozone treatment: catalyst selection. *Catalysts*, 10, 6, 611, 2020.
66. Basrık, C., Kutlay, K., Değirmenci, N., Boyama atık suyunun ozonlanarak boyamada tekrar kullanılması, 2021.
67. Lotito, A.M., Fratino, U., Bergna, G., Di Iaconi, C., Integrated biological and ozone treatment of printing textile wastewater. *Chem. Eng. J.*, 195, 261–269, 2012.
68. Ledakowicz, S., Żyłła, R., Paździor, K., Wrębiak, J., Sójka-Ledakowicz, J., Integration of ozonation and biological treatment of industrial wastewater from dyehouse. *Ozone: Sci. Eng.*, 39, 5, 357–365, 2017.
69. Eren, S., Eren, H.A., Bayaçlı, G., Özyurt, I., Color Removal Of Dısperse Dyeıng Waste Water By Ozone In An Example Dyehause. *Tekst. Müh.*, 30, 131, 260–263, 2023.
70. Ciardelli, G., Capannelli, G., Bottino, A., Ozone treatment of textile wastewaters for reuse. *Water Sci. Technol.*, 44, 5, 61–67, 2001.
71. Sundrarajan, M., Vishnu, G., Joseph, K., Decolorisation of exhausted reactive dye bath using ozonator for reuse. *Int. J. Environ. Sci. Technol.*, 4, 263–270, 2007.

72. Yin, H., Qiu, P., Qian, Y., Kong, Z., Zheng, X., Tang, Z., Guo, H., Textile wastewater treatment for water reuse: a case study. *Processes*, 7, 1, 34, 2019.
73. Bilińska, L., Blus, K., Gmurek, M., Ledakowicz, S., Coupling of electrocoagulation and ozone treatment for textile wastewater reuse. *Chem. Eng. J.*, 358, 992–1001, 2019.
74. Abdel-Halim, E.S., Simple and economic bleaching process for cotton fabric. *Carbohydr. Polym.*, *88*, 4, 1233–1238, 2012.
75. Eren, H.A. and Ozturk, D., The evaluation of ozonation as an environmentally friendly alternative for cotton preparation. *Text. Res. J.*, 81, 5, 512–519, 2011.
76. Prabaharan, M. and Rao, J.V., Combined desizing, scouring and bleaching of cotton using ozone, *Indian J. Fibre Text. Res.*, 28, 4, pp. 437–443, 2003.
77. Yigit, I., Eren, S., Eren, H.A., Ozone utilisation for discharge printing of reactive dyed cotton. *Color. Technol.*, 134, 1, 13–23, 2018.
78. Gedik, G., Aydın Kızılkaya, Y.M., Uyak, V., Koluman, A., Simultaneous Eco-friendly Bleaching and Retting Wastewater Treatment of Hemp Fiber with Ozone Application. *Fibers Polym.*, 24, 1, 57–72, 2023.
79. Prabaharan, M. and Rao, J.V., Study on ozone bleaching of cotton fabric-process optimisation, dyeing and finishing properties. *Color. Technol.*, 117, 2, 98–103, 2001.
80. Perincek, S.D., Duran, K., Korlu, A.E., Bahtiyari, I.M., An investigation in the use of ozone gas in the bleaching of cotton fabrics. *Ozone: Sci. Eng.*, 29, 5, 325–333, 2007.
81. Eren, H.A., Avinc, O., Erişmiş, B., Eren, S., Ultrasound-assisted ozone bleaching of cotton. *Cellulose*, 21, 4643–4658, 2014.
82. Arooj, F., Ahmad, N., Chaudhry, M.N., A pilot-scale application of ozone to bleach raw cotton fabric using various additives. *Ozone: Sci. Eng.*, 37, 3, 203–215, 2015.
83. Prabaharan, M., Nayar, R.C., Kumar, N.S., Rao, J.V., A study on the advanced oxidation of a cotton fabric by ozone. *Color. Technol.*, 116, 3, 83–86, 2000.
84. Bahtiyari, M.I. and Benli, H., Comparison of ozone-based cold bleaching processes with conventional pretreatment of cotton. *Ozone: Sci. Eng.*, 42, 5, 450–460, 2020.
85. Kurban, M., Yavas, A., Avinc, O., Eren, H.A., Nettle biofibre bleaching with ozonation. *Ind. Textila*, 67, 1, 46–54, 2016.
86. Maqsood, H.S., Bashir, U., Wiener, J., Puchalski, M., Sztajnowski, S., Militky, J., Ozone treatment of jute fibers. *Cellulose*, 24, 1543–1553, 2017.
87. Eren, H.A., Avinc, O., Uysal, P., Wilding, M., The effects of ozone treatment on polylactic acid (PLA) fibres. *Text. Res. J.*, 81, 11, 1091–1099, 2011.
88. Avinc, O., Eren, H.A., Uysal, P., Wilding, M., The effects of ozone treatment on soybean fibers. *Ozone: Sci. Eng.*, 34, 3, 143–150, 2012.
89. Perincek, S., Bahtiyari, M.I., Körlü, A.E., Duran, K., Ozone treatment of Angora rabbit fiber. *J. Cleaner Prod.*, 16, 17, 1900–1906, 2008.

90. Ul-Haq, N. and Nasir, H., Cleaner production technologies in desizing of cotton fabric. *J. Text. Inst.*, 103, 3, 304–310, 2012.
91. Turhan, Y. and Soydaş, S., The effects of ozone bleaching and ozone desizing method on whiteness and water absorption of 100% cotton terry fabrics. *Int. J. Mater. Sci. Appl.*, *7*, 3, 85–94, 2018.
92. Yiğit, I. and Eren, S., Haşil Sökme İşlemine Alternatif Bir Metot; Ozon Uygulamasi. *Uludağ Üniversitesi Mühendislik Fakültesi Derg.*, *22*, 1, 139–148, 2017.
93. Kan, C.W., Cheung, H.F., Kooh, F.M., An investigation of color fading of sulfur-dyed cotton fabric by plasma treatment. *Fibers Polym.*, 18, 767–772, 2017.
94. Turhan, K. and Ozturkcan, S.A., Decolorization and degradation of reactive dye in aqueous solution by ozonation in a semi-batch bubble column reactor. *Water Air Soil Pollut.*, *224*, 1–13, 2013.
95. Kıcık, H. and Eren, H.A., Application of ozone gas for the stripping of fabric ink-jet-printed with reactive dyes. *Color. Technol.*, 133, 6, 485–490, 2017.
96. Eren, S., Gümüs, B., Eren, H.A., Colour stripping of reactive-dyed cotton by ozone treatment. *Color. Technol.*, 132, 6, 466–471, 2016.
97. Atav, R., Yüksel, M.F., Özkaya, S., Buğdaycı, B., The use of ozone technology for color stripping and obtaining vintage effect on reactive dyed cotton fabrics. *J. Nat. Fibers*, 19, 13, 6648–6658, 2022.
98. Hu, L., Zhou, X., Chen, J., Zhang, X., Chen, G., Chang, W., Yi, C., Investigation of color fading and fabric-touch test for jeans through ozonation. *Ozone: Sci. Eng.*, 43, 3, 276–283, 2021.
99. Zulfiqar, A., Arooj, F., Aftab, M., Rashid, M., Luqman, M., Kashif, S.U.R., Naseer, R., A Sustainable Approach to Dyed Cotton Fabric Stripping Using Ozone. *Sustainability*, 15, 9, 7467, 2023.
100. Güneş, Y., Atav, R., Namırtı, O., Effectiveness of ozone in decolorization of reactive dye effluents depending on the dye chromophore. *Text. Res. J.*, *82*, 10, 994–1000, 2012.
101. Powar, A., Perwuelz, A., Behary, N., vinh Hoang, L., Aussenac, T., Loghin, C., Chen, G., Investigation into the color stripping of the pigment printed cotton fabric using the ozone assisted process: A study on the decolorization and characterization. *J. Eng. Fibers Fabr.*, 16, 1558925021992757, 2021.
102. İvedi, I., Yağcı, K., Tağaç, E., Development Of Ecological Denim Bleaching Methods, in: *International Conference on Textile and Connected R&D Domains (TEXTEHX)*, pp. 81–87, 2021.
103. Garcia, B., Reduced water washing of denim garments, in: *Denim*, pp. 405–423, Woodhead Publishing, Cambridge, UK, 2015.
104. Ben Fraj, A. and Jaouachi, B., Effects of ozone treatment on denim garment properties. *Color. Technol.*, 137, 6, 678–688, 2021.
105. Nayak, R., George, M., Jajpura, L., Khandual, A., Panwar, T., Laser and ozone applications for circularity journey in denim manufacturing-A developing country perspective. *Curr. Opin. Green Sustainable Chem.*, 38, 100680, 2022.

106. Atav, R., Gündüz, O., Yaz, S., Çakan, G., Sevgili, B., Eco-Friendly Ozone Process for Denim Garments as an Alternative to Conventional Bleaching. *Fibers Polym.*, 1–7, 2024.
107. Körlü, A., Use of ozone in the textile industry, in: *Textile Industry and Environment*, pp. 1–23, IntechOpen, Rijeka, Croatia, 2018.
108. Sarker, U.K., Kawser, N., Rahim, A., Al Parvez, A., Shahid, M.I., Superiority of sustainable ozone wash over conventional denim washing technique. *Int. J. Curr. Eng. Technol.*, 11, 5, 516–522, 2021.
109. Ben Hmida, S. and Ladhari, N., Study of parameters affecting dry and wet ozone bleaching of denim fabric. *Ozone: Sci. Eng.*, 38, 3, 175–180, 2016.
110. Eren, H.A., Afterclearing by ozonation: a novel approach for disperse dyeing of polyester. *Color. Technol.*, 122, 6, 329–333, 2006.
111. Eren, H.A. and Anis, P., Surface trimer removal of polyester fibers by ozone treatment. *Text. Res. J.*, 79, 7, 652–656, 2009.
112. Avinc, O., Eren, H.A., Uysal, P., Ozone applications for after-clearing of disperse-dyed poly (lactic acid) fibres. *Color. Technol.*, 128, 6, 479–487, 2012.
113. Eren, H.A., Gundogan, S., Eren, S., Kocaman, B., Substitution of reduction clearing step by ozone treatment at disperse dyeing of polyester. *Ozone: Sci. Eng.*, 35, 3, 196–200, 2013.
114. Ahmad, J. and Zhou, Z., Mechanical properties of natural as well as synthetic fiber reinforced concrete: a review. *Constr. Build. Mater.*, 333, 127353, 2022.
115. Michael, M.N., El-Zaher, N.A., Ibrahim, S.F., Investigation into surface modification of some polymeric fabrics by UV/ozone treatment. *Polym. Plast. Technol. Eng.*, 43, 4, 1041–1052, 2004.
116. Osbeck, S., Bradley, R.H., Liu, C., Idriss, H., Ward, S., Effect of an ultraviolet/ozone treatment on the surface texture and functional groups on polyacrylonitrile carbon fibres. *Carbon*, 49, 13, 4322–4330, 2011.
117. Wang, Y., Wiener, J., Militky, J., Mishra, R., Zhu, G., Ozone effect on the properties of aramid fabric. *Autex Res. J.*, 17, 2, 164–169, 2017.
118. Fattahi, F.S., Khoddami, A., Rahmatinejad, J., Nanoscale roughness on the surface of polyester fibers through ultraviolet/ozone treatment. *Iran. J. Polym Sci. Technol.*, 32, 457–473, 2020.
119. Yiğit, I., Eren, S., Özenç, A., Saka Dinç, Z., Surface treatment of hemp materials. *Oral presentation at the Autex 2024*, Czechia, 2024, June 17–19.
120. Rahmatinejad, J., Khoddami, A., Mazrouei-Sebdani, Z., Avinc, O., Polyester hydrophobicity enhancement *via* UV-Ozone irradiation, chemical pretreatment and fluorocarbon finishing combination. *Prog. Org. Coat.*, 101, 51–58, 2016.
121. Teklal, F., Djebbar, A., Allaoui, S., Hivet, G., Joliff, Y., Kacimi, B., A review of analytical models to describe pull-out behavior–Fiber/matrix adhesion. *Compos. Struct.*, 201, 791–815, 2018.
122. Fu, X., Lu, W., Chung, D.D.L., Ozone treatment of carbon fiber for reinforcing cement. *Carbon*, 36, 9, 1337–1345, 1998.

123. Downey, M.A. and Drzal, L.T., Toughening of carbon fiber-reinforced epoxy polymer composites utilizing fiber surface treatment and sizing. *Composites, Part A*, 90, 687–698, 2016.
124. Wang, L., Gao, S., Wang, J., Wang, W., Zhang, L., Tian, M., Surface modification of UHMWPE fibers by ozone treatment and UV grafting for adhesion improvement. *J. Adhes.*, 94, 1, 30–45, 2018.
125. Rajendran Royan, N.R., Sulong, A.B., Yuhana, N.Y., Chen, R.S., Ab Ghani, M.H., Ahmad, S., UV/O3 treatment as a surface modification of rice husk towards preparation of novel biocomposites. *PLoS One*, 13, 5, e0197345, 2018.
126. Sánchez, M.L., Patino, W., Cardenas, J., Physical-mechanical properties of bamboo fibers-reinforced biocomposites: Influence of surface treatment of fibers. *J. Build. Eng.*, 28, 101058, 2020.
127. Choi, H., Kim, Y.Y., Lim, H., Cho, J., Kang, J.W., Kim, K.S., Oxidation of polycyclic aromatic hydrocarbons by ozone in the presence of sand. *Water Sci. Technol.*, 43, 5, 349–356, 2001.
128. Kim, E.M. and Jang, J., Surface modification of meta-aramid films by UV/ozone irradiation. *Fibers Polym.*, 11, 677–682, 2010.
129. Elnagar, K., Abou Elmaaty, T., Raouf, S., Dyeing of polyester and polyamide synthetic fabrics with natural dyes using ecofriendly technique. *J. Text.*, 2014, 1, 363079, 2014.
130. Gabardo, R.S., de Carvalho Cotre, D.S., Lis Arias, M.J., Moisés, M.P., Martins Ferreira, B.T., Samulewski, R.B., Bezerra, F.M., Surface modification of polyester fabrics by ozone and its effect on coloration using disperse dyes. *Materials*, 14, 13, 3492, 2021.
131. Eren, S., Öztürk, M., Yiğit, I., The effect of surface modification with ozone in waterless (ScCO2) dyeing o polyester. *J. Innov. Eng. Nat. Sci.*, 4, 2, 314–325, 2023.
132. Atav, R. and Yurdakul, A., Effect of the ozonation process on the dyeability of mohair fibres. *Color. Technol.*, 127, 3, 159–166, 2011.
133. Anam, W., Akhtar, K.S., Muhammad, M., Sardar, S., Saleem, I., Development of novel and sustainable ozone based dyeing processes for cotton fabric. *Cellulose*, 1–15, 2024.
134. Bhattacharyya, D. and Baheti, V., Ozone assisted modification and pulverization of plant fibres. *Ind. Crops Prod.*, 199, 116711, 2023.
135. Sargunamani, D. and Selvakumar, N., A study on the effects of ozone treatment on the properties of raw and degummed mulberry silk fabrics. *Polym. Degrad. Stab.*, 91, 11, 2644–2653, 2006.
136. Balcı, N., Ömeroğulları, Z., Kut, D., Eren, H., Effects of plasma and ozone treatments on tensile and whiteness properties of 100% silk. *Uludağ Üniversitesi Mühendislik Fakültesi Derg.*, 20, 2, 43–56, 2015.
137. Eren, S., Yiğit, I., Kutlay, K., Kaya, Z., Basrık, C., Eren, H.A., A New Sight of Ozone Usage in Textile: Improving Flame Retardant Properties. *Polymers*, 16, 6, 735, 2024.

# 16

# Coating Technologies in Surface Modification of Textiles: Performance-Enhancing Textile Coatings

**A. Selcen Altınok**

*DOSETEX R&D, Tekirdag, Türkiye*

***Abstract***

In recent years, there have been significant changes in the coating field in terms of coating techniques, properties of coating materials, production techniques, and application areas. In the field of technical textiles, coating provides many functional properties to fabrics with different techniques. The techniques used vary according to the properties of the coating material. In recent years, modern techniques such as plasma and sol-gel have also been used.

The properties of coating materials vary depending on the area of use, desired properties, chemical, environmental, mechanical requirements, price, and application. Adding them to the final product has a significant impact on the desired properties. Weight, thickness, tensile strength, burst damage resistance, abrasion resistance, dimensional stability, adhesion, etc., tests are applied to determine the performance of the fabric and the coating material used.

In recent years, as the demand for technical textiles has increased, the demand for coated fabrics has also increased. This section provides a comprehensive overview of coating techniques, coating polymers used, fibers used in coated fabrics, desired properties of final products, areas of use of coated fabrics, tests applied to coated fabrics, and standards of these tests. Finally, future prospects and insights in this field are also discussed.

***Keywords***: Coating, coating in technical textiles, coating materials, coating polymers, surface modification in coating, binders, functionality in textiles

*Email*: selcenaltinok@dosetex.com

Aminoddin Haji and Çağlar Sivri (eds.) Surface Modification and Functionalization of Textiles, (395–436) 

## 16.1 Introduction

Coated and laminated textile products, resulting from coating and lamination processes, form a significant part of today's textile industry. The increasing demand for technical textiles, especially in recent years, has also increased the demand for coated fabrics. In the last 5 years, the global coated fabrics market has seen significant growth. The value of the market, which was approximately 21.6 billion USD dollars in 2023, is expected to reach approximately 29.8 billion USD dollars by 2030. In this case, approximately 6.2% growth is estimated. Key drivers include increasing demand from sectors such as transportation, furniture, and protective clothing, as well as a growing focus on sustainability.

Transportation remains the largest application segment, accounting for nearly 29.5% of the market in 2022, while industrial applications and protective clothing also play significant roles. Geographically, the Asia-Pacific region leads in demand due to rapid industrialization, especially in automotive manufacturing, whereas North America and Europe are also major markets [18, 19].

Coating is a process applied to fabrics to achieve special effects that cannot be obtained through standard finishing processes. Depending on the desired effect, chemicals in liquid, paste, or powder form are transferred to the fabric in dust, paste (pat), or foam form to create a film layer on the fabric. Some of the properties expected from textile materials, such as waterproofing, heat insulation, and aesthetic appearance, can be achieved by coating fabrics with polymers.

Coated and laminated fabrics, produced to protect against external factors, have many applications ranging from agricultural textiles to medical textiles, construction, and protective clothing. The coating material used, the applied production technique, and the structure and properties of the textile surface directly affect the functional properties of fabrics.

Coating in textiles refers to applying a layer of material onto the surface of a fabric to enhance its properties or add new functionalities. This coating can provide functional features on fabrics such as the following:

Waterproofing: Makes the fabric resistant to water penetration, often used in rainwear, outdoor gear, and protective clothing.
UV Protection: Blocks or absorbs harmful ultraviolet rays, often used in outdoor textiles like tents or umbrellas.
Chemical Resistance: Protects against exposure to hazardous chemicals, useful in medical and industrial environments.

Flame Retardancy: Reduces the fabric's flammability, making it safer for applications like firefighting gear or upholstery.
Breathability: Some coatings allow moisture to escape while blocking liquids, which is important for comfort in activewear.
Antibacterial: Prevents the growth of bacteria or fungi, common in medical textiles and sportswear.
Abrasion Resistance: Strengthens the fabric to withstand wear and tear, used in automotive, military, and heavy-duty applications.

The coatings are usually made from polymers, such as polyurethane (PU), polyvinyl chloride (PVC), silicone, or fluoropolymers, and are applied *via* techniques like knife coating, transfer coating, or dip coating. This process allows textiles to serve specific functional roles in industries like medical, military, and agriculture.

In recent years, the field of textile coatings has experienced significant innovations, driven by advancements in materials science, nanotechnology, and smart textiles.

Some studies on coating applications in textiles in literature:

In the study conducted by Ferziger, a tightly woven glass fiber fabric with at least one surface coated with one or more layers of flame-retardant foam was produced. Carbon black was considered suitable for the light-blocking fabric layer [1].

In a study by Hakansson and colleagues, the reflection, transmission, and absorption values of textile materials coated with polypyrrole were measured. The measurements were carried out using the free space test method in the frequency range of 1–18 GHz. Polypyrrole was polymerized in solution and coated on the polyester fabric surface. It was determined that shielding effectiveness increased in direct proportion to the polymerization time [2].

In the study presented by Mondal and Hu, it was aimed to provide high protection by applying hydrophilic coatings containing multi-walled carbon nanotubes (MWNTs) to protect cotton fabrics from UV radiation. Multi-walled carbon nanotubes were dispersed in a hydrophilic PU solution and applied to the fabrics using the solution knife coating technique. Three different MWNT concentrations were used: 0.25%, 1%, and 2.5%. Hydrophilic PU coatings containing nanotubes provided very high protection against UV radiation. The ultraviolet protection factor (UPF) value was 174 for 1% MWNT, and this value increased to 421 for 2.5% MWNT [3].

Renhby and colleagues applied coating on polyester fabrics with conductive polymers like polyaniline, polythiophene, and polypyrrole in the presence of an acrylic binder to obtain conductive textile surfaces. They worked on surface resistance, surface analysis, surface appearance, and the behavior of the coating paste. The concentration of the conductive polymer, the number of coated layers, and the desired conductivity on the surface, which can withstand aging and mechanical stress, varied [4].

Asker and colleagues conducted an experimental study on woven and knitted fabrics with different characteristics on the front and back sides. In the study, four types of conventional fabrics were selected as the base, two woven and two knitted. The raw material content of all selected fabrics was 100% cotton. The front surfaces of these four different base fabrics were coated using conventional methods, and three different chemical finishing agents were applied to the fabric using the impregnation method after the coating process.

The front surfaces of the coated fabrics became hydrophobic, whereas the back sides remained hydrophilic. Later, the chemical finishing process was applied to the hydrophilic side of the fabric. As a result, hybrid fabrics with different properties on each side were produced.

Durability, abrasion resistance, color measurement, and fastness tests were applied to the fabrics. When the results of these tests were examined, it was determined that the hybrid fabric production stages did not change these properties of the sample fabrics. While these properties remained unchanged, the fabrics achieved water repellency, hydrophilicity, and antibacterial-antifungal activity as desired. The results showed that hybrid fabric applications might yield better results in woven fabrics than in knitted fabrics. Additionally, when woven fabrics were examined internally, it was concluded that hybrid activity could be more easily achieved in thicker and higher-weight fabrics [5].

In a study by Negru and colleagues, the aim was to develop a fabric with electrical conductivity produced from cotton yarns coated with a carbon black-based solution. Different recipe results prepared using the tested carbon black particles, and other components of the coating mixture were presented to obtain the best results regarding electrical conductivity. Measurements were made according to the AATCC 76-1995 test method, and the surface electrical resistance was measured using a multimeter (1906 programming multimeter TTI). As a result, it was found that the increase in the concentration of carbon black particles led to a decrease in electrical resistance, whereas the concentration of other components of the coating material had less effect on the electrical properties of the conductive fabric [6].

Padleckiene and colleagues examined the effect of friction on mass loss and air permeability in breathable PU-coated fabrics. The study concluded that the friction force and the number of friction cycles affected the air permeability and mass loss parameters due to the mechanical deformation of the applied PU coating [7].

Gouda and Keshk studied the antibacterial properties of cotton fabrics coated with chitosan. The antibacterial properties were investigated using the AATCC 100-1999 test method with *S. aureus* and *E. coli* bacteria. The reduction percentages in the number of *S. aureus* and *E. coli* bacteria were found to be 80% and 70.5%, respectively, for chitosan-coated fabric; 86% and 84% for fabric coated with chitosan and titanium dioxide; and 97% and 95% for fabric coated with chitosan and zirconium oxide. For the fabric where both metal oxides were used with chitosan, a bacterial reduction of 96% and 93% was observed [8].

In their experimental study, Asanovic and colleagues worked on antibacterial coatings for medical purposes applied to knitted surfaces. Chitosan gel was applied to the knitted fabric, and antibacterial activity was measured at the end of the study [9].

In a study by Paul and colleagues, the UV resistance of cotton fabrics was improved by applying ZnO and $TiO_2$ nanoparticles. ZnO and $TiO_2$ nanoparticles were applied to both dyed and undyed fabrics. The application of ZnO particles to cotton yarns was performed according to the exhaust principle. The UV protection factor obtained as a result of the ZnO application to undyed fabrics was 30.03, whereas a value of 47.91 was obtained for dyed fabrics. $TiO_2$ nanoparticles were applied to bleached and reactively dyed fabrics using two different methods: sol-gel and a binding agent. In the binding agent method, two different $TiO_2$ phases—rutile and anatase—were used, and rutile was found to be more effective for UV protection. According to the sol-gel method application, a UPF value of 50+ was achieved in undyed fabrics both before and after washing, while this result was not obtained in dyed ones. Comparing $TiO_2$ and ZnO particles, it can be said that $TiO_2$ showed better UV protection performance [10].

In a study by Ledakowicz and colleagues, a sol-gel coating was applied to fabric with nano-sized $TiO_2$ and $TiO_2$ modified with aminosilane. The microstructures of the nano-structured titanium dioxide-coated fabrics were evaluated using high-resolution SEM and TEM electron microscopes. The absorption spectra of the fabrics were determined using the UV-Vis Jasco V-550 apparatus. According to EN 13758-1:2002, the UPF of the fabrics was also measured with the same device. It was determined that the UPF values of the fabrics treated with $TiO_2$ and aminosilane-modified $TiO_2$ were above 50 [17].

Gouda and Keshk applied a chitosan film to cotton fabric by coating it with zirconium oxide and titanium dioxide in different proportions. While the chitosan-titanium dioxide mixture film showed a very high interaction with cotton fabric, the same was not achieved with the zirconium-titanium-chitosan mixture film. According to the test results, the UV protection factor was 50 and above for the fabrics treated with titanium dioxide, while this value ranged between 45 and 50 with the addition of zirconium oxide. In both cases, using chitosan with metal oxides resulted in a UV protection value of 50 and above [8].

## 16.2 Coating Techniques

When determining the coating technique, the properties of the fabric to be coated, the fabric's construction, the form in which the coating material is applied, the viscosity range being worked with, the coating's precision and adhesion strength, the customer's quality criteria, and the economic feasibility of the process are the main factors to consider.

The coating material should have a viscosity that allows it to spread on the yarn and fiber surfaces, and the fabric surface should be smooth and even after coating. In every technique, the fabric should be fully opened to its width before coating, and tension-controlled feeding should be used. After coating, the coated fabric should be processed in a dryer to evaporate and remove the solvents present in the fabric before cooling and winding.

Coating methods can be categorized into three sections: methods where the coating material is liquid, methods where the material is solid, and modern coating methods such as plasma and sol-gel technology.

### 16.2.1 Liquid Coating Methods

In the coating technique where the coating material is liquid, factors such as viscosity, surface tension, and drying come to the fore. Liquid coating techniques are generally used in areas where uniform, thin coatings are needed. Liquid coating techniques are examined in two groups.

#### *16.2.1.1 Pre-Dosed Coating*

Pre-dose coating is a practice in which the exact amount of coating material is pre-applied in a controlled manner onto a substrate to be coated, in applications where the amount of coating material is critical. This technique is preferred in various applications in materials where high performance is required, especially technical textiles, where uniformity, quality, and efficiency come to the fore. The aim of the pre-dosed coating method, which allows controlled dosing, is to achieve consistent thickness, improved adhesion and targeted functionality.

Advantages of pre-dosed coating method:

- In applications where precision comes to the fore, pre-dosed coating ensures extremely precise application of the coating material, minimizing unnecessary material use. With controlled material use, it ensures that the required amount of material is used.
- In areas where uniformity, that is, stability and homogeneity, is desired, material can be applied evenly throughout the ground layer, as the coating amount and thickness are measured in advance. This method is especially suitable for protective textiles, medical textiles, electronic textiles, etc., where even distribution of the coating layer is required. It is preferred in applications.
- Material waste that may occur during the coating process is minimized and excess material usage is prevented by controlling the material dosage. With this advantage of the process, more sustainable and cost-effective application opportunities are offered.
- Reproducibility of the process is possible by dosing the material in a controlled manner. In applications made with this technique, it maintains consistency between different batches during quality control in the automotive, defense and health industries.
- It can be used in coating applications such as waterproofing, UV protection, and thermal insulation that require pre-dosing due to the advantage of being able to apply the coating material in a controlled manner. Special applications can be made with special prescriptions.

From the perspective of application techniques:

Depending on the materials and desired properties, there are different methods of applying pre-dosed coatings:

Roll Coating: The coating process is carried out by applying a predetermined amount of coating material to the surface to be coated through rollers. Because the material is transferred to the surfaces to be coated *via* a roller, this method is very suitable and useful for flexible surfaces such as technical textiles.

Spray Coating: The coating process is carried out by controlled spraying of the coating material in predetermined dosing amounts onto the surface to be coated. It is generally preferred on sensitive floors that are damaged by mechanical pressure. It is also used in complex, 3D coatings.

Dip Coating: The coating process is carried out by dipping the surface or material to be coated into a pre-dosed solution, that is, prepared according

to a certain recipe, and then the drying process is carried out. It provides a consistent coating layer, which is widely used in products such as gloves and socks, where coating in open width fabric is not possible.
Screen Printing: A patterned surface is obtained by dosing the coating material onto the surface to be coated in the desired amount using a digital screen with predefined patterns.

Advantages in the field of technical textiles:

- Textiles coated with pre-dosed methods have very good performance properties. High strength, resistance to harsh weather conditions, good flame retardancy, antimicrobiality, etc. It is possible to add advanced features.
- Using less amount of coating material provides an advantage in terms of energy efficiency. It reduces energy consumption during production as a controlled process is required.
- Reducing the amount of waste material and increasing the coating process efficiency reduces the overall costs in production.

Pre-dosed coating methods are used in many areas. For smart fabrics that integrate electronics or sensors in the field of wearable technologies, pre-dosed coating enables the precise application of conductive layers or protective elements without compromising flexibility. In the development of wound care products or implantable fabrics in the field of medical textiles, precision coating increases biocompatibility and reduces infection risks.

Although pre-dosed coating methods are advantageous, they also have difficulties. In cases where there are incompatibilities between the coating material and the coated surface, the desired product properties may not be achieved. This situation can be determined and prevented with pre-application studies. In addition, the smoothness and evenness of the coating surface is an important detail that increases the coating quality. Coating applications with pre-dosed processes are widely preferred, especially in applications that require great precision and where performance criteria are a priority. It is also a very suitable method for environmentally friendly and sustainable applications.

Pre-dosed coating is becoming increasingly popular in industries where precision and performance are critical, offering a tailored and sustainable approach to coating processes.

### *16.2.1.2 Post-Dosed (On-the-Go) Coating*

Post-dosed coating method is the application of coating material to the underside of the textile surface after the dyeing and fabric production processes. This method is preferred if the fabric is to be given additional functional and operational properties. Below are key details about the process and its applications:

Stages of the post-dosed coating process

1) Before the coating process, the fabric is cleaned by passing through washing, scrubbing, and corona discharge pre-treatments. Thus, the bonding and surface smoothness on the surface are ensured.
2) The coating process is applied by determining the desired thickness and surface properties according to the coating technique selected depending on the application.
3) After the coating process, the fabric is passed through a drying chamber and the drying process is performed. Thus, the solvents and excess moisture on the coating material are removed.
4) A curing process with heat or UV light is performed to cross-link the polymers used as coating material to the fabric surface and to increase the strength and adhesion of the coating.
5) As a final finishing process, additional processes can be applied to obtain a smooth or fluffier appearance on the coated fabric surface.

In the final dosing coating processes, a coating material is applied to the fabric immediately after the first production of the base fabric. It is especially preferred in products where high-performance properties such as waterproofing, flame resistance and improved durability are desired. You can see general information about coating techniques after dosing below.

Knife-Roll Coating: The coating material is spread on the fabric surface by passing it under a blade on the fabric roll with the help of a roller. The thickness of the coating is determined by the gap between the fabric surface and the blade.

Roll Coating: The coating process is carried out by spreading the coating material on the fabric surface with the help of a roller. This method is preferred in production applications that require high speed.

Coating by Dipping Method: The coating material is made ready for application in a bath and the fabric to be coated is completely passed through this bath and the coating process is carried out by removing excess material with the help of rollers. Although equal coating homogeneity is provided with this method, sometimes higher coating thicknesses due to excessive absorption of the coating material may be encountered.

Spray Coating: The coating material is sprayed onto the fabric of the surface to be coated. It is preferred in applications where sensitive and local uniformity and surface aesthetics are important.

In the pad dry cure coating method, a bath containing the coating material is prepared and the fabric is passed through this bath. The excess coating liquid remaining on the fabric is removed by passing through successive pads and the coating liquid on the fabric surface is dried, hardened and bonded to the fabric. While polymers such as PU, PVC, silicone, and acrylic are widely preferred in coating processes, thermoplastic and thermoset polymers are generally preferred to provide functional properties such as waterproofing, durability, and flexibility to the fabric and to create protective, shiny, and functional surfaces.

Nanocoating processes should be preferred to provide more advanced functional properties such as self-cleaning, antimicrobial properties, and advanced UV protection.

Advantages of post-dosed coatings

High Durability: With coatings applied after production, resistance to abrasion and wear can be created in order to withstand harsh conditions in automotive and protective textiles.

Waterproofness/Water Resistance: This is a very desired feature especially in protective clothing and outdoor textiles, and this feature is provided with coating techniques applied after fabric production.

Breathability: It is a very necessary and very frequently desired functional feature in wearable textiles and medical textiles.

UV Protection: It is a preferred feature especially in outdoor textiles, outerwear, and agricultural textiles. Textile surfaces that can absorb UV rays can be developed with the applied coatings.

Flame Retardancy: This feature can be provided with both coating and yarns used in textile fabrics. Flame retardant chemicals are coated on the fabric surface after production with coating and used to increase safety in textiles used in defense, automotive, and interior furniture.

Aesthetics and Surface Changes: With different coatings, a decorative effect can be created on the textile surface and a shiny or matte appearance can be achieved, and a visually more impressive surface can be obtained in the fabric.

Post-dosing coating method is used in the different fields of technical textiles:

Coatings used in automotive textiles provide resistance to oil, heat, and abrasion in the car's interior, upholstery, and airbags.

Textile fabrics that are flame retardant, resistant to chemicals, and water resistant functional properties are highly preferred in the defense industry and industrial textiles.

Textile fabrics that are given antimicrobial and liquid repellent properties with coatings are frequently preferred in medical textiles and masks.

Waterproof, breathable coatings are very important for sports textiles as they can provide comfort and performance in harsh weather conditions. Resistance to weather conditions, UV protection, and breathability are highly preferred in agricultural textiles.

In the field of wearable technologies, conductive coatings that can have a structure that is both conductive and flexible, is comfortable, and allows sensor integration on the textile surface are very important.

Coatings that generally allow the addition of aesthetic properties such as shine or texture while preserving the drape of the basic fabric are highly preferred in the field of fashion and ready-to-wear.

In the field of healthcare, it is generally used in antimicrobial and liquid barrier coatings used in medical devices, gowns, and other protective clothing.

In recent years, with environmental sustainability coming to the forefront, advances in material technology have fed this area and minimized environmental damage, and water-based high value-added coating materials have been developed. It is anticipated that developments in this area will continue to increase.

Smart and functional coating integration of conductive materials and sensors into the coating: It has enabled the development of smart textiles that monitor temperature, heart rate, or other biological parameters. In recent years, there have been serious advances, especially in the field of medical textiles. Figure 16.1 shows wearable technologies where coating techniques are widely used to integrate electronic materials into conductive textile structures. It is predicted that these products will increase in the coming years in order to make our lives easier and to monitor our health more closely.

In the field of nanotechnology, nanomaterials are increasingly used in coatings to improve fabric functionality, such as creating self-cleaning or superhydrophobic surfaces. Nanocoatings are used especially in agricultural applications in fruit and vegetable cultivation to protect the product from pests and climatic conditions. In this field, it is possible to obtain light, flexible materials that are breathable and can transfer sunlight (Figure 16.2).

In order to enable agriculture on other planets, studies have been carried out on vegetable cultivation under laboratory conditions and positive

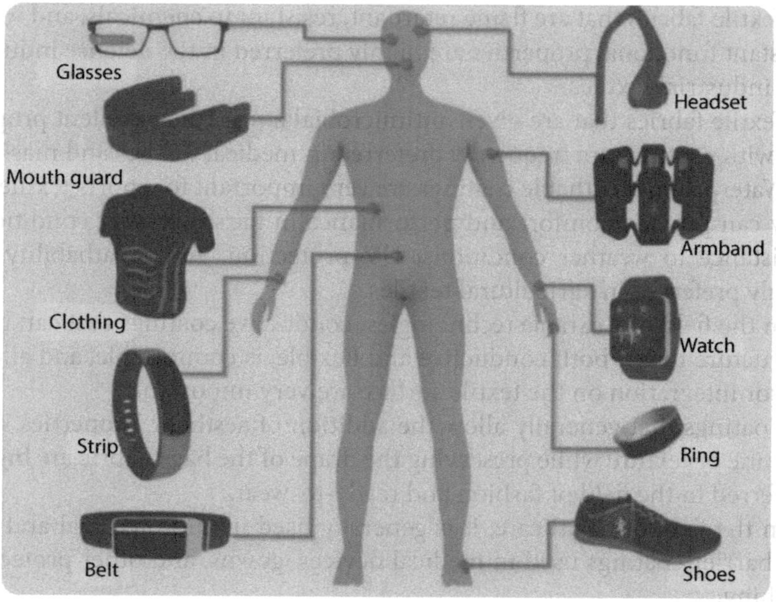


**Figure 16.1** Wearable technologies [18, 19].

**Figure 16.2** Nanocoated smart fabric for smart agriculture without pests.

results have been obtained. Coated textile surfaces to protect the plant are also used in this field. Figure 16.3 shows growing tomatoes on Mars and the Moon with coated textile surfaces, exhibited at the 2017 Frankfurt Technical Textiles Fair.

**Figure 16.3** Growing tomatoes on Mars and Moon with coated textile surfaces.

Coated fabrics are used extensively in spacesuits to provide different functional features such as thermal comfort, fire retardation, and protection from external environmental conditions Figure 16.4 shows coated fabrics for space suits that were first exhibited at the 2017 Technical Textiles Fair in Frankfurt.

Nanocoatings have an important role in creating electrically conductive textile surfaces. Especially in organic photovoltaics, flexible photovoltaic surfaces can be obtained on large surfaces at low costs. Advances in this field will facilitate the widespread use of organic solar cells. Figure 16.5 shows a textile-based nanocoated solar cell.

**Figure 16.4** Coated fabrics for spacesuits.

**Figure 16.5** Textile-based nanocoated solar cell.

Challenges of the post-dosed coating are as follows:

The adhesion of the coating material to the fabric surface is the most difficult and critical stage that determines the quality of the coating. It is necessary to ensure that the coating adheres well to the fabric substrate without compromising the flexibility of the fabric. Preservation of Fabric Properties: Although the desired functional properties are given to the fabric surface after the coating process, care must be taken not to lose the basic properties of the fabric such as drape, soft touch, shine, and breathability.

Therefore, the preparation of the formulation of the coating material is a very delicate process.

Environmental Effects: The risk of VOC emission by solvent organic compounds used in the coating material can create problems in terms of environment and safety. Therefore, although it affects the fabric quality in terms of performance, water-based or UV-cured coatings are preferred. In recent years, continuous innovative products are being developed in coating technologies due to the emphasis on environmental sustainability and multifunctionality.

### 16.2.2 Solid Coating Methods

In order to increase the durability of the textile surface and to provide different functional properties, solid coatings are coated on the lower layer of the textile fabric using materials such as polymers and resins with the following methods.

It is applied in the form of solid films and layers. Detailed insights into various solid coating methods, their processes, applications, and advantages are as follows:

Hot Melt coating:

Hot melt coating is made by applying a molten polymer or resin onto a textile substrate. The coating material is first heated until it melts. The liquid material is applied homogeneously to the fabric surface. In order for the coating material to solidify and form a smooth layer on the fabric, it is cooled and after cooling, it is rapidly solidified and bonded to the surface. Polyethylene (PE), polypropylene (PP), ethylene vinyl acetate, PU, and thermoplastic elastomers are very frequently preferred as coating materials. When we look at the areas where hot melt coating material is coated in general, in medical textiles, it is preferred for non-woven medical fabrics for masks, gowns, and bandages.

In the field of automotive textiles, it is used in airbags, seat covers and sound insulation materials. In addition, thermoplastic PU (TPU) coatings are highly preferred for shoe soles and upper materials in the shoe industry.

Being a solvent-free, environmentally friendly process, high adhesion to fabric surfaces and rapid solidification, excellent durability, and resistance to water and chemicals are the advantages of this method.

Powder Coating:
In this coating method, the coating material is adhered to the textile surface to be coated in the form of very small solid powder particles and the adhesion of these powders to the fabric is actually achieved by utilizing electrostatic charging. The fabric surface is electrostatically charged, fine-grained powders are melted in the oven or with infrared radiation and coated on the fabric surface, and a cooling process is performed to ensure a firm adhesion to the surface. Epoxy, polyester, PU and acrylic are preferred more as coating materials due to their thermoplasticity and smooth adhesion to the surface.

These coatings are used for fire-resistant and chemical-resistant fabrics in the field of protective textiles. It is used in decorative products, especially in metallic or colored coatings for bags, shoes, or interior textiles. It is preferred for conveyor belts and filter fabrics with high abrasion and chemical resistance in industrial textiles.

Creating a thick, smooth, and durable coating layer and surface with powder coating is an important advantage. High mechanical resistance to wear and tear also makes it a highly preferred coating method. The fact that volatile organic compounds (VOCs) are not used as solvents makes it stand out as an environmentally friendly coating method.

Extrusion Coating:
In this coating method, the polymer is passed through an extruder and melted, creating a thin film layer on the surface of the fabric moving on a moving belt. The polymer film formed solidifies as it cools and adheres to the fabric surface. Depending on the speed of the moving belt, surfaces of

the desired thickness can be created during the extrusion process, thin or thick. The most preferred coating materials such as PE, PP, and PVC are used in shade, vegetation and greenhouse film coatings, tarpaulins, tents, and protective covers in the agricultural field and in barrier fabrics used in medical gowns and bedding in the medical textile field.

This coating method has a high-speed process suitable for mass production, provides excellent moisture and chemical resistance, and is an ideal process for large-scale, continuous coating operations.

Laminate Coating:
In the lamination coating method, the coating process is carried out by laminating a pre-formed solid film or layer surface to the surface of a textile by applying heat and pressure. The formed solid film layer is bonded to the textile substrate to create a coated fabric.

Lamination is the simple bonding and coating of an adhesive or solid polymer to the textile surface.

Breathable and waterproof membranes that can be easily laminated with coating materials such as TPU and PE are used more frequently in sportswear and outerwear to provide functional properties such as flame retardancy in firefighter clothing and industrial protective clothing as protective textiles and in automotive textiles to provide functional properties such as sound insulation in upholstery and vehicle interiors.

With these coatings, high-performance fabrics with certain functional properties such as water-resistant, breathable, and protective can be produced and the thickness and smoothness of the coating can be controlled. Durable and long-lasting fabrics can be obtained.

Film Casting and Heat Transfer:
A solid polymer film is ensured to adhere to the textile surface by heating and pressing it onto the textile substrate with an upper pressure and pouring it homogeneously onto the fabric layer, or by melting the film slightly using heat transfer techniques.

It is then bonded to the fabric surface by solidifying the film during the cooling process. PU, PVC, and silicone films are preferred as coating materials.

It is used in shiny or metallic coatings on clothing and accessories in fashion textiles. This application is used in barrier textiles that provide resistance against liquids and pollutants in medical textiles. Weather-resistant fabrics for tarpaulins, tents, and banners in outdoor and industrial applications can be produced with this application.

The ability to adjust the film thickness precisely and consistently is one of the most important advantages of this method. High aesthetics can be achieved for decorative effects. Good resistance to abrasion, UV, and environmental factors can be achieved.

Calendaring:
It is a process in which the fabric is passed through heated rollers, along with a solid polymer or coating material. The pressure and heat from the rollers force the solid material to bond with the textile, creating a smooth, thin, and uniform coating.

The desired coating thickness can be achieved by passing the fabric through rollers multiple times.

It is used PVC, PU, and other thermoplastic resins as coating materials.

Considering the Application Areas:
In the field of technical textiles, in coated fabrics for industrial applications such as conveyor belts and inflatable structures, in the field of home textiles, and in waterproof and decorative coatings for flooring and curtains, it is used in automotive textiles for seating, flooring, and door panel coverings.

The prominent advantages of the application are that it is possible to produce extremely smooth coatings.

It is fast and cost-effective for large production processes, and high gloss or matte surfaces can be obtained.

Foam Coating:
In foam coating, a foam-like structure containing a solid polymer is applied to the fabric. The foam collapses on the fabric surface, forming a uniform solid layer upon drying or curing.

This process allows for controlled coating deposition, especially for lightweight fabrics.

PU, acrylics, and other resins are used as coating materials.

Considering the Application Areas:
In the field of apparel and fashion, lightweight waterproof coatings are used in rainwear and jackets. In medical textiles, breathable but waterproof coatings are used for medical drapes and protective clothing. In home textiles, fire-retardant coatings are used for furniture upholstery and bedding.

Advantages of application are as follows: It allows for lightweight and flexible coatings without adding bulk to the fabric. The coating is precisely controlled to adjust the thickness and ensure homogeneous distribution. It is a suitable method for delicate and light fabrics.

Screen Printing With Solid Coatings:
This method involves using a screen to apply a solid coating in specific patterns or areas on the fabric. The solid material (in paste or gel form) is applied through a screen onto the fabric, and, then, it is dried or cured to form a solid layer.

Silicone, rubber-based compounds, and thermoplastic materials are used in these coating methods.

In the field of fashion and design, this method can be used to create embossed designs on clothing, accessories, and interior fabrics. In the field of wearable technologies, functional coatings in precise patterns can be made for wearable sensors. The advantages of screen printing with solid coating are as follows:

- Highly precise placement of coatings.
- Enables decorative and functional applications simultaneously.
- It is the right choice especially when the number of colors and layers is high.

The solid coating method is preferred in technical textiles due to its advantages of providing flexibility to the surface and the ability to create sensitive and durable surfaces, especially in the automotive and medical textiles fields, by providing functional properties such as waterproofing, fire resistance, breathability, and resistance to chemicals.

### 16.2.3 Modern Coating Methods

Sol-Gel Coating Technology:
The prepared colloidal suspension is subjected to different chemical reactions to obtain a gel-like coating material. This gel-like material, consisting of metal alkoxides (such as silicon, titanium, and aluminum) dissolved in a solvent, is coated on the fabric surface by undergoing hydrolysis and condensation reactions. The coated surface is exposed to heat to ensure stronger adhesion to the surface.

Solution Preparation: Metal precursors are mixed in a solvent.

Deposition: The prepared colloidal suspension solution is applied to the fabric surface by coating methods such as dip coating, spin coating, and spray coating.

Gelation: The suspension transitions to a gel as the coating begins to solidify.

Drying/Curing: Heat treatment solidifies the gel into a durable coating.

The sol-gel coating method has many advantages. With this coating method, customized coatings with many properties such as hydrophobicity, UV protection, or antimicrobial activity can be created. It has the flexibility to be applied not only to textiles but also to many surfaces such as ceramics and glass. The possibility of precisely controlling the sol-gel process, coating thickness, and nanostructure as desired enables its use in the field of smart textiles. The coating material bonds strongly with the fabric after the curing process, providing excellent resistance to wear and tear.

The sol-gel coating method is used in different application areas. Self-cleaning properties can be provided to textile surfaces by using water-repellent and oil-repellent materials with the sol-gel coating technique. Sol-gel coatings containing silver nanoparticles are preferred in medical textiles because they provide antimicrobial properties for therapeutic purposes. It is also possible to see sol-gel coatings in energy storage applications on photovoltaic surfaces.

It is also necessary to consider some disadvantages of sol gel coating. In the sol gel coating process, the process is complex. Careful control of reaction procedures is necessary to prevent the coating from cracking or icing. The sol-gel process is more costly than traditional coatings, especially for large changeover applications. Sol-gel storage compartments must be divided and stored separately. This can sometimes make the process difficult.

Plasma Coating Technology:

Plasma is an ionized gas composed of ions, electrons, and neutral atoms. The plasma coating method is particularly preferred when very thin and functional surface coatings are desired on textile surfaces. Plasma is produced under low pressure using gases such as oxygen, nitrogen, argon, and these gases accumulate energy and interact with the reactive materials desired to adhere to the textile surface and create a thin coating surface, providing an effective coating process. In plasma coating, plasma is first produced. A high-energy field ionizes the gas, creating plasma.

By activating the textile surface, active areas are created for the coating material to adhere to the surface. The plasma environment enables the deposition of thin, uniform coatings on the nanometer scale on the fabric surface.

Plasma coating is applied in three different ways.

Plasma Polymerization Technique: Organic monomers are given to the plasma, polymer layers are deposited on the fabric, and the coating process is carried out.

Plasma Etching Technique: Plasma cleans the contaminants on the surface or activates it by changing the surface properties of the fabric before the coating is applied.

Plasma-Assisted Deposition Technique: Plasma enables special coatings with hydrophobic or conductive layers to be more easily bonded to the textile surface by deposition technique.

Advantages of the Plasma Coating Method:
Plasma coating allows the coating process to be carried out without requiring solvents or harmful chemicals; therefore, it is an environmentally friendly alternative method to traditional methods.

It allows surface modification without major changes. Plasma process allows the coating process by affecting only the surface of the textile without changing the volumetric properties of the fabric.

Plasma coatings create an extremely thin surface and allow precise control of surface properties without affecting the texture or flexibility of the material.

It has improved adhesion properties. Plasma treatment improves the functional properties of textile surfaces by improving the adhesion of other layers and increasing their durability.

Plasma coating is used in different areas; textiles with plasma technology applied are used on antibacterial surfaces. Textile surfaces can be further functionalized with antimicrobial properties using plasma polymerization. Plasma coatings are preferred to give hydrophobic properties to textiles and make them resistant to water, oil, and dirt. Plasma coatings are used in the field of electronics and smart textiles to deposit conductive layers on fabrics, integrating them with sensors or electronic circuits. In medical textiles, textile surfaces treated with plasma technology are used to create biocompatible surfaces for implants or wound dressings.

In addition to the advantages of the plasma coating method, it is necessary to consider its disadvantages. Because the cost of the equipment required for the plasma coating process is high, it cannot be used for the production of large-scale, low-cost textile fabrics. Plasma process is a complex and costly method because it requires special equipment such as vacuum chambers and sensitive control systems. Because plasma treatments are applied by activating only the surface layer of the textile, they may not be suitable for applications that require the coating material to penetrate into the lower layers.

Sol-gel and plasma coating methods are very high-level technologies for developing functional textile surfaces. While sol-gel technology offers very thin, optionally controllable, durable coating solutions, plasma offers sensitive coating solutions focused on environmental sustainability. However, due to the complexity of the application of these high technologies and their high costs, they are preferred to be used in products with high added value.

In conclusion, the choice of coating technique and material is determined by the desired properties of the coated product and the application requirements. Modern methods like plasma and sol-gel coatings are advancing the field, providing more precise control and expanding the possibilities for specialized coatings.

## 16.3 Coating Materials

In the field of technical textiles, the coating material applied to the fabric surface is generally viscous liquid. After this liquid is applied to the fabric surface, drying and curing processes are applied to ensure hardening after the coating process in order for it to adhere to the surface. The polymers used in the coatings are substances that can be applied as a melt or solution and form a solid film through solvent evaporation or cooling.

Thermoplastic polymers with cross-linking properties are used as coating materials in the field of technical textiles. These coated polymeric materials directly determine the strength and performance properties of the final product.

Environmental, mechanical, and chemical needs, desired cost and application area play an important role in the selection of coating materials. To put it more concretely, the widely known Lubrizol develops high-performance coating technologies in areas where flame retardancy, abrasion resistance, durability, and water repellency properties are desired. These coating technologies are used in industries such as automotive, work clothes, and healthcare, taking into account functional and aesthetic features [20].

Additionally, Topp Textil has introduced solvent-free thermoplastic polymer coatings, including polyesters and polyolefins. Their modern extrusion technology allows energy-efficient coating processes, enhancing functionalities like breathability, UV protection, and material reinforcement [21].

Innovations developed in recent years do not only focus on environmental sustainability but also prioritize functionality. In other words, the functional properties of environmentally friendly coating materials developed based on water are now very important. At this point, the focus is now on higher-level products [22]. Coating materials, including liquids, pastes, foams, and polymers, are selected according to specific usage requirements such as chemical resistance, mechanical strength, environmental conditions, and cost. After coating materials, especially polymer-based ones, are coated on the textile surface, different properties are added to the final product and used in the automotive, packaging, and construction sectors.

Polyvinyl Chloride (PVC):
PVC, the most widely used polymer coating material, stands out with its excellent durability, chemical, moisture and abrasion resistance, very good electrical insulation, and flame retardancy properties.

PVC: Upholstery, protective coatings, automotive interiors, and textile coatings for upholstery and raincoats are common areas of use.

The most important disadvantage to consider is that PVC can become brittle at low temperatures and release toxic chemicals during production and disposal.

Polyurethane (PU):
PU stands out as a coating material due to its durability against external weather conditions, flexibility, and resistance to abrasion. In addition, because it is possible to obtain very soft touch surfaces with PU, it is highly preferred in wearable applications.

PU Coatings: It is preferred in breathable coating applications for textiles, shoes, automotive seats, and protective films. PU foams are highly preferred in products that require insulation and fullness. However, low moisture resistance and high cost are the disadvantages of these materials.

Natural Rubber:
Natural rubber has high elasticity, flexibility, and excellent resistance to abrasion, flexibility, and tear strength.

Natural Rubber Coatings: It is preferred in gloves, tires, and medical products. These coatings are used in waterproof and stretchable textile fabrics.

It is important to keep in mind that this material has low resistance to UV light and is sensitive to oxidation and ozone during use.

Styrene Butadiene Rubber (SBR):
It is particularly preferred as a coating material due to its advantages of high abrasion resistance and high aging stability. An important advantage is that they have good resistance to water and some chemicals.

SBR Coatings:
It is preferred in outdoor fabrics, floor coverings, and tire coatings in the automotive industry.

SBR has limited resistance to oils, fuels and high temperatures.

Butyl Rubber:
Butyl rubber coated textiles are very durable surfaces because they have low moisture and gas permeability and high resistance to chemicals. Due to their high flexibility, they do not lose their flexibility even at low temperatures. Due to this high strength, they are used in roof coverings, waterproofing membranes and protective clothing. However, due to their low heat resistance, they are not preferred to be used at high temperatures.

Neoprene (Polychloroprene Rubber):
It is a chemically stable coating material. It is also highly preferred due to its flame retardancy, resistance to adverse weather conditions and ozone resistance. The durability of neoprene coatings allows it to work in wide temperature ranges. Due to these features, it is used in many areas from diving suits to gloves, from seals and cables to protective clothing. However, its high cost and low tensile strength make it not preferred in some areas.

Silicone:
It is used in non-stick coatings, medical applications, and outdoor fabrics such as tents and awnings, foams, sealants, and insulation materials due to its superior thermal stability, resistance to high and low temperatures, and high resistance to water, UV light, and chemicals.

However, its high cost and the need for an extra process to adhere to the surface limit its use.

The overall performance of the final product depends on the properties of the coating material. The barrier properties, adhesion, mechanical properties, thermal stability, and viscosity of the coating material will directly reflect on the final product. In applications performed at high temperatures, the thermal stability and strength of the coating material are very important. The properties of the coating material can be improved and controlled with surface preparation techniques. Thus, desired quality bonding to the fabric surface and smooth surfaces can be obtained. The barrier properties of the coating material also ensure that it fulfills its barrier function against moisture, chemicals, and gases in many areas.

Many criteria such as chemical requirements, environmental requirements, and mechanical properties are very important in the selection of coating materials.

In industrial conditions where resistance to oil, acid, and alkali chemicals is required, the selection of durable coating materials is very important. In addition, in difficult environmental conditions, resistance to humidity, high temperature, and UV light comes to the fore. If the final product is to be used in areas such as automotive, ready-made clothing, and construction, then properties such as abrasion resistance, flexibility, strength, and elongation at break are very important. The selection of coating material should be made depending on the application area and the price of the final product. Low-cost materials such as SBR can be used in areas that do not require very high quality and durability. When very high performance is needed, silicone and PU coating materials come to the forefront. In addition, coating materials that meet the requirements for environmental sustainability, low VOC emissions, and do not contain harmful chemicals should be preferred. In some special applications, value-added products are developed using multifunctional coating materials.

## 16.4 Fibers Used in Coated Fabrics

The fabric to be coated can be either woven or knitted fabric, as well as nonwoven fabrics directly produced from fibers, and coating can also be applied in yarn form. For a good coating process to be performed, certain basic properties are expected from the chosen fabric. These include durability; resistance to acids and chemicals; adhesion; low cost; a clean, smooth, even surface; and a compact structure, dimensional stability.

The fibers most preferred in base fabrics and the fabrics produced from these fibers have certain advantages and disadvantages in terms of the coating process. Fibers play a key role in defining the fabric's base properties. Common fibers include the following:

In textile coatings, the base fiber or substrate is critical as it provides the structural foundation for the coating material. The choice of base fiber can significantly influence the final product's properties, including strength, flexibility, breathability, durability, and resistance to chemicals, heat, or abrasion.

Table 16.1 shows the properties, application areas, and advantages of the most commonly used natural fibers in coated fabrics.

Table 16.2 shows the properties, application areas, and advantages of the most commonly used synthetic fibers in coated fabrics.

Table 16.3 shows the properties, application areas, and advantages of the most commonly used high-performance fibers in coated fabrics.

**Table 16.1** Natural fibers used in coated fabrics.

| Fiber | Properties | Applications | Advantages | Disadvantages |
|---|---|---|---|---|
| Cotton | Soft, breathable, and absorbent | Protective clothing, home textiles, and casual wear | Biodegradable, sustainable, and comfortable | Lower durability, prone to shrinking and wrinkling, and requires special coatings to enhance its strength and resistance to chemicals or water |
| **Wool** | Excellent insulation, and good moisture absorption | Winter wear, carpets, and protective gear | Naturally flame retardant and insulating | Expensive, less durable, and requires treatment to avoid shrinkage and degradation |
| **Silk** | Smooth texture, lightweight, and high strength-to-weight ratio | High-end fashion, medical textiles, and some protective applications | Natural sheen and high tensile strength | Expensive, sensitive to light, and requires special coatings for durability |

Blended fibers combine the properties of different materials to achieve a balance between comfort, strength, and durability. Some common blends are included in Table 16.4.

Properties affecting the selection of basic fibers to be used in the coating process as follows:

Fiber Strength: Selection should be made by considering the force to which the fiber will be exposed depending on the coating process to be applied. If very high strength and performance is desired, then fibers such as Kevlar and carbon should be used.

Resistance to Heat: In applications where very high heat resistance is desired, aramid and glass fibers that generally show very good performance are used.

Fiber Flexibility: This feature is very important especially in wearable products. Good performance can be achieved by using PU fibers.

**Table 16.2** Synthetic fibers used in coated fabrics.

| Fiber | Application | Properties | Advantages | Disadvantages |
|---|---|---|---|---|
| **Polyester** | Sportswear, automotive fabrics, medical textiles, and industrial applications | Strong, durable, and resistant to stretching and shrinking | High tensile strength, moisture resistance, and abrasion resistance | Poor breathability, may hold odors, and non-biodegradable |
| **Nylon (polyamide)** | Outdoor gear, military textiles, parachutes, and upholstery | High strength, good elasticity, and lightweight | Excellent abrasion resistance, high elasticity, and quick-drying | Sensitive to UV degradation and less breathable than natural fibers |
| **Polypropylene** | Nonwoven fabrics for medical, hygiene, and packaging uses | Lightweight, moisture-wicking, and chemically resistant | High chemical resistance, moisture resistance, and affordable | Poor UV resistance, flammable, and non-biodegradable |
| **Polyurethane (PU)** | Performance textiles, medical wear, sportswear, and artificial leather | High flexibility, stretchability, and elasticity | Waterproof, flexible, and durable | Sensitive to heat, can degrade over time, and non-biodegradable |

Fiber Resistance to Chemicals: Especially in applications where environmental conditions are difficult such as medical applications, PP, and polyester with high chemical resistance are preferred.

Fiber Breathability: Natural fibers such as wool and cotton are frequently used in products where breathability is a priority. However, when synthetic fibers are required for technical uses, microporous coating applications can be preferred to increase air flow. Natural fibers are more suitable for applications where comfort and sustainability are at the forefront, whereas synthetic and high-performance fibers are more suitable for technical applications

**Table 16.3** High-performance fibers used in coated fabrics.

| Fibers | Properties | Applications | Advantages | Disadvantages |
|---|---|---|---|---|
| **Kevlar (aramid fiber)** | Extremely strong, heat-resistant, and lightweight | Bulletproof vests, protective clothing, and aerospace textiles | High tensile strength, and heat and flame resistance | Expensive, stiff, and not flexible for fashion or comfort applications |
| **Carbon fiber** | Very high strength-to-weight ratio and rigid | Aerospace, automotive, sports equipment, and technical textiles | High stiffness, lightweight, and excellent heat resistance | Very expensive, brittle, and difficult to recycle |
| **Glass fiber** | High heat resistance and strong | Insulation materials, fireproof clothing, and reinforcement fabrics | Non-flammable and high thermal resistance | Brittle and can irritate skin, and low elasticity |
| **Dyneema (ultrahigh-molecular-weight polyethylene)** | Extremely strong, lightweight, and high abrasion resistance | Protective gear, ballistic protection, and cut-resistant fabrics | High impact strength, buoyant, and moisture-resistant | Expensive and low UV resistance |

**Table 16.4** Fiber blends used in coated fabrics.

| Fibers | Properties | Application | Advantages | Disadvantages |
|---|---|---|---|---|
| **Cotton/ polyester blends** | Combine the comfort of cotton with the durability and moisture resistance of polyester | Clothing, home textiles, and medical textiles | Better strength and durability than cotton alone | Less biodegradable than 100% cotton |
| **Nylon/ spandex blends** | Combine the strength and stretch of nylon with the elasticity of spandex | Sportswear, activewear, and compression garments | High stretchability and flexibility | Less breathability and retains odors |

such as protective clothing, medical textiles, and industrial uses. If it is desired to achieve better performance by combining the good properties of different fibers, then mixing different fibers is also a good option.

## 16.5 Production Techniques

The coating process is usually the coating of a polymeric material to provide many properties such as flame retardancy, water repellency, high strength, electrical conductivity, and thermal insulation to the textile surface. Coating techniques may vary depending on the coated textile surface, coating material, and application. The most commonly used coating techniques as follows:

Direct Coating:
In this coating method, the polymer or compound is directly coated on the fabric surface as the coating material. This technique is frequently used when it is desired to provide protective and functional properties to the textile surface to be coated.

The polymer as the coating material is spread on the fabric surface by means of a knife or a roller, and a layer is formed and coated. Controlling the thickness of the coating is achieved by precisely adjusting the gap between the knife and the roller. Curing and drying processes are performed to bond and solidify the layer formed on the fabric surface. The most commonly used coated polymers are as follows: PVC, PU, silicone, and acrylic. It is preferred in outdoor applications such as protective clothing, tarpaulins, and tents.

Transfer Coating:
In the transfer coating technique, the coating material is first transferred to a release paper or a film. In the second stage, this layer is transferred to the fabric and the coating process is carried out (Figure 16.6). After the drying process is carried out under heat and pressure, the release paper is removed from the surface and separated.

Advantages of this method are the production of smoother coatings with better adhesion and even thickness.

Application areas of transfer coating are synthetic leather, automotive interiors, and breathable fabrics in textile industry.

Lamination:
It involves bonding two or more layers together. The coating layer is laminated to the textile surface. The two layers are combined with heat and

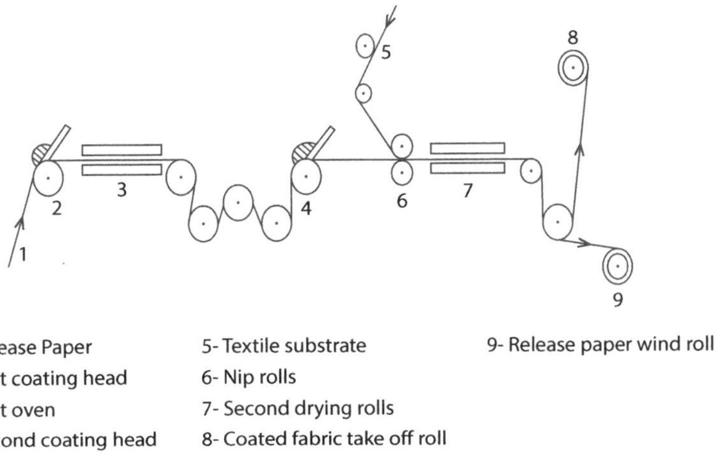


**Figure 16.6** Transfer coating process [16].

pressure. In some cases, adhesives are used to make the bond stronger. This application is highly preferred in waterproof and breathable fabrics such as Goretex, automotive, and medical textiles.

Foam Coating:
This coating technique is carried out by mixing a polymer material with air to obtain foam and then applying this foam material to the textile surface. After the foam material is mixed in a mixer, it is coated on the fabric directly or with the help of a knife. With the foam coating technique, less material is used while homogeneous distribution and thickness control are important advantages. Drying and curing processes are performed in order for the coating material to bond strongly to the fabric. It is frequently preferred in protective clothing, upholstery fabrics, and carpets where lightness is at the forefront.

Gravure Coating:
In the gravure coating technique, the coating process is carried out by applying the coating material precisely to the fabric surface with the help of engraved cylinders. The coating material is poured onto the gravure cylinders with patterns on them and the textile surface is passed over the cylinders and the coating material is transferred to the textile surface. The coating weight and thickness can be controlled as desired by means of gravure cylinders (Figure 16.7).

Application areas of gravüre coating are lightweight, finely coated fabrics used in high-tech applications such as medical and filtration textiles.

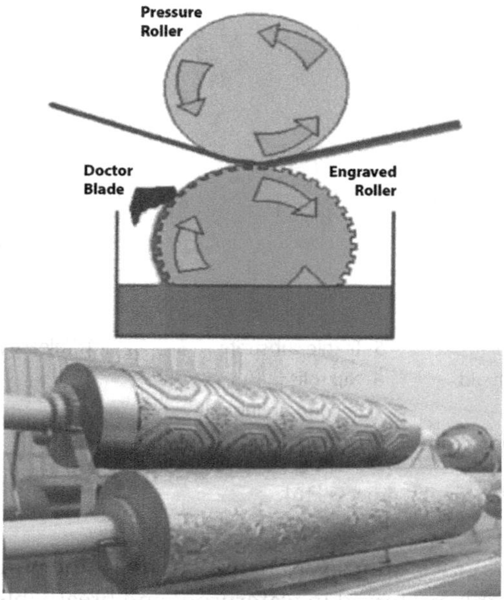


**Figure 16.7** Gravure coating [16].

Knife-Over-Roll Coating:
In this coating technique, the coating process is carried out by spreading the coating material on the fabric with the help of a knife while the fabric passes over a cylinder (Figure 16.8). The thickness can be controlled by adjusting the position of the knife during coating. This coating technique is highly preferred in many areas such as architectural applications, rubber coatings, and water- and oil-repellent fabrics.

Application areas of knife-over-roll coating are waterproof fabrics, rubber-coated textiles, and coated fabrics for architectural use.

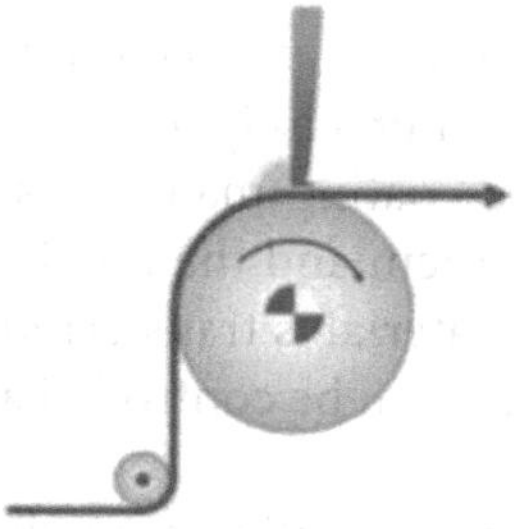

**Figure 16.8** Knife over roller [16].

Calendering:
It is a technique in which the fabric is passed through heated cylinders to make the fabric surface more smooth and add different features to the surface during the coating process. Thus, a bright, smooth surface can be obtained, and different patterns can be applied on the surface. This technique is mostly preferred in synthetic leather production, special film coatings, and technical textiles.

Dip Coating:
In this coating technique, the fabric to be coated is completely immersed in a coating bath containing coating solution and the coating process is carried out by impregnating the coating material on the fabric surface. After coating, the fabric is passed through the cylinders to remove excess coating liquid on the surface and a drying and curing process is applied to bond the material to the fabric (Figure 16.9). This technique is mostly preferred in protective clothing, composites, and filters.

Hot Melt Coating:
In this coating technique, the granular coating material is first melted and coated on the fabric surface in molten form with the help of a coating knife or roller. A cooling process is performed to ensure that the coated material adheres to the fabric. This technique is mostly used in laminated fabrics.

Spray Coating:
In this coating technique, the coating process is performed by spraying the atomized polymer or coating solution onto the textile surface to be coated

**Figure 16.9** Dip coating [16].

**Figure 16.10** Spray coating [16].

using high-pressure air (Figure 16.10). A drying process is performed to ensure that the coating material adheres to the fabric surface.

Application areas of spray coating are lightweight, breathable coatings for medical or outdoor textiles, where precise control over coating distribution is essential.

Screen Printing Coating:
Screen printing applies the coating in a patterned or selective manner, usually for decorative or functional purposes.

A stencil or screen is used to apply the coating material to specific areas of the fabric. The coated fabric is then dried and cured.

The screen printing technique is generally preferred in fashion, technical textiles, and some industrial fabrics where the entire fabric surface is not covered but only regionally coated. As in every coating technique, the coating material, coating surface, and coating technique directly affect the coating quality.

- Coating Material: The properties of the coating polymer to be used, such as flexibility and breathability, will also determine the properties of the final product.
- Surface Properties: The type of fabric (woven, non-woven, and knitted) affects how the coating will adhere and perform.
- Application Method: Different methods are suitable for specific applications depending on the precision and distribution needed.
- Coating Thickness: Thicker coatings provide better protection but may reduce flexibility and breathability.

- Curing and Drying: Proper curing (with heat or UV) ensures that the coating adheres well and remains durable under stress or environmental conditions.

Each coating technique serves unique purposes, depending on the final product's intended functionality and performance.

## 16.6 Functional Features

Waterproofing and Breathability:
This feature should be provided especially in medical textiles, sports textiles, and outerwear fabrics. With the technologies developed in recent years, highly comfortable coated fabrics have been developed that are resistant to water while allowing the water vapor produced by sweating to be expelled. Goretex technology, which is used especially in sportswear, is one of the leading technologies that allows the vapor to be expelled with micropores on the coating surface and prevents water from entering.

Thermal Insulation and Regulation:
Thermally stable materials are used to provide protection against temperature changes. In order to achieve the desired thermal performance, phase change materials (PCMs) and aerogels that provide excellent insulation properties due to the high air void ratio inside are added to polymer materials. PCMs absorb heat when the temperature rises and release heat when the temperature drops, keeping the temperature constant within certain ranges [11].

Electrical Conductivity:
Textile surfaces coated with nano-sized materials such as silver, carbon, and graphene have electrical conductivity properties, and these conductivity properties are used especially in wearable technologies, smart textiles, electrical circuits, and technical textiles where electrical conductivity is desired.

These coatings allow the development of flexible sensors, electronic textile surfaces, and integrated circuits used in clothing thanks to their flexible structures, washability, and antibacterial advantages.

Conductive threads used in wearable electronic devices form circuits integrated into garments. Graphene-based textiles are used to create flexible sensors [12].

UV and Chemical Resistance:
In particular, technical textile surfaces used outdoors and in industrial environments require resistance to UV and chemicals, and UV stabilizers and chemical-resistant coating materials are used to provide these properties. These coating materials prevent damage caused by UV and chemicals used in products used mostly in agriculture and construction [13].

Antimicrobial and Self-Cleaning:
Coating materials containing silver or titanium dioxide nanoparticles are used to provide protection against fungi, bacteria and microbes. Because these materials have hydrophobic properties, they provide self-cleaning properties by allowing water, oil, and dirt to flow and roll off the fabric surface without adhering to it. Fabrics coated with titanium dioxide have the ability to break down dirt on the surface under sunlight, thus providing self-cleaning properties under sunlight. Silver-containing coatings, which have antibacterial and infection-preventing properties, are also very frequently used, especially in the production of socks [14].

Energy Storage:
Recent developments include coating materials that enable energy storage capabilities in textiles. It is possible to see these innovative coating methods in wearable electronics, health monitoring systems, and smart textiles.

Textiles coated with conductive polymers or carbon nanotubes can function as batteries or supercapacitors [15].

These properties are used extensively not only in modern technical textile applications but also in different industries.

## 16.7 Commercial Usage Areas

Due to the superior properties of technical textiles, the demand in the industry is constantly increasing and as this demand increases, the demand for coated fabrics will also increase. Areas where coated fabrics are used as follows:

Automotive Textiles:
Because durability, functional properties, and aesthetic appearance are at the forefront in the automotive sector, coated fabrics are in great need. They are used in ceiling tiles, door, and hood heat and sound insulation, seat covers, and airbags in automobiles. In recent years, with the increase in electric vehicle production, functional properties such as lightness, flame

retardancy, and sustainability are being added to the fabric with coating techniques and used.

Defense and Protective Textiles:
Coated fabrics are added to the fabric with coating techniques for functional properties such as resistance to chemicals, ballistic protection, water resistance, and thermal comfort, which are especially demanded in military uniforms.

Sports and Activewear:
Because comfort is very important in sports textiles, different coating techniques are used to provide the fabric with breathability, flexibility, and waterproof properties. In addition, coating processes are applied not only for sportswear but also for sports equipment such as tents and backpacks to provide UV protection and high-performance properties.

Medical and Wearable Textiles:
Because high performance is required in medical textiles such as surgical gowns, bedding, wound care products, and wearable devices, many properties such as antimicrobial, moisture resistance, breathability, hydrophilicity, and hydrophobicity are provided with coating. Conductive coatings are now widely used in smart textiles and wearable technologies that we use in many areas of our lives.

Home Textiles:
Coated fabrics are preferred in home textiles such as curtains and upholstery to provide functional properties such as flame retardancy, dirt and stain repellency, and waterproofness. The rise of functional home textiles is expected to increase the use of coated fabrics in home textiles with the developing technology as these functional features provide both aesthetics and practicality to human life.

Industrial and Construction Textiles:
Weather resistance and UV protection properties are features that must be added to textile surfaces used in industrial and construction areas. Coated textiles are widely used in industrial areas as protective covers, filtration, conveyor belts, and in construction areas as geotextiles, soil stabilization, roof coverings, and insulation materials.

Agricultural and Space Technical Textiles:
In agriculture, coated textiles, especially those with features such as UV protection, breathability, temperature regulation, and water resistance, have been widely preferred in protective covers, greenhouses, and soil fertility in recent years. Because weight reduction is extremely important in textile surfaces used in space, thin and multifunctional coating surfaces are used. Lightweight and heat-resistant protective textile surfaces in space vehicles are products with high added value.

Photovoltaics and Conductive Textiles:
In recent years, it is an important development that organic solar panels can be produced with a larger surface area and industrially low cost and are easier by coating the textile surface. The use of conductive coated textile surfaces in solar panels and energy-collecting fabrics is increasing. In addition, textile surfaces with electrically conductive coatings are highly preferred in the development of smart systems, especially from wearable technologies to smart home products, due to the positive properties of the textile structure, and it is predicted that this increase in use will continue. Because coatings will now further increase functionalities such as self-cleaning and antibacterial properties, in parallel with developments in nanotechnology, they have become increasingly important in textile applications. Demands for environmentally friendly coatings that reduce the environmental impact of textile production and end use will increase, especially in sectors such as automotive, agriculture, and home textiles.

Personalization and multifunctionality will come to the fore. In particular, moisture management and thermal regulation are the most preferred features in coated fabrics. Due to the increasing focus on environmental sustainability, increasing efficiency, and safety, coating processes that provide fabrics with multiple functions will be increasingly applied.

## 16.8 Tests Applied to Coated Fabrics

Tests are performed to determine whether the coating material meets the requirements of the area in which it will be used and to measure its performance.

These tests measure both the coating material and the properties of the base textile fabric and determine whether it complies with the standards determined for the application area. It is seen the purpose for which these tests are performed in the application areas and their importance in terms of application areas in Table 16.5. Among these tests, especially

**Table 16.5** Tests applied for coated fabrics.

| Test | Purpose | Importance |
|---|---|---|
| **Weight test** | Determines the fabric's weight per unit area, typically expressed in grams per square meter ($g/m^2$) | Helps assess the density of the material, which affects durability, comfort, and functionality |
| **Thickness test** | Measures the total thickness of the coated fabric | Thickness impacts the fabric's mechanical properties, including flexibility and insulation capabilities |
| **Breaking strength** | Measures the maximum force the fabric can withstand before it breaks | Important for determining the fabric's durability under tensile loads |
| **Tear strength** | Evaluates the fabric's resistance to tearing | Helps assess the material's suitability for harsh environments where mechanical damage might occur |
| **Blast damage resistance** | Tests the fabric's ability to withstand explosive or high-energy impacts | Relevant in military and industrial applications |
| **Pilling resistance** | The purpose of this test is to measure the tendency of the fabric to form small, unwanted lumpy particles on its surface due to abrasion. | Ensures aesthetic and functional performance, especially for wearables or consumer products |
| **Abrasion resistance** | Assesses how well the fabric withstands friction and wear | Critical for long-lasting fabrics in high-friction applications, such as workwear or upholstery |
| **Dimensional stability** | Measures how much the fabric changes in size when subjected to different conditions like washing or humidity | Ensures the fabric retains its shape and size over time |
| **Water vapor permeability** | The purpose of this test is to measure the tendency of water vapor to pass through the fabric. | Important for breathable materials used in outdoor gear and smart textiles |

(*Continued*)

**Table 16.5** Tests applied for coated fabrics. (*Continued*)

| Test | Purpose | Importance |
|---|---|---|
| **Waterproofness and water repellency** | Evaluates how well the fabric resists water penetration | Essential for protective clothing, outdoor gear, or agricultural applications |
| **Low-temperature bending and resistance** | Tests the flexibility and performance of the fabric at low temperatures | Crucial for materials exposed to cold climates or extreme environments |
| **Adhesion test** | Measures the bonding strength between the fabric and the coating material | This test is especially critical because it determines how well the coating adheres to the base fabric. Poor adhesion can lead to peeling, delamination, and failure of the fabric in its intended use |

the adhesion test is an extremely important test in terms of evaluating the bonding strength between the fabric and the coating material. Therefore, special attention was paid to the adhesion test in more detail.

Because it is very important for the coating material to adhere to the fabric surface, the necessary tests must be carried out. This test determines whether the necessary force is within the desired range with different tests such as adhesion, peeling, and shearing to separate the coating material from the fabric surface. With the peel test, the force required to peel the coating material from the fabric bottom layer at an angle of 90° or 180° is determined in the opposite direction to the force indicating the presence of the adhesive. The force required to maintain peeling is measured and recorded.

In the shear test method, the fabric is stretched to ensure sliding between the surface and coating material layers. The force required to initiate and maintain sliding is measured and recorded. This force indicates the force required to separate the adhesive from the fabric and depends on many factors such as the formulation of the coating material, environmental effects, surface preparation, and the curing process. In the preparation phase before the bonding process, the fabric surface can be cleaned and the adhesion can be made stronger with a roughening process.

During the formulation of the coating, the chemistry of the coating material (e.g., polymers and additives) affects how well it will bond to the fabric.

At the curing process stage, curing conditions (temperature, time, and environment) when applying the coating can affect adhesion. Humidity, mechanical stress, environmental factors, and UV light can cause deformation of the adhesion over time.

The Importance of Adhesion in Smart Fabric Applications:
Because the adhesive material used in smart fabrics holds the coating material and the fabric together, it should not be affected by environmental factors such as humidity and sunlight. In addition, if the adhesive holds the taste elements such as sensors together on the smart fabric, then it is desired that its strength is high so that the integrity between the elements is not disrupted. Attention should be paid to the selection of the adhesive so that performance does not decrease over time.

Attention should be paid to the adhesion test in order for the coated fabric to meet commercial performance requirements.

Different international standards are used for the tests of coated fabrics. ISO (International Organization for Standardization) standards are mostly used for chemical resistance, flame retardancy and mechanical properties, American Society for Testing and Materials standards are for physical tests of abrasion and tensile strength, and National Fire Protection Association standards are mostly used in protective clothing. In addition, European Norms standards are preferred in the performance tests of coated fabrics in the EU market.

## 16.9 Future Expectations and Insights

Future expectations for coatings in technical textiles are shaped by innovation, environmental sustainability, and multifunctionality.

Focusing on environmental sustainability, water-based, biodegradable, and recyclable coating materials are being developed by taking into account the requirements desired in the world. In particular, products developed using natural materials such as starch, vegetable oils, chitosan will increase even more. Multifunctional, high value-added products will continue to be developed by combining environmental sustainability with functionality. Water-based coating materials will be preferred more to reduce harmful emissions (VOCs) and increase environmental safety compared to traditional solvent-based materials that are very harmful to the environment. Because the recyclability of the entire textile surface is important in terms of environmental impact, the coating material used must also be recyclable, and the demand for these materials will increase in terms of circular economy.

Smart and functional coatings will become more prominent, especially to provide different functional features to technical textile surfaces. Lightweight, flexible, and functional surfaces will continue to be developed.

With the increasing demand for protective equipment and medical textiles due to pandemics and epidemics that have increased greatly in recent years and can affect the whole world, silver ions and titanium dioxide coatings will come to the forefront, especially to provide antibacterial and self-cleaning properties to the fabric surface. With the advancement of technology in this field, it is also possible to produce solutions with wearable technologies. At this point, textile surfaces with electrically conductive coatings will be in greater demand. Sensor integration, which will allow textile products to monitor body signals or environmental conditions, will become more common with these coatings. The use of these applications will increase especially in health, sports, and military applications.

Coatings for energy production and storage, such as technical textiles containing photovoltaic or conductive coatings, smart textiles that can charge devices or provide heat, will make further progress.

Industries such as automotive, defense, and heavy-duty sportswear require improved durability and protection, which increases textile durability and enables them to last longer under mechanical stress. It will need wear-resistant coatings.

Fire-resistant and thermal-insulating coatings will become more prominent as innovations in fire-resistant coatings are vital for protective clothing, defense applications, and home textiles. Coatings that will provide better thermal regulation or protection against extreme temperatures will continue to be developed.

Demand for waterproof yet breathable coatings will continue to increase for outdoor sportswear, military textiles, and medical applications where comfort and protection are crucial.

It is anticipated that coatings in agricultural textiles will focus more on providing UV resistance, controlling light transmission, and improving plant growth conditions. Bioactive coatings are also expected to play a role in pest control or nutrient distribution in smart agriculture applications.

Developments in nanotechnology will allow for more homogeneous, lightweight, and very thin coating applications while also providing functional effects at the nano level. Therefore, nanocoatings will continue to develop even more, and demand will increase in many areas, especially in medical textiles, agriculture, automotive, conductive textiles, defense, space, and aviation.

Because graphene and carbon-based coatings are advanced materials, they will be in even greater demand for wearable technology, defense, space, and aviation applications due to their properties such as thermal resistance, strength and electrical conductivity. It will be important to develop coating materials that meet comfort needs such as temperature and humidity control for the user, and the demand for these materials will increase even more.

As digital and 3D printing technologies advance, coatings will become more customizable and efficient. It can be used in applications requiring special performance.

In summary, for future developments in coated technical textiles, it will focus on sustainability, functionality, and durability, with innovations that appeal to a wide range of sectors, including medicine, defence, space, sports, agriculture, and wearable electronics.

**AI disclosure statement:** The author confirms that no artificial intelligence (AI)-generated content or AI-assisted writing tools were used in the preparation of this chapter (including summaries and the abstract). All text, analysis, and conclusions are the result of the author's original work.

## References


1. D. Ferziger and J. Lippman, Foam Coated Fabrics. US Patent 4,677,016, 1987.
2. Hakansson, E., Amiet, A., Kaynak, A., Electromagnetic shielding properties of polypyrrole/polyester composites in the 1–18 GHz frequency range. *Synth. Met.*, 156, 917–925, 2006.
3. Mondal, S. and Hu, J.L., A novel approach to excellent UV protecting cotton fabric with functionalized MWNT containing water vapor permeable PU coating. *J. Appl. Polym. Sci.*, 103, 5, 3370–3376, 2007.
4. Rehby, W., Coating of textile fabrics with conductive polymers for smart textile applications. *Ambience 08 International Scientific Conference*, Borås, Sweden, 2008.
5. Asker, G., Balcı, O., Topçal, F., Determination of performance properties of hybrid fabrics produced by combination of coating and impregnation application methods. *Text. Eng.*, 18, 82, 6–18, 2011.
6. Negru, D., Buda, C., Avram, D., Electrical conductivity of woven fabrics coated with carbon black particles. *Fibres Text. East. Eur*, 20, 1, 53–56, 2012.
7. Padleckiene, I. and Petrulis, D., Effect of abrasion on the air permeability & mass loss of breathable-coated fabrics. *Fibres Text. East. Eur*, 17, 2–73, 50–54, 2009.
8. Gouda, M. and Keshk, S.M., Evaluation of multifunctional properties of cotton fabric based on metal/chitosan film. *Carbohydr. Polym.*, 80, 504–512, 2010.

9. Asanovic, K., Mihailovic, T., Skundric, P., Simovic, L., Some properties of antimicrobial coated knitted textile material evaluation. *Text. Res. J.*, 80, 16, 1665–1674, 2010.
10. Ying, K.-L., Hsieh, T.-E., Hsieh, Y.-F., Colloidal Dispersion of Nano-scale ZnO Powders Using Amphibious and Anionic Polyelectrolytes. *Ceram. Int.*, 35, 3, 1165–1171, 2009.
11. Mondal, S., Phase Change Materials for Smart Textiles – An Overview. *Appl. Therm. Eng.*, 28, 11, 1536–1550, 2008.
12. Stoppa, M. and Chiolerio, A., Wearable Electronics and Smart Textiles: A Critical Review. *Sensors*, 14, 7, 11957–11992, 2014.
13. La Mantia, F.P. and Morreale, M., Green Composites: A Brief Review. *Composites, Part A*, 42, 6, 579–588, 2011.
14. Rai, M., *et al.*, Silver Nanoparticles as a New Generation of Antimicrobials. *Biotechnol. Adv.*, 27, 1, 76–83, 2009.
15. Lee, J.H., *et al.*, Recent Advances in Flexible Energy Storage Devices Based on Carbon Nanomaterials. *Adv. Energy Mater.*, 8, 4, 1701749, 2018.
16. Sabat, P.K., Textile Coating Presentation, Texport Syndicate Ltd., 2024.
17. Pazdzior, K., Klepacz-Smółka, A., Ledakowicz, S., Sojka-Ledakowicz, J., Mrozińska, Z., Żyłła, R., Photocatalytic decomposition of formaldehyde in air using $TiO_2$ modified textiles. *Chemosphere*, 75, 2, 250–255, 2009.
18. Medical Textiles, https://med.stanford.edu/snyderlab/news/2022-our-wearable-future-part-1-what-will-new-tech-look-like-.html, 2022.
19. Coated Fabric Market, https://www.grandviewresearch.com/industry-analysis/coated-fabric-market, 2024.
20. Technical Textiles Coatings, https://www.lubrizol.com/Coatings/Markets/Textiles/Technical-Textile-Coatings, 2024.
21. Coating, https://www.topp-textil.de/en/expertise/coating, 2024.
22. Technical Textiles, https://www.technicaltextile.net/articles/coating-of-technical-textiles-4851, 2024.

# 17

# Textile Functionalization by Skincare Finishing

**Angela Danila**

*Department of Chemical Engineering in Textiles and Leather, Faculty of Industrial Design and Business Management, "Gheorghe Asachi" Technical University of Iasi, Iasi, Romania*

***Abstract***

The concern for a healthy lifestyle has become a priority for many consumers who no longer choose a textile product solely for its aesthetic and protective functions but also for a range of beneficial effects on the human body. Therefore, functional textiles that can be tailored to individual needs through a variety of obtained effects are becoming increasingly popular. In the production of functional textiles, the skincare finishes are pretty used to provide new functionalities (present a range of beneficial effects on the human body such as antimicrobial, aromatic, UV protection, insecticidal, anti-inflammatory, antioxidant, and antiaging), adding value to the final product, thus leading to increased consumer interest and market competitiveness.

This chapter reviews the current uses of textiles in skincare, types of textiles used for skincare finishes, methods for embedding active compounds into protective carriers, the materials involved in their production (including substances used as carriers and those used as bioactive compounds), methods for applying active systems onto textile substrates, release mechanisms of active compounds, analysis methods for skincare textiles, and some commercial types of textiles functionalized by skincare finishes.

***Keywords*:** Functionalization, textile, skincare, UV protection, antiaging, antimicrobial, anti-cellulite

*Email*: angela.danila@academic.tuiasi.ro

Aminoddin Haji and Çağlar Sivri (eds.) Surface Modification and Functionalization of Textiles, (437–472) © 2026 Scrivener Publishing LLC

## 17.1 Introduction

Functional textiles, in addition to their basic functions (protection from external factors, comfort, and aesthetics functions), offer a range of other functionalities with beneficial effects on the human body. Functional properties are integrated at various stages of production, including the fiber, yarn, fabric, or finished product stages. The materials used often include chemical agents capable of providing antibacterial, UV protection, anti-crease, fluorescent whitening, antimicrobial, water-repellent, self-cleaning, flame-retardant, etc., properties (Figure 17.1). The main challenge in producing functional textiles lies in the careful selection of textile materials and active compounds, as well as in the manufacturing processes [1]. By introducing innovative finishes that allow textiles to have multiple functionalities, they gain added value, and the production process is simplified. Textiles can now achieve superior performance characteristics, making them versatile and highly functional for various applications [2]. Functional textiles can be classified based on their intended application (Figure 17.1) [3].

The skincare industry is increasingly focusing on environmentally friendly ingredients for a healthier lifestyle, moving away from synthetic

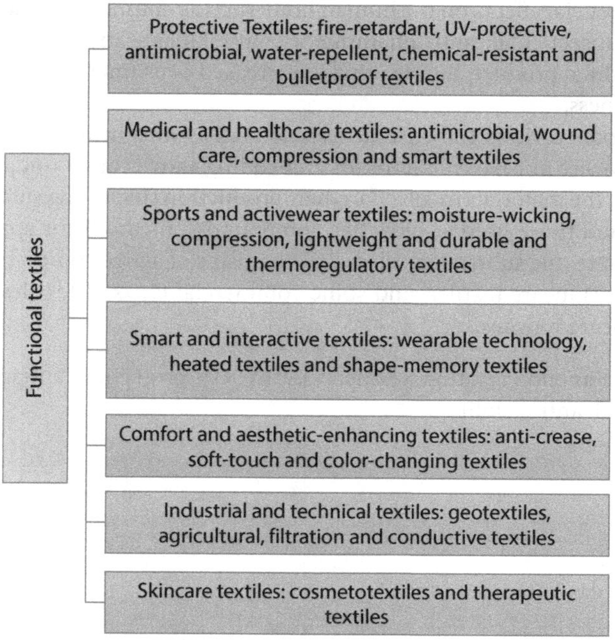


**Figure 17.1** Functional textiles classification (adapted from [3]).

chemicals that can harm health and the environment. Synthetic ingredients have been linked to skin irritations, allergies, asthma, cancer, and endocrine disruption. Lately, there is a growing preference for skincare products made with natural plant-based ingredients, which offer benefits like antioxidative, antimicrobial, antiaging, photoprotective, anti-cellulite, and anti-inflammatory properties. Skincare is essential for health and well-being, shifting from the use of active compounds in the form of emulsions/suspensions to their prolonged/controlled release from textile substrates. By applying skincare products, the skin is exposed to the action of eco-friendly products such as vitamins, minerals, and plant-derived metabolites (flavonoids, polyphenols, phenolic acids, and tannins), which have recently become increasingly popular for their health benefits [4]. By applying skincare active compounds on textile supports, there is a need to respect some rules:

- maintaining a healthy balance of skin microorganisms, which can prevent issues like acne, eczema, and other skin conditions;
- active compounds should support the skin's barrier function, leading to better protection and hydration;
- ensure that the skin's natural processes are supported, not disrupted [5].

The skin is the interface between the body and the external environment, acting as a protective barrier. It is constantly exposed to various factors that can lead to its damage (Table 17.1) [6].

The barrier function of the skin is primarily maintained by the stratum corneum, the outermost layer of the epidermis. Its unique structure, consisting of keratin-filled corneocytes and a lipid matrix, ensures protection against external aggressors, prevents water loss, and maintains overall skin health. Understanding and maintaining the integrity of the stratum

**Table 17.1** External factors that can lead to skin damage (adapted from [6]).

| External factors | Examples |
|---|---|
| Chemical agents | Corrosive, irritating, and allergenic substances |
| Physical factors | Ultraviolet radiation, air pollution, temperature variations, humidity variations, and cigarette smoke |
| Biological agents | Bacteria, viruses, and fungi |

corneum is essential for effective skincare and overall health [7]. The skin consists of three main layers: the epidermis, dermis, and hypodermis, along with appendages such as hair follicles and sweat glands. The epidermis ranges in thickness from 0.1 to 1.5 mm and includes the stratum corneum (SC), stratum lucidum, stratum granulosum, stratum spinosum, and stratum basale. The SC is the outermost layer, 10–40 μm thick, made up of 15 to 25 layers of dead, non-nucleated corneocytes, serving as the primary barrier. At the bottom of the epidermis, the basement membrane zone connects it to the dermis and acts as a secondary barrier, restricting the passage of larger molecules. The dermis, ranging from 1.5 to 4 mm in thickness, contains a capillary network that connects to the systemic circulation. It consists of the papillary and reticular layers, forming dense connective tissue and serving as the final barrier to external substances. These layers collectively protect the body from environmental harm while allowing essential nutrients and metabolites to pass through [8]. Substances that facilitate the penetration of active compounds are used to overcome the skin barrier and allow active compounds to penetrate within. These substances include fatty acids, dimethyl sulfoxide, surfactants, n-decyl methyl sulfoxide, pyrrolidones, and short-chain alcohols [9].

Protecting and maintaining the skin's natural barrier is essential to prevent water loss, reduce sensitivity, and avoid chronic skin conditions. Using gentle skincare products and minimizing exposure to harsh chemicals are key strategies for preserving healthy skin [10].

Among the products with topical application are skincare textiles that refer to textile material embedded with active ingredients, providing skincare benefits through contact with the skin. These textiles are broadly categorized into cosmetotextiles and therapeutic textiles. Cosmetotextiles (UV protection, antiaging, antimicrobial, anti-cellulite, moisturizing, and anti-wrinkle) are textiles infused with cosmetic substances that provide various skincare benefits when worn. Therapeutic textiles (burns, acne, skin wounds, microbial infection, inflammation, psoriasis, vitiligo, and fungal infection) are designed to deliver medical benefits, aiding in the treatment and management of various skin conditions [11].

The topical application of active compounds aims to maintain/treat the skin and its layers on a specific area of the body (epidermis and dermis) [12]. When skincare textiles are applied to the skin, active ingredients embedded in the fabric can penetrate the skin through several pathways (Figure 17.2): intercellular, intracellular, and transappendageal (follicular) [13].

These pathways are crucial for delivering the beneficial properties of the skincare ingredients. Its partition coefficient largely influences the primary pathway for a compound's penetration. Hydrophilic compounds are more

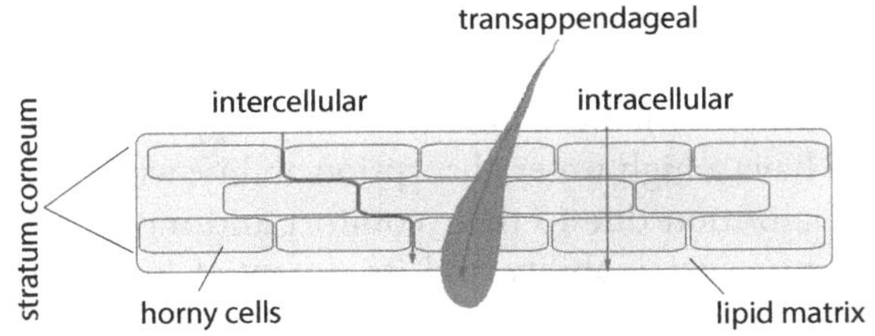


**Figure 17.2** Routes of active compound delivery *via* the skin (adapted from [13]).

likely to partition into the intracellular domains (given that corneocytes contain a hydrophilic core). At the same time, the intercellular pathway is ideal for the passage of highly hydrophobic molecules, as these molecules traverse the lipid-rich spaces between corneocytes. However, despite a compound's affinity for a particular route, most molecules simultaneously penetrate the skin through both pathways [8].

## 17.2 Textile Materials Used for Skincare Functionalization

To be used in the production of skincare textiles, textile substrates must be safe for direct skin contact, not cause irritation or allergic reactions, be comfortable to wear, allow the skin to breathe, and be non-carcinogenic, non-toxic, and biodegradable. The best textile materials for producing skincare textiles are both natural and regenerated. Some textile supports are as follows:

*Bamboo* is a sustainable alternative for textiles due to its rapid growth, minimal resource needs, and low environmental impact. It requires less water, does not require the use of pesticides, and absorbs significant carbon dioxide. Bamboo fibers are breathable, moisture-wicking, and possess antibacterial properties that remain active after multiple washes. They also offer natural UV protection and are easy to dye. These qualities make bamboo an eco-friendly choice for textile functionalization [14].

*Cotton* and its derivatives have been widely used in the healthcare textile sector for both medical and biomedical products. These include external applications like surgical clothing, drapes, and bedding, as well as internal uses such as wound dressings, tissue engineering, drug delivery, surgical procedures, and dental treatments. Since the 19th century, cotton has been used for wound coverage, providing warmth and bacterial protection. Numerous cotton-based and modified gauze products have been developed to meet biomedical needs. Cotton's large surface area, favorable mechanical properties, porosity, and permeability to air, water, and water vapor make it ideal for medical and biomaterial applications [15].

*Hemp* fibers have a high water absorption ability, which is beneficial for textile functionalization due to their comfort and moisture management properties. Hemp's sustainability and its potential in creating bio-based products have led to increased interest in hemp textiles. Challenges remain in the spinning process due to the need for specialized machinery. Hemp textiles offer benefits such as UV protection, antibacterial properties, and environmental sustainability [16].

*Silk* fibers are ideal for healthcare applications due to their strength, toughness, and ductility. They are used in wound dressings, drug delivery systems, and hydrogels. For example, silk textiles functionalized with silver nanoparticles provide antibacterial properties and support wound healing.

Silk's mechanical properties and softness make it suitable for hospital materials like patient drapes and gowns. It is also used for obtaining protective masks with titanium dioxide nanoparticles. Silk's hydrophobicity and antimicrobial qualities make it valuable for protective textiles. Additionally, silk dyed with natural extracts offers UV protection [17].

*Soy* protein fibers present a promising opportunity by utilizing low-value soy processing byproducts, benefiting agriculture, and green economies [18]. Soy fibers offer several skin benefits thanks to their isoflavones and polyphenols. They help reduce wrinkles and improve skin elasticity and hydration due to their antioxidant properties. Additionally, their anti-inflammatory and antimicrobial effects effectively manage acne and skin irritation. Soy-based products also improve skin hydration and strengthen the skin barrier. Moreover, soy extracts protect against UV damage from the sun. Overall, soy-based products can enhance skin health, reduce signs of aging, and offer protective benefits through topical use [19].

*Wool* fibers have outstanding properties such as UV resistance, biodegradability, biocompatibility, heat and sound insulation, odor and water absorption, durability, and breathability properties [20].

*Corn* fibers are eco-friendly, cheap, and very comfortable and have remarkable permeability, moisture absorbency, and UV protection [21].

*Banana* fibers are extracted from the stem (depending on the part of the stem from which they are extracted, banana fibers can be smooth and fine or coarse) and are composed of cellulose, hemicellulose, and lignin. Banana fiber is eco-friendly, strong, durable, light, and fire-resistant; has excellent mechanical properties and high moisture absorption; and is biodegradable [22].

*Casein* fibers, derived from milk proteins, are environmentally friendly, strong, and finer than synthetic fibers. They are compatible with human skin thanks to their pH of 6.8, and they are biodegradable and suitable for medical textiles and skincare. Casein fibers provide comfort, excellent air permeability and water transport properties [23].

## 17.3 Classification of Skincare Compounds

Typically, skincare systems intended for topical application consist of the active compound, the encapsulating compound, and other ingredients that may contribute to the effect of the final product. The role of the embedding compound (shell) is to ensure the prolonged/controlled release of the active compound (core) and to maintain its chemical, physical, and microbiological stability. The most commonly used forms of skincare systems are nanomaterials (organic and inorganic nanoparticles). Organic nanoparticles include lipid- and surfactant-derived nanoparticles, polymeric nanoparticles, and nanocrystals. Inorganic nanoparticles include metal nanoparticles, carbon-based nanoparticles, and nanotubes [24].

*<u>Organic Nanoparticles</u>*

*Vesicular Nanoparticles*

Liposomes are phospholipids (like phosphatidylcholines and phosphatidylserines) spherical vesicles that can embed and deliver active compounds to the upper layers of the skin. Their disadvantage lies in the fact that the transport of active compounds to the deeper layers of the skin is limited [9]. Liposomes protect the active compounds from external factors (water, oxygen, and light) and enhance their stability (in the case of volatile active compounds) and biological activity [25]. Figure 17.3 shows the classification of liposomes according to the number of double layers. For skincare delivery, the most used are large unilamellar vesicles (>50 nm) [26].

Ethosomes were first obtained in 1997 by Touitou and his collaborators. They are soft lipid vesicles that primarily contain phospholipids, 2%–45% alcohol, and water. In addition to these components, ethosomes may also contain polyglycol (for enhancing skin penetration), cholesterol (which provides stability to the vesicular membrane), edge activators (which increase skin permeability), dye (for characterization), vehicle (for gel formation), and other ingredients (e.g., dicetyl phosphate to prevent aggregation). Ethosomes are used for the transdermal delivery of active compounds. Due to their high deformability, ethosomes can easily penetrate the layers of the skin [27].

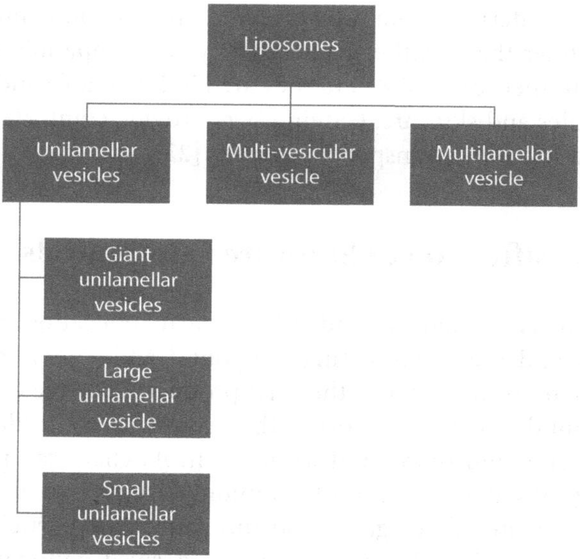


**Figure 17.3** Classification of liposomes (adapted from [26]).

Transferosomes are highly deformable vesicles that contain at least one aqueous compartment. Although transferosomes morphologically resemble liposomes, they are sufficiently deformable to pass through skin pores that are much smaller than their size [28]. The advantages of transferosomes are their low toxicity, ability to encapsulate hydrophilic and lipophilic compounds, and biodegradability. The transport pathways of transferases are either intracellular or transcellular, allowing them to reach the stratum corneum [29].

Niosomes are vesicular systems that contain lipidic components, amphiphilic non-ionic surfactants, and charge-inducing agents. Non-ionic surfactants used in niosome's formulation are alkyl ethers (alkyl glycerol ethers and oxyethylene alkyl ethers), alkyl esters (sorbitan fatty acid esters and polyoxyethylene sorbitan fatty acid esters), alkyl amides (alkyl galactosides and alkyl glucosides), block copoly (poloxamer/pluron), fatty alcohols (stearyl alcohol, acetyl alcohol, and myristyl alcohol), fatty acids (stearic, palmitic, and myristic acids). The lipidic component is cholesterol [30].

Transethosomes are vesicular systems for encapsulating active compounds that combine the advantages of transferosomes and ethosomes. Transethosome's composition is similar to that of ethosomes, but penetration enhancer or an edge activator [31].

Catanionic vesicles are obtained by mixing oppositely charged surfactants in water, without the need for external forces. This type of vesicle requires an excess of positive or negative charge to avoid precipitation and exhibits good stability and dispersion in aqueous solutions. Research on their use in transdermal delivery of active compounds is limited [32].

*Non-Vesicular Nanoparticles*

Solid-lipid nanoparticles are a colloidal system used for active compound delivery and are made from solid lipids (particle sizes: 50–500 nm), active compounds, and a surfactant layer [33]. Lipids used for solid lipid nanoparticles are triglycerides (such as tripalmitin, tricaprin, and trilaurin), fatty acids (such as myristic acid, stearic acid, and palmitic acid), monoglycerides (such as glyceryl hydroxyl stearate) and waxes (such as beeswax and carnauba wax) [34].

Nanostructured lipid carriers have been developed to improve the stability of active compounds. Nanostructured lipid carriers consist of liquid lipids, solid lipids and surfactants (hydrophilic emulsifiers, lipophilic emulsifiers, and amphiphilic emulsifiers). As liquid lipids can be used, Vitamin E, propylene glycol dicaprylocaprate, oleic acid, and isopropyl myristate. As solid lipids can be used, glyceryl behenate, glyceryl palmitostearate, fatty acids (e.g., stearic acid), triglycerides (e.g., tristearin), steroids (e.g., cholesterol), and waxes (e.g., cetylpalmitate). Surfactants help in reducing interfacial tension. The most commonly used surfactants are hydrophilic, lipophilic, and amphiphilic [35].

Nanoemulsions are colloidal systems formed from two immiscible phases and can be stabilized using a surfactant. Due to their droplet sizes (<100 nm), nanoemulsions can easily penetrate the layers of the skin. Nanoemulsions can be simple (O/W or W/O) or complex (O/W/O or W/O/W). To obtain nanoemulsions, 3%–10% surfactants are needed, whereas, for microemulsions, a surfactant concentration of ≥20% is required [36].

*Polymeric Nanoparticles*

Polymeric nanoparticles are carrier systems with diameters lower than 1 μm and can be homogeneous matrices (nanospheres) or have core-shell structures (nanocapsules).

Nanospheres are amorphous or crystalline spherical particles in which the active compound can be dissolved, incorporated or attached to the polymer matrix. Due to their small size (diameter varies between 10 and 200 nm), nanospheres can easily penetrate the layers of the skin [37]. In general, the polymers (in which the active compound is uniformly distributed) used to obtain the nanospheres can be synthetic homopolymers: e.g., poly(acrylate) and poly(methacrylate), poly(lactide-co-glycolide), poly(n-butylcyanoacrylate), poly(lactide), and poly methyl methacrylate; natural polymers, e.g., chitosan, sodium alginate, gelatine, and agar; copolymers, e.g., poly(ethylene glycol) [38].

Nanocapsules are carrier systems that have a typical core-shell structure. The inclusion compound of the active compound (polymer matrix) has an important role in protecting it from degradation and external factors and for prolonged/controlled release. The polymer used ensures stability, embedding efficiency and the degree of release of the active compound. The most used polymers are chitosan and alginate (natural polymers), poly (lactic-co-glycolic acid) and polyethylene glycol (synthetic polymers) [39].

*Nanocrystals* first appeared in the 19th century and are obtained through the nanocrystallization of poorly soluble active compounds [40]. Among the active compounds with topical applications used in the form of nanocrystals are caffeine, rutin, curcumin, quercetin, and others [41].

#### Inorganic Nanoparticles

##### *Metallic Nanoparticles*

Among the nanoparticles used in skincare applications, metallic nanoparticles occupy an important role. Metallic nanoparticles can be synthesized by chemical, physical, and biological methods. Among these, the most used are the biological ones because the other methods are quite expensive, the reducing agents used are toxic to the environment and also require the use of stabilization compounds that cause environmental problems. On the other hand, the biological method of obtaining metal nanoparticles uses natural and ecological reducing agents (extracts from plants or microorganisms). Bioactive compounds from plant extracts act as reducing and stabilizing agents [42].

##### *Carbon-Based Nanoparticles*

Carbon-based nanomaterials used for topical applications are fullerenes, nanotubes, and nanodiamonds [25].

Fullerenes have a spherical structure with 12 pentagonal rings and 20 hexagonal rings. Despite all the double bonds are conjugated, fullerenes behave more like alkenes, which are electron deficient and not like aromatic systems, which are electron rich [43]. Through the efficient penetration of the layers of the skin, they present a series of beneficial effects on the human body (antibacterial properties and antioxidants reduce sebum production) [44].

Nanodiamonds (diamond nanoparticles) are monocrystalline diamonds with high physical and chemical properties. Depending on the size of the particles, nanodiamonds are classified into nanocrystalline particles (1–150 nm) and ultra-nanocrystalline particles (2–10 nm) [45].

The nanotubes are of two types (single-walled and multi-walled) and show high loading efficiency with the active compound, high stability, biocompatibility, and prolonged/controlled release efficiency of the active compound. Despite the beneficial effects that nanotubes present, they pose problems of biodegradability and toxicity [46].

## 17.4 Examples of Skincare Functionalization

### 17.4.1 Skincare Textiles with UV Protection

Through functionalization, textiles can provide effective protection against harmful ultraviolet radiation (due to the ability to reflect, absorb and scatter the sun's radiation). When UV radiation falls on the textile surface, part of the radiation is reflected by it, another part of the radiation is absorbed by the textile material and converted into heat, and another part of the radiation passes through the textile material and reaches the surface of the skin, thus causing several problems. Textiles with UV protection can significantly reduce the transmission of UV radiation that affects the skin's surface [47]. Inorganic nanoparticles such as $TiO_2$, $SiO_2$, and ZnO are used in treating textile fibers, mainly due to their capacity to improve antibacterial and UV protection properties.

Emam *et al.* obtained functional textiles with UV protection by synthesis and obtaining cotton support nanocomposite-based coatings (chitosan–rice husk silica nanoparticles–zinc oxide nanoparticles). The obtained results showed that the cotton functionalized with developed nanocomposites presents a high level of UV protection with an ultraviolet protection factor value of 167.8 [48]. Mondal has described a series of nanomaterials used for obtaining UV-protective textiles: aloe vera nanoparticles applied on cotton fabrics (coating with chitosan); $SiO_2$, $TiO_2$, and ZnO nanoparticles applied on the cotton/PET blends (applied by pad-dry-cure method); Azadirachta indica nanoparticles applied on cotton fabrics (coating with chitosan); $SiO_2$-ZnO nanoparticles applied on nylon knitted fabrics (applied by pad-dry-cure method); graphene oxide in chitosan applied on cotton fabrics (five cycles of layer-by-layer); multi-walled carbon nanotube applied on cotton fabrics (Figure 17.4); nano-MIL (Ti and In) metal-organic framework applied on cotton and silk fabrics

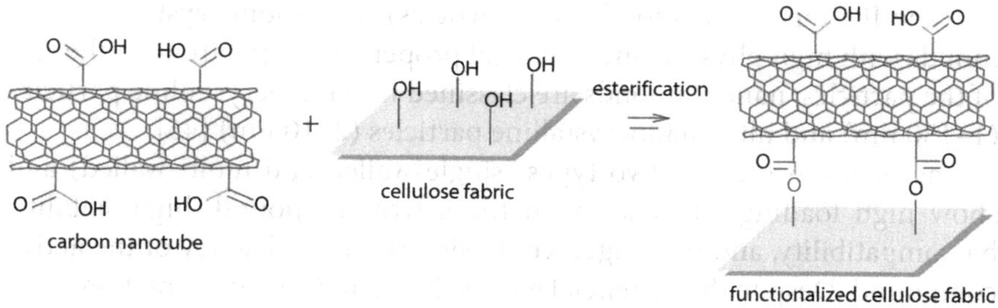


**Figure 17.4** Cellulose functionalization with carbon nanotube.

(by *in situ* synthesis process); and gold nanoparticle in chitosan applied on soybean protein–based knitted rib fabrics [49].

Metal nanoparticles can be applied to the textile support through several mechanisms:

1. Polymer and biosurfactant stabilization (Figure 17.5).

Polysaccharides (e.g., alginate, chitosan, cyclodextrin, starch, hyaluronic acid, pullulan, and dextran) are often used to stabilize metal nanoparticles and fix them on the textile support. These are natural macromolecules composed of monosaccharide units covalently linked by glycosidic bonds to form polymer chains and are also used for the effects they present, such as biodegradability, biocompatibility, and a series of beneficial effects for the human organism (for example, anti-inflammatory, antioxidant, and antimicrobial effect) [50]. Another method for stabilizing metal nanoparticles is the use of biosurfactants, which are non-toxic and biodegradable. In general, these are produced from plants and microorganisms (bacteria,

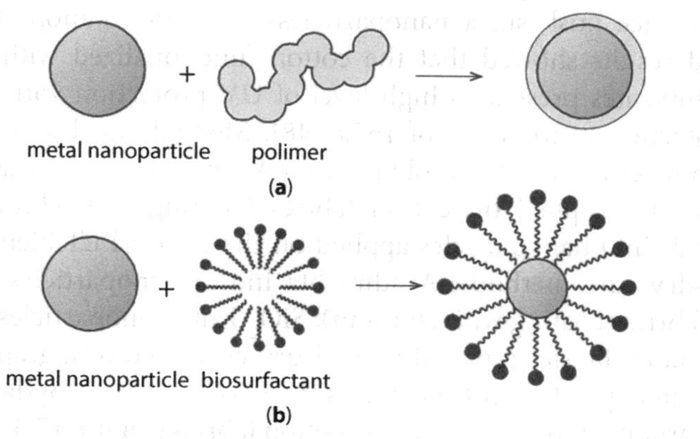


**Figure 17.5** Stabilization of nanoparticles using (a) surfactants and (b) polymers.

yeasts, and filamentous fungi). Their use as coupling agents for bioactive metal nanoparticles is a relatively new method. Biosurfactants produced by microorganisms are the most used in metal nanoparticle stabilization. Some examples of such biosurfactants are glycolipids (rhamnolipids, trehalose esters, sophorolipids, etc.), phospholipids and fatty acids, lipopeptides and lipoproteins (globomycin, amphomycin, gramicidins, fengycin, lichenysin G, etc.), polymeric biosurfactants (emulsan, liposan, alasan, etc.), and particulate biosurfactants (membrane vesicles, fimbriae, and whole cell) [51].

2. Physical and chemical adsorption (Figure 17.6). Metal nanoparticles can be directly applied to textile substrates using one of the finishing methods (e.g., exhaustion and padding).

In this case, the active compounds and the textile substrate are bound through physical forces (mostly van der Waals bonds). In the case of chemical fixation, covalent bonds are formed between the metal nanoparticles and the textile substrate. Activation of the textile surface with plasma or electrospinning is the method of obtaining functional textiles [52]. For example, Reznickova *et al.* used two techniques to create a chemical link between metal nanoparticles and the textile substrate (e.g., polyethylene terephthalate): 1. Grafting the textile substrate with dithiol (biphenyl-4,4′-dithiol), followed by the deposition of metal nanoparticles; 2. Coating the metal nanoparticles with biphenyl-4,4′-dithiol, followed by their deposition on polyethylene terephthalate surface activated by Ar plasma discharge [53]. Another study by Thinkohkaew *et al.* describe the fixation of metal nanoparticles ($TiO_2$) on polypropylene spunbond

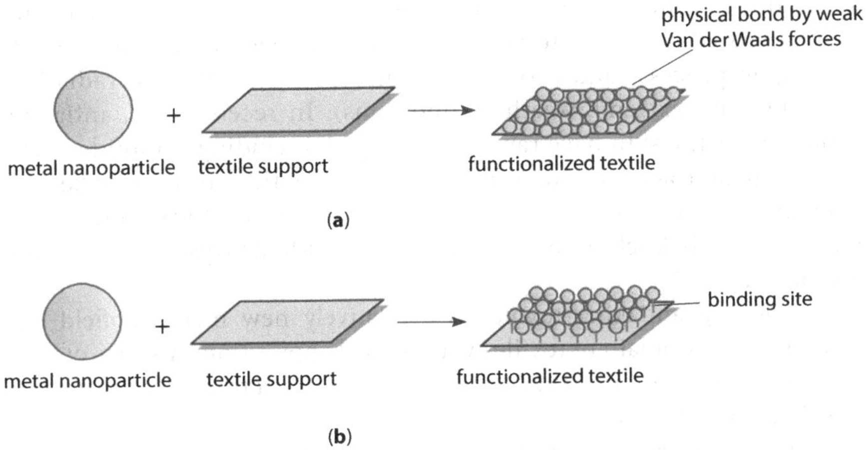


**Figure 17.6** The mechanism of (a) physical and (b) chemical adsorption.

non-woven fabric. In the first phase, the metallic nanoparticles were modified with maleic anhydride (after the modification, the structure of the $TiO_2$ crystal and its properties did not change), after which they were applied to the textile support [54].

3. *In Situ* Synthesis. This method involves obtaining metal nanoparticles directly on the textile support. For example, Asmat-Campos *et al.* obtained a functionalized textile support (70% cotton and 30% polyester) by immersing the textile sample in copper sulfate pentahydrate solution. Sodium hydroxide was added dropwise with vigorous stirring and a reducing agent. In the final phase, the textile support is removed from the tartar bath, rinsed and dried [55].

### 17.4.2 Skincare Textiles with Antiaging Properties

Aging is a natural and complex process that has been extensively studied in recent years. It is characterized by a degenerative process followed by a series of issues such as telomere shortening, tissue degeneration, chronic pathologies, dementia, cognitive deficits, and functional impairments. Several theories suggest that aging is a programmed process due to the activation/deactivation of certain genes and telomere shortening, the activity of the biological clock regulated by endocrine mechanisms, and the reduced functionality of the immune system. On the other hand, some theories attribute the aging process to external and internal factors. One of the most important compounds in aging is free radicals (reactive oxygen species), which lead to the alteration of normal cell function [56]. The skin also shows signs of aging (e.g., wrinkles and skin relaxation) due to cellular aging and decreased or degraded collagen content. The factors contributing to skin aging are endogenous (due to the natural aging process) and exogenous (harmful chemicals, UV radiation, smoking, and extreme weather conditions). In recent years, antiaging products for the skin have rapidly developed, including ceramides, retinol, hyaluronic acid, astaxanthin, collagen tripeptide, acetyl zinc ketone, epigallocatechin gallate, vitamin C, melatonin, protocatechuic acid, nature-based bakuchiol, poly L lactic acid, 10-hydroxystearic acid, and glycolic acid [57].

Developing antiaging textiles is a relatively new niche subfield and belongs to the domain of textiles with topical application. A series of antioxidant compounds are used to prevent the effects of free radicals that lead to skin aging (Table 17.2).

Polyphenols can be grafted onto the textiles by enzyme catalysis (Figure 17.7) and chemical catalysis (Figure 17.8) [67, 68].

**Table 17.2** Antioxidant compounds used for textile functionalization.

| Antioxidant compound | Method of applying to the textile substrate | References |
|---|---|---|
| Red pigment extracted from the isolated fungal strain *T. albobiverticillius* | cotton fabrics and fibers dyeing by using $CuSO_4$ and $FeSO_4$ mordants. | [58] |
| Resveratrol | Cotton (by hydrogen-bonding) and polyamide (by hydrophobic interactions) functionalization with resveratrol. | [59] |
| Blue bio-colorant phycocyanin (PC) | PC was applied on silk fabric *via* the cross-linking agent, such as genipin. | [59] |
| Gallic acid | The active compound was encapsulated in microspheres prepared from poly-ε-caprolactone and applied on polyamide fabric *via* the bath exhaustion method. | [60] |
| Silver nanoparticles obtained by using honeysuckle extract as a reducing agent | Silver nanoparticle mixture was applied to the silk fabric by the bath exhaustion method. | [61] |
| Phenolic extracts obtained from the outer skin of the onion (*Allium cepa L.*) | Unmordanted and mordanted (alum, tin chloride and tannic acid) cellulosic fabrics (cotton, linen, bamboo, hemp, and nettle) have died with dyeing extracts. | [62] |
| Extract spent coffee grounds; extract spent black tea leaves | Silk and wool fabrics were dyed without mordanting with these extracts. | [63] |
| Naturally phenolic compounds (e.g., pyrogallol, resorcinol, phloroglucinol, and pyrocatechol) | Cotton fabric was functionalized with phenolic compounds using a butane-1,2,3,4-tetracarboxylic acid cross-linker *via* a pad-dry-cure method. | [64] |
| Vitamin E | Knitted cotton fabrics were coated with nanoparticles containing vitamin E *via* the pad-cure process. | [65] |
| The various phenolic substances of pomegranate peel | Polyphenols were oxidized by an enzyme (laccase) to form colored quinone that bound to textiles (wool fabric). | [66] |

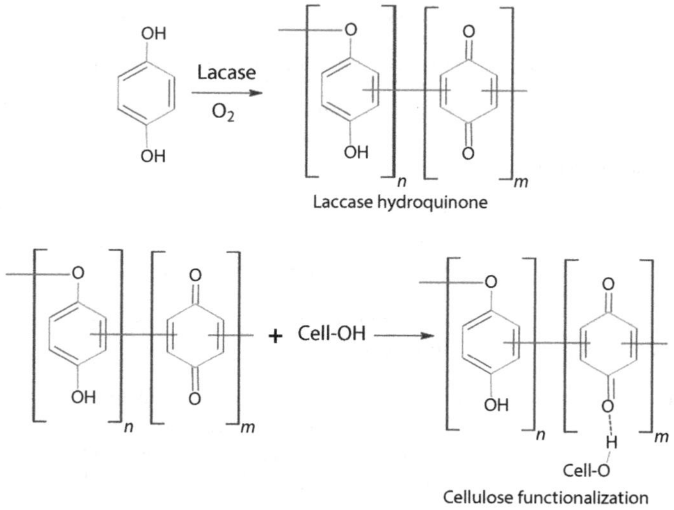


**Figure 17.7** Example of textile functionalization with phenolic compounds *via* enzyme catalysis.

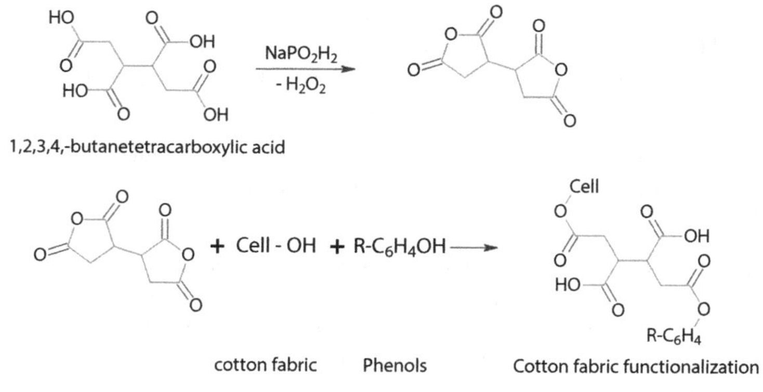


**Figure 17.8** Example of textile functionalization with phenolic compounds *via* chemical catalysis.

### 17.4.3 Skincare Textiles with Antimicrobial Properties

In recent years, a series of studies focused on developing textiles with antimicrobial properties. The ability of textile materials to absorb moisture leads to the growth of microorganisms, resulting in a range of undesirable effects for the consumer (unpleasant odor) and for the textile material (discoloration and reduced mechanical strength). Although there is a wide variety of

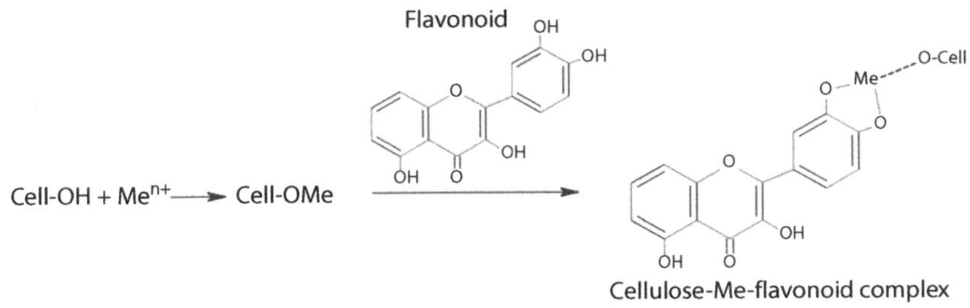


**Figure 17.9** Proposed mechanism of cellulose–metal complex formation.

antimicrobial compounds, a very important criterion in choosing the right compound is that it should not affect the skin's microbiome, which plays a crucial role in maintaining skin health. In addition to this criterion, antimicrobial compounds must be non-toxic and should not cause allergies or irritation to the skin. The most commonly used antimicrobial compounds in the development of functional textiles are synthetic (e.g., metals and metallic salts, triclosan, quaternary ammonium compounds, N-halamine compounds, and chlorhexidine) or natural (e.g., chitosan, essential oils, plant extracts, hydrogen peroxide, some natural dyes, and mordants) [69–71]. The development of nanotechnologies has led to the use of nanoparticles as antimicrobial agents (metal nanoparticles, metal oxide nanoparticles, and carbon-based nanoparticles), and they are considered an alternative to conventional antimicrobial compounds [72]. An example of the schematic representation of the *in situ* synthesis of metal (e.g., Ag, Cu, Zn, and Mg) flavonoid complex on cellulosic fabric is presented in Figure 17.9.

### 17.4.4 Skincare Textiles with Anti-Cellulite Properties

Cellulite (edematous fibrosclerotic panniculopathy) is a skin condition in lower body areas such as hips, thighs, and buttocks. The skin areas affected by cellulite have an orange-peel appearance with dimples and uneven texture [73, 74]. One of the causes of cellulite is the accumulation of fat in adipose tissue [75]. Currently, various anti-cellulite treatments are available, including radiofrequency devices, subcision, fillers, injectable biologic treatments, acoustic wave therapy, laser/light devices, and topical agents. The topical application of different anti-cellulite compounds is increasingly used. Natural remedies are widely used in topical administration due to the advantages they present [76]. For example, there are several studies on the use of caffeine in the treatment of cellulite, due to the stimulation of lipolytic activity. Caffeine-loaded nanoparticles can cross the skin barrier and present a good topical delivery [75]. Other topical applications for treating cellulite include seaweed extracts. The most commonly used are brown seaweeds that contain flavonoids, quercetin, and phlorotannins (e.g., *Fucus vesiculosus*, *Furcellaria lumbricalis*, *Laminaria digitata*, and *Pelvetia wrightii*). By stimulating blood flow, these algae help the body eliminate excess fluids [77]. Hegazi *et al.* demonstrated the anti-cellulite effect of *A. graveolens* and *P. crispum* fruit extracts (with main compounds apigenin and glycosides). Other plant-derived compounds that show promising anti-cellulite activity include forskolin (*Coleus forskohlii*), rutin (*Ruta graveolens*), asiaticoside (*Centella asiatica*), and escin (*Aesculus hippocastanum*) [78]. Other studies demonstrated the anti-cellulitic effect of the flowers of *Chrysanthemum indicum* L., which contains flavonoid and lignin metabolites that can be responsible for lipolytic activity. Another major sources of anti-cellulite active ingredients used in cosmeceuticals are the members of the family Lamiaceae, rich in polyphenols, terpenoids, and organic acids in their roots and flowers [79]. Abosabaa *et al.* studied the anti-cellulite effect of green tea extract incorporated into chitosan-sodium tripolyphosphate nanoparticles using lecithin *via* the ionic gelation method. The results demonstrated a significant reduction in adipocyte size and fat layer thickness, indicating that green tea nanoparticles can be successfully used for cellulite reduction [80].

Most active anti-cellulite compounds are applied to textile supports through encapsulation. Encapsulation is the process by which the active agent is enclosed within a homogeneous or heterogeneous matrix, on a nano (<1 μm) or microscopic (1–5,000 μm) scale. The encapsulated compound is called the core, and the surrounding material is the embedding substance. Ideally, the embedding material should be low hygroscopic, low-viscosity, highly soluble, inexpensive, inert, non-toxic, biodegradable, and biocompatible. The most commonly used encapsulating materials include polysaccharides (e.g., starch, gum arabic, alginate, pectin, chitosan, and agar-agar), phospholipids (e.g., lecithin, phosphatidylethanolamine, and phosphatidylcholine), and proteins (albumin, casein, and gluten) [81].

Microcapsules can be applied to textiles using various finishing methods such as padding, coating, spraying, or exhaustion. In the tartar bath, in addition to microcapsules and other finishing agents, a binder is also added to fix the microcapsules to the textile support and ensure their durability over time [82].

### 17.4.5 Fragranced Textiles

The sense of smell is much more active than the other senses, and the reaction to a particular scent is immediate. In recent years, fragranced textiles have gained popularity due to their beneficial effects on well-being (they help achieve a positive sensory and emotional response). The purpose of using fragranced textiles is to provide the consumer with a sense of well-being through a range of effects for the body (e.g., sedation, curing hypertension, curing rheumatism, relieving cough, curing diabetes, and being good for the stomach) and mind (e.g., desolation, irritability, hysteria, tension, and anxiety) and to eliminate unpleasant odors [83]. The most widely used bioactive compounds used to obtain fragranced textiles are essential oils. Essential oils are derived from plants (they are considered the life force of plants) and are used in various fields (food, cosmetics, pharmaceutical, textile, and leather industries) [84]. The basic compounds of essential oils are alcohols, aldehydes, acids, phenols, esters, and terpenes. The fragrance of essential oils plays a critical role in their biological activity, leading to beneficial effects on the central nervous system (e.g., revitalization, and stress reduction). Fragranced textiles obtained through the use of essential oils have beneficial effects on mood disorders, such as sleep promotion, and the reduction of anxiety and depression.

Anxiety is an unpleasant state that generates fear and nervousness, and the result of this state can lead to changes in certain physiological parameters (hypertension, tachycardia, nausea, aggressiveness, increased body temperature, and excessive sweating) [85]. Some essential oils that reduce anxiety are extracted from *Satureja brevicalyx* and *Satureja boliviana* [86], Lavender, citrus, rose [85], and *Citrus aurantium*.

Recently, sleep disorders have become increasingly common, and their impact on health is significant. A series of studies have demonstrated the promising effects of essential oils in improving sleep quality through their action on the nervous system. Although several studies show the effectiveness of essential oils in enhancing sleep quality, the exact mechanism by which they influence sleep is still unknown. For example, it has been shown that using a mix of essential oils (*Lavandin Super*, *Citrus bergamia*, and *Cananga odorata*) through inhalation significantly improves sleep quality [87].

Depression is a mental disorder and a significant problem that increasingly affects people worldwide. According to some theories, the symptoms of depression result from a serotonin deficiency. Several essential oils, such as lavender, sweet orange, bergamot, *Asarum heterotropoides*, ylang-ylang, rose, patchouli, lemongrass, sandalwood, valerian, and lemon, have demonstrated beneficial effects in reducing depressive symptoms [88, 89].

It is interesting how essential oils act on the body. They can be administered through three routes: topical, oral, or inhalation. Essential oils applied to textile supports (fragranced textiles) penetrate the body either topically or through inhalation. If essential oils are administered through inhalation, then there are two pathways by which they act on the body: the olfactory route and the respiratory route. In the olfactory route, the olfactory nerve is activated, sending signals to the brain. The olfactory neurons transmit electrical impulses to the limbic and hypothalamic regions of the brain. The topical application of essential oils involves their penetration into the skin layers, where they dissolve into the lipid component of the cell membrane. The degree of penetration depends on the chemical composition of the essential oils. The smaller the molecules of the essential oil, the more easily they reach the brain [85]. When applied as is, essential oils have several disadvantages, such as variable and uneven molecular sizes, which can result in improper penetration into the brain. They are also vulnerable to oxidation, and their biological activity can decrease due to decomposition and isomerization under high temperatures and humidity. To overcome these disadvantages, essential oils are encapsulated (e.g., microcapsules or nanocapsules) [84]. Encapsulation of essential oils includes several important stages: (1) formation of the wall around the active compound (the core); (2) stability of the core/wall system; and (3) controlled/prolonged release of the active compound (essential oil) from the encapsulation system. In the last few years, nanotechnology has been advancing rapidly. The choice of polymer system affects drug loading, release mechanisms, and overall effectiveness. Polymeric nanoparticles are classified based on the following:

- Structure: nanocapsules, nanospheres;
- Formulation: micelles, vesicles;
- Biodegradability: biodegradable (shell material is derived from natural sources such as plants, animals, bacteria, and fungi) and non-biodegradable (shell material is made from materials like poly lactic-co-glycolic acid and polylactide). Biodegradable nanoparticles use biopolymers for encapsulating active compounds (essential oils) and are preferred due to their reduced or absent risk of toxicity [90]. Natural biopolymers used for essential oil embedding are polysaccharides (e.g., chitosan, maltodextrin, alginate, and pectin), lipids (e.g., cholesterol, soy lecithin, and cocoa butter) and proteins (e.g., zein, caseins, and gelatin) [91].

Essential oils are embedded in polymer materials by three methods:

- Physical methods (e.g., spray drying, spray coating, ionic gelation, lyophilization, and crystalization);
- Physical methods (e.g., coacervation and evaporation);

- Chemical methods (e.g., interphase polymerization) [92].

The obtained embedding systems can have various forms, as shown in Figure 17.10.

Textile materials functionalized with microcapsules/nanocapsules can be for single-use (e.g., moisturizing gloves or socks, textile facial masks, and patches for various areas of the body) or for multiple uses (e.g., antimicrobial or fragrant bed linens, scented clothing, curtains, towels or sofa covers, and sportswear). In the case of obtaining textiles for multiple uses, it is important to fix the active micro/nanoparticles on the textile substrate to ensure their effectiveness even after washing and friction. There are two main methods for fixing micro/nanoparticles to the textile substrate:

- The use of polymer resins applied as a film that forms strong bonds with micro/nanoparticles and with textile substrates;
- The use of multifunctional cross-linking agents [93].

The fixation of polymer resins or cross-linking agents on the textile substrate is most commonly done using heat-curing or ultraviolet curing (Figure 17.11).

The disadvantage of using heat-curing is the reduction of the effect over time of the functionalized material, due to the heat that affects the active compound (essential oils). On the other hand, the use of ultraviolet curing

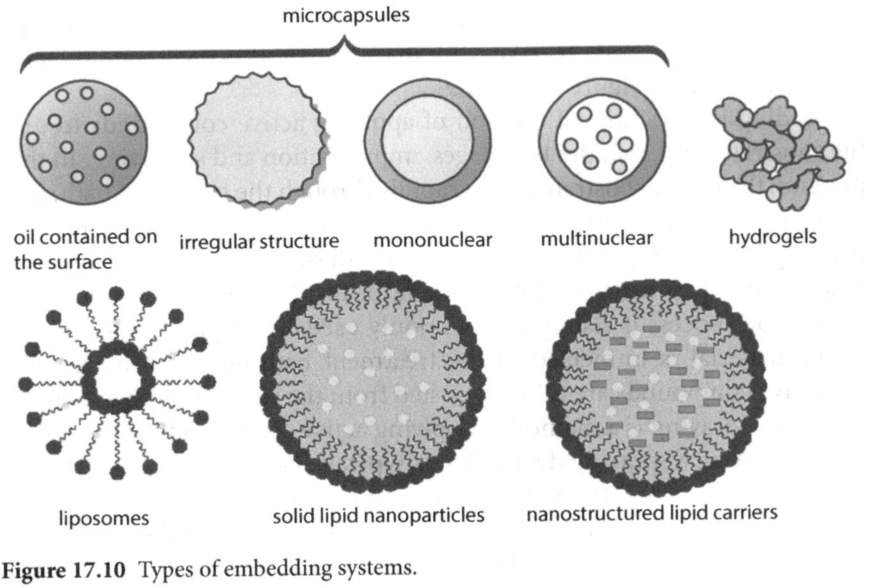


**Figure 17.10** Types of embedding systems.

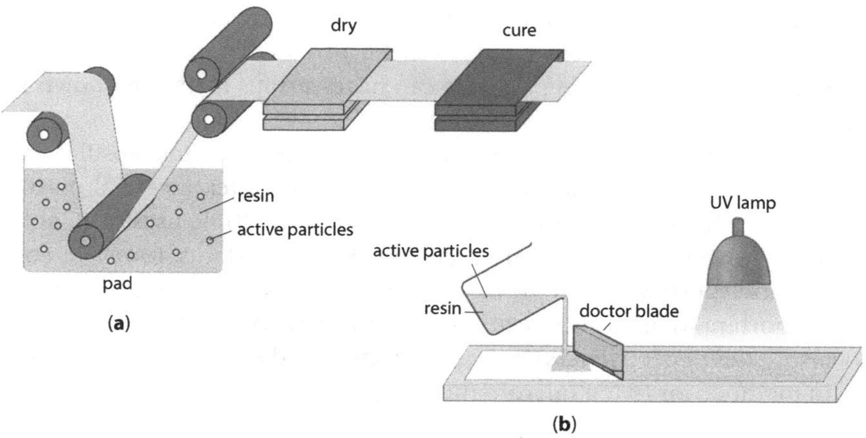


**Figure 17.11** A schematic of micro/nanoparticles fixation on textile support using (a) heat-curing and (b) UV curing.

offers a series of advantages, including low working temperature, quick process, energy saving, and reduction of pollution [94].

## 17.5 Methods for Micro/Nanoparticles Applying on Textile Support

The most used methods for micro/nanoparticles applying on textile support are padding, spraying, coating, exhaust method, sol-gel method, and plasma treatment (Figure 17.12) [69].

Padding is a traditional method of applying active compounds to textile substrates and includes two stages: impregnation and squeezing. In this method, the textile substrate passes rapidly through the treatment bath, and the fixation of the active compound occurs in a subsequent stage. Through padding, active compounds are applied to both sides of the textile substrate. Like the padding method, the treatment of textile substrates through the exhaustion process is also quite commonly used. Through exhaustion, the textile material is immersed in the treatment bath and kept there until the active compounds are fully exhausted from the solution onto the textile substrate. Spraying is a method of applying active compounds to the textile substrate using a spraying device that is typically coupled with compressed air. With this method, the active compound can be applied to specific areas of the textile substrate, ensuring a high level of uniformity and saving water, energy, and chemicals [95]. The coating method involves covering the

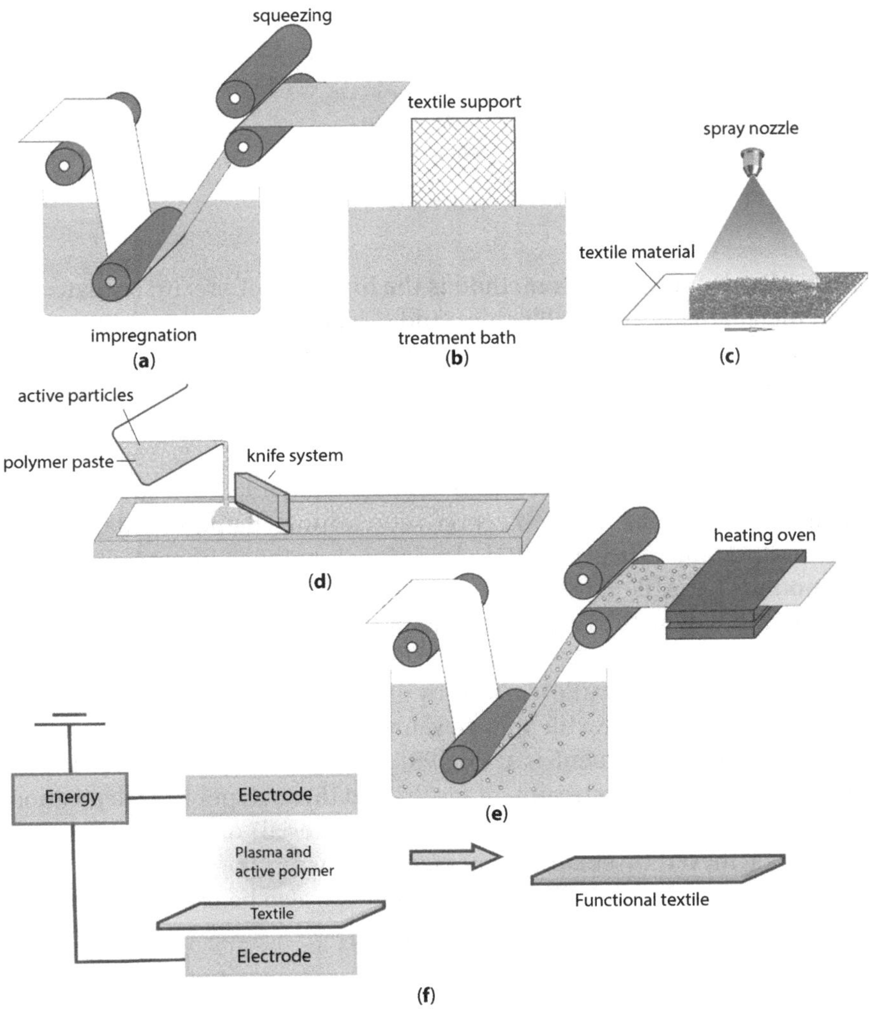


**Figure 17.12** Micro/nanoparticles applied on textile support by (a) padding, (b) exhaust method, (c) spraying, (d) coating, (e) sol-gel method, and (f) plasma treatment.

textile substrate with the active compound from a polymer paste or foam and is carried out using a coating system (a knife system, roller application, or with the help of a rotary printing screen) [96]. Due to its advantages (chemical savings, low working temperatures, ease of application, no need for special equipment, and the ability to obtain multifunctional textiles in a single step, which leads to water and energy savings), the sol-gel method is quite commonly applied to functionalized textile materials.

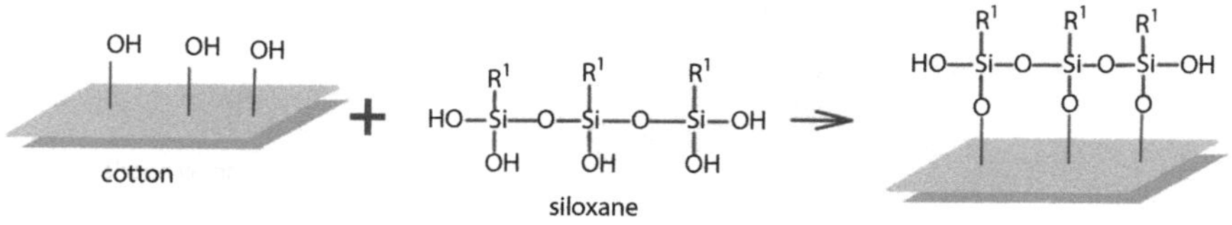


**Figure 17.13** The reactions of siloxane with cotton support.

The disadvantage of this method is the high cost of precursor materials (e.g., metallic salts, metal alkoxides, and silicon-based precursors) used in the sol-gel method [97]. For example, siloxanes (silicon-based precursors) can be used as precursor materials that undergo hydrolysis and condensation reactions in the sol-gel process, allowing them to be applied to various textile substrates (Figure 17.13).

The sol-gel method is used to obtain functional textiles such as anti-wrinkle, antibacterial, UV-protection, self-cleaning, flame-retardance, and water, oil, or soil repellence [97]. Traditional methods of applying active compounds to textile substrates require a large amount of water, use chemicals, and thus generate harmful pollutants. This disadvantage can be eliminated by applying an ecological alternative treatment, such as plasma, which modifies the textile surface by ionizing gas using an electric or magnetic field. In the textile industry, low-temperature and atmospheric-pressure plasma treatment is used [98]. The functionalization of textile substrates using plasma treatment is achieved through polymer deposition (applying them as a thin film on the textile material) or through polymer grafting (achieved by chemical interaction between the active compound and the textile substrate). Plasma treatment can also be used as a pretreatment of the active substrate to activate it [99].

## 17.6 Release Mechanisms of Active Compounds

The most commonly used semi-empirical and empirical models (Figure 17.14) for the release of active compounds are as follows:

Zero-order model

$$M_t = M_0 + k_0 t$$

Where:

$M_0$ - the initial amount of encapsulated compound;

$M_t$ - the amount of active compound released at the time *t*;

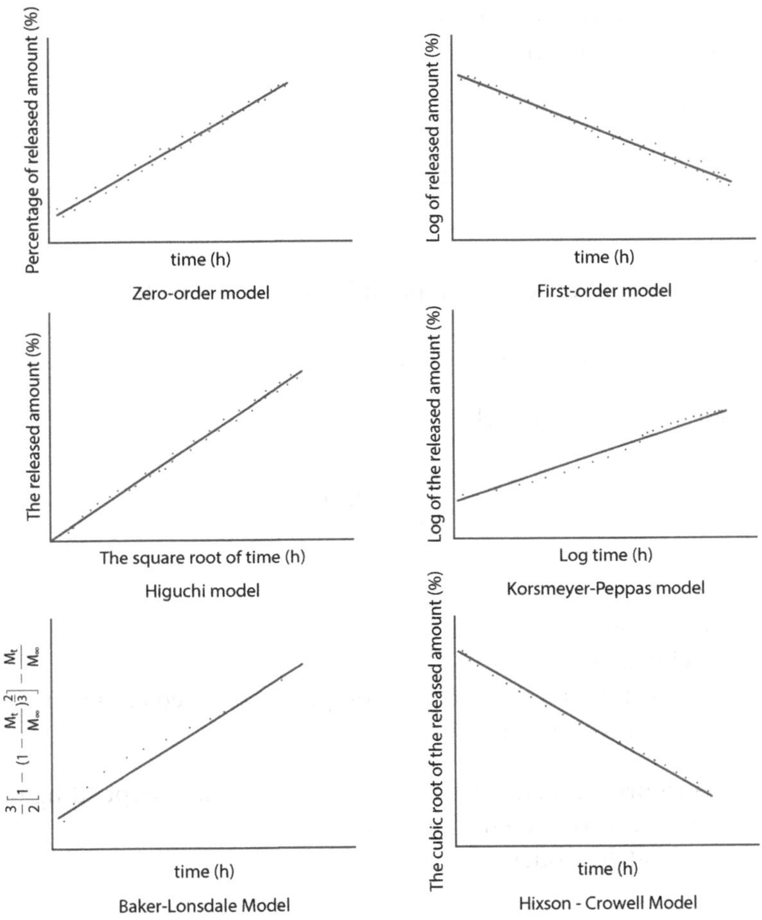


**Figure 17.14** The release graphs for the release models.

$k_0$ - the zero-order release constant;
t - the release time (h).

First-order model

$$M_t = M_0 e^{-k_1 t}$$

Where:
$M_0$ - the initial amount of encapsulated compound;
$M_t$ - the amount of active compound released at the time t;

$K_1$ - the first-order release constant;
t - the release time (h).

Higuchi model

$$M_t = k_H \sqrt{t}$$

Where:
$M_t$ - the amount of active compound released at time t;
t - the release time (h);
$K_H$ - the Higuchi constant [100].

Korsmeyer-Peppas model

$$\frac{M_t}{M_\infty} = k \cdot t^n$$

Where:
K - the Peppas release constant;
t - the release time (h);
$M_t/M_\infty$ - the fraction of the active compound released at the time *t*;
n - the release exponent.

The transport mechanisms of the active compound, depending on the release exponent, are presented in Table 17.3.
Baker-Lonsdale Model

$$\frac{3}{2}\left[1-\left(1-\frac{M_t}{M_\infty}\right)^{\frac{2}{3}}\right]-\frac{M_t}{M_\infty} = kt$$

**Table 17.3** Transport mechanisms of active principles [101].

| Release exponent (n) | Transport mechanism of active compounds | dMt/dt as a function of time |
|---|---|---|
| n = 0.5 | Fickian diffusion | $t^{-0.5}$ |
| 0.5 < n < 1 | Non-Fickian transport | $t^{n-2}$ |
| n = 1 | Case II transport | Zero-order release |
| n > 1 | Super Case II transport | $t^{n-1}$ |

where:

$M_t$ - The amount of active compound released at the time *t*;

$M_\infty$ - the amount of active compound released at infinite time;

K - the release constant, which corresponds to the graph when plotted as $[d(M_t/M_\infty)/dt]$ versus the inverse square root of time;

t - the release time (h).

Hixson-Crowell Model

$$\sqrt[3]{M_0} - \sqrt[3]{M_t} = k_{HC} \cdot t$$

Where:

$K_{HC}$ - Hixson-Crowell constant;

$M_0$ - initial concentration of the active compound;

$M_t$ - concentration of the active compound at time t;

t - release time (h).

## 17.7 Analysis Methods for Skincare Textiles

The most common methods for evaluating the effectiveness and safety of skincare textiles are:

- Sensory analysis involves testing skincare textiles based on potential consumers' tactile and olfactory sensations.
- Physico-chemical analyses: FTIR (Fourier transform infrared spectroscopy) is used to analyse the functional groups present in the textile material; TGA (thermogravimetric analysis) is used for measuring the weight variations of material depending on the applied temperature; XPS (X-ray photoelectron spectroscopy) provides information about the surface composition and the oxidation state of the elements in the textile material; and SEM-EDS (scanning electron microscopy–energy-dispersive X-ray spectroscopy) provides high-resolution images of the functionalized textile surface and offers information about the elemental composition of the investigated area.
- Microbiological analysis is carried out to ensure the safety and hygiene of products that come into direct contact with the skin and is performed according to the diffusion method adapted for natural plant extracts [102].

- The analysis of the effect of skincare textiles—depending on the desired effect, it must be tested accordingly.
- The main comfort indices' analysis includes the evaluation of hygroscopicity, air permeability, and water vapor permeability.

## 17.8 Commercial Functional Textiles

Due to changing consumer perceptions regarding the functionality of textile products and improvements in lifestyle, a wide range of functional textile materials has been available on the market in recent years.

Some brands that create clothing products with UV protection are "Sun With Love" (the textile used is regenerated ECONYL® nylon), "Coolibar" (uses ZnO particles to block UVA and UVB radiations), and "Columbia Sportswear." Some brands that create moisture-wicking and breathable textile products are "Under Armour" (HeatGear® and ColdGear® fabrics) and DuPont™ (uses Coolmax® technology). "Polygiene" is a brand that uses silver-based technologies to create anti-microbial and odor-resistant textiles (inhibiting the growth of different microorganisms). "Organic Basics" developed SilverTech 2.0 technology that integrates silver ions into fabrics to offer various functional benefits. The "Gore-Tex" brand created waterproof and water-resistant textiles. Another company that creates waterproof and water-resistant textiles is "The North Face" (uses DryVent™ technology). Some brands that create clothing products with flame-resistant and protective properties are "DuPont," which uses Nomex® technology, and "Carhartt," which creates flame-resistant workwear designed for industrial safety.

## 17.9 Conclusions

Due to the changing consumer perception of the functionality of textile products and the growing awareness of the importance of a healthy lifestyle, skincare products with various beneficial effects on the body have significantly developed in recent years. Skincare textiles are part of the functional textile category and are classified into cosmetotextiles and textiles with therapeutic effects. The skincare effect (e.g., anticellulite, emollient, antiaging, and anti-inflammatory) is achieved by applying various active compounds to the textile substrate, which are embedded in different protective systems that ensure their prolonged/controlled release.

**AI disclosure statement:** The author acknowledges the use of AI-based tools for minor language editing or grammatical correction. However, all intellectual contributions, analyses, and conclusions presented in this chapter are entirely the author's own.

## References

1. Haule, L.V. and Nambela, L., Sustainable application of nanomaterial for finishing of textile material, in: *Green nanomaterials for industrial applications*, U. Shanker, C.M. Hussain, M. Rani (Eds.), pp. 177–206, Elsevier, Amsterdam, Netherlands, 2022.
2. Yashovardhan, I., Deshpande, R.H., Atul, P., Multi-functional finishing of textiles. *Asian Dyer*, 10, 44–50, 2014.
3. Somkuwar, V., Introductory chapter: Functional textiles, in: *Textiles for functional applications*, B. Kumar (Ed.), IntechOpen, London, United Kingdom, 2021.
4. Alias, A.H.D., Abdul Razak, N.Q., Md Yusoff, M.H., Chin, K., Kamal, M.L., Uyup, N.H., Abdullah, S., Ridzuan, N.S., Saaid, M., Shafie, M.H., Antioxidant and antimicrobial activities of essential oils extracted from bamboo (Bambusa vulgaris) leaves and its application in skincare products: A review. *Biocatal. Agric. Biotechnol.*, 54, 102930, 2023.
5. McEwen, L., Déchelette, C., Fauverghe, S., Ecobiology: More than just a word, a philosophy emerging from skincare. *Ethics Med. Public Health*, 31, 100954, 2023.
6. Kouassi, M., Grisel, M., Gore, E., Multifunctional active ingredient-based delivery systems for skincare formulations: A review. *Colloids Surf., B*, 217, 112676, 2022.
7. Lotos, E., Danila, A., Vasiliu, A., Rosca, I., Stroian, D., Simionescu, B.C., Mihai, M., The potential emulsions of xanthan gum and Daucus carota macerated oil in functional textiles for skincare applications: Formulation, characterization, and performance evaluation. *Colloids Surf., A*, 682, 132960, 2024.
8. Chen, Q., Yi, S., Yang, L., Zhu, L., Penetration pathways, influencing factors and predictive models for dermal absorption of exobiotic molecules: A critical review. *Sci. Total Environ.*, 927, 172390, 2024.
9. Natsheh, H. and Touitou, E., Phospholipid vesicles for dermal/transdermal and nasal administration of active molecules: The effect of surfactants and alcohols on the fluidity of their lipid bilayers and penetration enhancement properties. *Molecules*, 25, 13, 2020.
10. Draelos, Z., *Cosmetic dermatology. Products and procedures*, 2nd ed., Wiley-Blackwell, Hoboken, NJ, USA, 2016.

11. Shetty, S. and Shetty, S., Cubosome-based cosmeceuticals: A breakthrough in skincare. *Drug Discovery Today*, 28, 7, 103623, 2023.
12. Massella, D., Argenziano, M., Ferri, A., Guan, J., Giraud, S., Cavalli, R., Barresi, A.A., Salaün, F., Bio-functional textiles: Combining pharmaceutical nanocarriers with fibrous materials for innovative dermatological therapies. *Pharmaceutics*, 11, 8, 403, 2019.
13. Atanasova, D., Staneva, D., Grabchev, I., Textile materials modified with stimuli-responsive drug carrier for skin topical and transdermal delivery. *Materials*, 14, 930, 2021.
14. Amjad, A.I., Bamboo fibre: A sustainable solution for textile manufacturing. *Adv. Bamboo Sci.*, 7, 100088, 2024.
15. Shahriari-Khalaji, M., Alassod, A., Nozhat, Z., Cotton-based health care textile: A mini review. *Polym. Bull.*, 79, 10409–10432, 2022.
16. Zimniewska, M., Hemp fibre properties and processing target textile: A review. *Materials*, 15, 1901, 2022.
17. Edison, O., Lucy, C., Edwin, K., Applications of silk in biomedical and healthcare textiles, in: *Textiles for functional applications*, IntechOpen, London, United Kingdom, 2021.
18. Tahir, M., Li, A., Moore, M., Ford, E., Theyson, T., Seyam, A., Development of eco-friendly soy protein fiber: A comprehensive critical review and prospects. *Fibers*, 12, 4, 31, 2024.
19. Natarelli, N., Gahoonia, N., Maloh, J., Sivamani, R.K., Clinical efficacy of topical or oral soy supplementation in dermatology: A systematic review. *J. Clin. Med.*, 12, 12, 4171, 2022.
20. Kilinc, M., Korkmaz, G., Kilinc, N., Kut, D., The use of wool fiber in technical textiles and recent developments, in: *The Wool Handbook: Morphology, Structure, Properties, Processing, and Applications*, pp. 441–465, The Textile Institute Book Series, Manchester, UK, 2024.
21. Dhama, A., New innovation in textiles: Corn fibre. *RJPSSs*, 43, 1, 152–160, 2017.
22. Rajput, A. and Khot, A., Functional aspect of eco-friendly banana fibre as textile. *Int. J. Home Sci.*, 8, 3, 01–06, 2022.
23. Pavan, M. and Samant, L., Regenerated milk fiber: An approach towards green textiles. *Just Agric.*, 3, 4, 2022.
24. Salvioni, L., Morelli, L., Ochoa, E., Labra, M., Fiandra, L., Palugan, L., Prosperi, D., Colombo, M., The emerging role of nanotechnology in skincare. *Adv. Colloid Interface Sci.*, 293, 102437, 2021.
25. Hedayati, S., Tarahi, M., Iraji, A., Hashempur, M.H., Recent developments in the encapsulation of lavender essential oil. *Adv. Colloid Interface Sci.*, 331, 103229, 2024.
26. Castañeda-Reyes, E.D., Perea-Flores, M.J., Davila-Ortiz, G., Lee, Y., de Mejia, E.G., Development, characterization and use of liposomes as amphipathic transporters of bioactive compounds for melanoma treatment and reduction of skin inflammation: A review. *Int. J. Nanomed.*, 15, 7627–7650, 2020.

27. Ambekar, A.W., Nagaraju, R., Lakshman, S.R., Ethosomes: A tool for transdermal drug delivery. *Curr. Trends Biotechnol. Pharm.*, 5, 972–981, 2011.
28. Rai, S., Pandey, V., Rai, G., Transfersomes as versatile and flexible nano-vesicular carriers in skin cancer therapy: The state of the art. *Nano Rev. Exp.*, 8, 1, 2017.
29. Umamaheshwari, D., Risvana, I., Vimal, M., Najiya, N., Safeetha, S., Transferosomes as a novel therapeutic delivery system: A review. *J. Pharm. Res. Int.*, 2021.
30. Liga, S., Paul, C., Moaca, E.A., Péter, F., Niosomes: Composition, formulation techniques, and recent progress as delivery systems in cancer therapy. *Pharmaceutics*, 16, 223, 2024.
31. Verma, S. and Utreja, P., Vesicular nanocarrier based treatment of skin fungal infections: Potential and emerging trends in nanoscale pharmacotherapy. *Asian J. Pharm. Sci.*, 14, 2, 117–129, 2019.
32. Richard, C., Cassel, S., Blanzat, M., Vesicular systems for dermal and transdermal drug delivery. *RSC Adv.*, 11, 1, 442–451, 2020.
33. Akanda, M., Mithu, M.S.H., Douroumis, D., Solid lipid nanoparticles: An effective lipid-based technology for cancer treatment. *J. Drug Delivery Sci. Technol.*, 86, 104709, 2023.
34. Pandey, S., Shaikh, F., Gupta, A., Tripathi, P., Yadav, J.S., A recent update: Solid lipid nanoparticles for effective drug delivery. *Adv. Pharm. Bull.*, 17–33, 2022.
35. Palekar-Shanbhag, P., Chandra, R., Rane, D., Lande, S., Deshmukh, P., Patil, P., Gawade, T., Nanostructured lipid carriers in cosmeceuticals. *World J. Pharm. Pharm. Sci.*, 8, 6, 1397–1422, 2019.
36. Che Marzuki, N.H., Wahab, R.A., Abdul Hamid, M., An overview of nanoemulsion: Concepts of development and cosmeceutical applications. *Biotechnol. Biotechnol. Equip.*, 33, 1, 779–797, 2019, https://doi.org/10.1080/13102818.2019.1620124.
37. Guterres, S.S., Alves, M.P., Pohlmann, A.R., Polymeric nanoparticles, nanospheres and nanocapsules for cutaneous applications. *Drug Target Insights*, 2, 147–157, 2007.
38. Ibrahim, N., Krishnakumar, K., Dineshkumar, B., Nair, S.K., An overview on nanosphere drug delivery. *Eur. J. Pharm. Med. Res.*, 5, 4, 192–198, 2018.
39. Deng, S., Gigliobianco, M.R., Censi, R., Di Martino, P., Polymeric nanocapsules as nanotechnological alternative for drug delivery system: Current status, challenges and opportunities. *Nanomaterials*, 10, 5, 847, 2020.
40. Jarvis, M., Krishnan, V., Mitragotri, S., Nanocrystals: A perspective on translational research and clinical studies. *Bioeng. Transl. Med.*, 4, 1, 5–16, 2018.
41. Parmar, P.K., Wadhawan, J., Bansal, A.K., Pharmaceutical nanocrystals: A promising approach for improved topical drug delivery. *Drug Discovery Today*, 26, 10, 2329–2349, 2021.
42. Alqaraleh, S.Y., Metallic nanoparticles applications in cosmetology: A comprehensive review. *Nanomed. Nanotechnol.*, 8, 3, 000238, 2023.

43. Bakry, R., Vallant, R.M., Najam-ul-Haq, M., Rainer, M., Szabo, Z., Huck, C.W., Bonn, G.K., Medicinal applications of fullerenes. *Int. J. Nanomed.*, 2, 4, 639–649, 2007.
44. Parinaz, N., Shohreh, N., Howard, M., Fullerene nanoparticles: Power skin penetration, antioxidant defense, 2022.
45. Chauhan, S., Jain, N., Nagaich, U., Nanodiamonds with powerful ability for drug delivery and biomedical applications: Recent updates on *in vivo* study and patents. *J. Pharm. Anal.*, 10, 1, 1–12, 2020.
46. Zare, H., Ahmadi, S., Ghasemi, A., Ghanbari, M., Rabiee, N., Bagherzadeh, M., Karimi, M., Webster, T.J., Hamblin, M.R., Mostafavi, E., Carbon nanotubes: Smart drug/gene delivery carriers. *Int. J. Nanomed.*, 16, 1681–1706, 2021.
47. Čuden, A.P. and Urbas, R., Advances in ultraviolet (UV) ray blocking textiles, in: *Functional and Technical Textiles*, pp. 213–273, The Textile Institute Book Series, Manchester, UK, 2023.
48. Emam, R.H., El-Sayed, I.E., Xin, J., Lu, X., Attia, N.F., Innovative coating of cotton textile fabrics for integrating outstanding flame hazard safety, UV protection, tensile strength strengthening and antibacterial properties. *Ind. Crops Prod.*, 216, 118752, 2024.
49. Mondal, S., Nanomaterials for UV protective textiles. *J. Ind. Text.*, 51, 4_suppl, 5592S–5621S, 2022.
50. Marta, F., Jorge, P., Ana, R., Rui, F., Liliana, M., Talita, N., Behnaz, M., Cátia, A., Rui, R., Andrea, Z., Polysaccharides and metal nanoparticles for functional textiles: A review. *Nanomaterials*, 12, 1006, 2022, https://doi.org/10.3390/nano12061006.
51. Femina, C. and Kamalesh, T., Advances in stabilization of metallic nanoparticle with biosurfactants—A review on current trends. *Heliyon*, 10, e29773, 2024.
52. Wang, D., Li, K., Zhou, C., Lei, L., Rancourt de Mimérand, Y.D., Jin, X., Guo, J., $Bi_2MoO_6$ and Ag nanoparticles immobilized on textile by plasma-derived innovative techniques to generate antimicrobial activity. *Appl. Surf. Sci.*, 585, 152591, 2022.
53. Reznickova, A., Novotna, Z., Kolska, Z., *et al.*, Immobilization of silver nanoparticles on polyethylene terephthalate. *Nanoscale Res. Lett.*, 9, 305, 2014.
54. Thinkohkaew, K., Piroonpan, T., Jiraborvornpongsa, N., Potiyaraj, P., Development of multifunctional polypropylene nonwoven fabric by radiation induced grafting of $TiO_2$ nanoparticles and trifluoroethyl methacrylate for protective textile applications. *Materialia*, 21, 101355, 2022.
55. Asmat-Campos, D., De Oca-Vásquez, G.M., Rojas-Jaimes, J., Delfín-Narciso, D., Juárez-Cortijo, L., Nazario-Naveda, R., Batista Menezes, D., Pereira, R., De la Cruz, M.S., $Cu_2O$ nanoparticles synthesized by green and chemical routes, and evaluation of their antibacterial and antifungal effect on functionalized textiles. *Biotechnol. Rep*, 37, e00785, 2023.
56. Bjørklund, G., Shanaida, M., Lysiuk, R., Butnariu, M., Peana, M., Sarac, I., Strus, O., Smetanina, K., Chirumbolo, S., Natural compounds and products from an anti-aging perspective. *Molecules*, 27, 20, 7084, 2021.

57. He, X., Wan, F., Su, W., Xie, W., Research progress on skin aging and active ingredients. *Molecules*, 28, 14, 5556, 2022.
58. Umesh, M., Suresh, S., Santosh, A.S., Prasad, S., Chinnathambi, A., Al Obaid, S., Jhanani, G., Shanmugam, S., Valorization of pineapple peel waste for fungal pigment production using Talaromyces albobiverticillius: Insights into antibacterial, antioxidant and textile dyeing properties. *Environ. Res.*, 229, 115973, 2023.
59. Wu, M., Zhou, Y., Tang, R., Bridging phycocyanin onto silk by genipin towards durable colouristic, antioxidant and UV protective properties: A sustainable strategy for fully bio-based textile. *Chem. Eng. J.*, 477, 146808, 2023.
60. Alonso, C., Martí, M., Barba, C., Lis, M., Rubio, L., Coderch, L., Skin penetration and antioxidant effect of cosmeto-textiles with gallic acid. *J. Photochem. Photobiol., B*, 156, 50–55, 2016.
61. Zhou, Y. and Tang, R., Facile and eco-friendly fabrication of AgNPs coated silk for antibacterial and antioxidant textiles using honeysuckle extract. *J. Photochem. Photobiol., B*, 178, 463–471, 2017.
62. Bartolini, D., Pallottelli, L., Sgargetta, D., Varfaj, I., Macchiarulo, A., Galli, F., Romani, A., Sardella, R., Clementi, C., Natural cellulosic biofunctional textiles from onion (Allium cepa L.) skin extracts: A sustainable strategy for skin protection. *Ind. Crops Prod.*, 212, 118295, 2024.
63. Xia, W., Li, Z., Tang, Y., Li, Q., Sustainable recycling of café waste as natural bio resource and its value adding applications in green and effective dyeing/bio finishing of textile. *Sep. Purif. Technol.*, 309, 123091, 2023.
64. Hong, K.H., Preparation and properties of phenolic compound/BTCA treated cotton fabrics for functional textile applications. *Cellulose*, 22, 2129–2136, 2015.
65. Ghaheh, F.S., Khoddami, A., Alihosseini, F., Jing, S., Ribeiro, A., Cavaco-Paulo, A., Silva, C., Antioxidant cosmetotextiles: Cotton coating with nanoparticles containing vitamin E. *Process Biochem.*, 59, 46–51, 2017.
66. Li, J., Song, N., Wang, Y., Chen, L., Liang, Z., Jia, W., Bio-coloration and Antibacterial Function of Wool grafted with pomegranate peel polyphenols catalyzed by laccase. *Mater. Today Commun.*, 40, 109910, 2024.
67. Kim, S., Biopolyphenolics in textile, in: *Advances in Textile Biotechnology*, 2nd ed., A. Cavaco-Paulo, V.A. Nierstrasz, Q. Wang (Eds.), pp. 159–183, The Textile Institute Book Series, Manchester, United Kingdom, 2018.
68. Bai, R., Yu, Y., Wang, Q., Shen, J., Yuan, J., Fan, X., Chitosan-templated bio-coloration of cotton fabrics *via* laccase-catalyzed polymerization of hydroquinone. *Eng. Life Sci.*, 19, 9, 643–654, 2019.
69. Broadhead, R., Craeye, L., Callewaert, C., The Future of Functional Clothing for an Improved Skin and Textile Microbiome Relationship. *Microorganisms*, 9, 6, 1192, 2021.
70. Gulati, R., Sharma, S., Sharma, R.K., Antimicrobial textile: recent developments and functional perspective. *Polym. Bull.*, 79, 5747–5771, 2022.

71. Tanasa, F., Teaca, C., Nechifor, M., Ignat, M., Duceac, I.A., Ignat, L., Highly Specialized Textiles with Antimicrobial Functionality—Advances and Challenges. *Textiles*, 3, 2, 219–245, 2023.
72. Swetha, A., Satheesh Kumar, B., Jaison, J., Murugesan, M., Emerging nanomaterials for antibacterial textile fabrication. *Naunyn-Schmiedeberg's Arch. Pharmacol.*, 394, 2021.
73. Sadick, N., Treatment for cellulite. *Int. J. Womens Dermatol.*, 5, 1, 68–72, 2019.
74. Małgorzata, M., Wilk, E., Zieliński, J., Seersucker Woven Fabrics with Therapeutic Properties. *Fibres Text. East. Eur*, 26, 5, 54–58, 2018.
75. Santos, EMD., Macedo, L.M.D., Tundisi, L.L., Ataide, J.A., Camargo, G.A., Alves, R.C., Oliveira, M.B.P., Mazzola, P.G., Coffee by-products in topical formulations: A review. *Trends Food Sci. Technol.*, 111, 280–291, 2021.
76. Ngamdokmai, N., Waranuch, N., Chootip, K., Jampachaisri, K., Scholfield, C.N., Ingkaninan, K., Cellulite Reduction by Modified Thai Herbal Compresses; A Randomized Double-Blind Trial. *J. Evid.-Based Integr. Med.*, 23, 2515690X18794158, 2018.
77. Bedoux, G., Hardouin, K., Burlot, A.S., Bourgougnon, N., Bioactive Components from Seaweeds: Cosmetic Applications and Future Development. *Adv. Bot. Res.*, 71, 345–378, 2013.
78. Hegazi, N.M., Salem, M.A., Saad, H.H., Aborehab, N.M., El Bishbishy, M.H., Ezzat, S.M., Untargeted metabolomics-based molecular networking for chemical characterization of selected Apiaceae fruit extracts in relation to their antioxidant and anti-cellulite potentials. *Fitoterapia*, 173, 2024.
79. Charles Dorni, A., Amalraj, A., Gopi, S., Varma, K., Anjana, S., Novel cosmeceuticals from plants—An industry guided review. *J. Appl. Res. Med. Aromat. Plants*, 7, 1–26, 2017.
80. Sara, A., Mona, A., Aliaa, E., Hybrid chitosan-lipid nanoparticles of green tea extract as natural anti-cellulite agent with superior *in vivo* potency: full synthesis and analysis. *Drug Delivery*, 28, 2160–2176, 2021.
81. Sundar, S. and Parikh, J.K., Advances and trends in encapsulation of essential oils. *Int. J. Pharm.*, 635, 122668, 2023.
82. Siler Marinkovic, S., Bezbradica, D., Škundrić, P., Microencapsulation in the Textile Industry. *Chem. Ind. Chem. Eng. Q.*, 12, 2006.
83. Astuti, K. and Gaurav, K., Surveying consumer acceptance of textile products with fragrance finish: A detailed review. *Tekst. Ind.*, 69, 37–45, 2022.
84. Huo, Y., Deng, W., Sun, X., Zhou, L., Zhang, Q., Hu, J., Extract toolkit for essential oils: State of the art, trends, and challenges. *Food Chem.*, 461, 140854, 2024.
85. Vora, L.K., Gholap, A.D., Hatvate, N.T., Naren, P., Khan, S., Chavda, V.P., Balar, P.C., Gandhi, J., Khatri, D.K., Essential oils for clinical aromatherapy: A comprehensive review. *J. Ethnopharmacol.*, 330, 118180, 2024.
86. Soto-Vásquez, M.R. and Alvarado-García, P.A.A., Aromatherapy with two essential oils from Satureja genre and mindfulness meditation to reduce anxiety in humans. *J. Tradit. Complement. Med.*, 7, 1, 121–125, 2016.

87. Carlos, N., Enhancing Sleep Quality through Aromatherapy: Updating the Therapeutic Benefits of Essential Oils. *Braz. J. Health Aromather. Essent. Oil*, 1, 2024.
88. Fung, T.K.H., Lau, B.W.M., Ngai, S.P.C., Tsang, H.W.H., Therapeutic Effect and Mechanisms of Essential Oils in Mood Disorders: Interaction between the Nervous and Respiratory Systems. *Int. J. Mol. Sci.*, 22, 4844, 2021.
89. Fonseca, E.C., Ferreira, L.R., Figueiredo, P.L., Maia, C.D., Setzer, W.N., Da Silva, J.K., Antidepressant Effects of Essential Oils: A Review of the Past Decade (2012–2022) and Molecular Docking Study of Their Major Chemical Components. *Int. J. Mol. Sci.*, 24, 11, 9244, 2022.
90. Adeyemi, S.B., Akere, A.M., Orege, J.I., Ejeromeghene, O., Orege, O.B., Akolade, J.O., Polymeric nanoparticles for enhanced delivery and improved bioactivity of essential oils. *Heliyon*, 9, 6, e16543, 2023.
91. Yammine, J., Chihib, N.E., Gharsallaoui, A., *et al.*, Advances in essential oils encapsulation: development, characterization and release mechanisms. *Polym. Bull.*, 81, 3837–3882, 2024.
92. Cândido de Souza, W.F., Almeida, F., Melo, A., Soares, A., Forte, M., Janser, R., Sato, H., Immobilization Techniques on Bioprocesses: Current Applications Regarding Enzymes, Microorganisms, and Essential Oils. *Food Bioprocess Technol.*, 15, 1449–1476, 2022.
93. Xiao, Z., Liu, H., Zhao, Q., Niu, Y., Chen, Z., Zhao, D., Application of microencapsulation technology in silk fibers. *J. Appl. Polym. Sci.*, 139, 25, e52351, 2022
94. Ghayempour, S. and Montazer, M., Tragacanth nanocapsules containing Chamomile extract prepared through sono-assisted W/O/W microemulsion and UV cured on cotton fabric. *Carbohydr. Polym.*, 170, 234–240, 2017.
95. Spraying, Retrieved from https://uptti.ac.in/classroom-content/data/5_6161180347603091731.pdf [Accessed 03.09.2024].
96. Coating, Retrieved from https://www.drpetry.de/en/products/product-overview/coating/ [Accessed 03.09.2024].
97. Nurhan, OC. and Buket, A., Sol-Gel Applications in Textile Finishing Processes, in: *Recent Applications in Sol-Gel Synthesis*, U. Chandra (Ed.), IntechOpen, London, United Kingdom, 2017.
98. Ssekasamba, H., Tebyetekerwa, M., Haodong, C., *et al.*, Review of Plasma Surface Engineering Technology Toward Sustainable Textile Materials. *Mater. Circ. Econ.*, 6, 27, 2024.
99. Naebe, M., Haque, ANMA., Haji, A., Plasma-Assisted Antimicrobial Finishing of Textiles: A Review. *Engineering*, 12, 2021.
100. Sarfaraz, M.D. and Joshi, V.G., Development and Evaluation of Enteric Coated Herbal Drug Delivery System for Treatment of Asthma. *IOSR J. Pharm.*, 4, 12, 34–46, 2014.
101. Maleki, M., Amani-Tehrana, M., Latifia, M., Mathur, S., Drug release profile in core–shell nanofibrous structures: a study on Peppas equation and

artificial neural network modelling. *Comput. Methods Programs Biomed.*, 113, 92–100, 2014.
102. Danila, A., Mureşan, E.I., Ibănescu, S., Popescu, A., Danu, M., Zaharia, C., Türkoğlu, G.C., Erkan, G., Staras, A., Preparation, characterization, and application of polysaccharide-based emulsions incorporated with lavender essential oil for skin-friendly cellulosic support. *Int. J. Biol. Macromol.*, 191, 405–413, 2021.

# 18

# Future Perspectives of Surface Modifications and Functionalization of Textiles

**Arnob Dhar Pranta[1,2] and Md. Reazuddin Repon[3]***

*[1]ZR Research Institute for Advanced Materials, Sherpur, Bangladesh*
*[2]Department of Textile Engineering, Mawlana Bhashani Science and Technology University, Santosh, Tangail, Bangladesh*
*[3]Department of Bioproducts and Biosystems, School of Chemical Engineering, Aalto University, Espoo, Finland*

***Abstract***
The performance of textiles can be improved through surface modifications and functionalization. There are different processes that can impart additional characteristics like antimicrobial activity, water repellency, UV protection, and flame resistance to textiles. Advanced processes like chemical treatments, nanotechnology, and plasma technology have increased the potential uses of textile in different sectors. These approaches provide precise control over textile properties. Surface modification and functionalization of textiles have opened doors in the recent advancement of nanotechnology and artificial intelligence. This chapter explores the future prospects of surface modification and functionalization. It also covers the existing challenges regarding advanced surface modification and functionalization of textiles.

***Keywords***: Smart textiles, surface modification, functional textiles, advanced textiles

## 18.1 Introduction

Surface modifications and functionalization are now essential methods for enhancing the sustainability and performance of textiles. These processes aim

**Corresponding author*: reazmbstu.te@gmail.com

Aminoddin Haji and Çağlar Sivri (eds.) Surface Modification and Functionalization of Textiles, (473–488) © 2026 Scrivener Publishing LLC

to enhance the textile functionality by altering the surface attributes to make it more comfortable, antimicrobial, water-repellent, and UV-protective. Among the techniques that have been employed to accomplish particular functional characteristics are chemical treatments, plasma technology, and nanotechnology. Advanced surface modifications help to achieve different functional characteristics within the textiles without reducing the fabric durability [1]. Imparting functional traits in textiles is significant for the ever-growing and requirement for high-performance textiles in industries like automobile, healthcare, sports, military, and many more. The world is shifting toward eco-friendly technology integration in every industry. The use of bio-based and environmentally sustainable surface treatment is increasing day by day. Advances in surface modification now allow textiles to serve multiple purposes with a wide range of applications. Functional coatings with nanomaterials are expected to significantly impact textile applications.

The growing interest in surface modification and functionalization of textiles has the potential to greatly improve fabric properties and performance. Different surface treatments can enhance textiles to meet the requirements of modern industries. These modifications involve altering the outer layer of the fabric to add features such as mechanical properties and increased durability [2]. There are many approaches available for the surface modification and functionalization of textiles such as chemical, physical, and biological techniques [3, 4]. The performance enhancement and the adaptability of textures for different practical applications require the surface modification and functionalization. Characteristics like water resistance, UV protection, and antimicrobial activity can be achieved by using these processes.

This chapter aims to explore the future prospects of surface modification and functionalization of textiles. It will deeply look into the emerging areas like sustainability and artificial intelligence (AI) integration in surface modification. Apart from that, the role of surface modification in the area of smart textiles along with the possible advances in biotechnology will also be explored. The challenges regarding surface modification and functionalization of textiles will also be discussed.

## 18.2 Future Perspectives

### 18.2.1 Sustainability in Textile Functionalization

The integration of sustainable practices in textile functionalization has emerged as a critical frontier to address the environmental challenges associated with conventional textile production and finishing processes. The future of textile functionalization is expected to revolve around greener technologies, resource efficiency, and the use of eco-friendly raw materials.

#### *18.2.1.1 Eco-Friendly Chemical Functionalization*

Traditional methods of functionalization have frequently made use of harsh chemicals that are known to cause harm to both humans and the environment [5]. A more sustainable alternative is the growing trend of using chemical agents that are bio-based and environmentally friendly. More and more studies are looking into the possibility of using natural polymers like chitosan, alginate, and starch derivatives to impart functionality without leaving harmful residues [6]. The antimicrobial, biodegradable, and renewable chitosan, for instance, comes from crustacean shells. Additional UV protection, flame retardancy and dye fixation capabilities have been demonstrated by tannins, lignin, and pectin derived from agricultural waste [7]. In order to make bio-based functional finishes last longer, scientists are looking into new ways to cross-link materials that do not use formaldehyde or other VOCs. Enzymatic processes also show promise because they allow functionalization under mild conditions which means less energy is needed. For example, a long-lasting finish can be achieved with little environmental effect by grafting functional groups onto textile fibers using laccase-based treatments [8].

#### *18.2.1.2 Waterless Functionalization Technologies*

Due to the high volumes of wastewater and chemical discharge produced by traditional wet processing methods, the textile industry ranks among the top water users. Revolutionary solutions are offered by waterless technologies like plasma treatments, laser surface modifications, and supercritical $CO_2$ finishing [9]. For example, plasma treatments can make fabrics much more reactive, which means they can have antimicrobial or hydrophobic qualities added to them without using chemicals. Another technology that shows promise for functionalization is supercritical $CO_2$ that does not require solvents. It is possible to treat textiles consistently without using water or drying time by dissolving functional agents in supercritical $CO_2$ [10]. Dyeing, antimicrobial finishes, and UV protection have all found success with this method. Dry and precise, laser surface modification uses concentrated light to change the surface characteristics of fabrics. Improved coating adhesion or direct functionalities, like better moisture management, can be achieved through the formation of micro- and nano-structures [11].

#### *18.2.1.3 Functionalization with Recycled and Biodegradable Materials*

Functionalizing textiles with recycled and biodegradable materials helps with two important parts of sustainability: lowering trash and making sure that they break down at the end of their life. Functional coatings are finding more and more applications on recycled polymers like post-consumer polyethylene terephthalate [12]. For instance, coatings made of polylactic acid offer resistance to moisture and antimicrobial qualities; when composted in an industrial setting, they break down into harmless byproducts. The potential of proteins and peptides derived from waste biomass to impart self-cleaning or anti-odor functionalities is being investigated [13]. Using bio-composites that are functional textiles made of a combination of natural fibers and biodegradable matrices is another potential route.

#### *18.2.1.4 Low-Energy Functionalization Processes*

Energy efficiency is one of the significant factors in terms of sustainable textile functionalization. It is important because the conventional methods frequently require a drying cycle with a long time and high-temperature curing. Photonic techniques and nanostructure materials are some possible substitutes that have low energy are currently under development. Nanoparticles such as silver and titanium dioxide help to achieve functional effects with much lower concentrations and active energies [14, 15]. Ultrasonic and microwave technologies are getting attention in the recent times. As a result, nanomaterials are becoming the primary choice for researchers as these materials have a very strong potential to improve chemical reaction at reduced temperatures.

#### *18.2.1.5 Circular Functionalization Approaches*

The main purpose of circularity in textile functionalization is to develop sustainable methods and materials. Sustainable methods and materials can be used again and again. Functional coating that can be recycled in modular units is a great idea. Advanced technologies allow the removal of coatings without damaging the main structure of the textiles [16]. Biomass of waste such as lignin nanoparticles or cellulose nanocrystals can be another way to use as functional agents. The future research can be focused on integrating bio-circular materials into larger production processes [17]. Targeted re-functionalization and increased material life expectancy are both made possible by these technologies' real-time data on fabric composition and condition.

#### 18.2.1.6 *Legislation-Driven Innovations*

Stricter environmental regulations and consumer demand for transparency are expected to drive innovations in sustainable functionalization. Compliance with standards such as OEKO-TEX®, ZDHC (Zero Discharge of Hazardous Chemicals), and REACH (Registration, Evaluation, Authorization, and Restriction of Chemicals) will necessitate the elimination of hazardous chemicals and the adoption of greener alternatives.

#### 18.2.1.7 *Consumer-Centric Functionalization*

Consumer preferences for sustainable and high-performing textiles are anticipated to shape the future of textile functionalization. There will be an increasing demand for functionality coupled with sustainability, such as organic antimicrobial finishes or biodegradable waterproof coatings [18]. Smart labeling systems that highlight the eco-friendly attributes of functionalized textiles, supported by blockchain-based traceability.

### 18.2.2 Advanced Nanotechnology Applications

Textile surface modifications and functionalization could be revolutionized by nanotechnology. By incorporating nanomaterials into textile substrates, novel fabrics with exceptional characteristics can be produced which were previously impossible using traditional techniques.

#### 18.2.2.1 *Multifunctional Nano-Coatings*

An important step toward improving fabrics' performance and functionality has been the creation of multifunctional nano-coatings for use in textiles. Fabrics can be treated with nanoparticles such as silver, zinc oxide, or titanium dioxide to make them water-repellent, antimicrobial, and UV protector all at once [19]. Efforts are being made by researchers to make these nano-coatings more washable and long-lasting, so that their multifunctional properties can withstand environmental factors and repeated washings. Furthermore, research into biocompatible and biodegradable nanoparticles has been prompted by the need to reduce environmental impact [20].

#### *18.2.2.2 Carbon-Based Nanomaterials for Textile Enhancement*

It is highly anticipated that graphene, carbon nanotubes, and nanodiamonds, all carbon-based nanomaterials, can be incorporated into clothing. The use of graphene and carbon nanotubes in conductive textiles is gaining popularity in wearable electronics. These materials are primarily used for energy storage devices, heating elements, and sensors [21]. This technology opens new opportunities for smart textiles. The development of lightweight and flexible textiles that can store energy and track physiological movements. The tensile strength of textiles remarkably increases with the addition of carbon nanotubes. These nanomaterials can be uniformly incorporated into the textile fibers without compromising the primary features of the apparel. Hence, carbon-based nanomaterials utilized in textiles, particularly in high-performance textiles [22].

#### *18.2.2.3 Integration of Nanotechnology in Smart Textiles*

The application of nanotechnology makes it possible for the textiles to respond in real time to the outside environment. Real-time sensing can be performed through the integration of nanotechnology. It helps the textile to sensing, moisture control, and thermal regulation adaptation [23]. One example is the use of piezoelectric nanomaterials, which can convert mechanical movement into electrical energy. The creation of autonomous smart clothing that track vital signs or environmental factors might be radically altered by this. Integrating nano-sensors into textiles is another application of nanotechnology that allows for the real-time monitoring of a variety of physiological parameters, including temperature and heart rate [24].

#### *18.2.2.4 Nanotechnology in Dyeing and Finishing Processes*

The use of nanotechnology has the potential to significantly improve the efficiency and environmental friendliness of the textile finishing and dyeing industries. A lot of water and chemicals are used in traditional dyeing methods, which is bad for the environment. By facilitating more effective dye fixation at the nanoscale, nanoparticle-assisted dyeing provides a novel approach that improves colorfastness while decreasing chemical and water consumption [25]. More vibrant and long-lasting colors, even after multiple washes, are another benefit of this method. Reducing the need for harsh chemicals and significantly minimizing water consumption compared to traditional dyeing methods, these nanotechnology-driven processes unquestionably benefit the environment [26].

#### 18.2.2.5 *Integration with AI and 3D Printing*

Nanotechnology, 3D printing, and AI are coming together to create new opportunities for functionalizing textiles. Nanostructures optimized for particular textile applications can be designed and optimized using AI algorithms, enhancing the functionality and precision of nanomaterials [27]. AI can optimize the deposition process to guarantee performance and uniformity which simplifies the integration of nanomaterials into textile production. Furthermore, with the development of 3D printing technology, tailored nanomaterial coatings or structures can now be precisely manufactured into textiles [28]. As the traditional manufacturing techniques and processes do not give the opportunity for the mass customization and highly specialized functional textile production, these technologies help to overcome these challenges.

### 18.2.3 Surface Modification through AI Integration

The addition of AI-based surface modification of textiles is a novel approach that has the potential to revolutionize the textile sector. Functionalization processes with the integration of AI have the potential to enhance the performance of textiles.

#### 18.2.3.1 *Data-Driven Approaches in Surface Modification*

The most efficient surface modification methods can be predicted by AI algorithms. Technologies like machine learning and deep learning can analyze large datasets [29]. Surface modification can optimize the desired results. It helps to improve functionality and durability. It can process the information about a wide range of textile characteristics like fabric type the composition of the fiber and fabric and the condition in which the fabric will be treated [30].

#### 18.2.3.2 *Automation of Surface Modification Processes*

The production environments and the parameters are being revolutionized by the automation driven AI. It enables real-time control and adjustment of surface modification processes. AI-driven automatic system can precisely monitor, control, and modify the variables like environmental conditions like temperature, pressure, and concentration [31]. Functional textiles with large-scale production at the industry level would be convenient with the help of AI-controlled robots and machines.

### *18.2.3.3 Personalization of Functional Textiles*

The most significant advantage of AI integration is the personalized and individualized functionalization. It can be predicted how various surface treatment will affect our properties of textiles through the use of machine learning tools [32]. Hence, highly specialized products can be produced. Textiles with specific features like water resistance, antimicrobial qualities, or UV protection can be imported to meet the consumer demands. Customized surface modifications can be done by manufacturers through AI.

### *18.2.3.4 Enhanced Material Discovery*

AI has a role in surface modification that goes beyond finding new agents and properties. It can predict new materials for surface modification that can enhance textile performance using machine learning tools. It can also discover new materials with better features and bio-sourced. It is highly challenging to do the same task in conventional research techniques [33].

### *18.2.3.5 Sustainability and Energy Efficiency*

The sustainability of surface modification processes can be greatly enhanced by the addition of AI. It will help to optimize energy, resources, and chemicals. This cutting-edge technology will allow to suggest changes in the total procedure to reduce the waste. The energy consumption can be much more controlled optimized and minimized. It can be done by the meticulous observation of the circumstances surrounding the surface treatment [34, 35]. AI can, for instance, adjust the parameters of plasma treatment or coating application in order to reduce the amount of chemical that is used excessively or the number of steps that require a lot of energy. AI-powered solutions will assist in lowering the environmental impact of surface modification processes, which is becoming increasingly important for the textile industry as it faces increasing pressure to meet sustainability standards.

### *18.2.3.6 Predictive Maintenance and Performance Monitoring*

AI can predict particular surface modifications. It will give information about the imported coating degradation before it takes place. With the use of the predictive analytics domain, it can and hence the functionality and durability of textiles. The longevity of the textile products will be increased by the use of AI. Also, the functional textile performance was tracked over time. This feature will allow to predict possible failures of the areas that need to be researched for further improvements [36, 37]. The ability to carry out condition based monitoring and predictive maintenance is important for high-performance textiles [38].

### 18.2.4 Smart and Wearable Textiles

Several sectors like fashion, sports, and the healthcare industry can be completely transformed through the advancement in wearable textiles [39]. Clothing will be transformed from passive objects into active devices that are capable of performing a wide range of functions as a result of the development of such textiles, which will enable a more seamless integration of technology into daily life through the transition [40]. In the field of smart and wearable textiles, one of the most important areas of research that will be conducted in the future is the development of materials that are biocompatible, flexible, and durable for the purpose of integrating electronic components. By incorporating conductive fibers into textiles, such as carbon nanotubes, graphene, and silver nanoparticles, it is possible for textiles to keep their inherent properties while also gaining the ability to sense environmental stimuli or perform physiological signal monitoring [41].

Smart textiles will revolutionize the upcoming health sector. It will be continuously used to merge vital signs like blood pressure, temperature hard heart rate, and even glucose levels. It will help to take the necessary actions to cure the patient. The incorporation of bio-active substances and compounds into the textile fibers will increase the advanced textiles performance even more [42]. Wearables that are equipped with sensors can monitor the movement of the player along with the muscle activity. It also determines the fatigue level of the athletes. With the early detection, it will help to detect the stress or the strain in the body [43]. The fashion industry is also looking forward to adopting smart textiles. It is manufacturing apparels that can adapt to the preferences of the wearer along with the environment [44].

### 18.2.5 Biotechnological Advancements

The surface modification and functionalization with the help of biotechnological development have gained attention in recent years. It is environmentally sustainable and allows to creation of innovative methods to enhance the features of textiles. It does not depend on hazardous chemicals or energy-intensive procedures [45, 46]. The use of enzymes is one of the promising developments to modify the surface of textiles. Enzymatic treatments like cellulase and laccase have already proved great results in developing the softness, absorption of dyes, and even biodegradability of fibers. In addition, it helps to reduce the environmental impact compared to the traditional chemical process apart from enhancing the textile qualities. It has been found that, with the control enzyme concentration the intended surface modification can be achieved. Besides, it has the least amount of environmental impact [47]. The genetic engineering of microorganisms has enabled surface modification of textiles with the help of bio-based coatings. Microorganisms like bacteria and fungi along with biopolymers like chitosan, polylactic acid, and others can be produced. These bio-based compounds have a substantial effect on the surface area of the textile. It helps to impart desired features to the surface. As a result, textiles come with additional features like antimicrobial, water-repellent, and UV-blocking properties [48]. Biopolymer-based coatings perform better compared to traditional synthetic proteins. Environmentally sustainable. Researchers have shown that these bio-derived agents provide effective protection against harmful pathogens [49]. The addition of specific proteins or enzymes to the textile matrix helps to provide unique features. The addition of spider silk protein to the textile can be a great example. These proteins are added to textures for genetic engineering to increase the elasticity, strength and resilience. By mimicking the naturally occurring qualities of silk, this bio-inspired method provides a sustainable substitute for synthetic fibers [50].

## 18.3 Challenges

### 18.3.1 Complexity of Processes

The surface modification of textiles and functionalization usually consists of a series of multistep procedures. It should have proper control over chemical concentration, temperature, and time. This procedure makes the optimization of the textile outcomes quite challenging. Even though developed techniques can produce high-quality results, there are still challenges to overcome when implementing these processes on an industrial scale. Challenges of maintaining constant performance and ensuring quality control become more evident.

### 18.3.2 Environmental Concerns

Environmental concerns are one of the key considerations in terms of surface modification. One of the major barriers that ecosystems and human health can be at risk from is the chemicals used in these processes. Catalyst, solvents, and other components can harm the environment in a massive way. Improper disposal of chemicals can lead to environmental pollution.

### 18.3.3 Durability and Stability of Functionalization

The durability and the stability of the functionalized coating are very significant when textiles are subjected to different conditions like mechanical stress, washing may decay or damage the coating, consequently losing the effectiveness. Environmental parameters like temperature, pressure, and ultraviolet light also play a crucial role in damaging the imported coatings on the surface of the textile. A product's lifespan can be shortened by repeated washings or regular use.

### 18.3.4 Material Compatibility

Material compatibility is also an important factor for the surface modification. One of them major charges the issues of compatibility between textile substrate and modification agent is a significant barrier. Natural and synthetic fiber have different properties. It can affect how each type of fiber responds to the functionalization process and agents. Sensitive chemical treatments on the natural fiber can produce unwanted issues or effects. Wool and cotton may face physical degradation along with a change in texture due to sensitive chemical treatment. On the other hand, synthetic fiber have uneven coating or poor adhesion issues during processing. As a result, these fibers require more intensive treatments.

### 18.3.5 Cost Considerations

The associated cost of advanced textile functionalization techniques is yet another significant consideration. It is very challenging for the manufacturers. In developing countries, the large-scale production through these technologies is very costly. Eventually, it will increase the production cost. Functionalization of textiles requires chemicals, machineries, tools, and many more. It requires an enormous amount of investment for large-scale production. Although techniques like sol-gel coating, nanoparticle deposition, electroless deposition and plasma treatment are effective, they are very high in cost.

### 18.3.6 Health and Safety Concerns

The potential hazardous effects of surface modification and functionalization of textiles is one of the major barriers to human health and the environment. There were many substances like catalyst, nanoparticles, and solvents that were highly toxic, which were used in the process. These compounds pose risks to the workers in the manufacturing process. The long exposure to these chemicals can bring issues in the respiratory system, allergic reactions, or skin irritation.

## 18.4 Conclusion

The functionalization and surface modification of textiles have already improved in the textile industry. Through surface modification process, it is possible to develop textiles and apparel to be more durable with multi-functional features. The addition of non-toxic chemicals and bio-sourced compounds also meets the consumer demands. Fabrics are now used for a wide range of multifunctional applications. It reduces the dependency on hazardous chemicals and the negative effects on the environment. The advancements in this area not only increase the purpose of textiles but also open new doors to sustainable and eco-friendly solutions. It is more likely that multidisciplinary efforts that bring together materials science, biotechnology, and environmental science will result in the development of innovative solutions that are able to satisfy the requirements of developing markets. The future of surface-modified textiles lies in their ability to improve functionality while also making a contribution to the sustainability of the environment and the well-being of consumers.

**AI disclosure statement:** The author acknowledges the use of AI-based tools for minor language editing or grammatical correction. However, all intellectual contributions, analyses, and conclusions presented in this chapter are entirely the author's own.

## References

1. Arslan, A., *et al.*, Surface Texture Manufacturing Techniques and Tribological Effect of Surface Texturing on Cutting Tool Performance: A Review. *Crit. Rev. Solid State Mater. Sci.*, 41, 6, 447–481, Nov. 2016, doi: 10.1080/10408436.2016.1186597.

2. Nemani, S.K., *et al.*, Surface Modification of Polymers: Methods and Applications. *Adv. Mater. Interfaces*, 5, 24, Dec. 2018, doi: 10.1002/admi.201801247.
3. Goddard, J.M. and Hotchkiss, J.H., Polymer surface modification for the attachment of bioactive compounds. *Prog. Polym. Sci.*, 32, 7, 698–725, Jul. 2007, doi: 10.1016/j.progpolymsci.2007.04.002.
4. Nadi, A., Boukhriss, A., Bentis, A., Jabrane, E., Gmouh, S., Evolution in the surface modification of textiles: a review. *Text. Prog.*, 50, 2, 67–108, Apr. 2018, doi: 10.1080/00405167.2018.1533659.
5. Gaba, J.M., Regulation of Toxic Pollutants Under the Clean Water Act: NPDES Toxics Control Strategies. *J. Air L. Com.*, 50, 761, 1984.
6. Vinod, A., Sanjay, M.R., Suchart, S., Jyotishkumar, P., Renewable and sustainable biobased materials: An assessment on biofibers, biofilms, biopolymers and biocomposites. *J. Cleaner Prod.*, 258, 120978, Jun. 2020, doi: 10.1016/j.jclepro.2020.120978.
7. Uddin, Md.N., Desai, F., Asmatulu, E., Engineered nanomaterials in the environment: bioaccumulation, biomagnification and biotransformation. *Environ. Chem. Lett.*, 18, 4, 1073–1083, Jul. 2020, doi: 10.1007/s10311-019-00947-0.
8. Salemdeeb, R., Saint, R., Pomponi, F., Pratt, K., Lenaghan, M., Beyond recycling: An LCA-based decision-support tool to accelerate Scotland's transition to a circular economy. *Resour. Conserv. Recyl. Adv.*, 13, 200069, May 2022, doi: 10.1016/J.RCRADV.2022.200069.
9. Banchero, M., Recent advances in supercritical fluid dyeing. *Color. Technol.*, 136, 4, 317–335, Aug. 2020, doi: 10.1111/cote.12469.
10. Khaw, K.Y., Parat, M.O., Shaw, P.N., Falconer, J.R., Solvent supercritical fluid technologies to extract bioactive compounds from natural sources: A review. *Molecules*, MDPI, Switzerland, Jul. 01, 2017, MDPI AG. doi: 10.3390/molecules22071186.
11. Hou, A., Chen, B., Dai, J., Zhang, K., Using supercritical carbon dioxide as solvent to replace water in polyethylene terephthalate (PET) fabric dyeing procedures. *J. Cleaner Prod.*, 18, 10–11, 1009–1014, Jul. 2010, doi: 10.1016/j.jclepro.2010.03.001.
12. Sahoo, S.K. and Dash, A.K., Recycled fibres from polyester and nylon waste, in: *Sustainable Fibres for Fashion and Textile Manufacturing*, pp. 309–332, Cambridge, United States, Jan. 2023, doi: 10.1016/B978-0-12-824052-6.00008-1.
13. Kumartasli, S. and Avinc, O., Recycled Thermoplastics: Textile Fiber Production, Scientific and Recent Commercial Developments, pp. 169–192, 2021, doi: 10.1007/978-981-16-3627-1_8.

14. Ramasubramanian, B., Sundarrajan, S., Rao, R.P., Reddy, M.V., Chellappan, V., Ramakrishna, S., Novel low-carbon energy solutions for powering emerging wearables, smart textiles, and medical devices. *Energy Environ. Sci.*, 15, 12, 4928–4981, 2022, doi: 10.1039/D2EE02695C.
15. Khan, M., *et al.*, Development of self-cleaning and antibacterial properties on cotton fabric using silver nanoparticles and PFOTS. *SPE Polym.*, 2024, doi: 10.1002/pls2.10143.
16. Rattalino, F., Circular advantage anyone? Sustainability-driven innovation and circularity at Patagonia, Inc. *Thunderbird Int. Bus. Rev.*, 60, 5, 747–755, Sep. 2018, doi: 10.1002/tie.21917.
17. Batool, G. and Nawab, Y., Regulation in Recycling and Circularity: Future Prospective, pp. 267–282, 2023, doi: 10.1007/978-3-031-49479-6_11.
18. Yoon, H.S. and Occeña, L.G., Influencing factors of trust in consumer-to-consumer electronic commerce with gender and age. *Int. J. Inf. Manage.*, 35, 3, 352–363, Jun. 2015, doi: 10.1016/J.IJINFOMGT.2015.02.003.
19. Sojoudi, H., *et al.*, Stable Wettability Control of Nanoporous Microstructures by iCVD Coating of Carbon Nanotubes. *ACS Appl. Mater. Interfaces*, 9, 49, 43287–43299, Dec. 2017, doi: 10.1021/acsami.7b13713.
20. Singh, P., Balayan, S., Sarin, R.K., Jain, U., Differential surface characterizations of cotton fibers coated with TiO2 and ZnO nanoparticles for nano-based analysis of fabrics. *J. Eng. Fiber Fabr.*, 16, 2021, doi: 10.1177/15589250211043833.
21. Abu Zaid, N.S.K., Nasser, M.S., Onaizi, S.A., Pickering Emulsions Stabilized by Metal-Organic Frameworks, Graphene-Based Materials, and Carbon Nanotubes: A Comprehensive Review. *J. Mol. Liq.*, 393, 123617, Jan. 2024, doi: 10.1016/j.molliq.2023.123617.
22. Rho, W.Y., Jeon, H., Kim, H.S., Chung, W.J., Suh, J.S., Jun, B.H., Recent progress in dye-sensitized solar cells for improving efficiency: TiO2 nanotube arrays in active layer. *J. Nanomater.*, 2015, 2015, doi: 10.1155/2015/247689.
23. Hu, C., Yang, L., Wang, Y., Recent advances in smart-responsive hydrogels for tissue repairing. *MedComm – Biomaterials and Applications*, John Wiley & Sons Australia Ltd., Milton, Queensland, Australia, Sep. 01, 2022, John Wiley and Sons Inc. doi: 10.1002/mba2.23.
24. Tien, J.M., Internet of Things, Real-Time Decision Making, and Artificial Intelligence. *Ann. Data Sci.*, 4, 2, 149–178, Jun. 2017, doi: 10.1007/s40745-017-0112-5.
25. Shah, S.S., Shaikh, M.N., Khan, M.Y., Alfasane, Md.A., Rahman, M.M., Aziz, Md.A., Present Status and Future Prospects of Jute in Nanotechnology: A Review. *Chem. Rec.*, 21, 7, 1631–1665, Jul. 2021, doi: 10.1002/tcr.202100135.
26. Syduzzaman, Md., Hassan, A., Anik, H.R., Akter, M., Islam, Md.R., Nanotechnology for High-Performance Textiles: A Promising Frontier for Innovation. *ChemNanoMat*, 9, 9, Sep. 2023, doi: 10.1002/cnma.202300205.

27. Dong, K., Peng, X., Wang, Z.L., Fiber/Fabric-Based Piezoelectric and Triboelectric Nanogenerators for Flexible/Stretchable and Wearable Electronics and Artificial Intelligence. *Adv. Mater.*, 32, 5, Feb. 2020, doi: 10.1002/adma.201902549.
28. Anik, H.R., Tushar, S.I., Mahmud, S., Khadem, A.H., Sen, P., Akter, M., Into the Revolution of NanoFusion: Merging High Performance and Aesthetics by Nanomaterials in Textile Finishes. *Adv. Mater. Interfaces*, Nov. 2024, doi: 10.1002/admi.202400368.
29. Zhang, W., Gu, X., Tang, L., Yin, Y., Liu, D., Zhang, Y., Application of machine learning, deep learning and optimization algorithms in geoengineering and geoscience: Comprehensive review and future challenge. *Gondwana Res.*, 109, 1–17, Sep. 2022, doi: 10.1016/j.gr.2022.03.015.
30. Cao, Y., Taghvaie Nakhjiri, A., Ghadiri, M., Different applications of machine learning approaches in materials science and engineering: Comprehensive review. *Eng. Appl. Artif. Intell.*, 135, 108783, Sep. 2024, doi: 10.1016/j.engappai.2024.108783.
31. Shaw, I., Ali, Y.S., Nie, C., Zhang, K., Chen, C., Xiao, Y., Integrating Artificial Intelligence and Microfluidics Technology for Psoriasis Therapy: A Comprehensive Review for Research and Clinical Applications. *Adv. Intell. Syst.*, Nov. 2024, doi: 10.1002/aisy.202400558.
32. Shi, J., *et al.*, Smart Textile-Integrated Microelectronic Systems for Wearable Applications. *Adv. Mater.*, 32, 5, Feb. 2020, doi: 10.1002/adma.201901958.
33. Bin Abu Sofian, A.D.A., *et al.*, Advances, Synergy, and Perspectives of Machine Learning and Biobased Polymers for Energy, Fuels, and Biochemicals for a Sustainable Future. *Energy Fuels*, 38, 3, 1593–1617, Feb. 2024, doi: 10.1021/acs.energyfuels.3c03842.
34. Dolšak, J., Hrovatin, N., Zorić, J., Analysing consumer preferences, characteristics, and behaviour to identify energy-efficient consumers. *Sustainability (Switzerland)*, 12, 23, 1–21, Dec. 2020, doi: 10.3390/su12239870.
35. Fu, K., *et al.*, Designing Textile Architectures for High Energy-Efficiency Human Body Sweat- and Cooling-Management. *Adv. Fiber Mater.*, 1, 1, 61–70, Sep. 2019, doi: 10.1007/s42765-019-0003-y.
36. Ahsan, M., Teay, S.H., Sayem, A.S.M., Albarbar, A., Smart Clothing Framework for Health Monitoring Applications. *Signals*, 3, 1, 113–145, Mar. 2022, doi: 10.3390/signals3010009.
37. Rani, P., Kaur, P., Jain, V., Shokeen, J., Nain, S., Blockchain-based IoT enabled health monitoring system. *J. Supercomput.*, 78, 15, 17284–17308, Oct. 2022, doi: 10.1007/s11227-022-04584-3.
38. Zhang, J., Xu, J., Lim, J., Nolan, J.K., Lee, H., Lee, C.H., Wearable Glucose Monitoring and Implantable Drug Delivery Systems for Diabetes Management. *Adv. Healthcare Mater.*, 10, 17, Sep. 2021, doi: 10.1002/adhm.202100194.

39. Rai, P.K., Seasonal monitoring of heavy metals and physicochemical characteristics in a lentic ecosystem of subtropical industrial region, India. *Environ. Monit. Assess.*, 165, 1–4, 407–433, Jun. 2010, doi: 10.1007/s10661-009-0956-z.
40. Caiazzo, B., Murino, T., Petrillo, A., Piccirillo, G., Santini, S., An IoT-based and cloud-assisted AI-driven monitoring platform for smart manufacturing: design architecture and experimental validation. *J. Manuf. Technol. Manage.*, 34, 4, 507–534, May 2023, doi: 10.1108/JMTM-02-2022-0092.
41. Miah, M.H., Chand, D.S., Malhi, G.S., Selected river pollution in Bangladesh based on industrial growth and economic perspective: a review. *Environ. Monit. Assess.*, 195, 1, 98, Jan. 2023, doi: 10.1007/s10661-022-10663-y.
42. M. N, S. S, E. S, H. P. S, M. A, and V. M, Implementation of Monitoring and Controlling of pH and TDS in Process Industry, in: *2023 2nd International Conference on Edge Computing and Applications (ICECAA)*, Jul. 2023, IEEE, pp. 1587–1592, doi: 10.1109/ICECAA58104.2023.10212189.
43. Alsahag, M., *et al.*, Preparation of carboxymethyl cellulose/polyvinyl alcohol wound dressing composite immobilized with anthocyanin extract for colorimetric monitoring of wound healing and prevention of wound infection. *Int. J. Biol. Macromol.*, 224, 233–242, Jan. 2023, doi: 10.1016/j.ijbiomac.2022.10.119.
44. Chan, M., Estève, D., Fourniols, J.Y., Escriba, C., Campo, E., Smart wearable systems: Current status and future challenges. *Artif. Intell. Med.*, 56, 3, 137–156, Nov. 2012, doi: 10.1016/j.artmed.2012.09.003.
45. Torah, R., *et al.*, Enabling platform technology for smart fabric design and printing. *J. Eng. Fiber Fabr.*, 14, Jan. 2019, doi: 10.1177/1558925019845903.
46. Singh Arora, D. and Kumar Sharma, R., Ligninolytic fungal laccases and their biotechnological applications. *Appl. Biochem. Biotechnol.*, 160, 6, 1760–1788, Mar. 2010, doi: 10.1007/s12010-009-8676-y.
47. de Oliveira, F., de Oliveira, A.C., Sánchez-Muñoz, S., Balbino, T.R., Santos-Ebinuma, V.C., da Silva, S.S., Sustainability feasibility of fungi-based biocolorants by biotechnological routes. *Chem. Eng. J.*, 152942, 2024.
48. Cruz-Casas, D.E., Aguilar, C.N., Ascacio-Valdés, J.A., Rodríguez-Herrera, R., Chávez-González, M.L., Flores-Gallegos, A.C., Enzymatic hydrolysis and microbial fermentation: The most favorable biotechnological methods for the release of bioactive peptides. *Food Chem.: Mol. Sci.*, 3, 100047, 2021, doi: 10.1016/j.fochms.2021.100047.
49. Dhruv Patel, D. and Bhatt, S., Environmental pollution, toxicity profile, and physico-chemical and biotechnological approaches for treatment of textile wastewater. *Biotechnol. Genet. Eng. Rev.*, 38, 1, 33–86, Jan. 2022, doi: 10.1080/02648725.2022.2048434.
50. Sánchez-Muñoz, S., *et al.*, Production of fungal and bacterial pigments and their applications, in: *Biotechnological Production of Bioactive Compounds*, pp. 327–361, Elsevier Ltd., Cambridge, United States, 2019, doi: 10.1016/B978-0-444-64323-0.00011-4.

# Index

Printed in the USA/Agawam, MA
September 16, 2026

910660.008